NONLINEAR THEORY OF

ELASTICITY

Applications in Biomechanics

NONLINEAR THEORY OF

ELASTICITY

Applications in Biomechanics

Larry A Taber

Washington University, USA

NEW JERSEY • LONDON • SINGAPORE • SHANGHAI • HONG KONG • TAIPEI • BANGALORE

Published by

World Scientific Publishing Co. Pte. Ltd.
5 Toh Tuck Link, Singapore 596224
USA office: Suite 202, 1060 Main Street, River Edge, NJ 07661
UK office: 57 Shelton Street, Covent Garden, London WC2H 9HE

British Library Cataloguing-in-Publication Data
A catalogue record for this book is available from the British Library.

NONLINEAR THEORY OF ELASTICITY
Applications in Biomechanics

ISBN 981-238-735-8

Printed in Singapore by World Scientific Printers (S) Pte Ltd

To my wife Char, my two-legged kids Jordan and Eli,
and my four-legged kids Seamus and Jamie

Preface

Problems involving the deformation of soft biological tissues are among the most challenging in biomechanics. Soft tissues are viscoelastic composite materials composed of cells and extracellular matrix, are frequently arranged in complex geometries, and routinely experience large strains under normal and pathological conditions. As the discipline of biomechanics matures, new problems continually appear that demand more and more sophisticated analytical and numerical solution techniques. For example, the field of tissue engineering, which deals with *in vitro* construction of natural replacement tissues, could benefit greatly from analyses that integrate nonlinear mechanical behavior at the molecular, cellular, tissue, and whole organ levels.

Although soft tissues actually are not elastic, investigators often incorporate elasticity as a simplifying assumption. In fact, many tissues exhibit approximately elastic (pseudoelastic) behavior under a fairly wide range of conditions. Hence, knowledge of the fundamental principles of nonlinear elasticity is crucial to understanding the biomechanics of soft tissues.

Unfortunately, however, many students trained in biomechanics are exposed only to relatively elementary concepts of nonlinear elasticity during the course of their studies. Later, these students may need to learn this daunting topic on their own. Having been in this boat myself at one time, I know just how difficult this task can be.

One reason this subject can be so difficult to learn is that the various methods, notations, and definitions used in the classical literature can be confusing. While this situation has improved in recent years, one goal of this book is to unify the principal approaches used by researchers in the field of nonlinear elasticity. Some authors, for example, work in convected coordinates, while others define separate coordinate systems for the undeformed and deformed bodies. This book uses and compares both methods.

A hallmark of this book is an extensive use of tensor and dyadic analysis in general curvilinear coordinates. Some may question the need for such complexity in this age of high-speed computers, especially in biomechanics, where most problems of interest require computational methods, in which case Cartesian coordinates often are sufficient. While this is true, much of the classical and even the current literature is littered with general curvilinear coordinates, and so reading and understanding these papers requires knowing the language of general tensor analysis.

The emphasis on dyadic notation was motivated by courses I took many years ago on linear elasticity and shell theory, respectively, from Professors George Herrmann and Charles R. Steele at Stanford University. These courses introduced me to the systematic elegance of dyadics. Over the years, my students and I have found that learning the intricacies of nonlinear elasticity is aided by routine use of dyadics, which form a link between direct and indicial notation. Direct notation lends insight and clarity to the basic physical principles, while indicial notation often makes algebraic manipulations easier (and is needed to solve problems). Judicious use of dyadics sometimes can clarify subtle differences, as well as similarities, between the various approaches used in solving nonlinear problems.

Throughout the book, I have tried to present as many details as possible for each derivation. While this feature may be tedious to some readers, I know how frustrating it can be when steps are omitted or glossed over. In some instances, however, derivations are left to exercises at the end of the chapter. Another possible source of irritation is the way many problems are solved using more than one approach. The purpose of doing this is to emphasize that there is more than one way to skin a cat, as well as to bring out some subtleties that may be missed otherwise.

Readers of this book would benefit by prior experience with linear elasticity. For those readers not already familiar with the linear theory, Appendix A provides an introduction. This book can be used as a textbook for a course on the nonlinear theory of elasticity, as a reference for a course on biomechanics, or as a reference for researchers trying to learn the subject on their own.

After a brief introduction in Chapter 1, Chapter 2 establishes the mathematical background that is needed to navigate the remainder of this book. This chapter covers general tensor and dyadic analysis, introducing the basic concepts of changes in coordinates and reference frames. Next, Chapters 3 and 4 develop and compare the basic measures of deformation and stress, and present the field equations of solid mechanics. Chapter 5 then discusses constitutive theory and develops stress-strain relations for nonlinear materials with various material symmetries. Finally, Chapter 6 solves several specific problems in soft tissue mechanics using the theory and techniques presented in the previous chapters.

The emphasis of this book is on learning the engineering fundamentals needed to solve nonlinear problems in biomechanics. Some mathematical and many biological details are omitted. For more in-depth modern treatments of some of these issues, I highly recommend the books by Gerhard A. Holzapfel and Jay D. Humphrey that are listed in the bibliography.

I am greatly indebted to Ashok Ramasubramanian, who diligently worked through most of the in-text derivations and the problems, pointing out numerous errors. I also would like to thank Millard F. Beatty and one of my students, Evan Zamir, who provided helpful comments on earlier versions of this book. Finally, I thank Charlene Taber for her valuable editorial assistance. Any remaining errors are my own, and I would appreciate any feedback.

Larry A. Taber
St. Louis, Missouri

Contents

Chapter 1

Introduction

Problems involving deformation of soft biological tissues are among the most challenging in applied mechanics. In addition to undergoing large strains, soft tissues exhibit nonlinear time-dependent behavior similar to that of viscoelastic materials (Fung, 1993). Moreover, biological tissues are not simply passive materials like those used in traditional engineering applications. Rather, they actively contract, grow, and remodel. Further complicating matters, problems in soft tissue biomechanics often involve complex three-dimensional geometry, dynamic loading conditions, and fluid-solid interaction. These problems are highly nonlinear in general.

This book deals with nonlinear analysis of *elastic* materials, i.e., materials that deform without loss of energy. At first glance, it may seem that limiting the discussion to elastic materials severely restricts the range of biomechanics applications. There is, however, a characteristic of many soft tissues that allows the use of nonlinear elasticity theory as a first approximation. This characteristic, called **pseudoelasticity**, was introduced by Y.C. Fung during the 1970s (Fung et al., 1979). Even today, with the rapidly increasing power of finite element and other computational methods, researchers take advantage of this feature to justify using elasticity theory to study the mechanics of soft tissues.

The concept of pseudoelasticity is based on the following experimental observation. During a tensile test, a soft tissue specimen ordinarily follows separate loading and unloading stress-strain curves, which form a hysteresis loop (Fig. 1.1). Such behavior is typical of viscoelastic materials. If the loading/unloading cycle is repeated at the same rate, the curves shift, presumably due to passive microstructural remodeling. After several loading cycles, however, the stress-strain curves become repeatable. Moreover, the response is relatively insensitive to loading

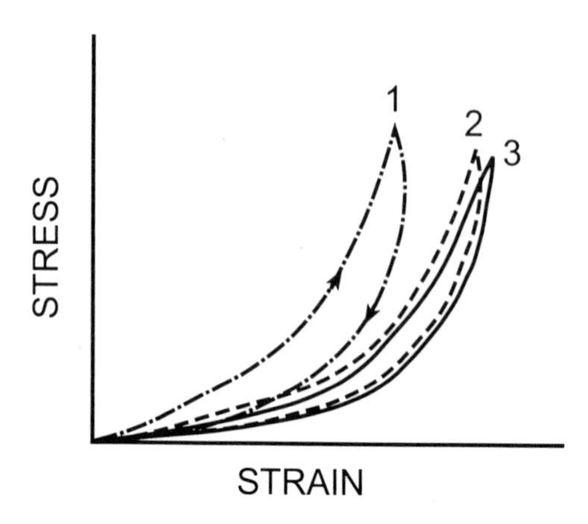

Fig. 1.1 Loading and unloading curves for a typical soft tissue. After the three cycles shown, the curves show relatively little additional change.

rate. To describe this behavior, Fung at al. (1979) suggested that the tissue can be treated as two separate elastic materials — one during loading and another during unloading, i.e., the tissue is pseudoelastic.

This is one possible approach, but separate loading and unloading constitutive relations have seldom been used in practice. Rather, most researchers either use viscoelasticity theory or, more commonly, simply assume that the material is elastic with the constitutive relation based only on the loading curve. This approach, although not strictly correct, has provided useful approximate results in a large number of problems. One feature that usually cannot be ignored, however, is nonlinear behavior, which is the focus of this book.

Early development of the nonlinear theory of elasticity is attributed largely to R.S. Rivlin, whose work on this topic began during the 1940s. Many of his early papers have been republished together as a collection (Rivlin et al., 1997). Rivlin initially worked without tensors, which are a staple of this book, but later came to embrace it as a convenient analytical tool. In addition, the classic work of Truesdell and Noll (1965), which helped put nonlinear elasticity on a firm theoretical foundation, exudes tensors. Today, it is difficult to imagine learning this subject without the use of tensor analysis.

In analyzing the mechanics of soft tissue, it is essential to have an in-depth understanding of the theory of nonlinear elasticity. Hence, Chapters 2 through 5 present the fundamental equations of the nonlinear theory. Chapter 2 presents the mathematical background needed to understand the material, and Chapters 3,

4, and 5 deal with the analysis of deformation, stress, and constitutive behavior, respectively. Some examples from biomechanics are discussed in these chapters, but the emphasis is on physical and mathematical concepts. Then, with the theoretical foundation in place, Chapter 6 examines applications to a relatively wide range of specific problems in soft tissue mechanics.

Chapter 2

Vectors, Dyadics, and Tensors

Physical quantities can be represented by tensors. Zeroeth-order tensors (scalars) have only a magnitude (e.g., temperature), first-order tensors (vectors) have a magnitude and direction (e.g., velocity), second-order tensors (dyadics) have a magnitude and two associated directions (e.g., stress[1]), and so on. The literature on nonlinear elasticity is filled with tensors, which provide a convenient tool for deriving the governing equations. For solving these equations, tensor quantities then can be expressed in terms of components relative to any convenient coordinate system. Because much of the historical literature on finite elasticity deals with general curvilinear coordinates, such generality is a main feature throughout this book.

2.1 Reference Frames and Coordinate Systems

Often the terms "reference frame" and "coordinate system" are used synonymously. There is, however, an important distinction. A **reference frame** is attached to an observer, and it may be in motion relative to other frames with other attached observers. In each frame, there can be one or more **coordinate systems** that can be used to locate a point in that frame. The coordinate systems *in a particular frame* are fixed relative to each other.

Consider a vector $\mathbf{a}$ fixed in a reference frame A (Fig. 2.1), which contains a Cartesian (x, y, z) and a cylindrical polar (r, θ, z) coordinate system (possibly with different origins). Relative to the Cartesian system, $\mathbf{a}$ has the components $\{a_x, a_y, a_z\}$, and, relative to the polar system, $\mathbf{a}$ has the components $\{a_r, a_\theta, a_z\}$.

[1] A stress component depends on its direction of action and the orientation of the area on which it acts.

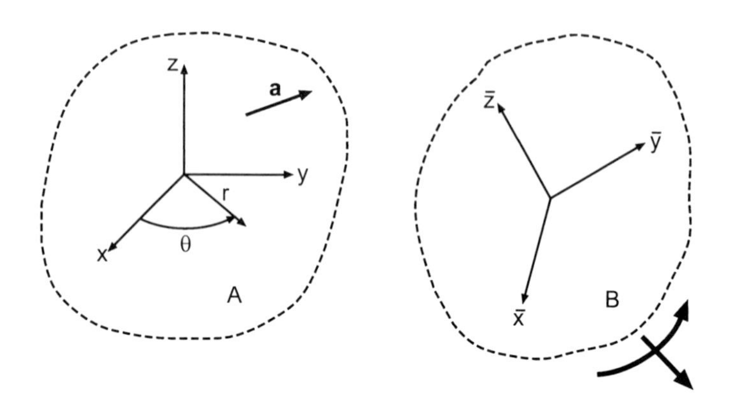

Fig. 2.1 Reference frames A and B with attached coordinate systems. Frame B is in motion relative to frame A.

These different components describe the *same* vector $\mathbf{a}$. The transformation of components from the Cartesian to the polar system is an example of a **coordinate transformation** .

Now, consider another reference frame B that is translating and rotating relative to A (Fig. 2.1). The vector $\mathbf{a}$ is fixed, i.e., **geometrically invariant**, in A, but it appears to be changing direction to an observer in B. Relative to a Cartesian system $(\bar{x}, \bar{y}, \bar{z})$ fixed in B, the components of $\mathbf{a}$ change with time. A coordinate transformation again can be used to relate the components of $\mathbf{a}$ from either coordinate system in A to the Cartesian system in B, but now the transformation coefficients are time-dependent.

Thus far, the only difference between reference frames and coordinate systems is the potential time dependency of the transformation of vector components. The distinction between the two becomes more significant in physical problems. For example, consider Newton's second law of motion $\mathbf{f} = m\mathbf{a}$, where $\mathbf{f}$ is the force vector acting on a particle of mass m and acceleration $\mathbf{a}$. If A is an inertial (nonrotating, nonaccelerating) reference frame, then this equation is valid in *any* coordinate system in A, although its components may change. Relative to another inertial frame B, $\mathbf{f} = m\mathbf{a}$ remains valid (if relativistic effects are ignored.) If B is a noninertial (rotating or accelerating) frame, however, the *form* of Newton's law changes. The reason for this lies in the acceleration vector $\mathbf{a}$. Transforming $\mathbf{f}$ from frame A to frame B involves only a time-dependent coordinate transformation, but the transformation of $\mathbf{a}$ is not so direct. For instance, if $\mathbf{f} = \mathbf{a} = \mathbf{0}$ in A,

$\mathbf{f}$ also is zero as seen by an observer in B, but the mass moves with $\mathbf{a} \neq \mathbf{0}$ relative to this accelerating observer. (Coriolis and other effects enter the analysis.) The force vector is an example of a **frame indifferent** or **objective** quantity, while the acceleration vector $\mathbf{a}$ is **coordinate invariant** but not frame indifferent.

Tensor analysis deals with coordinate transformations. In this chapter, we assume that we are dealing with various coordinate systems in a single frame of reference at a given time. Although the development in this chapter is valid for n-dimensional Euclidean space, we restrict our attention to three dimensions.

2.2 Vectors

A vector $\mathbf{a}$ has a magnitude and direction. Relative to a given coordinate system, $\mathbf{a}$ possesses a unique set of components. If the coordinate system in a given reference frame changes, the components of $\mathbf{a}$ change, but the vector $\mathbf{a}$ does not. A vector, therefore, is geometrically invariant, and the form of a vector equation is independent of any coordinate system in that frame.

Due in part to their invariance, vector formulations of physical problems often provide intuitive clarity. Solving these problems, however, usually requires expressing equations in terms of scalar components in a given coordinate system. Many problems simplify when the governing equations are written in a coordinate system other than Cartesian, e.g., polar or ellipsoidal coordinates. A general curvilinear coordinate system does not even have to be orthogonal. In this section, we examine how vectors can be described in various coordinate systems.

2.2.1 *Base Vectors*

We begin by introducing a notational convention. In rectangular Cartesian coordinates, we can write

$$\begin{aligned} \mathbf{a} &= \hat{a}^1\mathbf{e}_1 + \hat{a}^2\mathbf{e}_2 + \hat{a}^3\mathbf{e}_3 \\ &= \sum_{i=1}^{3} \hat{a}^i\mathbf{e}_i \equiv \hat{a}^i\mathbf{e}_i \end{aligned} \tag{2.1}$$

where the $\hat{a}^i$ are the components of $\mathbf{a}$ with respect to the orthogonal triad of unit vectors $\mathbf{e}_i$, which are tangent to the coordinate axes z^1, z^2, and z^3.[2] The second line of Eq. (2.1) defines the **summation convention**, with summation implied over 1,2,3 for each repeated index in a single term. Because any letter can replace it

[2]Henceforth, unless stated otherwise, the $\mathbf{e}_i$ represent unit vectors and z^i represent Cartesian coordinates, a special case of the general curvilinear coordinates x^i.

without changing the meaning of an expression, a repeated index also is called a "dummy" index. For example,

$$a_i^i = a_k^k = a_1^1 + a_2^2 + a_3^3,$$

while

$$a_i b_j^i = a_k b_j^k = \sum_{i=1}^{3} a_i b_j^i$$

gives a separate equation for each $j = 1, 2, 3$. Note that the number of upper and lower indices in an equation must "balance," i.e., after all summations are carried out, each term must be left with the same subscripts and superscripts. Moreover, a single term cannot contain, for example, three i's or even two i's on the same level; hence, $a_i b_i$ and a_{ii} are meaningless expressions. Throughout the remainder of this book, the summation convention holds unless stated otherwise or unless the indices are placed in parentheses, e.g., the terms $a_{(ii)}$ and $a_{(i)} b^{(i)}$ are not summed over i but rather represent separate expressions for each $i = 1, 2, 3$.

In a Cartesian coordinate system, the $\mathbf{e}_i$ are called **base vectors**. As shown by Eq. (2.1), *any* vector can be expressed as a linear combination of the base vectors. If the coordinates z_i and base vectors $\mathbf{e}_i$ undergo a rigid-body rotation to give a new coordinate system $\bar{z}^i$ with new base vectors $\bar{\mathbf{e}}_i$ (Fig. 2.2), then $\mathbf{a}$ can be resolved along these new base vectors as

$$\mathbf{a} = \bar{\hat{a}}^i \bar{\mathbf{e}}_i \tag{2.2}$$

where $\bar{\hat{a}}^i \neq \hat{a}^i$. A primary objective of tensor analysis is to relate components and base vectors in different coordinate systems.

Similarly, in **general curvilinear coordinates** (x^1, x^2, x^3), we introduce a set of base vectors $\mathbf{g}_i$, which are tangent to the x^i-curves at each point in space (Fig. 2.3). But now, as we will see, these vectors are not necessarily unit vectors or dimensionless. Moreover, in moving along the coordinate curves, the $\mathbf{g}_i$ can change in both magnitude and direction. In terms of these base vectors, similar to Eq. (2.1), the vector $\mathbf{a}$ can be written

$$\mathbf{a} = a^i \mathbf{g}_i \tag{2.3}$$

where the a^i are called the **contravariant components** (superscript i) of $\mathbf{a}$ relative to the **covariant base vectors** $\mathbf{g}_i$ (subscript i) of the x^i coordinate system.[3]

[3]Actually, any set of three linearly independent, i.e., noncoplanar, vectors can be used as "base vectors." The $\mathbf{g}_i$ can be considered the "natural base vectors" for a given coordinate system.

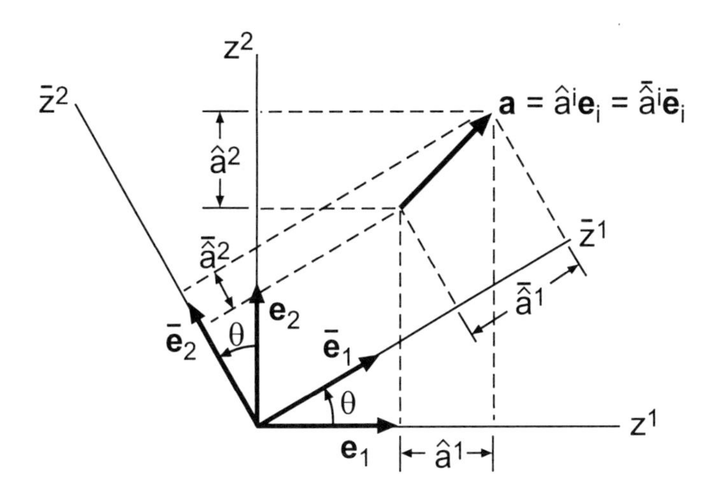

Fig. 2.2 Components of a vector $\mathbf{a}$ relative to two Cartesian coordinate systems, z^i and $\bar{z}^i$.

We now show how the covariant base vectors can be computed for a general x^i system. The **position vector** from the origin to any point $P(x^i)$ in space is (Fig. 2.3)

$$\mathbf{r} = \mathbf{r}(x^i) = r^i \mathbf{g}_i, \tag{2.4}$$

and the differential position vector is

$$d\mathbf{r} = \frac{\partial \mathbf{r}}{\partial x^i}\, dx^i = \mathbf{r}_{,i}\, dx^i \tag{2.5}$$

where $\mathbf{r}_{,i} \equiv \partial \mathbf{r}/\partial x^i$. (Note that a superscript in the denominator becomes a subscript in the numerator.) In the vicinity of P, a small movement of the tip of $\mathbf{r}$ along the x^1 coordinate curve corresponds to a change dx^1 in the coordinate, with similar results for the other directions. We take dx^i as the component of $d\mathbf{r}$ in the $\mathbf{g}_i$ direction. Then, resolving the vector $d\mathbf{r}$ in components relative to the basis $\{\mathbf{g}_i\}$ gives

$$\boxed{d\mathbf{r} = \mathbf{g}_i\, dx^i,} \tag{2.6}$$

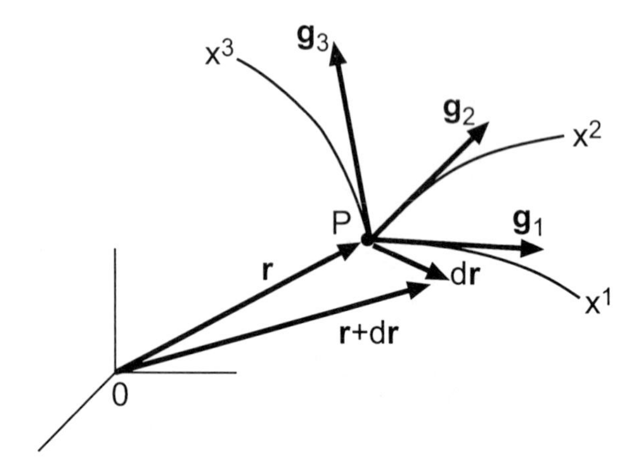

Fig. 2.3 Position vector $\mathbf{r}$ and differential position vector $d\mathbf{r}$ in curvilinear coordinate system with base vectors $\mathbf{g}_i$.

and comparison with Eq. (2.5) shows that

$$\boxed{\mathbf{g}_i = \mathbf{r}_{,i}\,.} \tag{2.7}$$

According to Eq. (2.6), the dx^i are the contravariant components of $d\mathbf{r}$ with respect to the $\mathbf{g}_i$. Note, however, that the x^i are *not* the contravariant components of $\mathbf{r}$ in general; only in Cartesian coordinates can we set $r^i = x^i$. Moreover, at a given time in general curvilinear coordinates, the base vectors and, therefore, the components of a vector are functions of position, while in Cartesian coordinates, the base vectors and the components of a vector are constant with respect to changes of position.

Example 2.1 Determine the covariant base vectors for each of the following coordinate systems:[4]

(1) Cartesian coordinates: $(x^1, x^2, x^3) = (x, y, z)$.
(2) Cylindrical polar coordinates: $(x^1, x^2, x^3) = (r, \theta, z)$ with $x = r\cos\theta$ and $y = r\sin\theta$ (Fig. 2.4a).

[4]To avoid confusion, powers are accompanied by parentheses. Thus, $(y)^2$ is "y squared," whereas y^2 is a contravariant vector component.

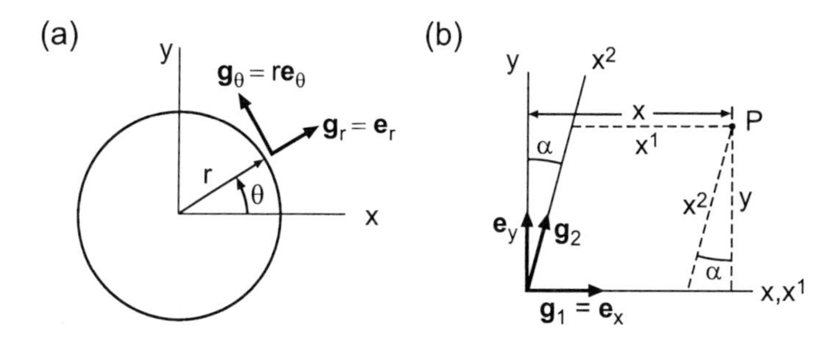

Fig. 2.4 Cylindrical polar (a) and skew (b) coordinates.

(3) Skew coordinates: (x^1, x^2, x^3) with $x = x^1 + x^2 \sin\alpha$, $y = x^2 \cos\alpha$, and $z = x^3$ (Fig. 2.4b).

Solution. Relative to these coordinate systems, the position vector to a given point P in space can be written in the forms

$$\begin{aligned} \mathbf{r} &= x\mathbf{e}_x + y\mathbf{e}_y + z\mathbf{e}_z \\ &= r\cos\theta\,\mathbf{e}_x + r\sin\theta\,\mathbf{e}_y + z\mathbf{e}_z \\ &= (x^1 + x^2\sin\alpha)\mathbf{e}_x + (x^2\cos\alpha)\mathbf{e}_y + x^3\mathbf{e}_z. \end{aligned} \tag{2.8}$$

Then, Eq. (2.7) gives the base vectors

$$\begin{aligned} \mathbf{g}_x &= \mathbf{r}_{,x} = \mathbf{e}_x \\ \mathbf{g}_y &= \mathbf{r}_{,y} = \mathbf{e}_y \\ \mathbf{g}_z &= \mathbf{r}_{,z} = \mathbf{e}_z \end{aligned} \tag{2.9}$$

in Cartesian coordinates,

$$\begin{aligned} \mathbf{g}_r &= \mathbf{r}_{,r} = \cos\theta\,\mathbf{e}_x + \sin\theta\,\mathbf{e}_y = \mathbf{e}_r \\ \mathbf{g}_\theta &= \mathbf{r}_{,\theta} = r(-\sin\theta\,\mathbf{e}_x + \cos\theta\,\mathbf{e}_y) = r\mathbf{e}_\theta \\ \mathbf{g}_z &= \mathbf{r}_{,z} = \mathbf{e}_z \end{aligned} \tag{2.10}$$

in cylindrical polar coordinates ($\mathbf{e}_r$ and $\mathbf{e}_\theta$ are unit vectors, see Fig. 2.4a), and

$$\begin{aligned} \mathbf{g}_1 &= \mathbf{r}_{,1} = \mathbf{e}_x \\ \mathbf{g}_2 &= \mathbf{r}_{,2} = \sin\alpha\,\mathbf{e}_x + \cos\alpha\,\mathbf{e}_y \\ \mathbf{g}_3 &= \mathbf{r}_{,3} = \mathbf{e}_z \end{aligned} \tag{2.11}$$

in skew coordinates. These sets of base vectors are illustrated in Fig. 2.4 (in two dimensions).

Here, we note several things. First, we have expressed $\mathbf{r}$ in terms of the Cartesian base vectors in each case; since these vectors are constant, this simplifies the algebra. Second, $\mathbf{g}_\theta$ is not a unit vector and it carries the dimension of length. Third, since $\mathbf{g}_x{\cdot}\mathbf{g}_y = \mathbf{g}_r{\cdot}\mathbf{g}_\theta = 0$, Cartesian and polar coordinates are orthogonal. (Obviously, all of these coordinate systems are orthogonal relative to the z-direction.) However, $\mathbf{g}_1{\cdot}\mathbf{g}_2 = \sin\alpha \neq 0$, indicating the nonorthogonality of skew coordinates (for $\alpha \neq 0$). ■

Example 2.2 For cylindrical polar coordinates, determine $\mathbf{g}_r$ and $\mathbf{g}_\theta$ without introducing a Cartesian system.

Solution. In polar coordinates, the position vector is

$$\mathbf{r} = r\,\mathbf{e}_r(\theta),$$

and differentiation yields

$$\begin{aligned} \mathbf{g}_r &= \mathbf{r}_{,r} = \mathbf{e}_r \\ \mathbf{g}_\theta &= \mathbf{r}_{,\theta} = r\frac{\partial \mathbf{e}_r}{\partial \theta}. \end{aligned}$$

The first expression agrees with Eq. $(2.10)_1$. To find $\partial\mathbf{e}_r/\partial\theta$ directly, we examine the geometry (Fig. 2.5). Consider two points in the $r\theta$-plane, P_1 and P_2, that are located the same distance from the origin but are separated by the small angle $d\theta$. In moving from P_1 to P_2, only the orientation of $\mathbf{e}_r$ changes. To a first approximation, this change is $(\partial\mathbf{e}_r/\partial\theta)\,d\theta$. The geometry of Fig. 2.5 gives

$$\left|\frac{\partial \mathbf{e}_r}{\partial \theta}\,d\theta\right| = |\mathbf{e}_r|\cdot d\theta$$

or

$$\left|\frac{\partial \mathbf{e}_r}{\partial \theta}\right| = 1,$$

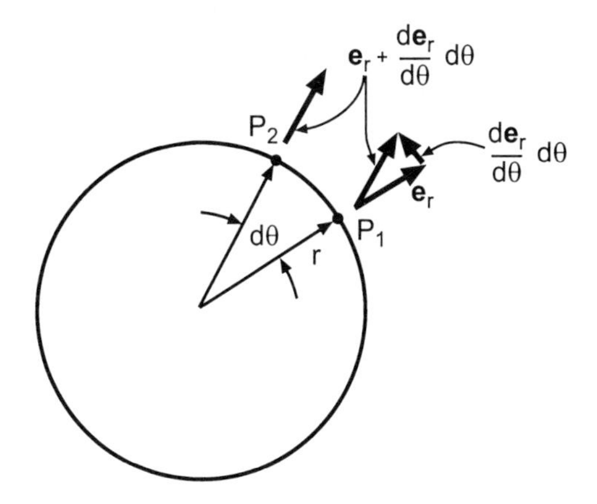

Fig. 2.5 Geometry for the differential change in the unit vector $\mathbf{e}_r$ with θ.

which shows that the derivative has unit magnitude. Moreover, as $d\theta \to 0$, the vector $\partial\mathbf{e}_r/\partial\theta$ becomes perpendicular to $\mathbf{e}_r$, i.e., it points in the direction of $\mathbf{e}_\theta$. Hence,

$$\frac{\partial \mathbf{e}_r}{\partial \theta} = \mathbf{e}_\theta$$

and so $\mathbf{g}_\theta = r\mathbf{e}_\theta$, in agreement with Eq. $(2.10)_2$. ■

2.2.2 *Reciprocal Base Vectors*

The dot product of the two vectors $\mathbf{a} = a^i\mathbf{g}_i$ and $\mathbf{b} = b^j\mathbf{g}_j$ is[5]

$$\begin{aligned}\mathbf{a}\cdot\mathbf{b} &= (a^i\mathbf{g}_i)\cdot(b^j\mathbf{g}_j) = a^ib^j\mathbf{g}_i\cdot\mathbf{g}_j \\ &= a^1b^1\mathbf{g}_1\cdot\mathbf{g}_1 + a^1b^2\mathbf{g}_1\cdot\mathbf{g}_2 + a^1b^3\mathbf{g}_1\cdot\mathbf{g}_3 \\ &\quad + a^2b^1\mathbf{g}_2\cdot\mathbf{g}_1 + a^2b^2\mathbf{g}_2\cdot\mathbf{g}_2 + a^2b^3\mathbf{g}_2\cdot\mathbf{g}_3 \\ &\quad + a^3b^1\mathbf{g}_3\cdot\mathbf{g}_1 + a^3b^2\mathbf{g}_3\cdot\mathbf{g}_2 + a^3b^3\mathbf{g}_3\cdot\mathbf{g}_3.\end{aligned} \tag{2.12}$$

If the coordinates are orthogonal, then $\mathbf{g}_i\cdot\mathbf{g}_j = 0$ for $i \neq j$, and this equation simplifies considerably. For general coordinates, however, this expression is quite

[5]Note that, in the representation for $\mathbf{b}$, we have replaced the i's by j's without altering the meaning. This procedure is necessary since more than two of any index is not allowed in any single term.

a mess.

To render this situation less painful, we introduce a new set of base vectors $\mathbf{g}^i$, which are called **reciprocal** or **contravariant base vectors** in the x^i coordinate system. [Simmonds (1994) calls the $\mathbf{g}_i$ and $\mathbf{g}^i$ "cellar" and "roof" base vectors, respectively, according to the location of the index.] The $\mathbf{g}^i$ are defined by the relation

$$\boxed{\mathbf{g}_i \cdot \mathbf{g}^j = \delta_i^j} \tag{2.13}$$

where

$$\delta_i^j = \begin{cases} 1 \text{ for } i = j \\ 0 \text{ for } i \neq j \end{cases} \tag{2.14}$$

is the **Kronecker delta**. Thus, $\mathbf{g}_1$ is orthogonal to both $\mathbf{g}^2$ and $\mathbf{g}^3$, and so on (Fig. 2.6).

Of course, a vector can be written in terms of components relative to *any* basis. Hence, if $\mathbf{b}$ is expressed relative to the contravariant basis, i.e., $\mathbf{b} = b_i \mathbf{g}^i$, where the b_i are the **covariant components** of $\mathbf{b}$, then

$$\begin{aligned} \mathbf{a} \cdot \mathbf{b} &= (a^i \mathbf{g}_i) \cdot (b_j \mathbf{g}^j) = a^i b_j (\mathbf{g}_i \cdot \mathbf{g}^j) = a^i b_j \delta_i^j \\ &= a^i b_i \\ &= a^1 b_1 + a^2 b_2 + a^3 b_3, \end{aligned} \tag{2.15}$$

which is a more palatable expression than Eq. (2.12). In this manipulation, we have used Eqs. (2.13) and (2.14). Note also that the effect of δ_i^j in a term is simply to replace j by i or, equivalently, i by j, with δ_i^j then removed.

Figure 2.6 illustrates a set of covariant and contravariant base vectors at a point in two-dimensional space. The vectors $\mathbf{g}^1$ and $\mathbf{g}^2$ are orthogonal to $\mathbf{g}_2$ and $\mathbf{g}_1$, respectively. Moreover, the magnitude and directions of $\mathbf{g}^1$ and $\mathbf{g}^2$ are determined by the requirement $\mathbf{g}^1 \cdot \mathbf{g}_1 = \mathbf{g}^2 \cdot \mathbf{g}_2 = +1$.

The $\mathbf{g}^i$ can be computed in terms of the $\mathbf{g}_i$ as follows. First, consider $\mathbf{g}^1$. Since it is orthogonal to both $\mathbf{g}_2$ and $\mathbf{g}_3$, we can write

$$\sqrt{g}\, \mathbf{g}^1 = \mathbf{g}_2 \times \mathbf{g}_3$$

according to the properties of the cross product. The scaling factor $\sqrt{g}$, found by dotting both sides of this equation with $\mathbf{g}_1$, is

$$\boxed{\sqrt{g} = \mathbf{g}_1 \cdot (\mathbf{g}_2 \times \mathbf{g}_3)} \tag{2.16}$$

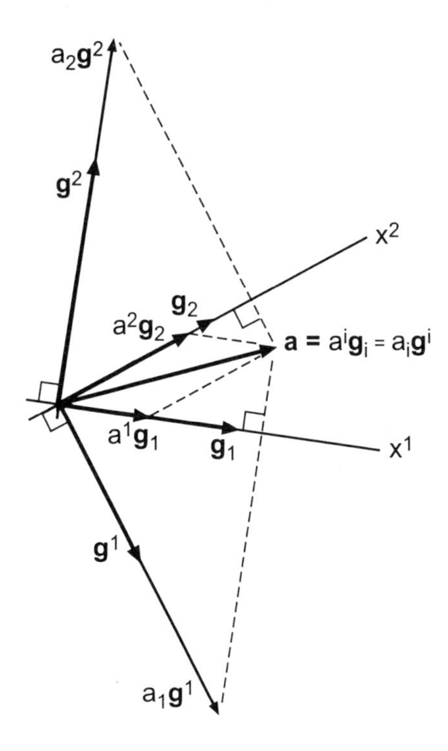

Fig. 2.6 Covariant and contravariant base vectors and components of a vector $\mathbf{a}$ in a two-dimensional nonorthogonal coordinate system.

which is the **scalar triple product** of the covariant base vectors. From elementary vector analysis, Eq. (2.16) implies that $\sqrt{g}$ represents the volume of the parallelepiped with the vectors $\mathbf{g}_i$ as edges. The reason for the square root is discussed later (Section 2.6.2). This and similar manipulations yield

$$\boxed{\mathbf{g}^1 = \frac{\mathbf{g}_2 \times \mathbf{g}_3}{\sqrt{g}}, \qquad \mathbf{g}^2 = \frac{\mathbf{g}_3 \times \mathbf{g}_1}{\sqrt{g}}, \qquad \mathbf{g}^3 = \frac{\mathbf{g}_1 \times \mathbf{g}_2}{\sqrt{g}}.} \tag{2.17}$$

Any vector can be expressed in terms of either its contravariant or covariant components (Fig. 2.6), i.e.,

$$\mathbf{a} = a^i \mathbf{g}_i = a_i \mathbf{g}^i. \tag{2.18}$$

The components of $\mathbf{a}$ can be extracted by dotting this relation with $\mathbf{g}^j$ or $\mathbf{g}_j$ to get

$$\begin{aligned}\mathbf{a}\cdot\mathbf{g}^j &= (a^i\mathbf{g}_i)\cdot\mathbf{g}^j = a^i\delta_i^j = a^j \\ \mathbf{a}\cdot\mathbf{g}_j &= (a_i\mathbf{g}^i)\cdot\mathbf{g}_j = a_i\delta_j^i = a_j.\end{aligned} \tag{2.19}$$

These equations show that dotting a vector with a base vector gives the component of the same variance.

The physical meaning of Eqs. (2.19) can be seen using elementary vector analysis (see Fig. 2.6). First, we can write $\mathbf{g}_j = |\mathbf{g}_j|\mathbf{e}_j$ (j not summed), where $|\mathbf{g}_j| = \sqrt{\mathbf{g}_j\cdot\mathbf{g}_j}$ is the magnitude of $\mathbf{g}_j$ and $\mathbf{e}_j$ is the unit vector in the direction of $\mathbf{g}_j$. Thus, $a_j = \mathbf{a}\cdot\mathbf{g}_j = |\mathbf{g}_j|(\mathbf{a}\cdot\mathbf{e}_j)$ (j not summed), which shows that a_j represents the product of $|\mathbf{g}_j|$ and the orthogonal projection of $\mathbf{a}$ in the direction of $\mathbf{g}_j$. Similarly, $a^j = \mathbf{a}\cdot\mathbf{g}^j = |\mathbf{g}^j|(\mathbf{a}\cdot\mathbf{e}^j)$ (j not summed), indicating that a^j is the product of $|\mathbf{g}^j|$ and the orthogonal projection of $\mathbf{a}$ in the direction of $\mathbf{g}^j$.

Judicious use of covariant and contravariant components and base vectors leads to numerous simplifications in vector and tensor analysis.

Example 2.3 Determine the contravariant base vectors for each of the coordinate systems of Example 2.1 (Fig. 2.4).

Solution. Recall that the scalar triple product of the vectors $\mathbf{a}$, $\mathbf{b}$, and $\mathbf{c}$ can be written

$$\mathbf{a}\cdot(\mathbf{b}\times\mathbf{c}) = \mathbf{b}\cdot(\mathbf{c}\times\mathbf{a}) = \mathbf{c}\cdot(\mathbf{a}\times\mathbf{b}) = \begin{vmatrix} a_1 & a_2 & a_3 \\ b_1 & b_2 & b_3 \\ c_1 & c_2 & c_3 \end{vmatrix} \tag{2.20}$$

where a_i, b_i, and c_i are *Cartesian components*. (The determinant representation for the scalar triple product is not valid in every coordinate system.) Thus, Eqs. (2.9)–(2.11), and (2.16) give

$$\begin{aligned}\sqrt{g} &= \mathbf{g}_x\cdot(\mathbf{g}_y\times\mathbf{g}_z) = \begin{vmatrix} 1 & 0 & 0 \\ 0 & 1 & 0 \\ 0 & 0 & 1 \end{vmatrix} = 1 \qquad \text{(Cartesian)} \\ &= \mathbf{g}_r\cdot(\mathbf{g}_\theta\times\mathbf{g}_z) = \begin{vmatrix} \cos\theta & \sin\theta & 0 \\ -r\sin\theta & r\cos\theta & 0 \\ 0 & 0 & 1 \end{vmatrix} = r \qquad \text{(cylindrical)} \\ &= \mathbf{g}_1\cdot(\mathbf{g}_2\times\mathbf{g}_3) = \begin{vmatrix} 1 & 0 & 0 \\ \sin\alpha & \cos\alpha & 0 \\ 0 & 0 & 1 \end{vmatrix} = \cos\alpha \qquad \text{(skew)}\end{aligned} \tag{2.21}$$

Now, Eqs. (2.9)–(2.11) and (2.17) give the reciprocal base vectors

$$\begin{aligned} \mathbf{g}^x &= \mathbf{g}_y \times \mathbf{g}_z = \mathbf{e}_y \times \mathbf{e}_z = \mathbf{e}_x \\ \mathbf{g}^y &= \mathbf{g}_z \times \mathbf{g}_x = \mathbf{e}_z \times \mathbf{e}_x = \mathbf{e}_y \\ \mathbf{g}^z &= \mathbf{g}_x \times \mathbf{g}_y = \mathbf{e}_x \times \mathbf{e}_y = \mathbf{e}_z \end{aligned} \tag{2.22}$$

in Cartesian coordinates,

$$\begin{aligned} \mathbf{g}^r &= \frac{1}{r}(\mathbf{g}_\theta \times \mathbf{g}_z) = \frac{1}{r}(r\mathbf{e}_\theta \times \mathbf{e}_z) = \mathbf{e}_r \\ \mathbf{g}^\theta &= \frac{1}{r}(\mathbf{g}_z \times \mathbf{g}_r) = \frac{1}{r}(\mathbf{e}_z \times \mathbf{e}_r) = \frac{\mathbf{e}_\theta}{r} \\ \mathbf{g}^z &= \frac{1}{r}(\mathbf{g}_r \times \mathbf{g}_\theta) = \frac{1}{r}(\mathbf{e}_r \times r\mathbf{e}_\theta) = \mathbf{e}_z \end{aligned} \tag{2.23}$$

in cylindrical polar coordinates, and

$$\begin{aligned} \mathbf{g}^1 &= (\cos\alpha)^{-1}(\mathbf{g}_2 \times \mathbf{g}_3) = (\cos\alpha)^{-1}(\cos\alpha\,\mathbf{e}_x - \sin\alpha\,\mathbf{e}_y) = \mathbf{e}_x - \tan\alpha\,\mathbf{e}_y \\ \mathbf{g}^2 &= (\cos\alpha)^{-1}(\mathbf{g}_3 \times \mathbf{g}_1) = (\cos\alpha)^{-1}\mathbf{e}_y \\ \mathbf{g}^3 &= (\cos\alpha)^{-1}(\mathbf{g}_1 \times \mathbf{g}_2) = (\cos\alpha)^{-1}(\cos\alpha\,\mathbf{e}_z) = \mathbf{e}_z \end{aligned} \tag{2.24}$$

in skew coordinates. Using Eqs. (2.9)–(2.11) and (2.22)–(2.24), we can show easily that Eq. (2.13) is satisfied in each case.

Note that, in orthogonal coordinates (e.g., Cartesian or cylindrical polar), $\mathbf{g}^i$ is in the same direction as $\mathbf{g}_i$ [see Eqs. (2.22) and (2.23)], but the magnitudes of these base vectors may differ. Moreover, in Cartesian coordinates, $\mathbf{g}_i = \mathbf{g}^i$, and so there is no difference between covariant and contravariant quantities. Because the placement of the index is immaterial in this case, the components of Cartesian tensors often are written with subscripts only. ■

2.3 Dyadics and Tensors

2.3.1 *Dyadics*

The **tensor product** of two vectors $\mathbf{a}$ and $\mathbf{b}$, denoted by $\mathbf{ab}$, is called a **dyad**.[6] Since each of the vectors in a dyad is geometrically invariant in a given frame

[6]Some authors use the notation $\mathbf{a} \otimes \mathbf{b}$ for the tensor product.

of reference, it follows that the dyad also is invariant. The order of the vectors composing a dyad is significant, and so the tensor product is not commutative, i.e., $\mathbf{ab} \neq \mathbf{ba}$. The same is true of higher-order **polyads**, such as $\mathbf{abc}$ or $\mathbf{abcd}$, which also are useful in some applications. Here, we are concerned primarily with dyads and linear combinations of dyads, which are called **dyadics**. Equations written in terms of vector and tensor components (e.g., $f_i = ma_i$) are written in **indicial notation**, whereas those written without components (e.g., $\mathbf{f} = m\mathbf{a}$) are said to be written in **direct notation**. As we will see later, dyadics link indicial and direct notation.

Algebraic manipulation of dyads is straightforward so long as care is taken to maintain the order of the component vectors. For example, the operations

$$\begin{aligned} \mathbf{ab}\cdot\mathbf{c} &= \mathbf{a}(\mathbf{b}\cdot\mathbf{c}) \\ \mathbf{c}\cdot\mathbf{ab} &= (\mathbf{c}\cdot\mathbf{a})\mathbf{b} \end{aligned} \tag{2.25}$$

show that the dot product of a dyad with a vector yields another vector. Similarly, the equations

$$\begin{aligned} \mathbf{ab}\cdot\mathbf{cd} &= \mathbf{a}(\mathbf{b}\cdot\mathbf{c})\mathbf{d} = (\mathbf{b}\cdot\mathbf{c})\,\mathbf{ad} \\ \mathbf{cd}\cdot\mathbf{ab} &= \mathbf{c}(\mathbf{d}\cdot\mathbf{a})\mathbf{b} = (\mathbf{d}\cdot\mathbf{a})\,\mathbf{cb} \end{aligned} \tag{2.26}$$

show that the **single-dot product** of two dyads produces another dyad. Furthermore, these equations show that the single-dot product of two dyads is not commutative in general. (Recall that the dot product of two vectors *is* commutative.)

Two types of **double-dot (scalar) product** are useful in manipulating dyads. The **vertical double-dot product** is defined by

$$\boxed{\begin{aligned} \mathbf{ab}:\mathbf{cd} &= (\mathbf{a}\cdot\mathbf{c})(\mathbf{b}\cdot\mathbf{d}) = \mathbf{c}\cdot(\mathbf{ab})\cdot\mathbf{d} \\ &= (\mathbf{c}\cdot\mathbf{a})(\mathbf{d}\cdot\mathbf{b}) = \mathbf{cd}:\mathbf{ab} \end{aligned}} \tag{2.27}$$

which shows that this operation is commutative for *dyads*. For polyads, however, this may not be the case, e.g.,

$$\begin{aligned} \mathbf{abc}:\mathbf{defg} &= \mathbf{a}(\mathbf{bc}:\mathbf{de})\mathbf{fg} \\ &= \mathbf{a}(\mathbf{b}\cdot\mathbf{d})(\mathbf{c}\cdot\mathbf{e})\mathbf{fg} \\ &= (\mathbf{b}\cdot\mathbf{d})(\mathbf{c}\cdot\mathbf{e})\mathbf{afg} \\ &\neq \mathbf{defg}:\mathbf{abc}. \end{aligned}$$

Table 2.1 Vector and polyadic formulas

$$
\begin{aligned}
\mathbf{a}+\mathbf{b} &= \mathbf{b}+\mathbf{a} \\
(\mathbf{a}+\mathbf{b})+\mathbf{c} &= \mathbf{a}+(\mathbf{b}+\mathbf{c}) \\
\mathbf{a}\cdot\mathbf{b} &= \mathbf{b}\cdot\mathbf{a} \\
\mathbf{a}\cdot(\mathbf{b}+\mathbf{c}) &= \mathbf{a}\cdot\mathbf{b}+\mathbf{a}\cdot\mathbf{c} \\
\phi(\mathbf{a}+\mathbf{b}) &= \phi\mathbf{a}+\phi\mathbf{b} \\
(\phi+\psi)\mathbf{a} &= \phi\mathbf{a}+\psi\mathbf{a} \\
\mathbf{a}\times\mathbf{b} &= -\mathbf{b}\times\mathbf{a} \\
\mathbf{a}\times(\mathbf{b}\times\mathbf{c}) &= (\mathbf{a}\cdot\mathbf{c})\mathbf{b}-(\mathbf{a}\cdot\mathbf{b})\mathbf{c} = \mathbf{a}\cdot(\mathbf{c}\mathbf{b}-\mathbf{b}\mathbf{c}) \\
(\mathbf{a}\times\mathbf{b})\times\mathbf{c} &= (\mathbf{a}\cdot\mathbf{c})\mathbf{b}-(\mathbf{b}\cdot\mathbf{c})\mathbf{a} = \mathbf{c}\cdot(\mathbf{a}\mathbf{b}-\mathbf{b}\mathbf{a}) \\
\mathbf{a}\cdot(\mathbf{b}\times\mathbf{c}) &= \mathbf{b}\cdot(\mathbf{c}\times\mathbf{a}) = \mathbf{c}\cdot(\mathbf{a}\times\mathbf{b}) \\
\phi(\mathbf{a}\mathbf{b}) &= (\phi\mathbf{a})\mathbf{b} = \mathbf{a}(\phi\mathbf{b}) = \phi\mathbf{a}\mathbf{b} \\
\mathbf{a}(\mathbf{b}\mathbf{c}) &= (\mathbf{a}\mathbf{b})\mathbf{c} = \mathbf{a}\mathbf{b}\mathbf{c} \\
\mathbf{a}(\mathbf{b}+\mathbf{c}) &= \mathbf{a}\mathbf{b}+\mathbf{a}\mathbf{c} \\
\phi(\mathbf{a}\mathbf{b}+\mathbf{c}\mathbf{d}) &= \phi\mathbf{a}\mathbf{b}+\phi\mathbf{c}\mathbf{d}
\end{aligned}
$$

Again, with care taken to maintain order, the algebra is straightforward.

The other type of double-dot product is the **horizontal double-dot product** defined by

$$
\boxed{\mathbf{a}\mathbf{b}\cdot\cdot\mathbf{c}\mathbf{d} = (\mathbf{a}\cdot\mathbf{d})(\mathbf{b}\cdot\mathbf{c}) = \mathbf{c}\mathbf{d}\cdot\cdot\mathbf{a}\mathbf{b}} \tag{2.28}
$$

which also is commutative for dyads. These operations can be extended to polyads by adding more dots, e.g., $\mathbf{abc}\vdots\mathbf{def} = (\mathbf{a}\cdot\mathbf{d})(\mathbf{b}\cdot\mathbf{e})(\mathbf{c}\cdot\mathbf{f})$. Table 2.1 contains some useful relations for vectors and polyads.

2.3.2 *Tensors*

A **linear vector function** $\mathbf{T}$ satisfies the relations

$$
\mathbf{T}\cdot(\mathbf{a}+\mathbf{b}) = \mathbf{T}\cdot\mathbf{a}+\mathbf{T}\cdot\mathbf{b} \qquad \text{and} \qquad \mathbf{T}\cdot(\phi\mathbf{a}) = \phi\mathbf{T}\cdot\mathbf{a} \tag{2.29}
$$

for any vectors $\mathbf{a}$ and $\mathbf{b}$ and scalar ϕ. A second-order **tensor** $\mathbf{T}$ can be defined as a linear vector function that, when dotted with a vector, transforms the vector into another vector, i.e.,[7]

$$\boxed{\mathbf{a} = \mathbf{T}\cdot\mathbf{b}.} \tag{2.30}$$

Since the vectors $\mathbf{a}$ and $\mathbf{b}$ are invariant, $\mathbf{T}$ also must be invariant. A *third-order* tensor is a linear vector function that transforms a vector into a second-order tensor through a single-dot product or a second-order tensor into a vector through a double-dot product, and so on. The equation

$$\begin{aligned}(\mathbf{ab})\cdot(\phi\mathbf{c} + \psi\mathbf{d}) &= \phi\mathbf{a}(\mathbf{b}\cdot\mathbf{c}) + \psi\mathbf{a}(\mathbf{b}\cdot\mathbf{d}) \\ &= [\phi(\mathbf{b}\cdot\mathbf{c}) + \psi(\mathbf{b}\cdot\mathbf{d})]\mathbf{a},\end{aligned} \tag{2.31}$$

with ϕ and ψ being scalars, shows that a dyad $\mathbf{ab}$ is a linear vector function that satisfies the definition of a second-order tensor. (Table 2.1 has been used in this equation.) In this way, we can show that a vector is a first-order tensor, a dyad is a second-order tensor, a triad is a third-order tensor, and so on.

Any second-order tensor can be represented by a dyadic. Because they span the space, base vectors are convenient for forming the component dyads. In three-dimensional space, for example, we can write

$$\begin{aligned}\mathbf{T} &= T^{11}\mathbf{g}_1\mathbf{g}_1 + T^{12}\mathbf{g}_1\mathbf{g}_2 + T^{13}\mathbf{g}_1\mathbf{g}_3 \\ &\quad + T^{21}\mathbf{g}_2\mathbf{g}_1 + T^{22}\mathbf{g}_2\mathbf{g}_2 + T^{23}\mathbf{g}_2\mathbf{g}_3 \\ &\quad + T^{31}\mathbf{g}_3\mathbf{g}_1 + T^{32}\mathbf{g}_3\mathbf{g}_2 + T^{33}\mathbf{g}_3\mathbf{g}_3 \\ &= T^{ij}\mathbf{g}_i\mathbf{g}_j\end{aligned} \tag{2.32}$$

where the T^{ij} are the contravariant components of the second-order tensor $\mathbf{T}$ with respect to the covariant basis $\{\mathbf{g}_i\mathbf{g}_j\}$. Like a vector, a tensor can be expressed in terms of components referred to any basis. Also like a vector, under a change of basis in a given frame of reference, a tensor does not change, but its components do. For a vector, the base vectors form a basis; for a tensor, the dyads (polyads) composed of base vectors form a basis.

In terms of natural base vectors, $\mathbf{T}$ has four fundamental component represen-

[7] In this chapter, lower-case bold letters denote first-order tensors (vectors), and upper-case bold letters denote higher-order tensors.

tations:

$$\boxed{\mathbf{T} = T^{ij}\mathbf{g}_i\mathbf{g}_j = T_{ij}\mathbf{g}^i\mathbf{g}^j = T^{i}_{\cdot j}\mathbf{g}_i\mathbf{g}^j = T_{i}^{\cdot j}\mathbf{g}^i\mathbf{g}_j.} \tag{2.33}$$

The T_{ij} are covariant components, and the $T^{i}_{\cdot j}$ and $T_{i}^{\cdot j}$ are **mixed components** of $\mathbf{T}$. By convention, the first index on a tensor component is associated with the first vector of the dyad, and the second index is associated with the second vector. Since $T^{i}_{\cdot j} \neq T_{j}^{\cdot i}$ in general, the dots in the mixed components are needed for clarity to keep the place of the first index. Again, we emphasize the importance of the order of the base vectors in Eq. (2.33).

The single-dot product of a vector with a base vector was used previously to extract the component of the vector with the same variance as the base vector [see Eqs. (2.19)]. Analogously, vertical-double-dotting a second-order tensor $\mathbf{T}$ with a dyadic basis gives the components of $\mathbf{T}$ with the same variance as the basis. For example, extending Eqs. (2.19) gives[8]

$$\begin{aligned} \mathbf{T}{:}\mathbf{g}_i\mathbf{g}^j &= (T_{k}^{\cdot l}\mathbf{g}^k\mathbf{g}_l){:}\mathbf{g}_i\mathbf{g}^j \\ &= T_{k}^{\cdot l}(\mathbf{g}^k{\cdot}\mathbf{g}_i)(\mathbf{g}_l{\cdot}\mathbf{g}^j) \\ &= T_{k}^{\cdot l}\delta^k_i\delta^j_l \\ &= T_{i}^{\cdot j} \end{aligned} \tag{2.34}$$

which is of the same variance as $\mathbf{g}_i\mathbf{g}^j$. The other components of $\mathbf{T}$ can be found similarly, giving

$$\boxed{\begin{aligned} \mathbf{T}{:}\mathbf{g}_i\mathbf{g}_j &= \mathbf{g}_i{\cdot}\mathbf{T}{\cdot}\mathbf{g}_j = T_{ij} \\ \mathbf{T}{:}\mathbf{g}^i\mathbf{g}^j &= \mathbf{g}^i{\cdot}\mathbf{T}{\cdot}\mathbf{g}^j = T^{ij} \\ \mathbf{T}{:}\mathbf{g}_i\mathbf{g}^j &= \mathbf{g}_i{\cdot}\mathbf{T}{\cdot}\mathbf{g}^j = T_{i}^{\cdot j} \\ \mathbf{T}{:}\mathbf{g}^i\mathbf{g}_j &= \mathbf{g}^i{\cdot}\mathbf{T}{\cdot}\mathbf{g}_j = T^{i}_{\cdot j}. \end{aligned}} \tag{2.35}$$

Note that the physical meaning of the projection of a tensor on a dyad is not intuitively obvious.

[8]Later, we will see that we could have used any of the four representations for $\mathbf{T}$ in Eq. (2.33) and ended up with the same result (see Problem **2–14**).

2.3.3 *Matrix Representation of Tensors*

Matrix algebra is useful for computing with the *components* of tensors. However, it must be kept in mind that these components are attached to base vectors. For example, $[T_{ij}]$ and $[T_i^{\cdot j}]$ are matrix representations of the same tensor $\mathbf{T}$, but their full meanings are not known without defining their bases.

In this book, we write the matrix form of a tensor $\mathbf{T}$ in three ways, e.g.,

$$\mathbf{T}_{(\mathbf{g}_i\mathbf{g}^j)} = [T^i_{\cdot j}] = \begin{bmatrix} T^1_{\cdot 1} & T^1_{\cdot 2} & T^1_{\cdot 3} \\ T^2_{\cdot 1} & T^2_{\cdot 2} & T^2_{\cdot 3} \\ T^3_{\cdot 1} & T^3_{\cdot 2} & T^3_{\cdot 3} \end{bmatrix}.$$

In this expression, the first term identifies the basis through a subscript on the tensor, the second term is shorthand form for the matrix of the indicated components, and the last term is used in calculations. Once the basis is specified, we may write simply $\mathbf{T}$ for the matrix of components in subsequent manipulations. In addition, we use the following conventions:

(1) For a second-order tensor, the first index (or base vector) corresponds to the row and the second to the column of the matrix.

(2) If we leave the place-holding dot out of the mixed components of a second-order tensor and write $T^i_{\cdot j} = T^i_j$ or $T_j^{\cdot i} = T^i_j$, then the contravariant index gives the row and the covariant index the column. [In Simmonds' terminology, the **r**oof and **c**ellar indices correspond to the **r**ow and **c**olumn, respectively (Simmonds, 1994).]

(3) If the basis is not specified, then components are associated with the natural base vectors $\mathbf{g}_i$ and $\mathbf{g}^i$.

2.3.4 *Some Properties of Tensors*

Tables 2.2 and 2.3 give some useful relations involving tensors. In this section, we focus on the transpose and inverse. Other properties are discussed in Section 2.5.

Transpose

The **transpose** of a tensor $\mathbf{T}$, denoted by $\mathbf{T}^T$, is defined by the relation

$$\boxed{\mathbf{b}\cdot\mathbf{T}\cdot\mathbf{a} = \mathbf{a}\cdot\mathbf{T}^T\cdot\mathbf{b}} \tag{2.36}$$

where $\mathbf{a}$ and $\mathbf{b}$ are arbitrary vectors. Because the dot product of a second-order tensor and a vector is a vector, we can write this equation in the form $\mathbf{b}\cdot(\mathbf{T}\cdot\mathbf{a}) =$

Table 2.2 Tensor formulas

$$
\begin{aligned}
\mathbf{T}+\mathbf{U} &= \mathbf{U}+\mathbf{T} \\
(\mathbf{T}+\mathbf{U})+\mathbf{V} &= \mathbf{T}+(\mathbf{U}+\mathbf{V}) \\
\mathbf{T}\cdot(\mathbf{U}+\mathbf{V}) &= \mathbf{T}\cdot\mathbf{U}+\mathbf{T}\cdot\mathbf{V} \\
\phi(\mathbf{T}+\mathbf{U}) &= \phi\mathbf{T}+\phi\mathbf{U} \\
(\phi+\psi)\mathbf{T} &= \phi\mathbf{T}+\psi\mathbf{T} \\
\mathbf{T}\cdot(\mathbf{U}\cdot\mathbf{V}) &= (\mathbf{T}\cdot\mathbf{U})\cdot\mathbf{V} \\
\mathbf{T}:\mathbf{U} &= \mathbf{T}^T:\mathbf{U}^T = \mathbf{U}:\mathbf{T} \\
\mathbf{T}:(\mathbf{U}+\mathbf{V}) &= \mathbf{T}:\mathbf{U}+\mathbf{T}:\mathbf{V} \\
\mathbf{T}:(\mathbf{U}\cdot\mathbf{V}) &= \mathbf{U}:(\mathbf{T}\cdot\mathbf{V}^T) = \mathbf{V}:(\mathbf{U}^T\cdot\mathbf{T}) \\
\phi(\mathbf{T}:\mathbf{U}) &= (\phi\mathbf{T}):\mathbf{U} = \mathbf{T}:(\phi\mathbf{U})
\end{aligned}
$$

$(\mathbf{a}\cdot\mathbf{T}^T)\cdot\mathbf{b} = \mathbf{b}\cdot(\mathbf{a}\cdot\mathbf{T}^T)$. Hence, Eq. (2.36) is equivalent to

$$\boxed{\mathbf{T}\cdot\mathbf{a} = \mathbf{a}\cdot\mathbf{T}^T.} \tag{2.37}$$

To find the component representation for $\mathbf{T}^T$, we substitute, for example, Eq. $(2.33)_1$ to get

$$
\begin{aligned}
\mathbf{T}\cdot\mathbf{a} &= (T^{ij}\mathbf{g}_i\mathbf{g}_j)\cdot\mathbf{a} = T^{ij}\mathbf{g}_i(\mathbf{g}_j\cdot\mathbf{a}) \\
&= T^{ij}(\mathbf{a}\cdot\mathbf{g}_j)\mathbf{g}_i = \mathbf{a}\cdot(T^{ij}\mathbf{g}_j\mathbf{g}_i) = \mathbf{a}\cdot\mathbf{T}^T,
\end{aligned} \tag{2.38}
$$

and so $\mathbf{T}^T = T^{ij}\mathbf{g}_j\mathbf{g}_i$. These and similar manipulations with the other component forms of $\mathbf{T}$ show that

$$\mathbf{T}^T = T^{ij}\mathbf{g}_j\mathbf{g}_i = T_{ij}\mathbf{g}^j\mathbf{g}^i = T^{i}_{\cdot j}\mathbf{g}^j\mathbf{g}_i = T_i^{\cdot j}\mathbf{g}_j\mathbf{g}^i. \tag{2.39}$$

Comparison with Eq. (2.33) reveals that the transpose of a second-order tensor is given by interchanging the base vectors in any of its dyadic representations (or by interchanging the rows and columns in a matrix representation).

A second-order tensor $\mathbf{T}$ is **symmetric** if $\mathbf{T} = \mathbf{T}^T$; if $\mathbf{T} = -\mathbf{T}^T$, the tensor is **antisymmetric** or **skew symmetric**. For a symmetric tensor, Eqs. (2.35) and

Table 2.3 Formulas for transpose and inverse

$$
\begin{aligned}
(\mathbf{ab})^T &= \mathbf{ba} \\
(\mathbf{T}+\mathbf{U})^T &= \mathbf{T}^T + \mathbf{U}^T \\
(\mathbf{T}^T)^T &= \mathbf{T} \\
(\mathbf{T}\cdot\mathbf{U})^T &= \mathbf{U}^T\cdot\mathbf{T}^T \\
(\mathbf{T}^T)^{-1} &= (\mathbf{T}^{-1})^T \equiv \mathbf{T}^{-T} \\
(\mathbf{T}\cdot\mathbf{U})^{-1} &= \mathbf{U}^{-1}\cdot\mathbf{T}^{-1} \\
(\mathbf{T}^{-1})^m &= (\mathbf{T}^m)^{-1} \equiv \mathbf{T}^{-m}
\end{aligned}
$$

(2.36) provide the following relations between the components:

$$
\begin{aligned}
\mathbf{g}^i\cdot\mathbf{T}^T\cdot\mathbf{g}^j = \mathbf{g}^j\cdot\mathbf{T}\cdot\mathbf{g}^i \quad &\rightarrow \quad T^{ij} = T^{ji} \\
\mathbf{g}_i\cdot\mathbf{T}^T\cdot\mathbf{g}_j = \mathbf{g}_j\cdot\mathbf{T}\cdot\mathbf{g}_i \quad &\rightarrow \quad T_{ij} = T_{ji} \\
\mathbf{g}^i\cdot\mathbf{T}^T\cdot\mathbf{g}_j = \mathbf{g}_j\cdot\mathbf{T}\cdot\mathbf{g}^i \quad &\rightarrow \quad T^i_{\cdot j} = T_j^{\cdot i}.
\end{aligned}
\tag{2.40}
$$

Note that symmetry of $\mathbf{T}$ does *not* imply $T^i_{\cdot j} = T^j_{\cdot i}$. Thus, in general, the *matrix* of mixed components with $T^i_{\cdot j}$ in the *ith* row and *jth* column is not equal to the transpose of the matrix with $T^j_{\cdot i}$ in the *jth* row and *ith* column. In other words, symmetry of the tensor $\mathbf{T}$ does not imply that the matrix of mixed components is symmetric.

Inverse

The **inverse** of a tensor is defined by

$$
\boxed{\mathbf{T}\cdot\mathbf{T}^{-1} = \mathbf{T}^{-1}\cdot\mathbf{T} = \mathbf{I}}
\tag{2.41}
$$

where $\mathbf{T}^{-1}$ is the inverse of $\mathbf{T}$ and $\mathbf{I}$ is the **identity tensor**, which satisfies $\mathbf{T}\cdot\mathbf{I} = \mathbf{T}$. It turns out that the mixed components of $\mathbf{I}$ are the Kronecker delta δ^i_j, and so

$$
\boxed{\mathbf{I} = \delta^i_j\mathbf{g}_i\mathbf{g}^j = \delta^i_j\mathbf{g}^j\mathbf{g}_i = \mathbf{g}_i\mathbf{g}^i = \mathbf{g}^i\mathbf{g}_i}
\tag{2.42}
$$

which shows that $\mathbf{I} = \mathbf{I}^T$. These representations can be verified, for example, by the manipulation

$$\begin{aligned} \mathbf{T}\cdot\mathbf{I} &= (T^{ij}\mathbf{g}_i\mathbf{g}_j)\cdot(\mathbf{g}^k\mathbf{g}_k) = T^{ij}\mathbf{g}_i\delta_j^k\mathbf{g}_k \\ &= T^{ij}\mathbf{g}_i\mathbf{g}_j = \mathbf{T}. \end{aligned} \tag{2.43}$$

A similar manipulation confirms that $\mathbf{a}\cdot\mathbf{I} = \mathbf{a}$.

2.4 Coordinate Transformation

Deriving governing equations for physical problems is usually easiest in Cartesian coordinates. When symmetry is present, however, it may be more convenient to solve a specific problem in some other coordinate system, such as cylindrical or spherical polar coordinates. Tensor analysis offers a way to take advantage of both situations. First, the equations can be derived in Cartesian coordinates. Next, they can be expressed in direct (coordinate-free) notation. Finally, the components relative to the specialized basis can be extracted. For this to be possible, we must be able to relate quantities in different bases (coordinate systems).

2.4.1 *Transformation of Base Vectors*

Coordinate Transformation Tensors

Let $\mathbf{g}_i$ and $\bar{\mathbf{g}}_i$ represent the covariant base vectors and $\mathbf{g}^i$ and $\bar{\mathbf{g}}^i$ the contravariant base vectors in the curvilinear coordinate systems x^i and $\bar{x}^i$, respectively. The base vectors can be converted from the unbarred to the barred system through the relations

$$\boxed{\bar{\mathbf{g}}_i = \mathbf{A}\cdot\mathbf{g}_i, \qquad \bar{\mathbf{g}}^i = \mathbf{B}\cdot\mathbf{g}^i} \tag{2.44}$$

where $\mathbf{A}$ and $\mathbf{B}$ are **coordinate transformation tensors**. Convenient mixed-component representations for these tensors are

$$\mathbf{A} = A^i_{\cdot j}\mathbf{g}_i\mathbf{g}^j, \qquad \mathbf{B} = B_i^{\cdot j}\mathbf{g}^i\mathbf{g}_j. \tag{2.45}$$

Substituting these equations into Eqs. (2.44) yields

$$\begin{aligned} \bar{\mathbf{g}}_i &= (A^j_{\cdot k}\mathbf{g}_j\mathbf{g}^k)\cdot\mathbf{g}_i = A^j_{\cdot k}\mathbf{g}_j\delta_i^k = A^j_{\cdot i}\mathbf{g}_j \\ \bar{\mathbf{g}}^i &= (B_j^{\cdot k}\mathbf{g}^j\mathbf{g}_k)\cdot\mathbf{g}^i = B_j^{\cdot k}\mathbf{g}^j\delta_k^i = B_j^{\cdot i}\mathbf{g}^j \end{aligned} \tag{2.46}$$

which show that $A^{j}_{.i}$ is the component of the vector $\bar{\mathbf{g}}_i$ in the direction of $\mathbf{g}_j$, and $B_j^{.i}$ is the component of $\bar{\mathbf{g}}^i$ in the direction of $\mathbf{g}^j$. Moreover, combining Eqs. (2.45) and (2.46) yields

$$\boxed{\mathbf{A} = \bar{\mathbf{g}}_i\mathbf{g}^i, \qquad \mathbf{B} = \bar{\mathbf{g}}^i\mathbf{g}_i,} \tag{2.47}$$

which also can be seen by directly substituting these relations into Eqs. (2.44).

In general, the only components of $\mathbf{A}$ and $\mathbf{B}$ we need are those indicated in Eqs. (2.46). Without confusion, therefore, we set

$$A^{j}_{.i} \equiv A^{j}_{i}, \qquad B_j^{.i} \equiv B^i_j \tag{2.48}$$

with the understanding that dyadic operations are to be carried out with $\mathbf{A}$ and $\mathbf{B}$ in the forms of Eqs. (2.45) or (2.47). Then, dotting Eqs. $(2.46)_{1,2}$ with $\mathbf{g}^k$ and $\mathbf{g}_k$, respectively, and changing dummy indices yields

$$\boxed{A^j_i = \bar{\mathbf{g}}_i\cdot\mathbf{g}^j, \qquad B^i_j = \bar{\mathbf{g}}^i\cdot\mathbf{g}_j}, \tag{2.49}$$

which can be considered defining relations for the mixed components of the transformation tensors. If the barred and unbarred systems coincide, Eqs. (2.13) and (2.49) show that $A^i_j = B^i_j = \delta^i_j$, and Eqs. (2.42) and (2.47) show that $\mathbf{A} = \mathbf{B} = \mathbf{I}$, as expected.

The transformation tensors $\mathbf{A}$ and $\mathbf{B}$ are related. Since the base vectors in any coordinate system must satisfy Eq. (2.13), then $\mathbf{g}_i\cdot\mathbf{g}^j = \bar{\mathbf{g}}_i\cdot\bar{\mathbf{g}}^j = \delta^j_i$. Thus, Eqs. (2.37) and (2.44) give

$$\begin{aligned} \mathbf{g}_i\cdot\mathbf{g}^j &= \bar{\mathbf{g}}_i\cdot\bar{\mathbf{g}}^j = (\mathbf{A}\cdot\mathbf{g}_i)\cdot(\mathbf{B}\cdot\mathbf{g}^j) \\ &= (\mathbf{g}_i\cdot\mathbf{A}^T)\cdot(\mathbf{B}\cdot\mathbf{g}^j) = \mathbf{g}_i\cdot(\mathbf{A}^T\cdot\mathbf{B})\cdot\mathbf{g}^j \end{aligned}$$

which shows that $\mathbf{A}^T\cdot\mathbf{B} = \mathbf{I}$ or

$$\boxed{\mathbf{B} = \mathbf{A}^{-T}} \tag{2.50}$$

with $\mathbf{A}^{-T} \equiv (\mathbf{A}^T)^{-1}$ as defined in Table 2.3. Similarly, Eqs. (2.13) and (2.46) yield

$$\bar{\mathbf{g}}_i\cdot\bar{\mathbf{g}}^j = (A^k_i\mathbf{g}_k)\cdot(B^j_l\mathbf{g}^l) = A^k_iB^j_l\delta^l_k = \delta^j_i$$

or

$$\boxed{A_i^k B_k^j = \delta_i^j.} \tag{2.51}$$

Example 2.4 Compute the tensors $\mathbf{A}$ and $\mathbf{B}$ for a transformation from Cartesian $[(x^1, x^2, x^3) = (x, y, z)]$ to cylindrical polar $[(\bar{x}^1, \bar{x}^2, \bar{x}^3) = (r, \theta, z)]$ coordinates (Fig. 2.4a).

Solution. Let $\mathbf{g}^i$ and $\bar{\mathbf{g}}_i$ represent base vectors in the Cartesian and polar coordinate systems, respectively. Then, Eq. (2.47)$_1$ gives

$$\begin{aligned}\mathbf{A} = \bar{\mathbf{g}}_i \mathbf{g}^i &= \bar{\mathbf{g}}_1 \mathbf{g}^1 + \bar{\mathbf{g}}_2 \mathbf{g}^2 + \bar{\mathbf{g}}_3 \mathbf{g}^3 \\ &= \bar{\mathbf{g}}_r \mathbf{g}^x + \bar{\mathbf{g}}_\theta \mathbf{g}^y + \bar{\mathbf{g}}_z \mathbf{g}^z.\end{aligned} \tag{2.52}$$

Substituting Eqs. (2.10) for the $\bar{\mathbf{g}}_i$ and (2.22) for the $\mathbf{g}^i$ yields

$$\begin{aligned}\mathbf{A} &= (\cos\theta\, \mathbf{e}_x + \sin\theta\, \mathbf{e}_y)\mathbf{e}_x + r(-\sin\theta\, \mathbf{e}_x + \cos\theta\, \mathbf{e}_y)\mathbf{e}_y + \mathbf{e}_z\mathbf{e}_z \\ &= \cos\theta\, \mathbf{e}_x\mathbf{e}_x - r\sin\theta\, \mathbf{e}_x\mathbf{e}_y + \sin\theta\, \mathbf{e}_y\mathbf{e}_x + r\cos\theta\, \mathbf{e}_y\mathbf{e}_y + \mathbf{e}_z\mathbf{e}_z,\end{aligned}$$

or, in matrix form,

$$\mathbf{A}_{(\mathbf{e}_i\mathbf{e}_j)} = [A_j^i] = \begin{bmatrix} \cos\theta & -r\sin\theta & 0 \\ \sin\theta & r\cos\theta & 0 \\ 0 & 0 & 1 \end{bmatrix}. \tag{2.53}$$

(Note that, for a Cartesian basis, the positions of the indices are immaterial.)

The tensor $\mathbf{B}$ can be found from Eq. (2.47) or Eq. (2.50). Here, we use the latter equation and first compute

$$\mathbf{A}^T_{(\mathbf{e}_j\mathbf{e}_i)} = \begin{bmatrix} \cos\theta & \sin\theta & 0 \\ -r\sin\theta & r\cos\theta & 0 \\ 0 & 0 & 1 \end{bmatrix}. \tag{2.54}$$

The inverse of a tensor $\mathbf{T}$ can be computed from

$$\mathbf{T}^{-1} = \frac{\operatorname{cof}\mathbf{T}}{\det\mathbf{T}} \tag{2.55}$$

where the components of $\operatorname{cof}\mathbf{T}$ are the cofactors of the matrix $[T^i_{\cdot j}]$, and $\det\mathbf{T}$ is the determinant of $[T^i_{\cdot j}]$ (see Section 2.5.2). Thus, we have

$$\mathbf{B} = \mathbf{A}^{-T} = (\mathbf{A}^T)^{-1} = \frac{1}{r}\begin{bmatrix} r\cos\theta & r\sin\theta & 0 \\ -\sin\theta & \cos\theta & 0 \\ 0 & 0 & r \end{bmatrix}. \tag{2.56}$$

These expressions for $\mathbf{A}$ and $\mathbf{B}$ can be used to check our results from Examples 2.1 and 2.3. For instance, using Eq. $(2.44)_1$, we can transform $\mathbf{g}_2 = \mathbf{g}_y$ to $\bar{\mathbf{g}}_2 = \bar{\mathbf{g}}_\theta$ by

$$\begin{aligned} \bar{\mathbf{g}}_\theta &= \mathbf{A}\cdot\mathbf{g}_y = \mathbf{A}\cdot\mathbf{e}_y = \begin{bmatrix} \cos\theta & -r\sin\theta & 0 \\ \sin\theta & r\cos\theta & 0 \\ 0 & 0 & 1 \end{bmatrix} \begin{bmatrix} 0 \\ 1 \\ 0 \end{bmatrix} \\ &= \begin{bmatrix} -r\sin\theta \\ r\cos\theta \\ 0 \end{bmatrix} = -r\sin\theta\,\mathbf{e}_x + r\cos\theta\,\mathbf{e}_y \end{aligned}$$

which agrees with $\mathbf{g}_\theta$ in Eq. (2.10). ■

Components of the Coordinate Transformation Tensors

In practice, the components of the transformation tensors usually are computed from the equations that relate the unbarred and barred curvilinear coordinates. For a unique mapping of one system to the other, we can write

$$\bar{x}^i = \bar{x}^i(x^j), \qquad x^i = x^i(\bar{x}^j). \tag{2.57}$$

For example, if $(x^1, x^2, x^3) = (x, y, z)$ represent a set of Cartesian coordinates and $(\bar{x}^1, \bar{x}^2, \bar{x}^3) = (r, \theta, z)$ are cylindrical polar coordinates, then Eqs. $(2.57)_2$ become $x^1 = \bar{x}^1 \cos\bar{x}^2$, $x^2 = \bar{x}^1 \sin\bar{x}^2$, and $x^3 = \bar{x}^3$.

The differential position vector, which is the same in both coordinate systems, can be written

$$d\mathbf{r} = \bar{\mathbf{g}}_i\, d\bar{x}^i = \mathbf{g}_i\, dx^i = \mathbf{g}_i \frac{\partial x^i}{\partial \bar{x}^j}\, d\bar{x}^j \tag{2.58}$$

where the chain rule for differentiation has been used. In this equation, dx^i and $d\bar{x}^i$ are the contravariant components of $d\mathbf{r}$ in the two coordinate systems. This equation gives the relation

$$\left(\bar{\mathbf{g}}_i - \mathbf{g}_j \frac{\partial x^j}{\partial \bar{x}^i}\right) d\bar{x}^i = \mathbf{0}.$$

For arbitrary $d\bar{x}^i$, the expression in parentheses must vanish, and comparison with

Eq. $(2.46)_1$ shows that

$$\boxed{A_i^j = \frac{\partial x^j}{\partial \bar{x}^i}.} \tag{2.59}$$

The components of $\mathbf{B}$ can be found from Eq. (2.51), i.e.,

$$A_i^k B_k^j = \frac{\partial x^k}{\partial \bar{x}^i} B_k^j = \delta_i^j.$$

Since the chain rule gives

$$\frac{\partial x^k}{\partial \bar{x}^i}\frac{\partial \bar{x}^j}{\partial x^k} = \frac{\partial \bar{x}^j}{\partial \bar{x}^i} = \delta_i^j,$$

comparing these expressions and changing indices yields

$$\boxed{B_j^i = \frac{\partial \bar{x}^i}{\partial x^j}.} \tag{2.60}$$

Often the x^i are known in terms of the $\bar{x}^j$ (or vice versa), and the relations between coordinates are difficult to invert. In this case, Eq. (2.59) can be used to compute A_i^j, and then Eq. (2.50) or (2.51) gives the B_j^i.

Example 2.5 Use Eqs. (2.59) and (2.60) to compute the components of the tensors $\mathbf{A}$ and $\mathbf{B}$ for a transformation from Cartesian to cylindrical polar coordinates (Fig. 2.4a).

Solution. Let $(x^1, x^2, x^3) = (x, y, z)$ and $(\bar{x}^1, \bar{x}^2, \bar{x}^3) = (r, \theta, z)$; then

$$x^1 = \bar{x}^1 \cos \bar{x}^2, \qquad x^2 = \bar{x}^1 \sin \bar{x}^2, \qquad x^3 = \bar{x}^3$$

$$\bar{x}^1 = \left[(x^1)^2 + (x^2)^2\right]^{\frac{1}{2}}, \qquad \bar{x}^2 = \tan^{-1}\frac{x^2}{x^1}, \qquad x^3 = \bar{x}^3.$$

We compute, for instance,

$$B_2^1 = \frac{\partial \bar{x}^1}{\partial x^2} = \frac{x^2}{[(x^1)^2 + (x^2)^2]^{\frac{1}{2}}} = \frac{\bar{x}^1 \sin \bar{x}^2}{\bar{x}^1} = \sin\theta$$

which agrees with Eq. (2.56) of Example 2.4. The other components of $\mathbf{A}$ and $\mathbf{B}$ can be determined similarly (see Problem **2–9**). ■

2.4.2 *Transformation of Vector Components*

The coordinate transformation relations for the components of a vector follow directly from those for the base vectors. Relative to the unbarred and barred coordinate systems, the vector $\mathbf{a}$ can be written

$$\mathbf{a} = a^i\mathbf{g}_i = a_i\mathbf{g}^i = \bar{a}^i\bar{\mathbf{g}}_i = \bar{a}_i\bar{\mathbf{g}}^i, \tag{2.61}$$

and applying Eqs. (2.19) and (2.49) gives the components

$$\begin{aligned}
\bar{a}^i &= \bar{\mathbf{g}}^i\cdot\mathbf{a} = \bar{\mathbf{g}}^i\cdot(a^j\mathbf{g}_j) = B^i_{j}a^j \\
\bar{a}_i &= \bar{\mathbf{g}}_i\cdot\mathbf{a} = \bar{\mathbf{g}}_i\cdot(a_j\mathbf{g}^j) = A^j_{i}a_j \\
a^i &= \mathbf{g}^i\cdot\mathbf{a} = \mathbf{g}^i\cdot(\bar{a}^j\bar{\mathbf{g}}_j) = A^i_{j}\bar{a}^j \\
a_i &= \mathbf{g}_i\cdot\mathbf{a} = \mathbf{g}_i\cdot(\bar{a}_j\bar{\mathbf{g}}^j) = B^j_{i}\bar{a}_j.
\end{aligned} \tag{2.62}$$

Comparison with Eqs. (2.46) reveals that the transformations of vector components parallel those of the base vectors.

2.4.3 *Transformation of Tensor Components*

Like those for vector components, the coordinate transformation relations for tensor components fall out naturally when dyadic notation is used. For example, consider the transformation of the covariant components of the tensor

$$\mathbf{T} = T_{ij}\mathbf{g}^i\mathbf{g}^j = \bar{T}_{ij}\bar{\mathbf{g}}^i\bar{\mathbf{g}}^j. \tag{2.63}$$

Extracting the components relative to the barred system yields [see Eq. (2.35)]

$$\bar{T}_{ij} = \bar{\mathbf{g}}_i\cdot\mathbf{T}\cdot\bar{\mathbf{g}}_j = \bar{\mathbf{g}}_i\cdot(T_{kl}\mathbf{g}^k\mathbf{g}^l)\cdot\bar{\mathbf{g}}_j.$$

With Eqs. (2.49), this and similar expressions for the other components give

$$\begin{aligned}
\bar{T}_{ij} &= A^k_{i}A^l_{j}T_{kl} \\
\bar{T}^{ij} &= B^i_{k}B^j_{l}T^{kl} \\
\bar{T}_i^{\cdot j} &= A^k_{i}B^j_{l}T_k^{\cdot l} \\
\bar{T}^i_{\cdot j} &= B^i_{k}A^l_{j}T^k_{\cdot l}.
\end{aligned} \tag{2.64}$$

Inspecting Eqs. (2.62) and (2.64) reveals a pattern. Transformation from the unbarred to the barred system requires an A^i_{j} for each covariant index and a B^i_{j} for each contravariant index. The pattern reverses when the transformation is the other way. It is more consistent with our approach, however, to memorize Eqs. (2.49) and simply dot the vector and dyadic representations with the appropriate base vectors. This is how we obtained Eqs. (2.62) and (2.64).

Before leaving our discussion of coordinate transformation, we point out an important feature of tensor analysis. As Eqs. (2.33) and (2.64) show, if the components of a tensor with respect to one basis vanish, then the components with respect to any basis vanish. *Thus, an equation written in tensor components, like an equation written in direct notation, holds in any coordinate system in a given reference frame.*

2.5 Tensor Invariants

2.5.1 *Trace*

The **trace** of a dyad is defined by

$$\boxed{\operatorname{tr}\mathbf{ab} \equiv \mathbf{a}\cdot\mathbf{b}.} \tag{2.65}$$

This operation also is called a **contraction** of the dyad $\mathbf{ab}$, since inserting a dot between the vectors lowers the order of $\mathbf{ab}$ by two, transforming it into the scalar $\mathbf{a}\cdot\mathbf{b}$. For a second-order tensor $\mathbf{T}$, the trace is defined in terms of *mixed components* as

$$\operatorname{tr}\mathbf{T} \equiv \operatorname{tr}(T^{i}_{\cdot j}\mathbf{g}_i\mathbf{g}^j) = T^{i}_{\cdot j}\mathbf{g}_i\cdot\mathbf{g}^j = T^{i}_{\cdot j}\delta^j_i = T^{i}_{\cdot i}.$$

Equivalently, the trace of $\mathbf{T}$ is the trace of the *matrix* of mixed components, i.e.,

$$\boxed{\operatorname{tr}\mathbf{T} = \operatorname{tr}[T^{i}_{\cdot j}] = T^{i}_{\cdot i}.} \tag{2.66}$$

In the barred coordinate system, Eqs. (2.51) and $(2.64)_4$ give

$$\bar{T}^{i}_{\cdot i} = B^i_k A^l_i T^{k}_{\cdot l} = \delta^l_k T^{k}_{\cdot l} = T^{k}_{\cdot k} = T^{i}_{\cdot i}. \tag{2.67}$$

Thus, $\operatorname{tr}\mathbf{T}$ does not change with a change of basis, and so the trace is called a **scalar invariant** of the tensor $\mathbf{T}$. Some useful properties of the trace are given in Table 2.4.

Here, we note that any tensor $\mathbf{T}$ can be decomposed as

$$\mathbf{T} = \mathbf{T}_S + \mathbf{T}_A \tag{2.68}$$

Table 2.4 Formulas for trace (n = dimension of the space)

$$
\begin{aligned}
\operatorname{tr}\mathbf{I} &= n \\
\operatorname{tr}\mathbf{T} &= \mathbf{T}{:}\mathbf{I} \\
\operatorname{tr}(\mathbf{T}\cdot\mathbf{U}^T) &= \operatorname{tr}(\mathbf{T}^T\cdot\mathbf{U}) = \mathbf{T}{:}\mathbf{U} \\
\operatorname{tr}\mathbf{T}^T &= \operatorname{tr}\mathbf{T} \\
\operatorname{tr}(\phi\mathbf{T}) &= \phi\operatorname{tr}\mathbf{T} \\
\operatorname{tr}(\mathbf{T}+\mathbf{U}) &= \operatorname{tr}\mathbf{T}+\operatorname{tr}\mathbf{U} \\
\operatorname{tr}(\mathbf{T}\cdot\mathbf{U}) &= \operatorname{tr}(\mathbf{U}\cdot\mathbf{T}) \\
\operatorname{tr}(\mathbf{T}\cdot\mathbf{U}\cdot\mathbf{V}) &= \operatorname{tr}(\mathbf{U}\cdot\mathbf{V}\cdot\mathbf{T}) = \operatorname{tr}(\mathbf{V}\cdot\mathbf{T}\cdot\mathbf{U})
\end{aligned}
$$

where

$$
\begin{aligned}
\mathbf{T}_S &= \tfrac{1}{2}(\mathbf{T}+\mathbf{T}^T) = \mathbf{T}_S^T \\
\mathbf{T}_A &= \tfrac{1}{2}(\mathbf{T}-\mathbf{T}^T) = -\mathbf{T}_A^T
\end{aligned}
\tag{2.69}
$$

are symmetric and antisymmetric tensors, respectively. These tensors have the properties

$$
\operatorname{tr}\mathbf{T}_S = \operatorname{tr}\mathbf{T}, \qquad \operatorname{tr}\mathbf{T}_A = 0. \tag{2.70}
$$

2.5.2 *Determinant*

The **determinant** of a tensor $\mathbf{T}$ is defined as the determinant of the matrix of mixed components, i.e.,

$$
\boxed{\det\mathbf{T} \equiv \det[T^i_{\cdot j}].} \tag{2.71}
$$

Table 2.5 contains some useful properties of the determinant. Equations (2.50), (2.64), (2.71), and Table 2.5 give

$$
\begin{aligned}
\det\bar{\mathbf{T}} &= \det[\bar{T}^i_{\cdot j}] = \det[B^i_k A^l_j T^k_{\cdot l}] \\
&= \det[B^i_k]\det[A^l_j]\det[T^k_{\cdot l}] = \det\mathbf{B}\,\det\mathbf{A}\,\det\mathbf{T} \\
&= \det\mathbf{A}^{-T}\,\det\mathbf{A}\,\det\mathbf{T} = \det\mathbf{A}^{-1}\,\det\mathbf{A}\,\det\mathbf{T} \\
&= \det(\mathbf{A}^{-1}\cdot\mathbf{A})\,\det\mathbf{T} = \det\mathbf{I}\,\det\mathbf{T} \\
&= \det\mathbf{T},
\end{aligned}
\tag{2.72}
$$

Table 2.5 Formulas for determinant (n = dimension of the space)

$$
\begin{aligned}
\det \mathbf{I} &= 1 \\
\det \mathbf{T}^T &= \det \mathbf{T} \\
\det(\phi \mathbf{T}) &= \phi^n \det \mathbf{T} \\
\det(\mathbf{T}\cdot\mathbf{U}) &= \det \mathbf{T} \det \mathbf{U} \\
\frac{\partial \det \mathbf{T}}{\partial \mathbf{T}} &= (\det \mathbf{T})\, \mathbf{T}^{-T}
\end{aligned}
$$

which shows that the scalar $\det \mathbf{T}$ is invariant under a change of basis.

2.5.3 *Eigenvalues and Eigenvectors*

In the eigenvalue problem for a tensor $\mathbf{T}$, we wish to find a vector $\mathbf{a}$ such that the transformation of $\mathbf{a}$ by $\mathbf{T}$ produces another vector in the direction of $\mathbf{a}$. Mathematically, this can be stated as

$$\mathbf{T}\cdot\mathbf{a} = \lambda \mathbf{a}$$

or

$$(\mathbf{T} - \lambda \mathbf{I})\cdot\mathbf{a} = \mathbf{0} \tag{2.73}$$

where λ is the **eigenvalue** (principal value) and $\mathbf{a}$ is the **eigenvector** (principal direction) of $\mathbf{T}$. For a nontrivial solution ($\mathbf{a} \neq \mathbf{0}$) in n-dimensional space, we must have

$$\det(\mathbf{T} - \lambda \mathbf{I}) = 0 \tag{2.74}$$

which gives the **characteristic equation** of order n to be solved for the λ_i ($i = 1, 2, \ldots, n$). With the λ_i known, Eq. (2.73) then provides the corresponding $\mathbf{a}_i$. Since Eqs. (2.73) and (2.74) are geometrically invariant, both λ_i and $\mathbf{a}_i$ are invariants of the tensor $\mathbf{T}$.

In mixed-component form, Eq. (2.73) can be written

$$(T^i_{\cdot j} \mathbf{g}_i \mathbf{g}^j - \lambda \delta^i_j \mathbf{g}_i \mathbf{g}^j)\cdot(a^k \mathbf{g}_k) = (T^i_{\cdot j} - \lambda \delta^i_j)\mathbf{g}_i \delta^j_k a^k = \mathbf{0}$$

or, for arbitrary $\mathbf{g}_i$,

$$(T^i_{\cdot j} - \lambda\delta^i_j)a^j = 0. \tag{2.75}$$

In three-dimensional space, expanding $\det[T^i_{\cdot j} - \lambda\delta^i_j] = 0$ gives the characteristic equation

$$\boxed{-\lambda^3 + I_1\lambda^2 - I_2\lambda + I_3 = 0} \tag{2.76}$$

for the second-order tensor $\mathbf{T}$, where

$$\boxed{\begin{aligned} I_1 &= \operatorname{tr}\mathbf{T} \\ I_2 &= \tfrac{1}{2}[(\operatorname{tr}\mathbf{T})^2 - \operatorname{tr}\mathbf{T}^2] \\ I_3 &= \det\mathbf{T} \end{aligned}} \tag{2.77}$$

in which $\mathbf{T}^2 = \mathbf{T}\cdot\mathbf{T}$. Because λ is invariant, it follows from Eq. (2.76) that I_1, I_2, and I_3 also are invariant, and they are called the **principal invariants** of the tensor $\mathbf{T}$. Of course, any function of these invariants is also invariant. A useful relation is given by the **Cayley-Hamilton theorem**, which states that a tensor satisfies its own characteristic equation, i.e.,

$$\boxed{-\mathbf{T}^3 + I_1\mathbf{T}^2 - I_2\mathbf{T} + I_3\mathbf{I} = \mathbf{0}.} \tag{2.78}$$

2.6 Special Tensors

2.6.1 *Metric Tensor*

The square of the length of a differential line element is

$$\boxed{ds^2 = d\mathbf{r}\cdot d\mathbf{r} = (dx^i\,\mathbf{g}_i)\cdot(dx^j\,\mathbf{g}_j) = g_{ij}\,dx^i\,dx^j} \tag{2.79}$$

where Eq. (2.58) has been used and

$$\boxed{g_{ij} \equiv \mathbf{g}_i\cdot\mathbf{g}_j.} \tag{2.80}$$

The scalar invariant given by Eq. (2.79) is called the **metric** of the space, and the components g_{ij} characterize a particular coordinate system in the space.

The quantities g_{ij} have a useful mathematical interpretation. Since $\mathbf{g}_i\mathbf{g}_j{:}\mathbf{I} = \mathbf{g}_i{\cdot}\mathbf{I}{\cdot}\mathbf{g}_j = \mathbf{g}_i{\cdot}\mathbf{g}_j = g_{ij}$, Eq. (2.35)$_1$ shows that the g_{ij} are the covariant components of the identity tensor. For this reason, **I** also is called the **metric tensor**. Like any second-order tensor, **I** has four representations in any coordinate system:

$$\boxed{\mathbf{I} = g_{ij}\mathbf{g}^i\mathbf{g}^j = g^{ij}\mathbf{g}_i\mathbf{g}_j = g^{i}_{\cdot j}\mathbf{g}_i\mathbf{g}^j = g_j^{\cdot i}\mathbf{g}^j\mathbf{g}_i.} \tag{2.81}$$

Extracting the components of **I** gives the relations

$$\boxed{\begin{aligned} \mathbf{g}_i{\cdot}\mathbf{I}{\cdot}\mathbf{g}_j &= \mathbf{g}_i{\cdot}\mathbf{g}_j = g_{ij} \\ \mathbf{g}^i{\cdot}\mathbf{I}{\cdot}\mathbf{g}^j &= \mathbf{g}^i{\cdot}\mathbf{g}^j = g^{ij} \\ \mathbf{g}^i{\cdot}\mathbf{I}{\cdot}\mathbf{g}_j &= \mathbf{g}^i{\cdot}\mathbf{g}_j = g^{i}_{\cdot j} \\ \mathbf{g}_j{\cdot}\mathbf{I}{\cdot}\mathbf{g}^i &= \mathbf{g}_j{\cdot}\mathbf{g}^i = g_j^{\cdot i} \end{aligned}} \tag{2.82}$$

which show that $g_{ij} = g_{ji}$, $g^{ij} = g^{ji}$, and $g^{i}_{\cdot j} = g_j^{\cdot i}$. Thus, $\mathbf{I} = \mathbf{I}^T$ by Eqs. (2.40). Moreover, comparison with Eq. (2.42) shows that

$$\boxed{g^{i}_{\cdot j} = g_j^{\cdot i} = \delta^i_j.} \tag{2.83}$$

For a physical interpretation of the components of the metric tensor, we dot **I** with the base vectors as follows:

$$\begin{aligned} \mathbf{g}_i &= \mathbf{g}_i{\cdot}\mathbf{I} = \mathbf{g}_i{\cdot}(g_{kj}\mathbf{g}^k\mathbf{g}^j) = g_{kj}\delta^k_i\mathbf{g}^j = g_{ij}\mathbf{g}^j \\ \mathbf{g}^i &= \mathbf{g}^i{\cdot}\mathbf{I} = \mathbf{g}^i{\cdot}(g^{kj}\mathbf{g}_k\mathbf{g}_j) = g^{kj}\delta^i_k\mathbf{g}_j = g^{ij}\mathbf{g}_j. \end{aligned} \tag{2.84}$$

These expressions show that g_{ij} is the component of $\mathbf{g}_i$ in the direction of $\mathbf{g}^j$, and g^{ij} is the component of $\mathbf{g}^i$ along $\mathbf{g}_j$. Another useful relation is given by

$$\delta^j_i = \mathbf{g}_i{\cdot}\mathbf{g}^j = (g_{ik}\mathbf{g}^k){\cdot}(g^{jl}\mathbf{g}_l) = g_{ik}g^{jl}\delta^k_l$$

or

$$\boxed{g_{ik}g^{jk} = \delta^j_i.} \tag{2.85}$$

In *orthogonal curvilinear coordinates*, Eqs. (2.82) show that $g^{ij} = g_{ij} = 0$ for $i \neq j$, and so Eq. (2.85) gives $g_{(ii)} = 1/g^{(ii)}$.

The covariant and contravariant components of $\mathbf{I}$ can be used to raise and lower indices. One example of this is illustrated by Eqs. (2.84), where the indices of the base vectors are raised or lowered. For vector components, Eqs. (2.18), (2.19), and (2.82) give

$$\begin{aligned} a^i &= \mathbf{g}^i \cdot \mathbf{a} = \mathbf{g}^i \cdot (a_j \mathbf{g}^j) = g^{ij} a_j \\ a_i &= \mathbf{g}_i \cdot \mathbf{a} = \mathbf{g}_i \cdot (a^j \mathbf{g}_j) = g_{ij} a^j. \end{aligned} \tag{2.86}$$

In general, raising an index requires a g^{ij} and lowering an index requires a g_{ij}. (The raised or lowered index also is replaced by i or j.) This applies also to tensor components. For example,

$$\begin{aligned} T^{i}_{\cdot j} &= \mathbf{g}^i \cdot \mathbf{T} \cdot \mathbf{g}_j = \mathbf{g}^i \cdot (T_k^{\cdot l} \mathbf{g}^k \mathbf{g}_l) \cdot \mathbf{g}_j \\ &= g^{ik} g_{lj} T_k^{\cdot l} \end{aligned} \tag{2.87}$$

in which g^{ik} raises one index, while g_{lj} lowers the other. Similarly, we can find

$$\begin{aligned} T_{ij} &= g_{ik} g_{lj} T^{kl} \\ T^{ij} &= g^{ik} g^{lj} T_{kl} \\ T_j^{\cdot i} &= g^{ik} g_{lj} T^{l}_{\cdot k}. \end{aligned} \tag{2.88}$$

Note that the components of the metric tensor $\mathbf{I}$ relate quantities in the *same* coordinate system, while the components of the coordinate transformation tensors $\mathbf{A}$ and $\mathbf{B}$ relate quantities in *different* coordinate systems [compare Eqs. (2.46) and (2.84)].

2.6.2 *Permutation Tensor*

According to Eq. (2.71), $\det \mathbf{T} = \det[T^{i}_{\cdot j}]$. In this section, we write $\det \mathbf{T} = \det[T^i_j]$ for convenience, where T^i_j is the component in the ith row and the jth column of the matrix; hence,

$$T = \det \mathbf{T} = \begin{vmatrix} T_1^1 & T_2^1 & T_3^1 \\ T_1^2 & T_2^2 & T_3^2 \\ T_1^3 & T_2^3 & T_3^3 \end{vmatrix}. \tag{2.89}$$

Expanding the determinant in the usual manner shows that it can be written

$$T = T_1^i T_2^j T_3^k e_{ijk} = T_i^1 T_j^2 T_k^3 e^{ijk} \tag{2.90}$$

or, equivalently,

$$e_{pqr}T = e_{ijk}T^i_pT^j_qT^k_r, \qquad e^{pqr}T = e^{ijk}T^p_iT^q_jT^r_k \tag{2.91}$$

where

$$\boxed{e_{ijk} = e^{ijk} = \begin{cases} +1 & \text{if } (i,j,k) \text{ is an even permutation} \\ -1 & \text{if } (i,j,k) \text{ is an odd permutation} \\ 0 & \text{if two or more indices are equal} \end{cases}} \tag{2.92}$$

are **permutation symbols**. (An even permutation also is called **cyclic** and an odd permutation is **anticyclic**.)

Permutation symbols are useful in manipulations involving cross products. For example, consider the cross product of the *unit vectors* given by

$$\begin{aligned} \mathbf{e}_i &= \frac{\mathbf{g}_i}{|\mathbf{g}_i|} = \frac{\mathbf{g}_i}{\sqrt{g_{(ii)}}} \\ \mathbf{e}^i &= \frac{\mathbf{g}^i}{|\mathbf{g}^i|} = \frac{\mathbf{g}^i}{\sqrt{g^{(ii)}}} \end{aligned} \tag{2.93}$$

where $|\mathbf{g}_i| = \sqrt{\mathbf{g}_{(i)}\cdot\mathbf{g}_{(i)}} = \sqrt{g_{(ii)}}$ is the magnitude of the base vector $\mathbf{g}_i$, with a similar meaning for $|\mathbf{g}^i|$. Since $\mathbf{e}^1$ is orthogonal to $\mathbf{e}_2$ and $\mathbf{e}_3$, we have $\mathbf{e}_2 \times \mathbf{e}_3 = \mathbf{e}^1$, $\mathbf{e}_3 \times \mathbf{e}_2 = -\mathbf{e}^1$, $\mathbf{e}_2 \times \mathbf{e}_2 = 0$, etc., or

$$\mathbf{e}_i \times \mathbf{e}_j = e_{ijk}\mathbf{e}^k \tag{2.94}$$

in general, which shows that

$$(\mathbf{e}_i \times \mathbf{e}_j)\cdot\mathbf{e}_k = e_{ijk}. \tag{2.95}$$

Extending this concept to general base vectors, we write

$$\begin{aligned} \mathbf{g}_i \times \mathbf{g}_j &= \epsilon_{ijk}\mathbf{g}^k \\ (\mathbf{g}_i \times \mathbf{g}_j)\cdot\mathbf{g}_k &= \epsilon_{ijk} \end{aligned} \tag{2.96}$$

and

$$\begin{aligned} \mathbf{g}^i \times \mathbf{g}^j &= \epsilon^{ijk}\mathbf{g}_k \\ (\mathbf{g}^i \times \mathbf{g}^j)\cdot\mathbf{g}^k &= \epsilon^{ijk}. \end{aligned} \tag{2.97}$$

The ϵ_{ijk} and ϵ^{ijk} are components of the third-order **permutation tensor**

$$\boxed{\boldsymbol{\epsilon} = \epsilon_{ijk}\, \mathbf{g}^i\mathbf{g}^j\mathbf{g}^k = \epsilon^{ijk}\, \mathbf{g}_i\mathbf{g}_j\mathbf{g}_k.} \tag{2.98}$$

In direct notation, Eqs. (2.96) and (2.97) can be written

$$\begin{aligned} \mathbf{g}_i{\times}\mathbf{g}_j &= \mathbf{g}_i\mathbf{g}_j{:}\boldsymbol{\epsilon} \\ (\mathbf{g}_i{\times}\mathbf{g}_j){\cdot}\mathbf{g}_k &= \mathbf{g}_i\mathbf{g}_j\mathbf{g}_k\,\dot{:}\,\boldsymbol{\epsilon} \end{aligned} \tag{2.99}$$

and

$$\begin{aligned} \mathbf{g}^i{\times}\mathbf{g}^j &= \mathbf{g}^i\mathbf{g}^j{:}\boldsymbol{\epsilon} \\ (\mathbf{g}^i{\times}\mathbf{g}^j){\cdot}\mathbf{g}^k &= \mathbf{g}^i\mathbf{g}^j\mathbf{g}^k\,\dot{:}\,\boldsymbol{\epsilon} \end{aligned} \tag{2.100}$$

as can be verified easily using Eq. (2.27) and its extension, the triple-dot product given by $\mathbf{abc}\,\dot{:}\,\mathbf{def} = (\mathbf{a}{\cdot}\mathbf{d})(\mathbf{b}{\cdot}\mathbf{e})(\mathbf{c}{\cdot}\mathbf{f})$.

In Cartesian coordinates, $\epsilon_{ijk} = e_{ijk}$, but since the $\mathbf{g}_i$ are not always unit vectors, the ϵ_{ijk} differ from the e_{ijk} in general. To find the relation between these components, we let the unit vectors $\mathbf{e}_i$ of Eqs. (2.93)–(2.95) be Cartesian base vectors ($\mathbf{e}_i = \mathbf{e}^i$) and set $\mathbf{e}_i = \bar{\mathbf{g}}_i$ and $e_{ijk} = \bar{\epsilon}_{ijk}$. Then, the permutation tensor can be written in terms of the Cartesian (barred) system and a general (unbarred) system as

$$\boldsymbol{\epsilon} = e_{ijk}\mathbf{e}^i\mathbf{e}^j\mathbf{e}^k = \bar{\epsilon}_{ijk}\, \bar{\mathbf{g}}^i\bar{\mathbf{g}}^j\bar{\mathbf{g}}^k = \epsilon_{ijk}\, \mathbf{g}^i\mathbf{g}^j\mathbf{g}^k. \tag{2.101}$$

Relative to the natural basis, the covariant components are

$$\begin{aligned} \epsilon_{ijk} &= \mathbf{g}_i\mathbf{g}_j\mathbf{g}_k\,\dot{:}\,\boldsymbol{\epsilon} \\ &= \mathbf{g}_i\mathbf{g}_j\mathbf{g}_k\,\dot{:}\,(\bar{\epsilon}_{lmn}\, \bar{\mathbf{g}}^l\bar{\mathbf{g}}^m\bar{\mathbf{g}}^n) \\ &= (\mathbf{g}_i{\cdot}\bar{\mathbf{g}}^l)(\mathbf{g}_j{\cdot}\bar{\mathbf{g}}^m)(\mathbf{g}_k{\cdot}\bar{\mathbf{g}}^n)\, \bar{\epsilon}_{lmn} \\ &= B_i^l B_j^m B_k^n\, e_{lmn} \end{aligned} \tag{2.102}$$

in which Eq. $(2.49)_2$ has been used. This transformation and Eq. $(2.99)_1$, which shows that $\boldsymbol{\epsilon}$ converts a dyad into a vector via the double-dot product, demonstrate the tensor character of $\boldsymbol{\epsilon}$.

The relation (2.102) between ϵ_{ijk} and e_{ijk} can be simplified as follows. First, for $(i, j, k) = (1, 2, 3)$, comparison of Eqs. (2.90) and (2.102) reveals that $\epsilon_{123} = \det[B^i_j]$, with a similar result for other even permutations of the indices. (The sign changes for odd permutations.) Next, we evaluate $\det[B^i_j] = \det \mathbf{B}$. In terms of unbarred and barred base vectors, the identity tensor is [see Eq. (2.81)]

$$\mathbf{I} = g_{ij}\mathbf{g}^i\mathbf{g}^j = \bar{g}_{ij}\bar{\mathbf{g}}^i\bar{\mathbf{g}}^j. \tag{2.103}$$

Extracting the covariant components and applying Eq. (2.49) yields

$$\begin{aligned} g_{ij} &= \mathbf{I}{:}\mathbf{g}_i\mathbf{g}_j = (\bar{g}_{kl}\bar{\mathbf{g}}^k\bar{\mathbf{g}}^l){:}\mathbf{g}_i\mathbf{g}_j \\ &= \bar{g}_{kl}(\bar{\mathbf{g}}^k{\cdot}\mathbf{g}_i)(\bar{\mathbf{g}}^l{\cdot}\mathbf{g}_j) \\ &= B^k_i B^l_j \bar{g}_{kl}. \end{aligned} \tag{2.104}$$

Therefore,

$$\det[g_{ij}] = \det[B^k_i]\det[B^l_j]\det[\bar{g}_{kl}] = (\det \mathbf{B})^2 \det[\bar{g}_{kl}]$$

or

$$g = (\det \mathbf{B})^2\, \bar{g} \tag{2.105}$$

where

$$\boxed{g \equiv \det[g_{ij}], \qquad \bar{g} \equiv \det[\bar{g}_{ij}].} \tag{2.106}$$

Finally, since $\bar{g} = 1$ in Cartesian coordinates, Eq. (2.105) gives

$$\det \mathbf{B} = \sqrt{g}. \tag{2.107}$$

Because $\epsilon_{123} = \det \mathbf{B}$, $\epsilon_{321} = -\det \mathbf{B}$, etc., this result and a similar analysis for ϵ^{ijk} imply the relations

$$\boxed{\begin{aligned} \epsilon_{ijk} &= \sqrt{g}\, e_{ijk} \\ \epsilon^{ijk} &= e^{ijk}/\sqrt{g}. \end{aligned}} \tag{2.108}$$

Note that, as Eqs. (2.96) and (2.108) give $(\mathbf{g}_1 \times \mathbf{g}_2){\cdot}\mathbf{g}_3 = \epsilon_{123} = \sqrt{g}$, the quantity $g = \det[g_{ij}]$ is the same as that in Eq. (2.16), i.e., it represents the volume of the parallelepiped with the $\mathbf{g}_i$ as edges. Equations (2.108), therefore, explain the square root in Eq. (2.16).

Now, we list a couple of useful formulas involving the permutation tensor. First, with Eqs. (2.96) and (2.99), the cross product of two vectors can be written

$$\begin{aligned} \mathbf{a}\times\mathbf{b} &= (a^i\mathbf{g}_i)\times(b^j\mathbf{g}_j) \\ &= a^i b^j \epsilon_{ijk}\mathbf{g}^k \\ &= a^i b^j \mathbf{g}_i\mathbf{g}_j{:}\boldsymbol{\epsilon}. \end{aligned} \tag{2.109}$$

Second, the ϵ-δ **identity** is

$$\boxed{\epsilon_{ijk}\epsilon^{irs} = \delta_j^r\delta_k^s - \delta_j^s\delta_k^r.} \tag{2.110}$$

2.6.3 *Orthogonal Tensors*

Consider the transformation of two vectors $\mathbf{a}$ and $\mathbf{b}$ according to the relations

$$\bar{\mathbf{a}} = \mathbf{Q}\cdot\mathbf{a}, \qquad \bar{\mathbf{b}} = \mathbf{Q}\cdot\mathbf{b} \tag{2.111}$$

where $\mathbf{Q}$ is an **orthogonal tensor** defined by

$$\boxed{\mathbf{Q}^T\cdot\mathbf{Q} = \mathbf{Q}\cdot\mathbf{Q}^T = \mathbf{I},} \tag{2.112}$$

i.e., $\mathbf{Q}^T = \mathbf{Q}^{-1}$. Then,

$$\begin{aligned} \bar{\mathbf{a}}\cdot\bar{\mathbf{b}} &= (\mathbf{Q}\cdot\mathbf{a})\cdot(\mathbf{Q}\cdot\mathbf{b}) = (\mathbf{a}\cdot\mathbf{Q}^T)\cdot(\mathbf{Q}\cdot\mathbf{b}) \\ &= \mathbf{a}\cdot(\mathbf{Q}^T\cdot\mathbf{Q})\cdot\mathbf{b} = \mathbf{a}\cdot\mathbf{I}\cdot\mathbf{b} \\ &= \mathbf{a}\cdot\mathbf{b} \end{aligned} \tag{2.113}$$

where Eq. (2.37) and associativity have been used. Thus, the angle between the vectors $\bar{\mathbf{a}}$ and $\bar{\mathbf{b}}$ is the same as that between $\mathbf{a}$ and $\mathbf{b}$. Moreover, because $\det(\mathbf{Q}^T\cdot\mathbf{Q}) = \det\mathbf{Q}^T\det\mathbf{Q} = \det\mathbf{Q}^2 = \det\mathbf{I} = 1$, we have $\det\mathbf{Q} = \pm 1$. If $\det\mathbf{Q} = |\mathbf{Q}| = +1$, $\mathbf{Q}$ is called a **proper orthogonal tensor**, and Eq. (2.111) gives $|\bar{\mathbf{a}}| = |\mathbf{Q}|\,|\mathbf{a}| = |\mathbf{a}|$, and similarly $|\bar{\mathbf{b}}| = |\mathbf{b}|$. In this case, therefore, $\mathbf{a}$ and $\mathbf{b}$ retain their lengths and relative positions in space, and so these vectors undergo a rigid-body rotation due to the transformation by $\mathbf{Q}$. For this reason, a proper orthogonal tensor also can be called a **rotation tensor**.

Example 2.6 Consider rigid-body rotation about the z-axis of the Cartesian base vectors $\mathbf{e}_x$ and $\mathbf{e}_y$ into the new unit vectors $\bar{\mathbf{e}}_x$ and $\bar{\mathbf{e}}_y$ [Fig. 2.2 (page 9) with $(\mathbf{e}_1, \mathbf{e}_2) = (\mathbf{e}_x, \mathbf{e}_y)$ and $(\bar{\mathbf{e}}_1, \bar{\mathbf{e}}_2) = (\bar{\mathbf{e}}_x, \bar{\mathbf{e}}_y)$]. According to Eqs. (2.111), the transformation can be expressed in the form

$$\bar{\mathbf{e}}_x = \mathbf{Q}\cdot\mathbf{e}_x, \qquad \bar{\mathbf{e}}_y = \mathbf{Q}\cdot\mathbf{e}_y, \qquad \bar{\mathbf{e}}_z = \mathbf{Q}\cdot\mathbf{e}_z. \tag{2.114}$$

Determine the Cartesian components of $\mathbf{Q}$.

Solution. In this relatively simple problem, we can write by inspection

$$\mathbf{Q} = \bar{\mathbf{e}}_x\mathbf{e}_x + \bar{\mathbf{e}}_y\mathbf{e}_y + \bar{\mathbf{e}}_z\mathbf{e}_z \tag{2.115}$$

where (see Fig. 2.2)

$$\begin{aligned} \bar{\mathbf{e}}_x &= \mathbf{e}_x \cos\theta + \mathbf{e}_y \sin\theta \\ \bar{\mathbf{e}}_y &= -\mathbf{e}_x \sin\theta + \mathbf{e}_y \cos\theta \\ \bar{\mathbf{e}}_z &= \mathbf{e}_z. \end{aligned} \tag{2.116}$$

Substitution of Eq. (2.115) into (2.114) verifies this representation. In fact, comparing Eqs. $(2.47)_1$ and (2.115) shows that this rotation can be considered a coordinate transformation from the unbarred to the barred base vectors. In matrix form, inserting Eqs. (2.116) into (2.115) yields

$$\mathbf{Q}_{(\mathbf{e}_i\mathbf{e}_j)} = \begin{bmatrix} \cos\theta & -\sin\theta & 0 \\ \sin\theta & \cos\theta & 0 \\ 0 & 0 & 1 \end{bmatrix}. \tag{2.117}$$

Finally, $\mathbf{Q}\cdot\mathbf{Q}^T = \mathbf{e}_x\mathbf{e}_x + \mathbf{e}_y\mathbf{e}_y + \mathbf{e}_z\mathbf{e}_z = \mathbf{I}$ can be verified easily. ■

2.7 Physical Components

In general curvilinear coordinates, the components of vectors and tensors do not always have a direct physical meaning. If the base vectors have different dimensions, for example, then the components have different dimensions, since their combination must be consistent with the dimensions of the vector or tensor. For instance, the velocity vector in cylindrical polar coordinates can be written

$$\mathbf{v} = v^r\mathbf{g}_r + v^\theta\mathbf{g}_\theta + v^z\mathbf{g}_z$$

where $\mathbf{g}_r = \mathbf{e}_r$, $\mathbf{g}_\theta = r\mathbf{e}_\theta$, and $\mathbf{g}_z = \mathbf{e}_z$ (see Example 2.1, page 10). Since $\mathbf{g}_r$ and $\mathbf{g}_z$ are dimensionless, v^r and v^z have units of velocity, but v^θ has units of $(\text{time})^{-1}$. Moreover, if dimensionless base vectors are not unit vectors, the magnitudes of the components are not consistent with the physical quantity actually observed. The **physical components** of a vector or tensor are components that have physically meaningful units and magnitudes. Often it is convenient to derive the governing equations for a problem in terms of tensor components but to solve the problem using physical components.

2.7.1 *Physical Components of Vectors*

The physical components of a vector $\mathbf{a}$ are the components of $\mathbf{a}$ relative to a set of unit vectors. With the unit vectors $\mathbf{e}_i$ defined by Eq. $(2.93)_1$, we have

$$\mathbf{a} = a^i\mathbf{g}_i = a^i\sqrt{g_{(ii)}}\,\mathbf{e}_i \equiv \hat{a}^i\mathbf{e}_i \tag{2.118}$$

where

$$\boxed{\hat{a}^i = a^i\sqrt{g_{(ii)}}} \tag{2.119}$$

are the physical components of $\mathbf{a}$. Since the $\mathbf{e}_i$ are dimensionless, the $\hat{a}^i$ have the same units as $\mathbf{a}$. Moreover, $\hat{a}^i$ is the length of the vector $a^{(i)}\mathbf{g}_{(i)}$ (see Fig. 2.6, page 15). In terms of the covariant components of $\mathbf{a}$, Eqs. $(2.86)_1$ and (2.119) give

$$\hat{a}^i = g^{ij}a_j\sqrt{g_{(ii)}}. \tag{2.120}$$

2.7.2 *Physical Components of Tensors*

For second-order tensors, physical components can be defined in various ways. Here, we base our definition on the mixed tensor components $T^i_{\cdot j}$ of $\mathbf{T}$, which were used to define the trace and determinant [Eqs. (2.66) and (2.71)]. The $T^i_{\cdot j}$ can be transformed to or from other components using the metric tensor to raise or lower indices.

Consider the transformation

$$\mathbf{b} = \mathbf{T}\cdot\mathbf{a}, \tag{2.121}$$

which gives the components

$$\begin{aligned} b^i &= \mathbf{g}^i \cdot \mathbf{b} = \mathbf{g}^i \cdot \mathbf{T} \cdot \mathbf{a} \\ &= \mathbf{g}^i \cdot (T^k_{\cdot l} \mathbf{g}_k \mathbf{g}^l) \cdot (a^j \mathbf{g}_j) \\ &= T^k_{\cdot l} a^j \delta^i_k \delta^l_j \\ &= T^i_{\cdot j} a^j . \end{aligned} \tag{2.122}$$

We define the physical components $\hat{T}^i_{\cdot j}$ of $\mathbf{T}$ through the analogous relation

$$\hat{b}^i \equiv \hat{T}^i_{\cdot j} \hat{a}^j \tag{2.123}$$

where Eq. (2.119) gives

$$\hat{a}^j = a^j \sqrt{g_{(jj)}}, \qquad \hat{b}^i = b^i \sqrt{g_{(ii)}}. \tag{2.124}$$

Since $\hat{b}^i$ and $\hat{a}^j$ possess physically meaningful units, $\hat{T}^i_{\cdot j}$ does also. Inserting Eqs. (2.124) into Eq. (2.122) yields

$$\hat{b}^i = T^i_{\cdot j} \sqrt{\frac{g_{(ii)}}{g_{(jj)}}}\, \hat{a}^j$$

and comparison with Eq. (2.123) gives

$$\boxed{\hat{T}^i_{\cdot j} = T^i_{\cdot j} \sqrt{\frac{g_{(ii)}}{g_{(jj)}}}.} \tag{2.125}$$

It is useful to note that Eq. (2.125) and Table 2.5 give

$$\begin{aligned} \hat{T}^i_{\cdot i} &= T^i_{\cdot i} = \operatorname{tr} \mathbf{T} \\ \det[\hat{T}^i_{\cdot j}] &= \det[T^i_{\cdot j}] = \det \mathbf{T}. \end{aligned} \tag{2.126}$$

Hence, the invariants given by Eq. (2.77) have the same values when expressed in terms of *either* tensor or physical components. This makes it easier to set up problems in terms of physical components, but it must be remembered that, in general, these components do not obey the rules for tensor transformation.

2.8 Vector and Tensor Calculus

In Cartesian coordinates, differentiation of vectors and tensors with respect to scalars is straightforward. Since the base vectors are constant, only differentiation of the components is required. In general curvilinear coordinates, however, the base vectors can change from point to point in space. Thus, differentiation of base vectors may enter the analysis, complicating the situation. In addition, we often need to differentiate quantities with respect to vectors and tensors. This section addresses these operations.

2.8.1 *Differentiation with Respect to Scalars*

First, we consider differentiation of vectors and tensors with respect to the coordinates x^i. For any function $\phi(x^i)$, this operation is denoted by a comma, i.e., $\partial\phi/\partial x^i \equiv \phi_{,i}$.

Differentiation of Base Vectors

Since the derivative $\mathbf{g}_{i,j}$ is a vector, it can be resolved into components relative to the base vectors as

$$\boxed{\mathbf{g}_{i,j} = \Gamma_{ijk}\mathbf{g}^k = \Gamma^k_{ij}\mathbf{g}_k} \tag{2.127}$$

where the components Γ_{ijk} and Γ^k_{ij} are called **Christoffel symbols**. It is important to note that these quantities are *not* components of a third-order tensor. Dotting this relation with the base vectors yields

$$\begin{aligned}
\mathbf{g}_{i,j}\cdot\mathbf{g}_k &= (\Gamma_{ijl}\mathbf{g}^l)\cdot\mathbf{g}_k = \Gamma_{ijl}\delta^l_k \\
\mathbf{g}_{i,j}\cdot\mathbf{g}^k &= (\Gamma^l_{ij}\mathbf{g}_l)\cdot\mathbf{g}^k = \Gamma^l_{ij}\delta^k_l
\end{aligned}$$

or

$$\boxed{\begin{aligned}
\Gamma_{ijk} &= \mathbf{g}_{i,j}\cdot\mathbf{g}_k \\
\Gamma^k_{ij} &= \mathbf{g}_{i,j}\cdot\mathbf{g}^k.
\end{aligned}} \tag{2.128}$$

Christoffel symbols enter the analysis whenever we differentiate general vectors and tensors.

The derivative of a contravariant base vector can be found by differentiating the relation

$$\mathbf{g}_i\cdot\mathbf{g}^j = \delta^j_i$$

with respect to x^k to get

$$\mathbf{g}_{i},_{k}\cdot\mathbf{g}^{j}+\mathbf{g}_{i}\cdot\mathbf{g}^{j},_{k}=0.$$

Substituting Eq. (2.128)$_2$ gives

$$\mathbf{g}^{j},_{k}\cdot\mathbf{g}_{i}=-\Gamma^{j}_{ik} \tag{2.129}$$

or

$$\boxed{\mathbf{g}^{i},_{j}=-\Gamma^{i}_{kj}\mathbf{g}^{k}} \tag{2.130}$$

which can be verified by dotting both sides of this relation with $\mathbf{g}_l$ and changing the indices.

Two useful properties of the Christoffel symbols warrant attention. First, since $\mathbf{g}_i=\mathbf{r},_i$, then $\mathbf{g}_i,_j=\mathbf{r},_{ij}=\mathbf{r},_{ji}=\mathbf{g}_j,_i$. Thus, from Eq. (2.127), we find

$$\boxed{\Gamma_{ijk}=\Gamma_{jik}, \qquad \Gamma^{k}_{ij}=\Gamma^{k}_{ji}.} \tag{2.131}$$

Second, dotting Eq. (2.127) with $\mathbf{g}_l$ gives

$$\Gamma_{ijk}\mathbf{g}^{k}\cdot\mathbf{g}_{l}=\Gamma^{k}_{ij}\mathbf{g}_{k}\cdot\mathbf{g}_{l}$$

or, since $\mathbf{g}^k\cdot\mathbf{g}_l=\delta^k_l$ and $\mathbf{g}_k\cdot\mathbf{g}_l=g_{kl}$,

$$\Gamma_{ijl}=\Gamma^{k}_{ij}g_{kl}. \tag{2.132}$$

Similarly, dotting both sides of Eq. (2.127) with $\mathbf{g}^l$ yields

$$\Gamma^{l}_{ij}=\Gamma_{ijk}g^{kl}. \tag{2.133}$$

Hence, although Γ_{ijk} and Γ^k_{ij} are not tensor components, the components of the metric tensor can be used to raise or lower the *third* index.

A useful formula for computing the Christoffel symbols in a given coordinate system can be found by differentiating the expression

$$g_{ij}=\mathbf{g}_{i}\cdot\mathbf{g}_{j}$$

with respect to x^k to obtain

$$\begin{aligned} g_{ij},_{k} &= \mathbf{g}_{i},_{k}\cdot\mathbf{g}_{j}+\mathbf{g}_{i}\cdot\mathbf{g}_{j},_{k} \\ &= \Gamma_{ikj}+\Gamma_{jki} \end{aligned}$$

where Eq. (2.128)$_1$ has been used. Permuting the subscripts and using Eqs. (2.131) yields the sequence of equations

$$\begin{aligned} g_{ij},_k &= \Gamma_{kij} + \Gamma_{jki} \\ g_{jk},_i &= \Gamma_{ijk} + \Gamma_{kij} \\ g_{ki},_j &= \Gamma_{jki} + \Gamma_{ijk}. \end{aligned} \tag{2.134}$$

Now, adding the last two equations and subtracting the first gives

$$\boxed{2\Gamma_{ijk} = g_{jk},_i + g_{ki},_j - g_{ij},_k \, .} \tag{2.135}$$

Thus, if the metric for a coordinate system is known, this equation gives the Γ_{ijk} and then Eq. (2.133) gives the Γ^k_{ij}. In Cartesian coordinates, because the base vectors are constant, all of the Christoffel symbols are zero.

Example 2.7 Compute the Christoffel symbols for cylindrical polar coordinates.

Solution. With $(x^1, x^2, x^3) = (r, \theta, z)$, Eqs. (2.10) and (2.23) give the base vectors

$$\begin{aligned} \mathbf{g}_1 &= \mathbf{g}_r = \mathbf{e}_x \cos\theta + \mathbf{e}_y \sin\theta = \mathbf{e}_r \\ \mathbf{g}_2 &= \mathbf{g}_\theta = r(-\mathbf{e}_x \sin\theta + \mathbf{e}_y \cos\theta) = r\mathbf{e}_\theta \\ \mathbf{g}_3 &= \mathbf{g}_z = \mathbf{e}_z \\ \mathbf{g}^1 &= \mathbf{g}^r = \mathbf{e}_r \\ \mathbf{g}^2 &= \mathbf{g}^\theta = r^{-1}\mathbf{e}_\theta \\ \mathbf{g}^3 &= \mathbf{g}^z = \mathbf{e}_z, \end{aligned} \tag{2.136}$$

and the nonzero components of the metric tensor ($g_{ij} = \mathbf{g}_i \cdot \mathbf{g}_j$, $g^{ij} = \mathbf{g}^i \cdot \mathbf{g}^j$) are

$$\begin{aligned} g_{11} &= 1, \qquad g_{22} = r^2, \qquad g_{33} = 1 \\ g^{11} &= 1, \qquad g^{22} = r^{-2}, \qquad g^{33} = 1. \end{aligned} \tag{2.137}$$

As the only nonzero derivative of the g_{ij} is $g_{22},_1 = 2r$, the only nonzero Christoffel symbols given by Eq. (2.135) are

$$\begin{aligned} \Gamma_{221} &= -\tfrac{1}{2} g_{22},_1 = -r \\ \Gamma_{122} &= \Gamma_{212} = \tfrac{1}{2} g_{22},_1 = r \end{aligned} \tag{2.138}$$

and from $\Gamma^l_{ij} = \Gamma_{ijk} g^{kl}$,

$$\begin{aligned}\Gamma^1_{22} &= \Gamma_{221}\, g^{11} = -r \\ \Gamma^2_{12} &= \Gamma^2_{21} = \Gamma_{122}\, g^{22} = \Gamma_{212}\, g^{22} = r^{-1}.\end{aligned} \tag{2.139}$$

As a check, we now compute

$$\mathbf{g}_{r,\theta} = \mathbf{g}_{1,2} = \Gamma_{12k}\mathbf{g}^k = \Gamma^k_{12}\mathbf{g}_k$$

as given by Eqs. (2.127). Substituting Eqs. (2.136), (2.138), and (2.139) yields

$$\begin{aligned}\mathbf{g}_{r,\theta} &= \Gamma_{122}\, \mathbf{g}^2 = r(r^{-1}\mathbf{e}_\theta) = \mathbf{e}_\theta \\ &= \Gamma^2_{12}\, \mathbf{g}_2 = r^{-1}(r\mathbf{e}_\theta) = \mathbf{e}_\theta\end{aligned}$$

which agree with the result given by taking the derivative of Eq. $(2.136)_1$ with respect to θ. ■

Example 2.8 Compute the derivative of $\sqrt{g} = \mathbf{g}_1\cdot(\mathbf{g}_2\times\mathbf{g}_3)$ with respect to x^i. Express the result in terms of g and Christoffel symbols.

Solution. Differentiating a product provides

$$\begin{aligned}(\sqrt{g})_{,i} &= \mathbf{g}_{1,i}\cdot(\mathbf{g}_2\times\mathbf{g}_3) + \mathbf{g}_1\cdot(\mathbf{g}_{2,i}\times\mathbf{g}_3) + \mathbf{g}_1\cdot(\mathbf{g}_2\times\mathbf{g}_{3,i}) \\ &= \mathbf{g}_{1,i}\cdot(\mathbf{g}_2\times\mathbf{g}_3) + \mathbf{g}_{2,i}\cdot(\mathbf{g}_3\times\mathbf{g}_1) + \mathbf{g}_{3,i}\cdot(\mathbf{g}_1\times\mathbf{g}_2)\end{aligned}$$

where the vectors have been permuted according to the scalar triple product formula in Table 2.1 (page 19). Next, substituting Eq. (2.127) gives

$$(\sqrt{g})_{,i} = \Gamma^j_{1i}\mathbf{g}_j\cdot(\mathbf{g}_2\times\mathbf{g}_3) + \Gamma^j_{2i}\mathbf{g}_j\cdot(\mathbf{g}_3\times\mathbf{g}_1) + \Gamma^j_{3i}\mathbf{g}_j\cdot(\mathbf{g}_1\times\mathbf{g}_2).$$

Since $\mathbf{g}_j\cdot(\mathbf{g}_2\times\mathbf{g}_3) \neq 0$ only for $j = 1$, and so on for the other terms, this expression can be written

$$\begin{aligned}(\sqrt{g})_{,i} &= \Gamma^1_{1i}\mathbf{g}_1\cdot(\mathbf{g}_2\times\mathbf{g}_3) + \Gamma^2_{2i}\mathbf{g}_2\cdot(\mathbf{g}_3\times\mathbf{g}_1) + \Gamma^3_{3i}\mathbf{g}_3\cdot(\mathbf{g}_1\times\mathbf{g}_2) \\ &= \sqrt{g}\,(\Gamma^1_{1i} + \Gamma^2_{2i} + \Gamma^3_{3i})\end{aligned}$$

or

$$\boxed{(\sqrt{g})_{,i} = \sqrt{g}\,\Gamma^j_{ji}.} \tag{2.140}$$

■

Differentiation of Vectors

Differentiating the vector $\mathbf{a}$ and using Eqs. (2.127) and (2.130) gives

$$\begin{aligned}\mathbf{a},_j = (a^i\mathbf{g}_i),_j &= a^i,_j\,\mathbf{g}_i + a^i\mathbf{g}_i,_j \\ &= a^i,_j\,\mathbf{g}_i + a^i\Gamma^k_{ij}\mathbf{g}_k \\ = (a_i\mathbf{g}^i),_j &= a_i,_j\,\mathbf{g}^i + a_i\mathbf{g}^i,_j \\ &= a_i,_j\,\mathbf{g}^i - a_i\Gamma^i_{jk}\mathbf{g}^k.\end{aligned}$$

Interchanging the dummy indices i and k in the terms involving Christoffel symbols yields

$$\boxed{\mathbf{a},_j = a^i|_j\,\mathbf{g}_i = a_i|_j\,\mathbf{g}^i} \tag{2.141}$$

where

$$\boxed{\begin{aligned}a^i|_j &\equiv a^i,_j + a^k\Gamma^i_{jk} \\ a_i|_j &\equiv a_i,_j - a_k\Gamma^k_{ij}\end{aligned}} \tag{2.142}$$

are called the **covariant derivatives** of a^i and a_i, respectively. Later, we will show that $a^i|_j$ and $a_i|_j$ are mixed and covariant components, respectively, of a second-order tensor (Section 2.8.3). In Cartesian coordinates, the vanishing of the Christoffel symbols implies that covariant and ordinary differentiation are the same. In general curvilinear coordinates, however, $a^i|_j = 0$ implies that the *vector* $\mathbf{a}$ is constant, but the *components* a^i are not ($a^i,_j \neq 0$). (Consider, for example, a vector that translates in a polar coordinate system.) This is consistent with the idea that a vector is geometrically invariant.

From Eqs. (2.130), (2.141), and (2.142), the second derivative of the vector $\mathbf{a}$ is

$$\begin{aligned}\mathbf{a},_{jk} &= (a_i|_j\mathbf{g}^i),_k = (a_i|_j),_k\,\mathbf{g}^i + (a_i|_j)\mathbf{g}^i,_k \\ &= (a_i,_j - a_l\Gamma^l_{ij}),_k\,\mathbf{g}^i - a_i|_j\Gamma^i_{kl}\mathbf{g}^l \\ &= (a_i,_{jk} - a_l,_k\,\Gamma^l_{ij} - a_l\Gamma^l_{ij},_k - a_l|_j\Gamma^l_{ik})\mathbf{g}^i \\ &= [a_i,_{jk} - a_l,_k\,\Gamma^l_{ij} - a_l\Gamma^l_{ij},_k - (a_l,_j - a_m\Gamma^m_{jl})\Gamma^l_{ik}]\mathbf{g}^i\end{aligned} \tag{2.143}$$

with a similar relation for $\mathbf{a},_{jk} = (a^i|_j\mathbf{g}_i),_k$. Thus, taking higher derivatives of a vector is straightforward, but labor-intensive. Symbolic manipulation computer programs, however, can make tensor calculus virtually painless.

Differentiation of Tensors

Differentiation of tensors also is straightforward when they are expressed in the form of dyadics (or polyadics). For example, for a second-order tensor in terms of mixed components, Eqs. (2.127) and (2.130) give

$$\begin{aligned}\mathbf{T},_k &= (T^i_{.j}\mathbf{g}_i\mathbf{g}^j),_k \\ &= T^i_{.j},_k\,\mathbf{g}_i\mathbf{g}^j + T^i_{.j}\mathbf{g}_i,_k\,\mathbf{g}^j + T^i_{.j}\mathbf{g}_i\mathbf{g}^j,_k \\ &= T^i_{.j},_k\,\mathbf{g}_i\mathbf{g}^j + T^i_{.j}\Gamma^l_{ik}\mathbf{g}_l\mathbf{g}^j - T^i_{.j}\mathbf{g}_i\Gamma^j_{kl}\mathbf{g}^l.\end{aligned}$$

On interchanging dummy indices, these and similar computations involving the other component representations of $\mathbf{T}$ produce

$$\boxed{\mathbf{T},_k = T_{ij}|_k\mathbf{g}^i\mathbf{g}^j = T^{ij}|_k\mathbf{g}_i\mathbf{g}_j = T^i_{.j}|_k\mathbf{g}_i\mathbf{g}^j = T_i^{.j}|_k\mathbf{g}^i\mathbf{g}_j} \tag{2.144}$$

where

$$\boxed{\begin{aligned}T_{ij}|_k &= T_{ij},_k - T_{lj}\Gamma^l_{ik} - T_{il}\Gamma^l_{jk} \\ T^{ij}|_k &= T^{ij},_k + T^{lj}\Gamma^i_{kl} + T^{il}\Gamma^j_{kl} \\ T^i_{.j}|_k &= T^i_{.j},_k + T^l_{.j}\Gamma^i_{kl} - T^i_{.l}\Gamma^l_{jk} \\ T_i^{.j}|_k &= T_i^{.j},_k - T_l^{.j}\Gamma^l_{ik} + T_i^{.l}\Gamma^j_{kl}\end{aligned}} \tag{2.145}$$

are the covariant derivatives of the indicated components.

It is important to note that taking the ordinary derivative of a vector or tensor is equivalent to taking the covariant derivative of the components while holding the base vectors constant [Eqs. (2.141) and (2.144)]. We again emphasize that *a vanishing covariant derivative implies that the vector or tensor is constant, but generally the components of the vector or tensor are not constant* [Eqs. (2.142) and (2.145)].

The tensor character of the covariant derivative can be used to derive an interesting feature of the components of the special tensors $\mathbf{I}$ and $\boldsymbol{\epsilon}$. In Cartesian coordinates, $g_{ij},_k = \epsilon_{ijk},_l = 0$, which can be written

$$g_{ij}|_k = \epsilon_{ijk}|_l = 0 \tag{2.146}$$

since the Christoffel symbols vanish. These last equations are in tensor form and, therefore, hold in any coordinate system, indicating that the tensors $\mathbf{I}$ and $\boldsymbol{\epsilon}$ are constants in Euclidean space.

2.8.2 *Differentiation with Respect to Vectors and Tensors*

The derivative of a scalar function $\phi(\mathbf{a}, \mathbf{T})$ with respect to the vector $\mathbf{a}$ is defined by

$$\boxed{d\phi = d\mathbf{a} \cdot \frac{\partial \phi}{\partial \mathbf{a}}} \tag{2.147}$$

which indicates that $\partial\phi/\partial\mathbf{a}$ is a vector. If the differential of $\mathbf{a}$ is written in terms of its contravariant and covariant components as

$$d\mathbf{a} = \mathbf{g}_i \, da^i = \mathbf{g}^i \, da_i, \tag{2.148}$$

then substitution into Eq. (2.147) shows that

$$\frac{\partial}{\partial \mathbf{a}} = \mathbf{g}^i \frac{\partial}{\partial a^i} = \mathbf{g}_i \frac{\partial}{\partial a_i}. \tag{2.149}$$

The first of these representations is verified by

$$\begin{aligned} d\phi &= d\mathbf{a} \cdot \frac{\partial \phi}{\partial \mathbf{a}} = (\mathbf{g}_i \, da^i) \cdot \left(\mathbf{g}^j \frac{\partial \phi}{\partial a^j} \right) \\ &= da^i \delta_i^j \frac{\partial \phi}{\partial a^j} = \frac{\partial \phi}{\partial a^i} da^i = d\phi \end{aligned}$$

where the chain rule for differentiation has been used. The second of (2.149) can be confirmed similarly.

Analogously, the derivative of ϕ with respect to the second-order tensor $\mathbf{T}$ is defined by

$$\boxed{d\phi = d\mathbf{T} : \frac{\partial \phi}{\partial \mathbf{T}}} \tag{2.150}$$

and differentiation with respect to higher-order tensors follows by adding more dots. If the differential of $\mathbf{T}$ is written, for example, in the form

$$d\mathbf{T} = dT^{ij} \mathbf{g}_i \mathbf{g}_j, \tag{2.151}$$

then comparing Eqs. (2.150) and (2.151) gives

$$\frac{\partial}{\partial \mathbf{T}} = \mathbf{g}^i \mathbf{g}^j \frac{\partial}{\partial T^{ij}} \tag{2.152}$$

as is confirmed by

$$\begin{aligned} d\phi &= d\mathbf{T} : \frac{\partial \phi}{\partial \mathbf{T}} = (dT^{ij}\mathbf{g}_i\mathbf{g}_j) : \left(\mathbf{g}^k\mathbf{g}^l \frac{\partial \phi}{\partial T^{kl}}\right) \\ &= dT^{ij} \frac{\partial \phi}{\partial T^{kl}} (\mathbf{g}_i \cdot \mathbf{g}^k)(\mathbf{g}_j \cdot \mathbf{g}^l) = dT^{ij} \frac{\partial \phi}{\partial T^{kl}} \delta_i^k \delta_j^l \\ &= \frac{\partial \phi}{\partial T^{ij}} dT^{ij} = d\phi. \end{aligned}$$

Equations (2.149) and (2.152) indicate that the component form of a derivative operator containing a vector or tensor is formed by moving the base vectors of the differential to the numerator and raising or lowering their indices. In summary, therefore, we can write

$$\boxed{\begin{aligned} \frac{\partial}{\partial \mathbf{a}} &= \mathbf{g}^i \frac{\partial}{\partial a^i} = \mathbf{g}_i \frac{\partial}{\partial a_i} \\ \frac{\partial}{\partial \mathbf{T}} &= \mathbf{g}^i\mathbf{g}^j \frac{\partial}{\partial T^{ij}} = \mathbf{g}_i\mathbf{g}_j \frac{\partial}{\partial T_{ij}} = \mathbf{g}^i\mathbf{g}_j \frac{\partial}{\partial T^i_{.j}} = \mathbf{g}_i\mathbf{g}^j \frac{\partial}{\partial T_i^{.j}}. \end{aligned}} \tag{2.153}$$

Note that the base vectors and the components in each of these expressions have the same variance.

The above observation is useful in differentiating vectors and tensors with respect to other vectors and tensors. The basic definitions, which follow from Eqs. (2.147) and (2.150), are

$$\boxed{\begin{aligned} d\mathbf{b} &= d\mathbf{a} \cdot \frac{\partial \mathbf{b}}{\partial \mathbf{a}} = d\mathbf{T} : \frac{\partial \mathbf{b}}{\partial \mathbf{T}} \\ d\mathbf{U} &= d\mathbf{a} \cdot \frac{\partial \mathbf{U}}{\partial \mathbf{a}} = d\mathbf{T} : \frac{\partial \mathbf{U}}{\partial \mathbf{T}} \end{aligned}} \tag{2.154}$$

for $\mathbf{b} = \mathbf{b}(\mathbf{a}, \mathbf{T})$ and $\mathbf{U} = \mathbf{U}(\mathbf{a}, \mathbf{T})$. In these expressions, the order of the factors is important, e.g.,

$$\frac{\partial \mathbf{b}}{\partial \mathbf{T}} = \frac{\partial}{\partial \mathbf{T}}(\mathbf{b}) = \frac{\partial}{\mathbf{g}_i\mathbf{g}_j \partial T^{ij}}(\mathbf{b}) = \mathbf{g}^i\mathbf{g}^j \frac{\partial \mathbf{b}}{\partial T^{ij}}. \tag{2.155}$$

2.8.3 *Gradient Operator*

The **gradient** ("del") **operator** $\boldsymbol{\nabla}$ is defined by

$$\boxed{d\phi = d\mathbf{r}\cdot\boldsymbol{\nabla}\phi} \tag{2.156}$$

where $d\phi$ is the differential of the scalar ϕ in the direction of the differential position vector $d\mathbf{r} = \mathbf{g}_i\, dx^i$. Comparison with Eqs. (2.147)–(2.149) shows that

$$\boxed{\boldsymbol{\nabla} \equiv \frac{\partial}{\partial \mathbf{r}} = \mathbf{g}^i \frac{\partial}{\partial x^i}} \tag{2.157}$$

with the invariance of $\boldsymbol{\nabla}$ indicated by the form $\partial/\partial\mathbf{r}$. This invariance also can be demonstrated by the manipulation

$$\begin{aligned}
\boldsymbol{\nabla} &= \mathbf{g}^i \frac{\partial}{\partial x^i} = \delta^i_j \mathbf{g}^j \frac{\partial}{\partial x^i} = (A^i_k B^k_j)\mathbf{g}^j \frac{\partial}{\partial x^i} \\
&= (B^k_j \mathbf{g}^j) \frac{\partial x^i}{\partial \bar{x}^k} \frac{\partial}{\partial x^i} \\
&= \bar{\mathbf{g}}^k \frac{\partial}{\partial \bar{x}^k} = \bar{\mathbf{g}}^i \frac{\partial}{\partial \bar{x}^i}
\end{aligned}$$

which involves Eqs. (2.46), (2.51), and (2.59) and the chain rule for differentiation. Equation (2.156) also holds if ϕ is replaced by a vector or a tensor.

Example 2.9 Using the results from Example 2.1 (page 10), write the gradient operator in Cartesian and cylindrical polar coordinates.

Solution. With $(x^1, x^2, x^3) = (x, y, z)$ and Eqs. (2.22), Eq. (2.157) gives

$$\boldsymbol{\nabla} = \mathbf{e}_x \frac{\partial}{\partial x} + \mathbf{e}_y \frac{\partial}{\partial y} + \mathbf{e}_z \frac{\partial}{\partial z} \tag{2.158}$$

in Cartesian coordinates. Similarly, with $(x^1, x^2, x^3) = (r, \theta, z)$ and Eq. (2.23), we get

$$\boldsymbol{\nabla} = \mathbf{e}_r \frac{\partial}{\partial r} + \frac{\mathbf{e}_\theta}{r} \frac{\partial}{\partial \theta} + \mathbf{e}_z \frac{\partial}{\partial z} \tag{2.159}$$

in cylindrical coordinates. ■

Gradient, Divergence, and Curl

Using the representation for $\boldsymbol{\nabla}$ given by Eq. (2.157), we can compute the gradient, divergence, and curl of a function in any coordinate system. The **gradient** of a scalar function ϕ is

$$\boldsymbol{\nabla}\phi = \mathbf{g}^i \frac{\partial \phi}{\partial x^i} = \mathbf{g}^i \phi_{,i}$$

or

$$\boxed{\boldsymbol{\nabla}\phi = \mathbf{g}^i \phi|_i} \tag{2.160}$$

since ordinary and covariant differentiation of a scalar are the same. The gradient of a vector $\mathbf{a}$ is

$$\boldsymbol{\nabla}\mathbf{a} = \mathbf{g}^j \frac{\partial \mathbf{a}}{\partial x^j} = \mathbf{g}^j \mathbf{a}_{,j} \,,$$

and substituting Eqs. (2.141) yields

$$\boxed{\boldsymbol{\nabla}\mathbf{a} = a^i|_j \mathbf{g}^j \mathbf{g}_i = a_i|_j \mathbf{g}^j \mathbf{g}^i,} \tag{2.161}$$

where care has been taken to maintain the order of the base vectors. This expression shows that the covariant derivatives $a^i|_j$ and $a_i|_j$ are components of the second-order tensor $\boldsymbol{\nabla}\mathbf{a}$.

The **divergence** of a vector $\mathbf{a}$, defined by $\boldsymbol{\nabla}\cdot\mathbf{a}$, can be obtained by a contraction of the gradient, i.e., by inserting a dot between the base vectors of Eqs. (2.161). Hence,

$$\begin{aligned} \boldsymbol{\nabla}\cdot\mathbf{a} &= a^i|_j \mathbf{g}^j \cdot \mathbf{g}_i = a^i|_j \delta_i^j = a^i|_i \\ &= a_i|_j \mathbf{g}^j \cdot \mathbf{g}^i = a_i|_j g^{ij} = a_i|^i \end{aligned}$$

or

$$\boxed{\boldsymbol{\nabla}\cdot\mathbf{a} = a^i|_i = a_i|^i} \tag{2.162}$$

where the contravariant components of the metric tensor have been used to raise one of the indices on the tensor component $a_i|_j$. The form of $a_i|^i$ appears to indicate *contravariant* differentiation, but this operation is of limited use since the components dx_i of $d\mathbf{r} = \mathbf{g}^i\, dx_i$ needed for $\partial/\partial x_i$ are not along the coordinate curves in general. The components $a_i|^i$, therefore, are considered shorthand for $a_i|_j g^{ij}$.

The **curl** of a vector $\mathbf{a}$ is

$$\begin{aligned}\nabla\times\mathbf{a} &= \left(\mathbf{g}^j\frac{\partial}{\partial x^j}\right)\times\mathbf{a} = \mathbf{g}^j\times\mathbf{a}_{,j}\\ &= a_i|_j\mathbf{g}^j\times\mathbf{g}^i\end{aligned}$$

where Eq. (2.141) has been used. With Eqs. $(2.97)_1$ and $(2.100)_1$, this expression assumes the forms

$$\boxed{\begin{aligned}\nabla\times\mathbf{a} &= a_i|_j\mathbf{g}^j\mathbf{g}^i{:}\boldsymbol{\epsilon}\\ &= a_i|_j\epsilon^{jik}\mathbf{g}_k = a_j|_i\epsilon^{ijk}\mathbf{g}_k.\end{aligned}} \tag{2.163}$$

Second Derivative

If we let

$$\mathbf{T} = (\nabla\mathbf{a})^T = a_i|_j\mathbf{g}^i\mathbf{g}^j \tag{2.164}$$

from Eq. (2.161), then $T_{ij} = a_i|_j$, and Eq. $(2.145)_1$ gives

$$\begin{aligned}T_{ij}|_k &= (a_i|_j)|_k \equiv a_i|_{jk}\\ &= a_i|_{j,k} - a_l|_j\Gamma^l_{ik} - a_i|_l\Gamma^l_{jk}.\end{aligned} \tag{2.165}$$

With Eq. (2.142), this relation becomes

$$a_i|_{jk} = (a_{i,j} - a_l\Gamma^l_{ij})_{,k} - (a_{l,j} - a_m\Gamma^m_{lj})\Gamma^l_{ik} - (a_{i,l} - a_m\Gamma^m_{il})\Gamma^l_{jk} \tag{2.166}$$

which is the **second covariant derivative** of a_i. It is important to note, by comparing Eqs. (2.143) and (2.166), that $\mathbf{a}_{,jk} \neq a_i|_{jk}\mathbf{g}^i$ in general. The difference is the last term in Eq. (2.166), which comes from differentiating the second base vector in the dyadic representation of Eq. (2.164).

The second covariant derivatives of Eq. (2.166) represent the components of a third-order tensor. This can be shown by combining Eqs. (2.144), (2.157), and (2.161) to produce the **curvature tensor**[9]

$$\nabla(\nabla\mathbf{a})^T = \mathbf{g}^k\frac{\partial}{\partial x^k}(a_i|_j\mathbf{g}^i\mathbf{g}^j) = a_i|_{jk}\mathbf{g}^k\mathbf{g}^i\mathbf{g}^j$$

[9]The name *curvature tensor* stems from the physical meaning of the second derivative.

or

$$\boxed{\nabla(\nabla \mathbf{a})^T = a_j|_{ki}\mathbf{g}^i\mathbf{g}^j\mathbf{g}^k.} \tag{2.167}$$

This tensor has important implications that depend on the geometry of the space. For example, interchanging j and k in Eq. (2.166) to produce $a_i|_{kj}$ and using Eqs. (2.131) gives

$$\begin{aligned} a_i|_{jk} - a_i|_{kj} &= a_m(\Gamma^m_{ik},_j - \Gamma^m_{ij},_k + \Gamma^m_{lj}\Gamma^l_{ik} - \Gamma^m_{lk}\Gamma^l_{ij}) \\ &\equiv a_m R^m_{.ijk} \end{aligned} \tag{2.168}$$

where $R^m_{.ijk}$ are components of the **Riemann-Christoffel tensor**. In Cartesian coordinates, since the Christoffel symbols are zero, $R^m_{.ijk} = 0$, and so $a_i|_{jk} = a_i|_{kj}$. The tensor character of $R^m_{.ijk}$ indicates that this relation holds in all coordinate systems in *three-dimensional (flat) Euclidean space*, and thus the order of covariant differentiation is interchangeable. On the two-dimensional space of a curved surface, however, a Cartesian system is not possible, and so $R^m_{.ijk} \neq 0$ in general. Hence, covariant differentiation is not always interchangeable on a curved surface.[10] This result is important in shell theory.

Next, we compute the **Laplacian** of a scalar ϕ by

$$\nabla^2\phi \equiv \nabla\cdot\nabla\phi. \tag{2.169}$$

Application of Eqs. (2.160) and (2.162) yields

$$\nabla^2\phi = \nabla\cdot(\mathbf{g}^i\phi|_i) = \phi|_i|^i$$

or

$$\boxed{\nabla^2\phi = \phi|_i^i = \phi|_{ij}g^{ij}.} \tag{2.170}$$

Substituting Eqs. (2.130) and (2.157) into (2.169) gives the explicit form

$$\nabla^2\phi = g^{ij}\phi,_{ij} - \Gamma^j_{ki}g^{ik}\phi,_j\,. \tag{2.171}$$

[10] A curved surface formed by curling an initially flat surface is an exception that allows interchangeable differentiation. This type of surface is called **developable.**

Operations with the Gradient Operator

With direct notation, many operations using the gradient operator are straightforward. For example, Eqs. (2.144) and (2.157) yield the gradient of the tensor $\mathbf{T} = T_{ij}\mathbf{g}^i\mathbf{g}^j$ as

$$\boxed{\nabla\mathbf{T} = \mathbf{g}^k\mathbf{T}_{,k} = T_{ij}|_k\mathbf{g}^k\mathbf{g}^i\mathbf{g}^j} \tag{2.172}$$

with Eq. $(2.145)_1$ providing $T_{ij}|_k$. This equation shows that the covariant derivatives of the second-order tensor components in Eqs. (2.145) are components of the third-order tensor $\nabla\mathbf{T}$.

Other operations can be a little tricky, however. Consider, for instance, the gradient of the dot product $\mathbf{a}\cdot\mathbf{b}$. Directly differentiating the product gives $\nabla(\mathbf{a}\cdot\mathbf{b}) = (\nabla\mathbf{a})\cdot\mathbf{b} + \mathbf{a}\cdot(\nabla\mathbf{b})$, which happens to be incorrect. The correct relation can be derived formally by first taking the differential

$$d(\mathbf{a}\cdot\mathbf{b}) = (d\mathbf{a})\cdot\mathbf{b} + \mathbf{a}\cdot(d\mathbf{b}). \tag{2.173}$$

Next, using Eq. (2.156) and treating ∇ as a vector gives

$$\begin{aligned} d(\mathbf{a}\cdot\mathbf{b}) &= (d\mathbf{r}\cdot\nabla\mathbf{a})\cdot\mathbf{b} + \mathbf{a}\cdot(d\mathbf{r}\cdot\nabla\mathbf{b}) \\ &= (d\mathbf{r}\cdot\nabla\mathbf{a})\cdot\mathbf{b} + (d\mathbf{r}\cdot\nabla\mathbf{b})\cdot\mathbf{a} \\ &= d\mathbf{r}\cdot(\nabla\mathbf{a})\cdot\mathbf{b} + d\mathbf{r}\cdot(\nabla\mathbf{b})\cdot\mathbf{a} \\ &= d\mathbf{r}\cdot[(\nabla\mathbf{a})\cdot\mathbf{b} + (\nabla\mathbf{b})\cdot\mathbf{a}] \\ &\equiv d\mathbf{r}\cdot\nabla(\mathbf{a}\cdot\mathbf{b}) \end{aligned} \tag{2.174}$$

and thus

$$\nabla(\mathbf{a}\cdot\mathbf{b}) = (\nabla\mathbf{a})\cdot\mathbf{b} + (\nabla\mathbf{b})\cdot\mathbf{a}. \tag{2.175}$$

Note that we have used the fact that $d\mathbf{r}\cdot\nabla\mathbf{b}$ is a vector in the second line of Eq. (2.174).

Another method, which often is easier to apply, involves using the component form of ∇ given by Eq. (2.157), which we write in shorthand form as (Drew, 1961)

$$\boxed{\nabla = \mathbf{g}^i\frac{\partial}{\partial x^i} \equiv \mathbf{g}\,\partial.} \tag{2.176}$$

In manipulating equations with this expression for $\boldsymbol{\nabla}$, we must take care to keep the derivatives on the proper terms and to preserve the order of the vectors. Thus, ∂ must remain in front of its proper operand at all times, but we *can* move ∂ (not g) across dots and crosses, e.g., $\mathbf{a}\cdot\partial\mathbf{b} = \mathbf{a}\partial\cdot\mathbf{b}$. Note also that, since ∂ is simply a scalar differentiator, products can be differentiated without ambiguity. Now,

$$\begin{aligned}
\boldsymbol{\nabla}(\mathbf{a}\cdot\mathbf{b}) &= \mathbf{g}\,\partial(\mathbf{a}\cdot\mathbf{b}) \\
&= \mathbf{g}[(\partial\mathbf{a})\cdot\mathbf{b} + \mathbf{a}\cdot(\partial\mathbf{b})] \\
&= \mathbf{g}[(\partial\mathbf{a})\cdot\mathbf{b} + (\partial\mathbf{b})\cdot\mathbf{a}] \\
&= (\mathbf{g}\,\partial\mathbf{a})\cdot\mathbf{b} + (\mathbf{g}\,\partial\mathbf{b})\cdot\mathbf{a} \\
&= (\boldsymbol{\nabla}\mathbf{a})\cdot\mathbf{b} + (\boldsymbol{\nabla}\mathbf{b})\cdot\mathbf{a}
\end{aligned} \tag{2.177}$$

which agrees with Eq. (2.175).

Example 2.10 Compute the gradient, divergence, and curl of the dyad $\mathbf{ab}$.

Solution. With care taken to preserve the order, the gradient is

$$\begin{aligned}
\boldsymbol{\nabla}(\mathbf{ab}) &= (\mathbf{g}\,\partial)(\mathbf{ab}) \\
&= \mathbf{g}[(\partial\mathbf{a})\mathbf{b} + \mathbf{a}(\partial\mathbf{b})] \\
&= (\mathbf{g}\,\partial\mathbf{a})\mathbf{b} + \mathbf{g}\mathbf{a}\,\partial\mathbf{b}. \\
&= \boldsymbol{\nabla}(\mathbf{ab}) = (\boldsymbol{\nabla}\mathbf{a})\mathbf{b} + (\mathbf{a}\boldsymbol{\nabla})^T\mathbf{b}.
\end{aligned} \tag{2.178}$$

Similar manipulations yield the divergence as

$$\begin{aligned}
\boldsymbol{\nabla}\cdot(\mathbf{ab}) &= (\mathbf{g}\,\partial)\cdot(\mathbf{ab}) \\
&= \mathbf{g}\cdot[(\partial\mathbf{a})\mathbf{b} + \mathbf{a}(\partial\mathbf{b})] \\
&= (\mathbf{g}\,\partial)\cdot\mathbf{ab} + \mathbf{g}\cdot\mathbf{a}\,\partial\mathbf{b} \\
&= (\mathbf{g}\,\partial\cdot\mathbf{a})\mathbf{b} + \mathbf{a}\cdot\mathbf{g}\,\partial\mathbf{b} \\
&= (\boldsymbol{\nabla}\cdot\mathbf{a})\mathbf{b} + \mathbf{a}\cdot(\boldsymbol{\nabla}\mathbf{b}),
\end{aligned} \tag{2.179}$$

and the curl is

$$\begin{aligned}
\boldsymbol{\nabla}\times(\mathbf{ab}) &= (\mathbf{g}\,\partial)\times(\mathbf{ab}) \\
&= \mathbf{g}\times[(\partial\mathbf{a})\mathbf{b} + \mathbf{a}(\partial\mathbf{b})] \\
&= (\mathbf{g}\,\partial\times\mathbf{a})\mathbf{b} + \mathbf{g}\times\mathbf{a}\,\partial\mathbf{b} \\
&= (\mathbf{g}\,\partial\times\mathbf{a})\mathbf{b} - \mathbf{a}\times(\mathbf{g}\,\partial\mathbf{b}) \\
&= (\boldsymbol{\nabla}\times\mathbf{a})\mathbf{b} - \mathbf{a}\times(\boldsymbol{\nabla}\mathbf{b}).
\end{aligned} \tag{2.180}$$

■

Table 2.6 Gradient formulas

$$
\begin{aligned}
\boldsymbol{\nabla}(\phi\psi) &= (\boldsymbol{\nabla}\phi)\psi + \phi(\boldsymbol{\nabla}\psi) \\
\boldsymbol{\nabla}(\phi\mathbf{a}) &= (\boldsymbol{\nabla}\phi)\mathbf{a} + \phi(\boldsymbol{\nabla}\mathbf{a}) \\
\boldsymbol{\nabla}(\mathbf{a}\cdot\mathbf{b}) &= (\boldsymbol{\nabla}\mathbf{a})\cdot\mathbf{b} + (\boldsymbol{\nabla}\mathbf{b})\cdot\mathbf{a} \\
\boldsymbol{\nabla}(\mathbf{a}\mathbf{b}) &= (\boldsymbol{\nabla}\mathbf{a})\mathbf{b} + (\mathbf{a}\boldsymbol{\nabla})^T\mathbf{b} \\
\boldsymbol{\nabla}(\mathbf{T}\cdot\mathbf{a}) &= (\boldsymbol{\nabla}\mathbf{T})\cdot\mathbf{a} + (\boldsymbol{\nabla}\mathbf{a})\cdot\mathbf{T}^T \\
\boldsymbol{\nabla}\mathbf{r} &= \mathbf{I} \\
\boldsymbol{\nabla}(\mathbf{r}\cdot\mathbf{r}) &= 2\mathbf{r}
\end{aligned}
$$

Example 2.11 Compute the divergence of $\mathbf{T}\cdot\mathbf{a}$.

Solution. Substituting Eq. (2.176) yields

$$
\begin{aligned}
\boldsymbol{\nabla}\cdot(\mathbf{T}\cdot\mathbf{a}) &= (\mathbf{g}\,\partial)\cdot(\mathbf{T}\cdot\mathbf{a}) \\
&= \mathbf{g}\cdot[(\partial\mathbf{T})\cdot\mathbf{a} + \mathbf{T}\cdot(\partial\mathbf{a})] \\
&= (\mathbf{g}\,\partial)\cdot\mathbf{T}\cdot\mathbf{a} + \mathbf{g}\cdot\mathbf{T}\cdot(\partial\mathbf{a}) \\
&= (\boldsymbol{\nabla}\cdot\mathbf{T})\cdot\mathbf{a} + \mathbf{T}:\mathbf{g}\,\partial\mathbf{a} \\
&= (\boldsymbol{\nabla}\cdot\mathbf{T})\cdot\mathbf{a} + \mathbf{T}:(\boldsymbol{\nabla}\mathbf{a})
\end{aligned}
\tag{2.181}
$$

where the relation $\mathbf{a}\cdot\mathbf{T}\cdot\mathbf{b} = \mathbf{T}:\mathbf{ab}$ has been used. Other formulas involving the gradient operator are given in Tables 2.6–2.8. ■

2.8.4 *Integral Relations*

In continuum mechanics, we often need to convert volume integrals into surface integrals, surface integrals into line integrals, and vice versa. This section considers three useful theorems for these conversions. Although these relations can be derived in general curvilinear coordinates, it is simpler to derive them in Cartesian coordinates, as we do here.[11] The results then are written in direct notation for later specialization to any coordinate system.

[11] Since we are working in Cartesian coordinates, we do not distinguish between subscripts and superscripts in this section.

Table 2.7 Divergence formulas

$$
\begin{aligned}
\boldsymbol{\nabla}\cdot\mathbf{a} &= \operatorname{tr}\boldsymbol{\nabla}\mathbf{a} \\
\boldsymbol{\nabla}\cdot(\phi\mathbf{a}) &= (\boldsymbol{\nabla}\phi)\cdot\mathbf{a} + \phi(\boldsymbol{\nabla}\cdot\mathbf{a}) \\
\boldsymbol{\nabla}\cdot(\mathbf{ab}) &= (\boldsymbol{\nabla}\cdot\mathbf{a})\mathbf{b} + \mathbf{a}\cdot(\boldsymbol{\nabla}\mathbf{b}) \\
\boldsymbol{\nabla}\cdot(\mathbf{T}\cdot\mathbf{a}) &= (\boldsymbol{\nabla}\cdot\mathbf{T})\cdot\mathbf{a} + \mathbf{T}:\boldsymbol{\nabla}\mathbf{a} \\
\boldsymbol{\nabla}\cdot(\mathbf{a}\times\mathbf{b}) &= (\boldsymbol{\nabla}\times\mathbf{a})\cdot\mathbf{b} - \mathbf{a}\cdot(\boldsymbol{\nabla}\times\mathbf{b}) \\
\boldsymbol{\nabla}\cdot(\boldsymbol{\nabla}\times\mathbf{a}) &= 0 \\
\boldsymbol{\nabla}\cdot\mathbf{r} &= 3
\end{aligned}
$$

Table 2.8 Curl formulas

$$
\begin{aligned}
\boldsymbol{\nabla}\times(\phi\mathbf{a}) &= (\boldsymbol{\nabla}\phi)\times\mathbf{a} + \phi(\boldsymbol{\nabla}\times\mathbf{a}) \\
\boldsymbol{\nabla}\times(\mathbf{ab}) &= (\boldsymbol{\nabla}\times\mathbf{a})\mathbf{b} - \mathbf{a}\times(\boldsymbol{\nabla}\mathbf{b}) \\
\boldsymbol{\nabla}\times(\mathbf{a}\times\mathbf{b}) &= \mathbf{a}(\boldsymbol{\nabla}\cdot\mathbf{b}) + \mathbf{b}\cdot(\boldsymbol{\nabla}\mathbf{a}) - \mathbf{a}\cdot(\boldsymbol{\nabla}\mathbf{b}) - \mathbf{b}(\boldsymbol{\nabla}\cdot\mathbf{a}) \\
\boldsymbol{\nabla}\times(\boldsymbol{\nabla}\times\mathbf{a}) &= \boldsymbol{\nabla}(\boldsymbol{\nabla}\cdot\mathbf{a}) - \boldsymbol{\nabla}^2\mathbf{a} \\
\boldsymbol{\nabla}\times(\boldsymbol{\nabla}\phi) &= \mathbf{0} \\
\boldsymbol{\nabla}\times\mathbf{r} &= \mathbf{0}
\end{aligned}
$$

Gradient Theorem

For simplicity, we consider a two-dimensional region A bounded by a contour C, such that any line through A parallel to either one of the coordinate axes intersects C in only two points (Fig. 2.7). The curve C is divided by its leftmost and rightmost points ($x = a, b$) into a lower segment C_1, described by $y = y_1(x)$, and an upper segment C_2, described by $y = y_2(x)$. With the position vector to a point P on C given by

$$\mathbf{r} = x\mathbf{e}_x + y\mathbf{e}_y, \tag{2.182}$$

the *unit* tangent ("base") vector at P is [see Eq. (2.7)]

$$\mathbf{t} = \frac{d\mathbf{r}}{ds} = \frac{dx}{ds}\mathbf{e}_x + \frac{dy}{ds}\mathbf{e}_y \tag{2.183}$$

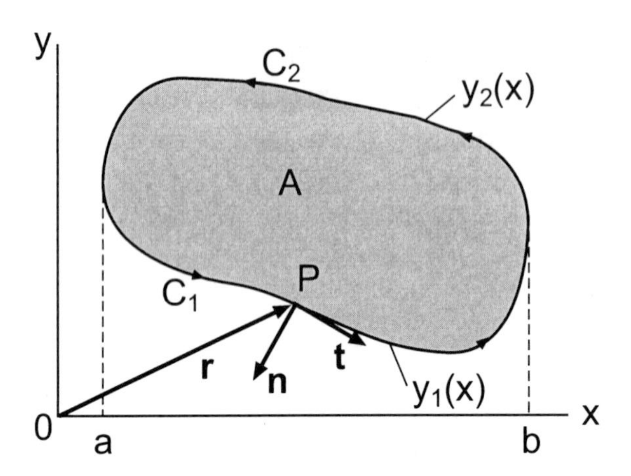

Fig. 2.7 Planar region A enclosed by curve $C = C_1 + C_2$.

where ds is the differential length along C, and the unit normal vector is

$$\mathbf{n} = \mathbf{t}\times\mathbf{e}_z = \frac{dy}{ds}\mathbf{e}_x - \frac{dx}{ds}\mathbf{e}_y = n_x\mathbf{e}_x + n_y\mathbf{e}_y. \tag{2.184}$$

For a function $\phi(x, y)$ defined in A, consider the area integral

$$\begin{aligned}\int_A \frac{\partial\phi}{\partial y}\, dA &= \int_a^b \int_{y_1(x)}^{y_2(x)} \frac{\partial\phi}{\partial y}\, dy\, dx \\ &= \int_a^b \{\phi[x, y_2(x)] - \phi[x, y_1(x)]\}\, dx \\ &= \int_a^b \{[\phi]_{C_2} - [\phi]_{C_1}\}\, dx. \end{aligned} \tag{2.185}$$

As indicated in Fig. 2.7, a positive contour integration corresponds to a counter-clockwise traversal of C. To make the first integral in Eq. (2.185) consistent with this convention, we write

$$\begin{aligned}\int_A \frac{\partial\phi}{\partial y}\, dA &= -\int_b^a [\phi]_{C_2}\, dx - \int_a^b [\phi]_{C_1}\, dx \\ &= -\int_C \phi\, dx = -\int_C \phi\, \frac{dx}{ds}\, ds.\end{aligned}$$

Finally, using Eq. (2.184) yields

$$\int_A \frac{\partial \phi}{\partial y}\, dA = \int_C \phi\, n_y\, ds \tag{2.186}$$

and a similar analysis gives

$$\int_A \frac{\partial \phi}{\partial x}\, dA = \int_C \phi\, n_x\, ds. \tag{2.187}$$

Equations (2.186) and (2.187) are the basic relations used to develop the integral theorems.

Now, multiplying Eqs. (2.186) and (2.187) by $\mathbf{e}_y$ and $\mathbf{e}_x$, respectively, and adding provides

$$\int_A (\mathbf{e}_x \frac{\partial \phi}{\partial x} + \mathbf{e}_y \frac{\partial \phi}{\partial y})\, dA = \int_C \phi\, (n_x \mathbf{e}_x + n_y \mathbf{e}_y)\, ds$$

or, since $\boldsymbol{\nabla} = \mathbf{e}_x \dfrac{\partial}{\partial x} + \mathbf{e}_y \dfrac{\partial}{\partial y}$ in two-dimensional Cartesian coordinates,

$$\int_A \boldsymbol{\nabla} \phi\, dA = \int_C \mathbf{n} \phi\, ds \tag{2.188}$$

with $\mathbf{n}$ given by Eq. (2.184). This relation, which is written in direct notation, is called the **gradient theorem** or **Green's theorem**. In three dimensions, it becomes

$$\boxed{\int_V \boldsymbol{\nabla} \phi\; dV = \int_A \mathbf{n} \phi\, dA} \tag{2.189}$$

where V is the volume enclosed by a surface of area A. Since it is written in coordinate-free form, this equation is valid for any coordinate system.

Divergence Theorem

Consider a vector $\mathbf{a} = a_x \mathbf{e}_x + a_y \mathbf{e}_y$. If we let $\phi = a_x$ in Eq. (2.187) and $\phi = a_y$ in Eq. (2.186) and add the results, we obtain

$$\int_A \left(\frac{\partial a_x}{\partial x} + \frac{\partial a_y}{\partial y} \right) dA = \int_C (a_x n_x + a_y n_y)\, ds.$$

In direct notation, this equation is

$$\int_A \boldsymbol{\nabla} \cdot \mathbf{a}\, dA = \int_C \mathbf{n} \cdot \mathbf{a}\, ds \tag{2.190}$$

which is called the **divergence theorem** or **Gauss's theorem**. In three dimensions, this relation becomes

$$\boxed{\int_V \boldsymbol{\nabla}\cdot\mathbf{a}\,dV = \int_A \mathbf{n}\cdot\mathbf{a}\,dA.} \tag{2.191}$$

Curl Theorem

If we set $\phi = a_y$ in Eq. (2.187) and $\phi = a_x$ in Eq. (2.186) and subtract the results, we get

$$\int_A \left(\frac{\partial a_y}{\partial x} - \frac{\partial a_x}{\partial y}\right) dA = \int_C (a_y n_x - a_x n_y)\, ds.$$

This equation can be written

$$\int_A \boldsymbol{\nabla}\times\mathbf{a}\,dA = \int_C \mathbf{n}\times\mathbf{a}\,ds \tag{2.192}$$

or, in three dimensions,

$$\boxed{\int_V \boldsymbol{\nabla}\times\mathbf{a}\,dV = \int_A \mathbf{n}\times\mathbf{a}\,dA} \tag{2.193}$$

which is called the **curl theorem**.

In summary, Eqs. (2.189), (2.191), and (2.193) can be written in the form

$$\boxed{\int_V \boldsymbol{\nabla} * \phi\,dV = \int_A \mathbf{n} * \phi\,dA} \tag{2.194}$$

where $*$ is a blank, a dot, or a cross; and ϕ can be a scalar, a vector, or a tensor, so long as the expression is meaningful. (For example, $\boldsymbol{\nabla}\times\phi$ is meaningless if ϕ is a scalar.)

2.9 Problems

2–1 Show the following:

(a) $\delta^i_j \delta^j_i = 3$.
(b) $\varepsilon_{ijk}\varepsilon^{ijk} = 6$.

(c) $\varepsilon_{mki}\varepsilon^{jki} = 2\delta^j_m$.

2–2 Evaluate the expression

$$\frac{\partial}{\partial x^i}\left(x^i x^j\right).$$

2–3 Write the following equation in index notation:

$$M = \left[(a_1)^2 + (a_2)^2 + (a_3)^2\right]\left[a_1 b^1 + a_2 b^2 + a_3 b^3\right].$$

2–4 Write the following equation in direct notation:

$$a_{ij} b^{jk} = c^k_i.$$

2–5 Let

$$\begin{aligned} \mathbf{a} &= 3\mathbf{e}_x + \mathbf{e}_y + \mathbf{e}_z \\ \mathbf{b} &= \mathbf{e}_x + 2\mathbf{e}_z \\ \mathbf{c} &= \mathbf{e}_x - 3\mathbf{e}_y + \mathbf{e}_z \end{aligned}$$

where the $\mathbf{e}_i$ are Cartesian base vectors. Compute $(\mathbf{ab})\cdot\mathbf{c}$ and $(\mathbf{ba})\cdot\mathbf{c}$.

2–6 Let $\mathbf{a}$ and $\mathbf{b}$ be vectors, while $\mathbf{A}$ and $\mathbf{B}$ are second-order tensors. Show the following:

(a) $(\mathbf{a}\times\mathbf{b})\cdot\mathbf{a} = 0$.
(b) $\mathbf{I} : \mathbf{A} = \mathrm{tr}\mathbf{A}$.
(c) $\mathbf{A} : \mathbf{B} = 0$ if $\mathbf{A}$ is symmetric and $\mathbf{B}$ is antisymmetric.
(d) $(\mathbf{A}^T)^{-1} = (\mathbf{A}^{-1})^T$.
(e) $(\mathbf{A}\cdot\mathbf{B})^{-1} = \mathbf{B}^{-1}\cdot\mathbf{A}^{-1}$.

2–7 If $\mathbf{T}$, $\mathbf{U}$, and $\mathbf{V}$ are second-order tensors:

(a) Write $\mathbf{T}\cdot\mathbf{U}$ in terms of the bases $\mathbf{g}_i\mathbf{g}_j$, $\mathbf{g}^i\mathbf{g}^j$, and $\mathbf{g}_i\mathbf{g}^j$.
(b) Write $(\mathbf{TU}) : \mathbf{V}$ in terms of the basis $\mathbf{g}^i\mathbf{g}_j$.
(c) Show that $\mathbf{T} : (\mathbf{U}\cdot\mathbf{V}) = \mathbf{U} : (\mathbf{T}\cdot\mathbf{V}^T)$.

2–8 Expressing the vectors $\mathbf{a}$, $\mathbf{b}$, and $\mathbf{c}$ in terms of general curvilinear components, show that $\mathbf{a}\times(\mathbf{b}\times\mathbf{c}) = (\mathbf{a}\cdot\mathbf{c})\mathbf{b} - (\mathbf{a}\cdot\mathbf{b})\mathbf{c}$. Use this relation to prove the ε-δ identity

$$\varepsilon_{ijk}\varepsilon^{irs} = \delta^r_j\delta^s_k - \delta^s_j\delta^r_k.$$

2–9 Cylindrical polar coordinates (r, θ, z) and Cartesian coordinates (x, y, z) are related by

$$\begin{aligned} x &= r\cos\theta \\ y &= r\sin\theta \\ z &= z. \end{aligned}$$

Using Eqs. (2.59) and (2.60), compute the components of the tensors **A** and **B** for a transformation from Cartesian to cylindrical coordinates.

2–10 Spherical polar coordinates (r, θ, ϕ) and Cartesian coordinates (x, y, z) are related by

$$\begin{aligned} x &= r\sin\theta\cos\phi \\ y &= r\sin\theta\sin\phi \\ z &= r\cos\theta. \end{aligned}$$

(a) Determine the base vectors $\mathbf{g}_{\bar{r}}$, $\mathbf{g}_{\theta}$, and $\mathbf{g}_{\phi}$.

(b) For a transformation from Cartesian to spherical coordinates, use Eq. (2.59) to compute the coordinate transformation components $A_i^{\cdot j}$.

(c) Show that $\bar{\mathbf{g}}_i = \mathbf{A} \cdot \mathbf{g}_i$, where $\bar{\mathbf{g}}_i = (\mathbf{g}_r, \mathbf{g}_\theta, \mathbf{g}_\phi)$, $\mathbf{g}_i = (\mathbf{e}_x, \mathbf{e}_y, \mathbf{e}_z)$, and **A** is the coordinate transformation tensor.

2–11 Parabolic cylindrical coordinates x^i and Cartesian coordinates z^i are related by

$$\begin{aligned} z^1 &= a(x^2 - x^1) \\ z^2 &= 2a(x^1x^2)^{1/2} \\ z^3 &= x^3 \end{aligned}$$

where a is a constant.

(a) Determine the base vectors $\mathbf{g}_i$ and $\mathbf{g}^i$ and compute the g_{ij} for parabolic coordinates.

(b) Write the physical components of the second-order tensor **T** in terms of the tensor components $T^i_{\cdot j}$ and the components T^{ij}.

2–12 If **T**, **U**, and **V** are orthogonal tensors, show that $\mathbf{T} \cdot \mathbf{U} \cdot \mathbf{V}$ is an orthogonal tensor.

2–13 Consider the vector $\mathbf{a} = 2\mathbf{e}_x - 5\mathbf{e}_y + 3\mathbf{e}_z$ and the base vectors

$$\begin{aligned} \mathbf{g}_1 &= \mathbf{e}_x + 2\mathbf{e}_y + \mathbf{e}_z \\ \mathbf{g}_2 &= \mathbf{e}_x - \mathbf{e}_z \\ \mathbf{g}_3 &= -3\mathbf{e}_x - 2\mathbf{e}_y + 4\mathbf{e}_z, \end{aligned}$$

where the $\mathbf{e}_i$ are Cartesian base vectors. Compute the $\mathbf{g}^i$ and the natural components a^i and a_i of $\mathbf{a}$.

2–14 If $\mathbf{T} = \mathbf{T}^{ij}\mathbf{g}_i\mathbf{g}_j$, show that $\mathbf{T} : \mathbf{g}_i\mathbf{g}^j = T_i^{\cdot j}$.

2–15 In Cartesian coordinates, a tensor has the matrix representation

$$\mathbf{T}_{(\mathbf{e}_i\mathbf{e}_j)} = \begin{bmatrix} -2 & 5 & 1 \\ 3 & 0 & -2 \\ 1 & -1 & 4 \end{bmatrix}.$$

(a) Relative to the base vectors in Problem **2–13**, compute the components T^{11}, T^{23}, T_{22}, T_{13}, $T_2^{\cdot 1}$, and $T_{\cdot 2}^{1}$.

(b) Compute the principal invariants of $\mathbf{T}$.

2–16 A tensor is defined by

$$\mathbf{A} = 2\mathbf{g}_1\mathbf{g}^1 + \mathbf{g}_2\mathbf{g}^2 - 5\mathbf{g}_3\mathbf{g}^3 + 3\mathbf{g}_2\mathbf{g}^3 + 3\mathbf{g}_3\mathbf{g}^2.$$

Compute the eigenvalues and eigenvectors of $\mathbf{A}$. Show that the eigenvectors are mutually orthogonal.

2–17 If $\mathbf{A}$, $\mathbf{B}$, and $\mathbf{C}$ are second-order tensors, show the following:

(a) $\dfrac{\partial}{\partial \mathbf{A}} \det \mathbf{A} = \det \mathbf{A} \cdot \mathbf{A}^{-T}$

(b) $\dfrac{\partial}{\partial \mathbf{A}} \mathrm{tr}\mathbf{A} = \mathbf{I}$

(c) $\dfrac{\partial}{\partial \mathbf{A}} \mathrm{tr}\mathbf{A}^2 = 2\mathbf{A}^T$

(d) $\dfrac{\partial}{\partial \mathbf{C}}(\mathbf{A} : \mathbf{B}) = \mathbf{A} : \dfrac{\partial \mathbf{B}}{\partial \mathbf{C}} + \mathbf{B} : \dfrac{\partial \mathbf{A}}{\partial \mathbf{C}}.$

2–18 Compute the Christoffel symbols for (a) spherical polar coordinates (Problem **2–10**) and (b) parabolic cylindrical coordinates (Problem **2–11**). Then, for each system, compute the derivatives $\mathbf{g}_{i,j}$ by taking straight derivatives of the $\mathbf{g}_i$ and also by using the Christoffel symbols.

2–19 Toroidal coordinates (r, θ, ϕ) and Cartesian coordinates (x, y, z) are related by

$$\begin{aligned} x &= (b + r\cos\phi)\cos\theta \\ y &= (b + r\cos\phi)\sin\theta \\ z &= r\sin\phi. \end{aligned}$$

(a) Derive the Christoffel symbols as listed in Appendix B.

(b) For axisymmetry ($\partial/\partial\theta = 0$), show that the Laplacian has the form

$$\nabla^2 = \frac{\partial^2}{\partial r^2} + \frac{1}{r}\frac{\partial}{\partial r} + \frac{1}{r^2}\frac{\partial^2}{\partial \phi^2} + \frac{\cos\phi}{(b + r\cos\phi)}\frac{\partial}{\partial r} - \frac{\sin\phi}{r(b + r\cos\phi)}\frac{\partial}{\partial \phi}.$$

2–20 Consider the vector

$$\mathbf{a} = \cos x^2 \mathbf{g}_1 - \frac{\sin x^2}{x^1}\mathbf{g}_2 + x^3 \mathbf{g}_3$$

in cylindrical polar coordinates with $(x^1, x^2, x^3) = (r, \theta, z)$. Compute the covariant derivatives $a^i|_j$.

2–21 Derive the expression

$$T^{ij}|_k = T^{ij}{}_{,k} + T^{lj}\Gamma^i_{kl} + T^{il}\Gamma^j_{kl}.$$

2–22 Using Eq. (2.157), write the gradient operator in spherical polar coordinates.

2–23 For the scalar function $\phi(x^i)$, use Eq. (2.169) to write $\boldsymbol{\nabla}^2\phi$ in cylindrical polar coordinates. Repeat the derivation using Eq. (2.170).

2–24 If $\mathbf{a}$ and $\mathbf{b}$ are vector functions and $\mathbf{T}$ is a second-order tensor function, show the following:

(a) $\boldsymbol{\nabla}(\mathbf{T} \cdot \mathbf{a}) = (\boldsymbol{\nabla}\mathbf{T}) \cdot \mathbf{a} + (\boldsymbol{\nabla}\mathbf{a}) \cdot \mathbf{T}^T$

(b) $\boldsymbol{\nabla} \cdot (\mathbf{a} \times \mathbf{b}) = (\boldsymbol{\nabla} \times \mathbf{a}) \cdot \mathbf{b} - \mathbf{a} \cdot (\boldsymbol{\nabla} \times \mathbf{b})$

(c) $\boldsymbol{\nabla} \times (\boldsymbol{\nabla} \times \mathbf{a}) = \boldsymbol{\nabla}(\boldsymbol{\nabla} \cdot \mathbf{a}) - \boldsymbol{\nabla}^2\mathbf{a}$

(d) $\boldsymbol{\nabla} \cdot (\boldsymbol{\nabla} \cdot \mathbf{a}\mathbf{I}) = \boldsymbol{\nabla}(\boldsymbol{\nabla} \cdot \mathbf{a})$

(e) $\boldsymbol{\nabla} \cdot (\boldsymbol{\nabla}\mathbf{a})^T = \boldsymbol{\nabla}(\boldsymbol{\nabla} \cdot \mathbf{a})$

(f) $\mathbf{I} : \boldsymbol{\nabla}\mathbf{a} = \boldsymbol{\nabla} \cdot \mathbf{a}.$

Chapter 3

Analysis of Deformation

The mechanics of hard tissues, such as bones, teeth, and horns, can be analyzed using the linear theory of elasticity, in which deformation is assumed to be "small." Under this condition, the distinction between the geometries of the undeformed and the deformed body can be ignored. Soft tissues, in contrast, often undergo "large," or "finite" deformation. In this case, **geometric nonlinearity** enters the analysis, even if the material properties (stress-strain relations) are linear. This chapter deals only with geometry. Nonlinear stress-strain relations characterize **material nonlinearity**, as discussed in Chapter 5.

One manifestation of geometric nonlinearity is nonlinear strain-displacement relations. Another can be seen through the example of a cantilever beam of length L with an end load P (Fig. 3.1). For small deflection, the peak bending moment is always approximately PL at the support. Classical linear beam theory is based on this assumption. When the deflection is large, however, the load moves toward the support by a significant distance, reducing the bending moment. Throughout most of this chapter, we make no restrictions on the magnitudes of deformation (i.e., strain), displacement, or rotation. Approximations for various restrictions are examined in Section 3.9.

Deformation of a body is measured relative to a **reference configuration**, which may or may not be stress free. In fact, most unloaded soft biological structures contain residual stress (Fung, 1990). Still, it often is convenient to *choose* the unloaded state as the reference configuration, even if it contains residual stress. In this chapter, we refer to the reference configuration as the *undeformed* configuration, keeping in mind that the term "undeformed" is relative. The term "deformed," then, refers to any other configuration.

When possible, we use upper-case letters to represent **Lagrangian** quantities,

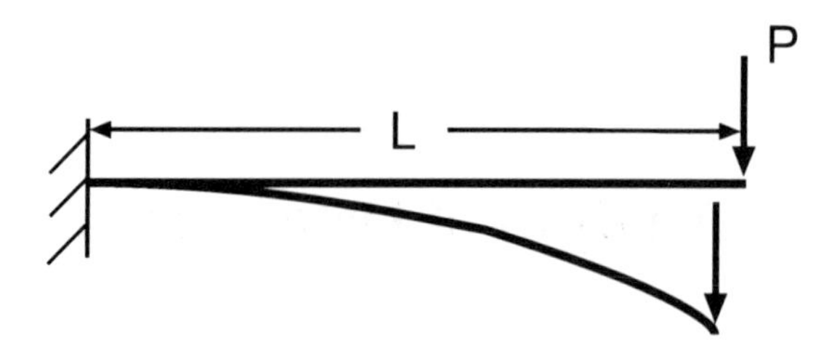

Fig. 3.1 Geometric nonlinearity in a cantilever beam.

referred to the undeformed configuration, and lower-case letters to represent **Eulerian** quantities referred to the deformed configuration. This convention also applies to indices, as we distinguish between I and i, so that $a_I^i \neq a_1^1 + a_2^2 + a_3^3$. Note that now some upper-case bold letters may be vectors, rather than tensors, a change from the convention in the previous chapter.

3.1 Deformation in One Dimension

To illustrate the basic measures used to describe the deformation of a solid body, we first consider uniaxial stretching of a bar (Fig. 3.2). We choose the unloaded bar of length L as the reference configuration. When subjected to end forces, the length of the bar changes to $\ell = L + \Delta L$, and the classical **engineering strain** or the **stretch** is defined as

$$E^* = \frac{\Delta L}{L} = \frac{\ell - L}{L} = \frac{\ell}{L} - 1. \tag{3.1}$$

In addition,

$$\lambda = \frac{\ell}{L} = 1 + E^* \tag{3.2}$$

is called the **stretch ratio**, being the ratio of the deformed to the undeformed (reference) length.

The deformation measures defined in Eqs. (3.1) and (3.2) are valid regardless of the magnitude of the deformation. When deformation is large, however, it generally is more convenient to define strain measures based on changes in *squared* length, rather than changes in length. A commonly used measure for large defor-

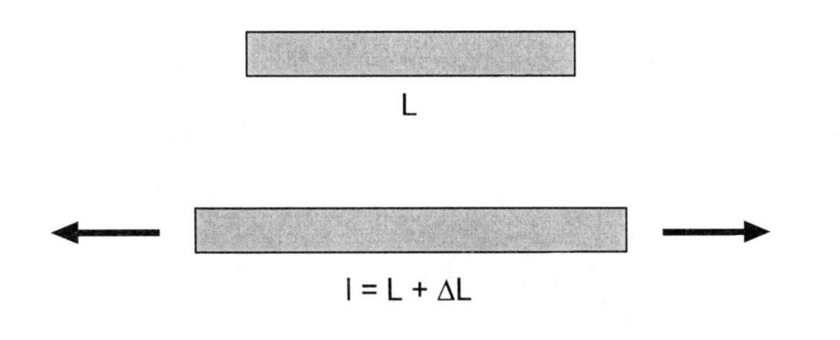

Fig. 3.2 Uniaxial stretching of a bar.

mation is the **Lagrangian strain**, which is defined to be

$$E = \frac{\ell^2 - L^2}{2L^2} \tag{3.3}$$

for uniaxial stretching of a bar. In terms of E^*, this relation can be written

$$E = E^*(1 + \tfrac{1}{2}E^*), \tag{3.4}$$

which implies

$$\begin{aligned} \lambda &= 1 + E^* = (1 + 2E)^{1/2} \\ E &= \tfrac{1}{2}(\lambda^2 - 1). \end{aligned} \tag{3.5}$$

Equation (3.4) can be verified by substituting (3.1).

The Lagrangian strain is referred to the undeformed configuration (length L of the bar). In some formulations, it is useful to define a strain measure relative to the deformed configuration. For the one-dimensional case, the **Eulerian strain** is defined as

$$e = \frac{\ell^2 - L^2}{2\ell^2}. \tag{3.6}$$

In terms of the other strain measures, this relation can be written

$$e = \tfrac{1}{2}(1 - \lambda^{-2}) = \frac{E}{1 + 2E} = \frac{E^*(1 + \frac{1}{2}E^*)}{(1 + E^*)^2}, \tag{3.7}$$

which can be confirmed by substituting Eqs. (3.2), (3.4), and (3.5).

Any one of the three strain measures (E^*, E, e) can be used to characterize a given deformation, although they differ numerically. If the deformation is small ($E^* << 1$), however, then Eqs. (3.4) and (3.7) show that $E \cong e \cong E^*$, i.e., all of

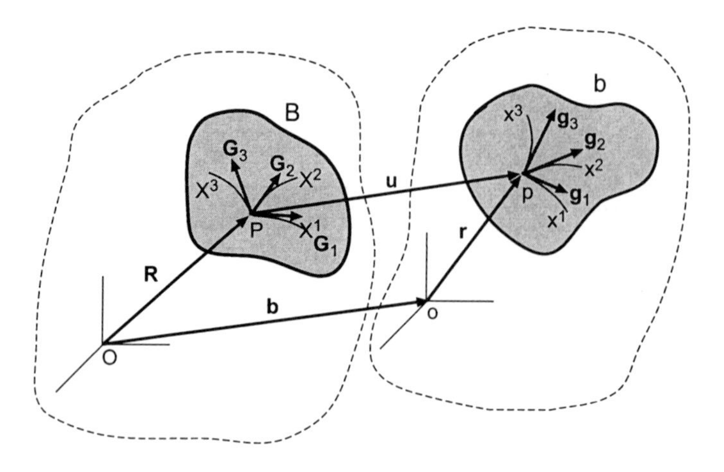

Fig. 3.3 Coordinates and base vectors in undeformed body B and deformed body b.

the strain measures are essentially equivalent. The development in the following sections extends these ideas to three dimensions.

3.2 Coordinate Systems and Base Vectors

Consider a deformation that sends a point P in the undeformed body B at time $t = 0$ into the point p in the deformed body b at time t (Fig. 3.3). The point P is located at the **material**, or **Lagrangian**, curvilinear coordinates X^I, and p is located at the **spatial**, or **Eulerian**, curvilinear coordinates x^i. In general, the coordinate systems X^I and x^i may be attached to two different reference frames, O and o, which may be in motion relative to each other (Fig. 3.3).

Some authors [e.g., Eringen (1962; 1980)] prefer to keep the X^I and x^i coordinate systems independent of each other. In this way, the coordinates can be chosen to take advantage of any symmetries in B and b. Other authors [e.g., Green and Zerna (1968)] assume that the X^I system is imbedded in B so that it deforms with the body, and the x^i system is then taken to coincide with the X^I system at all times during the deformation. Then, during the motion, each point keeps the same coordinate label, and the $x^i = X^I$ are called **convected coordinates**. Although the flexibility in choosing the deformed coordinate system is lost in this case, the basic equations assume a somewhat simpler form. In this chapter, we consider both choices.

Let $\mathbf{R}(X^I, t)$ and $\mathbf{r}(x^i, t)$ be the position vectors to P and p, respectively. Then, the mapping of P into p and the inverse mapping of p into P are described by

$$\boxed{\mathbf{r} = \mathbf{r}(\mathbf{R}, t), \qquad \mathbf{R} = \mathbf{R}(\mathbf{r}, t),} \tag{3.8}$$

respectively, which imply

$$\mathbf{r} = \mathbf{r}(x^i, t) = \mathbf{r}(X^I, t), \qquad \mathbf{R} = \mathbf{R}(X^I, t) = \mathbf{R}(x^i, t). \tag{3.9}$$

The covariant base vectors in B are

$$\boxed{\mathbf{G}_I = \frac{\partial \mathbf{R}}{\partial X^I} = \mathbf{R}_{,I}} \tag{3.10}$$

and, in b, two sets of covariant base vectors are given by

$$\boxed{\begin{aligned} \mathbf{g}_I &= \frac{\partial \mathbf{r}}{\partial X^I} = \mathbf{r}_{,I} \\ \mathbf{g}_i &= \frac{\partial \mathbf{r}}{\partial x^i} = \mathbf{r}_{,i} \,. \end{aligned}} \tag{3.11}$$

The $\mathbf{G}_I$ are tangent to the X^I coordinate curves at P, the $\mathbf{g}_i$ are tangent to the x^i coordinate curves at p (Fig. 3.3), and the $\mathbf{g}_I$ are tangent to the *convected* X^I coordinate curves at p.

Before proceeding further, we need to introduce a new notation. Since there are no upper- or lower-case numerical indices, we use an asterisk to denote an "upper-case number." This allows us to distinguish between $\mathbf{g}_I = \{\mathbf{g}_{1^*}, \mathbf{g}_{2^*}, \mathbf{g}_{3^*}\}$ and $\mathbf{g}_i = \{\mathbf{g}_1, \mathbf{g}_2, \mathbf{g}_3\}$. (The $\mathbf{g}_i$ are illustrated in Fig. 3.3.) If there is no confusion, however, the asterisk is omitted, i.e., $\mathbf{G}_I = \{\mathbf{G}_1, \mathbf{G}_2, \mathbf{G}_3\}$. Moreover, a similar confusion may arise when vector and tensor components are referred to the bases $\mathbf{G}_I$ and $\mathbf{g}_I$. Here again, and following Atluri (1984), we attach asterisks to indices associated with $\mathbf{g}_I$ or $\mathbf{g}^I$. For example, we write

$$\mathbf{T} = T_{IJ}\, \mathbf{G}^I \mathbf{G}^J = T_{I^*J^*}\, \mathbf{g}^I \mathbf{g}^J = T_{I^*J} \mathbf{g}^I \mathbf{G}^J .$$

The identity tensor, however, is written

$$\mathbf{I} = G_{IJ}\, \mathbf{G}^I \mathbf{G}^J = g_{IJ}\, \mathbf{g}^I \mathbf{g}^J = g_{ij}\, \mathbf{g}^i \mathbf{g}^j \tag{3.12}$$

since there is no confusion between G_{IJ} and g_{IJ}. Finally, note that the summation convention applies to repeated indices even if only one of them contains an asterisk, e.g., $a^I_{JI*} = a^1_{J1*} + a^2_{J2*} + a^3_{J3*}$.

Relations can be found that link the two sets of base vectors in b. Equations (3.10) and (3.11) give the differential length vectors

$$\boxed{\begin{aligned} d\mathbf{R} &= \mathbf{G}_I \, dX^I \\ d\mathbf{r} &= \mathbf{g}_I \, dX^I = \mathbf{g}_i \, dx^i. \end{aligned}} \tag{3.13}$$

If $x^i = X^I$ (convected coordinates), then $\mathbf{g}_i = \mathbf{g}_I$. Otherwise, Eq. $(3.13)_2$ gives

$$\boxed{\mathbf{g}_I = F^i_I \mathbf{g}_i, \qquad \mathbf{g}_i = F^I_i \mathbf{g}_I} \tag{3.14}$$

where

$$\boxed{F^i_I \equiv \frac{\partial x^i}{\partial X^I}, \qquad F^I_i \equiv \frac{\partial X^I}{\partial x^i}.} \tag{3.15}$$

These last two sets of equations can be considered a coordinate transformation between the x^i and the X^I coordinates, as given by Eqs. $(2.46)_1$ and (2.59). In this regard, using Eqs. $(2.46)_2$ and (2.60), we also can write

$$\boxed{\mathbf{g}^I = F^I_i \mathbf{g}^i, \qquad \mathbf{g}^i = F^i_I \mathbf{g}^I.} \tag{3.16}$$

Moreover, combining these relations gives

$$\begin{aligned} \delta^J_I &= \mathbf{g}_I \cdot \mathbf{g}^J = (F^i_I \mathbf{g}_i) \cdot (F^J_j \mathbf{g}^j) = F^i_I F^J_j \delta^j_i \\ \delta^j_i &= \mathbf{g}_i \cdot \mathbf{g}^j = (F^I_i \mathbf{g}_I) \cdot (F^j_J \mathbf{g}^J) = F^I_i F^j_J \delta^J_I \end{aligned}$$

or

$$\boxed{F^i_I F^J_i = \delta^J_I, \qquad F^I_i F^j_I = \delta^j_i,} \tag{3.17}$$

which also can be found from Eqs. (3.15) and the chain rule for differentiation, or from Eq. (2.51). Physical meanings for the F^i_I and F^I_i are discussed in the next section.

3.3 Deformation Gradient Tensor

In taking gradients, we must distinguish between gradients with respect to the undeformed and deformed configurations. With the gradient operator being $\boldsymbol{\nabla}$ in B and $\overline{\boldsymbol{\nabla}}$ in b, Eqs. (2.149), (2.157), and (3.13) yield

$$\begin{aligned} \boldsymbol{\nabla} \equiv \frac{\partial}{\partial \mathbf{R}} &= \mathbf{G}^I \frac{\partial}{\partial X^I} \\ \overline{\boldsymbol{\nabla}} \equiv \frac{\partial}{\partial \mathbf{r}} &= \mathbf{g}^i \frac{\partial}{\partial x^i} = \mathbf{g}^I \frac{\partial}{\partial X^I} \end{aligned} \tag{3.18}$$

since $\partial/(\mathbf{G}_I \partial X^I) = \mathbf{G}^I\, \partial/\partial X^I$, etc.

The **deformation gradient tensor** $\mathbf{F}$ of the deformed body b relative to the undeformed body B is defined to be the tensor that transforms a differential line element $d\mathbf{R}$ into $d\mathbf{r}$ (Fig. 3.4), i.e.,

$$d\mathbf{r} = \mathbf{F} \cdot d\mathbf{R} = d\mathbf{R} \cdot \mathbf{F}^T. \tag{3.19}$$

Similarly, the **spatial deformation gradient tensor** $\mathbf{F}^{-1}$ of B relative to b is defined by the inverse transformation

$$d\mathbf{R} = \mathbf{F}^{-1} \cdot d\mathbf{r} = d\mathbf{r} \cdot \mathbf{F}^{-T}. \tag{3.20}$$

Using these equations, we can write

$$\begin{aligned} \mathbf{F}^T &= \frac{\partial \mathbf{r}}{\partial \mathbf{R}} = \boldsymbol{\nabla} \mathbf{r} \\ \mathbf{F}^{-T} &= \frac{\partial \mathbf{R}}{\partial \mathbf{r}} = \overline{\boldsymbol{\nabla}} \mathbf{R}. \end{aligned} \tag{3.21}$$

These tensors also can be expressed in terms of base vectors alone. Substituting Eqs. (3.13) into (3.21) and noting Eq. (2.155) yields

$$\begin{aligned} \mathbf{F}^T &= \frac{\partial \mathbf{r}}{\partial \mathbf{R}} = \frac{\mathbf{g}_I\, dX^I}{\mathbf{G}_J\, dX^J} = \mathbf{G}^J \mathbf{g}_I \delta^I_J = \mathbf{G}^I \mathbf{g}_I \\ \mathbf{F}^{-T} &= \frac{\partial \mathbf{R}}{\partial \mathbf{r}} = \frac{\mathbf{G}_J\, dX^J}{\mathbf{g}_I\, dX^I} = \mathbf{g}^I \mathbf{G}_J \delta^J_I = \mathbf{g}^I \mathbf{G}_I \end{aligned}$$

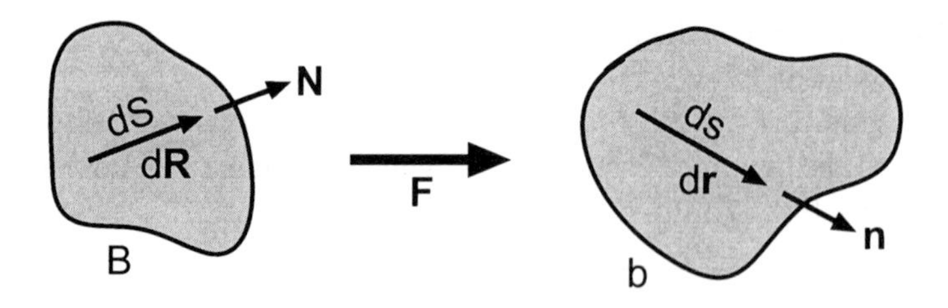

Fig. 3.4 Deformation of length element $d\mathbf{R}$ into length element $d\mathbf{r}$.

or, with Eqs. $(3.14)_1$, and $(3.16)_1$,

$$\boxed{\begin{aligned}\mathbf{F} &= \mathbf{g}_I\mathbf{G}^I = F^i_I\mathbf{g}_i\mathbf{G}^I \\ \mathbf{F}^{-1} &= \mathbf{G}_I\mathbf{g}^I = F^I_i\mathbf{G}_I\mathbf{g}^i.\end{aligned}} \tag{3.22}$$

The expression for $\mathbf{F}$ shows that the F^i_I are actually the mixed components $F^i_{\cdot I}$ of the deformation gradient tensor. We drop the dot with the understanding that the deformed base vector (lower case $\mathbf{g}$) always appears first in $\mathbf{F}$. Note that we also could derive $\mathbf{F}$ from Eqs. $(3.11)_1$, $(3.18)_1$, and $(3.21)_1$, i.e., $\mathbf{F}^T = \boldsymbol{\nabla}\mathbf{r} = \mathbf{G}^I\mathbf{r}_{,I} = \mathbf{G}^I\mathbf{g}_I$.

The mixed components $F^i_I = x^i{}_{,I}$ and $F^I_i = X^I{}_{,i}$ are called **deformation gradients**. Since the bases $\mathbf{G}_I$ and $\mathbf{g}_i$ are chosen independently of the deformation, the F^i_I or F^I_i contain the deformation part of the tensor $\mathbf{F}$. In convected coordinates, setting $x^i = X^J$ gives $F^J_I = X^J{}_{,I} = \delta^J_I$ and $\mathbf{g}_i = \mathbf{g}_J$, and so $\mathbf{F} = \delta^J_I\mathbf{g}_J\mathbf{G}^I = \mathbf{g}_I\mathbf{G}^I$, which is consistent with Eq. $(3.22)_1$. In this case, therefore, the deformation is shifted into the convected base vector $\mathbf{g}_I$. Tensor components such as F^i_I and F^I_i, which depend on base vectors in two different reference frames, are called **two-point tensor components**.

Equations (3.18) and (3.22) provide a useful identity. Since

$$\mathbf{F}^T\cdot\overline{\boldsymbol{\nabla}} = (\mathbf{G}^I\mathbf{g}_I)\cdot(\mathbf{g}^J\frac{\partial}{\partial X^J}) = \mathbf{G}^I\delta^J_I\frac{\partial}{\partial X^J} = \mathbf{G}^I\frac{\partial}{\partial X^I},$$

we have

$$\boxed{\boldsymbol{\nabla} = \mathbf{F}^T\cdot\overline{\boldsymbol{\nabla}}.} \tag{3.23}$$

3.4 Deformation and Strain Tensors

The deformation gradient tensor can serve as the primary deformation variable in a solid mechanics problem. It is relatively easy to compute, but it contains two features that may complicate analyses. First, as shown by Eqs. (3.22), $\mathbf{F}$ is not symmetric in general, and so all nine components relative to a given basis may be unknown in some problems. Second, as shown by Eq. (3.19), $\mathbf{F}$ *deforms and rotates* $d\mathbf{R}$ into $d\mathbf{r}$. Since rigid-body rotation of an element should produce no stresses, certain restrictions must be placed on how $\mathbf{F}$ appears in constitutive relations (see Section 5.2.5).[1]

One way around these problems is to define deformation variables based only on changes in length, and not orientation, of a line element. There are several ways to do this; here, we focus on the most common approaches.

3.4.1 *Deformation Tensors*

Consider an arbitrary differential element $d\mathbf{R}$ in B of length dS that deforms into $d\mathbf{r}$ of length ds in b (Fig. 3.4). In terms of $\mathbf{F}$, Eqs. (3.19) and (3.20) give

$$\begin{aligned} ds^2 &= d\mathbf{r}\cdot d\mathbf{r} = (d\mathbf{R}\cdot\mathbf{F}^T)\cdot(\mathbf{F}\cdot d\mathbf{R}) \equiv d\mathbf{R}\cdot\mathbf{C}\cdot d\mathbf{R} \\ dS^2 &= d\mathbf{R}\cdot d\mathbf{R} = (d\mathbf{r}\cdot\mathbf{F}^{-T})\cdot(\mathbf{F}^{-1}\cdot d\mathbf{r}) \equiv d\mathbf{r}\cdot\mathbf{B}^{-1}\cdot d\mathbf{r}. \end{aligned} \quad (3.24)$$

In these equations, since $(\mathbf{F}^{-T}\cdot\mathbf{F}^{-1})^{-1} = (\mathbf{F}^{-1})^{-1}\cdot(\mathbf{F}^{-T})^{-1} = \mathbf{F}\cdot\mathbf{F}^T$ (see Table 2.3, page 24),

$$\boxed{\begin{aligned} \mathbf{C} &= \mathbf{F}^T\cdot\mathbf{F} \\ \mathbf{B} &= \mathbf{F}\cdot\mathbf{F}^T \end{aligned}} \quad (3.25)$$

are the right and left Cauchy-Green **deformation tensors**, respectively. ($\mathbf{F}$ is on the right or left.) Because $\mathbf{C}^T = (\mathbf{F}^T\cdot\mathbf{F})^T = \mathbf{F}^T\cdot(\mathbf{F}^T)^T = \mathbf{F}^T\cdot\mathbf{F} = \mathbf{C}$ and similarly for $\mathbf{B}$, $\mathbf{C}$ and $\mathbf{B}$ are symmetric tensors.

[1] In certain instances, rotation information may be useful. One example is torsion of the left ventricle during the cardiac cycle (Arts et al., 1984; Hansen et al., 1988; Ingels et al., 1989; Taber et al., 1996), see Section 6.6.

3.4.2 *Strain Tensors*

The **Lagrangian strain tensor** $\mathbf{E}$ and the **Eulerian strain tensor** $\mathbf{e}$ are defined by the relations

$$\begin{aligned} ds^2 - dS^2 &= 2\,d\mathbf{R}\cdot\mathbf{E}\cdot d\mathbf{R} \\ &= 2\,d\mathbf{r}\cdot\mathbf{e}\cdot d\mathbf{r}. \end{aligned} \tag{3.26}$$

In the literature, various names are attached to these strain tensors. For instance, Green and St. Venant are often associated with $\mathbf{E}$, and Almansi and Hamel with $\mathbf{e}$. Equations (3.24) yield

$$\begin{aligned} ds^2 - dS^2 &= d\mathbf{R}\cdot\mathbf{C}\cdot d\mathbf{R} - d\mathbf{R}\cdot d\mathbf{R} = d\mathbf{R}\cdot(\mathbf{C} - \mathbf{I})\cdot d\mathbf{R} \\ &= d\mathbf{r}\cdot d\mathbf{r} - d\mathbf{r}\cdot\mathbf{B}^{-1}\cdot d\mathbf{r} = d\mathbf{r}\cdot(\mathbf{I} - \mathbf{B}^{-1})\cdot d\mathbf{r} \end{aligned}$$

and comparison with Eqs. (3.26) reveals

$$\boxed{\begin{aligned} 2\mathbf{E} &= \mathbf{C} - \mathbf{I} = \mathbf{F}^T\cdot\mathbf{F} - \mathbf{I} \\ 2\mathbf{e} &= \mathbf{I} - \mathbf{B}^{-1} = \mathbf{I} - \mathbf{F}^{-T}\cdot\mathbf{F}^{-1}. \end{aligned}} \tag{3.27}$$

Because $\mathbf{E}$ and $\mathbf{e}$ are tensors, the components of strain relative to any coordinate system can be obtained by double-dotting with the corresponding base dyad. First, substituting Eqs. (3.22) into (3.25) gives

$$\begin{aligned} \mathbf{C} &= \mathbf{F}^T\cdot\mathbf{F} = (\mathbf{G}^I\mathbf{g}_I)\cdot(\mathbf{g}_J\mathbf{G}^J) = g_{IJ}\mathbf{G}^I\mathbf{G}^J \\ &= (F_I^i\mathbf{G}^I\mathbf{g}_i)\cdot(F_J^j\mathbf{g}_j\mathbf{G}^J) = F_I^iF_J^jg_{ij}\mathbf{G}^I\mathbf{G}^J \end{aligned}$$

$$\begin{aligned} \mathbf{B}^{-1} &= \mathbf{F}^{-T}\cdot\mathbf{F}^{-1} = (\mathbf{g}^I\mathbf{G}_I)\cdot(\mathbf{G}_J\mathbf{g}^J) = G_{IJ}\mathbf{g}^I\mathbf{g}^J \\ &= (F_i^I\mathbf{g}^i\mathbf{G}_I)\cdot(F_j^J\mathbf{G}_J\mathbf{g}^j) = F_i^IF_j^JG_{IJ}\mathbf{g}^i\mathbf{g}^j. \end{aligned} \tag{3.28}$$

Next, inserting these equations, Eqs. (3.12), and the relations

$$\mathbf{E} = E_{IJ}\,\mathbf{G}^I\mathbf{G}^J, \qquad \mathbf{e} = e_{ij}\,\mathbf{g}^i\mathbf{g}^j \tag{3.29}$$

into Eqs. (3.27) and double-dotting with the appropriate dyads yields the strain components

$$\boxed{\begin{aligned} 2E_{IJ} &= C_{IJ} - G_{IJ} \\ 2e_{ij} &= g_{ij} - (B^{-1})_{ij} \end{aligned}} \tag{3.30}$$

where

$$\boxed{\begin{aligned} C_{IJ} &= F^i_I F^j_J g_{ij} = g_{IJ} \\ (B^{-1})_{ij} &= F^I_i F^J_j G_{IJ}. \end{aligned}} \tag{3.31}$$

These equations show that the components of the strain tensors reflect differences between the metrics of the deformed and undeformed bodies. Here, we also note that substituting Eqs. (3.13) and (3.29) into (3.26) gives the alternate definitions

$$\begin{aligned} ds^2 - dS^2 &= 2E_{IJ}\, dX^I dX^J \\ &= 2e_{ij}\, dx^i dx^j, \end{aligned} \tag{3.32}$$

for the Lagrangian and Eulerian strains.

Now, we pause to mention two important points. First, since $\mathbf{E}$ and $\mathbf{e}$ are tensors, all of the tools of tensor analysis in Chapter 2 can be used, including the computation of principal strains and strain invariants (see Section 3.7). To illustrate the second point, we examine the components of $\mathbf{e}$ defined by

$$\mathbf{e} = e_{I^*J^*}\, \mathbf{g}^I \mathbf{g}^J.$$

Equations (3.12), $(3.27)_2$, and $(3.28)_2$ give

$$2\mathbf{e} = \mathbf{I} - \mathbf{B}^{-1} = g_{IJ}\, \mathbf{g}^I \mathbf{g}^J - G_{IJ}\, \mathbf{g}^I \mathbf{g}^J$$

and, therefore,

$$2e_{I^*J^*} = g_{IJ} - G_{IJ}.$$

Comparison with Eqs. $(3.30)_1$ and $(3.31)_1$ shows that $E_{IJ} = e_{I^*J^*}$. Thus, although the strain tensors $\mathbf{E}$ and $\mathbf{e}$ are different, the components of $\mathbf{E}$ with respect to the basis $\{\mathbf{G}^I \mathbf{G}^J\}$ are equal to the components of $\mathbf{e}$ with respect to the basis $\{\mathbf{g}^I \mathbf{g}^J\}$. This equivalency of components of different tensors in appropriate bases will play a major role in our understanding of the relations between the components of the various stress tensors discussed in the next chapter.

Example 3.1 In Cartesian coordinates, the position of a point in an undeformed and deformed body are Z^I and z^i, respectively. Derive expressions for the Lagrangian and Eulerian strain components in terms of Z^I and z^i.

Solution. In Cartesian coordinates, we have $G_{IJ} = \delta_{IJ}$ and $g_{ij} = \delta_{ij}$. Thus, Eqs. (3.15), (3.30), and (3.31) give

$$
\begin{aligned}
E_{IJ} &= \frac{1}{2}\left(\frac{\partial z^k}{\partial Z^I}\frac{\partial z^k}{\partial Z^J} - \delta_{IJ}\right) \\
e_{ij} &= \frac{1}{2}\left(\delta_{ij} - \frac{\partial Z^K}{\partial z^i}\frac{\partial Z^K}{\partial z^j}\right).
\end{aligned}
\tag{3.33}
$$

■

3.4.3 *Physical Components of Strain*

As mentioned in Section 2.7.2, physical components of tensors can be defined in various ways, with one possibility given by Eq. (2.125). For strain, the situation is complicated further by the different bases that have been introduced. For stress and strain, our choice is guided by the defining relations for the tensors themselves.

We define the physical components of the Lagrangian strain tensor $\mathbf{E}$ by writing Eq. (3.32) in terms of physical quantities. Doing this yields

$$ds^2 - dS^2 = 2E_{IJ}\, dX^I dX^J = 2\hat{E}_{IJ}\, d\hat{X}^I d\hat{X}^J \tag{3.34}$$

where hat denotes a physical component. Since

$$d\mathbf{R} = \mathbf{G}_I\, dX^I = \mathbf{e}_I\, d\hat{X}^I \tag{3.35}$$

where

$$\mathbf{e}_I = \frac{\mathbf{G}_I}{\sqrt{G_{(II)}}} \tag{3.36}$$

is the unit vector in the direction of $\mathbf{G}_I$, we have

$$d\hat{X}^I = dX^I \sqrt{G_{(II)}}. \tag{3.37}$$

Inserting this relation into Eq. (3.34) shows that the physical Lagrangian strain components are given by

$$\hat{E}_{IJ} = \frac{E_{IJ}}{\sqrt{G_{(II)}G_{(JJ)}}}. \tag{3.38}$$

Here, we make an important observation. In terms of the reciprocal unit vectors

$$\mathbf{e}^I = \frac{\mathbf{G}^I}{\sqrt{G^{(II)}}}, \tag{3.39}$$

the Lagrangian strain tensor can be written

$$\mathbf{E} = E_{IJ}\mathbf{G}^I\mathbf{G}^J = \hat{E}_{IJ}\sqrt{G_{(II)}G_{(JJ)}G^{(II)}G^{(JJ)}}\mathbf{e}^I\mathbf{e}^J. \tag{3.40}$$

In orthogonal curvilinear coordinates, because $\mathbf{e}^I = \mathbf{e}_I$ and $G^{(II)} = 1/G_{(II)}$, this expression becomes

$$\mathbf{E} = \sum_{I=1}^{3}\sum_{J=1}^{3}\hat{E}_{IJ}\mathbf{e}_I\mathbf{e}_J. \tag{3.41}$$

This result is useful for setting up a problem in terms of physical components, *if* the undeformed body is referred to an orthogonal coordinate system. Otherwise, $\mathbf{E} = E_{IJ}\mathbf{G}^I\mathbf{G}^J$ should be used.

In the remainder of this section, we examine some simple deformations through examples.

Example 3.2 Soft biological tissues rarely are devoid of stress *in vivo*. In fact, most tissues experience ever-changing three-dimensional forces and deformations. To a first approximation, however, many tissues are subjected to simple tension or compression in one, two, or three dimensions. For example, papillary muscles stretch uniaxially to prevent the heart valves from inverting during systole, and articular cartilage in the knee is compressed between the femur and the tibia to provide a cushion between the surfaces of the bones. An example of multiaxial loading is that experienced by an epithelium, which is a thin layer of cells covering a tissue. When subjected to surface pressure, epithelial sheets undergo in-plane biaxial stretching and transverse compression. In addition, experiments designed to determine material properties commonly use uniaxial or biaxial loading protocols.

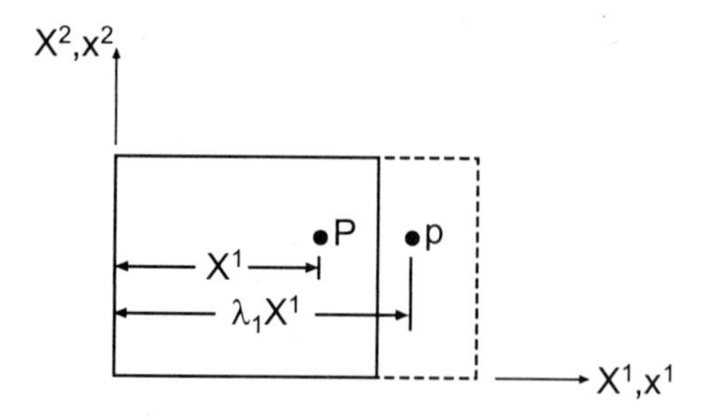

Fig. 3.5 Uniform extension in X^1-direction.

Thus, simple extension and compression of tissues is of fundamental importance in biomechanics.

Consider **uniform extension** of a rectangular block of material. With the coordinates X^I of P and x^i of p referred to the same Cartesian system, the deformation is described by (see Fig. 3.5)

$$x^1 = \lambda_1 X^1, \qquad x^2 = \lambda_2 X^2, \qquad x^3 = \lambda_3 X^3 \tag{3.42}$$

where the λ_I are stretch ratios in the coordinate directions. Compute the Cartesian components of $\mathbf{E}$ and $\mathbf{e}$ in terms of the λ_I.

Solution. The position vectors to P and p, respectively, are

$$\mathbf{R} = X^I \mathbf{e}_I, \qquad \mathbf{r} = x^i \mathbf{e}_i \tag{3.43}$$

where the $\mathbf{e}_I = \mathbf{e}_i$ are unit vectors along the coordinate axes. Thus, the base vectors are $\mathbf{G}_I = \mathbf{R}_{,I} = \mathbf{e}_I$ and $\mathbf{g}_i = \mathbf{r}_{,i} = \mathbf{e}_i$, and so $\mathbf{G}_{IJ} = \mathbf{e}_I \cdot \mathbf{e}_J = \delta_{IJ}$ and $g_{ij} = \mathbf{e}_i \cdot \mathbf{e}_j = \delta_{ij}$. In the following, we keep subscripts and superscripts distinct, although this is not really necessary, as $\mathbf{e}_I = \mathbf{e}^I$ in Cartesian coordinates.

Equations (3.18) and $(3.21)_1$ now give

$$\begin{aligned}
\mathbf{F}^T &= \nabla \mathbf{r} \\
&= \left(\mathbf{e}^1 \frac{\partial}{\partial X^1} + \mathbf{e}^2 \frac{\partial}{\partial X^2} + \mathbf{e}^3 \frac{\partial}{\partial X^3} \right) (\lambda_1 X^1 \mathbf{e}_1 + \lambda_2 X^2 \mathbf{e}_2 + \lambda_3 X^3 \mathbf{e}_3) \\
&= \lambda_1 \mathbf{e}^1 \mathbf{e}_1 + \lambda_2 \mathbf{e}^2 \mathbf{e}_2 + \lambda_3 \mathbf{e}^3 \mathbf{e}_3.
\end{aligned} \tag{3.44}$$

Therefore, $\mathbf{F} = \lambda_1\, \mathbf{e}_1 \mathbf{e}^1 + \lambda_2\, \mathbf{e}_2 \mathbf{e}^2 + \lambda_3\, \mathbf{e}_3 \mathbf{e}^3$, which shows that the stretch ratios are the only nonzero Cartesian components of $\mathbf{F}$. (Note that $\mathbf{F} = \mathbf{F}^T$ in this problem, since $\mathbf{e}^I = \mathbf{e}_I$.) In matrix form, we can write

$$\mathbf{F}_{(\mathbf{e}_i \mathbf{e}^I)} = \left[F_I^i\right] = \text{diag}\left[\lambda_1, \lambda_2, \lambda_3\right] \tag{3.45}$$

where

$$\text{diag}\left[a, b, c\right] \equiv \begin{bmatrix} a & 0 & 0 \\ 0 & b & 0 \\ 0 & 0 & c \end{bmatrix}. \tag{3.46}$$

The use of a Cartesian basis for both the undeformed and deformed coordinates allows us to employ matrix algebra without worrying about picking up extra factors that general base vectors may include. Thus, with all components referred to the unit base vectors $\mathbf{e}_I = \mathbf{e}^I = \mathbf{e}_i = \mathbf{e}^i$, Eqs. (3.25) give

$$\begin{aligned} \left[C_{IJ}\right] &= \left[F_I^i\right]^T \left[F_J^j\right] = \text{diag}\left[\lambda_1^2, \lambda_2^2, \lambda_3^2\right] \\ \left[B_{ij}\right] &= \left[F_I^i\right] \left[F_J^j\right]^T = \text{diag}\left[\lambda_1^2, \lambda_2^2, \lambda_3^2\right] \\ \left[B_{ij}\right]^{-1} &= \text{diag}\left[\lambda_1^{-2}, \lambda_2^{-2}, \lambda_3^{-2}\right]. \end{aligned} \tag{3.47}$$

It is easy to show that computing $\mathbf{C} = \mathbf{F}^T \cdot \mathbf{F}$ and $\mathbf{B} = \mathbf{F} \cdot \mathbf{F}^T$, with $\mathbf{F}$ given by Eq. (3.44), yields the same results. Finally, Eqs. (3.30) and (3.47) provide the strain components

$$\begin{aligned} \left[E_{IJ}\right] &= \tfrac{1}{2}\left[C_{IJ} - G_{IJ}\right] = \tfrac{1}{2}\left[C_{IJ} - \delta_{IJ}\right] \\ &= \text{diag}\left[\tfrac{1}{2}(\lambda_1^2 - 1), \tfrac{1}{2}(\lambda_2^2 - 1), \tfrac{1}{2}(\lambda_3^2 - 1)\right] \\ \\ \left[e_{ij}\right] &= \tfrac{1}{2}\left[g_{ij} - (B^{-1})_{ij}\right] = \tfrac{1}{2}\left[\delta_{ij} - (B^{-1})_{ij}\right] \\ &= \text{diag}\left[\tfrac{1}{2}(1 - \lambda_1^{-2}), \tfrac{1}{2}(1 - \lambda_2^{-2}), \tfrac{1}{2}(1 - \lambda_3^{-2})\right]. \end{aligned} \tag{3.48}$$

Note that the components of $\mathbf{E}$ and $\mathbf{e}$ are consistent with Eqs. (3.5) and (3.7). ■

Example 3.3 The importance of shear deformation in the mechanics of red blood cells is well known (Fung, 1993). However, the shear properties of most soft tissues have been largely unexplored. The reason is not that shear stress is considered relatively unimportant. The wall of the heart, for example, undergoes significant shearing during each heartbeat (Waldman et al., 1985), and flowing blood exerts shear stresses on vascular endothelial cells. A more likely reason that shear deformation has received relatively little attention is that testing in shear is more

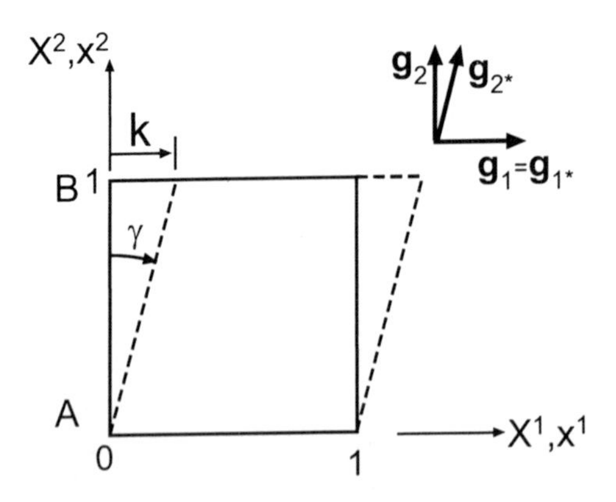

Fig. 3.6 Simple shear.

difficult technically than extension and compression. Recently, however, devices have been built specifically for shear testing of soft tissue (Sacks, 1999).

Consider **simple shear** of a unit cube (Fig. 3.6). During deformation, a point originally at (X^1, X^2, X^3) moves parallel to the X^1-axis to the new coordinates (x^1, x^2, x^3). With the coordinates X^I and x^i measured relative to the same Cartesian system, the undeformed and deformed position vectors are

$$\mathbf{R} = X^I \mathbf{e}_I, \qquad \mathbf{r} = x^i \mathbf{e}_i, \tag{3.49}$$

in which

$$x^1 = X^1 + kX^2, \qquad x^2 = X^2, \qquad x^3 = X^3. \tag{3.50}$$

In these equations, k is a constant and $\mathbf{e}_I = \mathbf{e}_i = \mathbf{G}_I = \mathbf{g}_i$.

(a) Using an analysis based on independently chosen x^i, compute $\mathbf{F}$ and then the $\mathbf{g}_I$.
(b) Using an analysis based on convected coordinates ($x^i = X^I$), compute the $\mathbf{g}_I$ and then $\mathbf{F}$.
(c) Determine the tensor components C_{IJ} and B_{ij}.

Solution. (a) For the independent x^i of Eqs. (3.50), the deformation gradients defined in Eq. (3.15)$_1$ are given by the matrix

$$\left[F^i_I\right] = \left[x^i{}_{,I}\right] = \begin{bmatrix} 1 & k & 0 \\ 0 & 1 & 0 \\ 0 & 0 & 1 \end{bmatrix}. \qquad (3.51)$$

Then, Eq. (3.22)$_1$ gives

$$\mathbf{F} = F^i_I \, \mathbf{g}_i \mathbf{G}^I = \mathbf{e}_1\mathbf{e}^1 + k\mathbf{e}_1\mathbf{e}^2 + \mathbf{e}_2\mathbf{e}^2 + \mathbf{e}_3\mathbf{e}^3, \qquad (3.52)$$

since $\mathbf{G}^I = \mathbf{e}^I$. Next, Eq. (3.14)$_1$ provides the convected base vectors

$$\begin{aligned} \mathbf{g}_{1^*} &= F^1_{1^*}\mathbf{g}_1 + F^2_{1^*}\mathbf{g}_2 + F^3_{1^*}\mathbf{g}_3 = \mathbf{e}_1 \\ \mathbf{g}_{2^*} &= F^1_{2^*}\mathbf{g}_1 + F^2_{2^*}\mathbf{g}_2 + F^3_{2^*}\mathbf{g}_3 = k\mathbf{e}_1 + \mathbf{e}_2 \\ \mathbf{g}_{3^*} &= F^1_{3^*}\mathbf{g}_1 + F^2_{3^*}\mathbf{g}_2 + F^3_{3^*}\mathbf{g}_3 = \mathbf{e}_3 \end{aligned} \qquad (3.53)$$

which are illustrated in Fig. 3.6. Note that the $\mathbf{g}_I = \{\mathbf{g}_{1^*}, \mathbf{g}_{2^*}, \mathbf{g}_{3^*}\}$ point along the edges of the *deformed* cube.

(b) In the convected-coordinate approach, the x^i coordinates do not enter the analysis directly. Rather, we compute the base vectors $\mathbf{g}_I$ directly from Eq. (3.11)$_1$ with $\mathbf{r}(X^I) = (X^1 + kX^2)\mathbf{e}_1 + X^2\mathbf{e}_2 + X^3\mathbf{e}_3$, i.e.,

$$\begin{aligned} \mathbf{g}_{1^*} &= \frac{\partial \mathbf{r}}{\partial X^1} = \mathbf{e}_1 \\ \mathbf{g}_{2^*} &= \frac{\partial \mathbf{r}}{\partial X^2} = k\mathbf{e}_1 + \mathbf{e}_2 \\ \mathbf{g}_{3^*} &= \frac{\partial \mathbf{r}}{\partial X^3} = \mathbf{e}_3, \end{aligned}$$

which agree with Eqs. (3.53). These relations and Eqs. (3.22)$_1$ yield

$$\begin{aligned} \mathbf{F} &= \mathbf{g}_I\mathbf{G}^I = \mathbf{g}_{1^*}\mathbf{G}^1 + \mathbf{g}_{2^*}\mathbf{G}^2 + \mathbf{g}_{3^*}\mathbf{G}^3 \\ &= \mathbf{e}_1\mathbf{e}^1 + (k\mathbf{e}_1 + \mathbf{e}_2)\mathbf{e}^2 + \mathbf{e}_3\mathbf{e}^3 \\ &= \mathbf{e}_1\mathbf{e}^1 + k\mathbf{e}_1\mathbf{e}^2 + \mathbf{e}_2\mathbf{e}^2 + \mathbf{e}_3\mathbf{e}^3, \end{aligned} \qquad (3.54)$$

which agrees with Eq. (3.52).

(c) As in the previous example, since the basis of $\left[F^i_I\right]$ is the set of Cartesian dyads $\{\mathbf{e}_i\mathbf{e}^I\}$, Eqs. (3.25) and elementary matrix operations give

$$[C_{IJ}] = \left[F_I^i\right]^T \left[F_J^j\right] = \begin{bmatrix} 1 & 0 & 0 \\ k & 1 & 0 \\ 0 & 0 & 1 \end{bmatrix} \begin{bmatrix} 1 & k & 0 \\ 0 & 1 & 0 \\ 0 & 0 & 1 \end{bmatrix} = \begin{bmatrix} 1 & k & 0 \\ k & 1+k^2 & 0 \\ 0 & 0 & 1 \end{bmatrix}$$

$$[B_{ij}] = \left[F_I^i\right] \left[F_J^j\right]^T = \begin{bmatrix} 1 & k & 0 \\ 0 & 1 & 0 \\ 0 & 0 & 1 \end{bmatrix} \begin{bmatrix} 1 & 0 & 0 \\ k & 1 & 0 \\ 0 & 0 & 1 \end{bmatrix} = \begin{bmatrix} 1+k^2 & k & 0 \\ k & 1 & 0 \\ 0 & 0 & 1 \end{bmatrix}. \tag{3.55}$$

Note that $C_{IJ} = g_{IJ} = \mathbf{g}_I \cdot \mathbf{g}_J$ [see Eq. $(3.31)_1$] also could be computed directly using Eqs. (3.53). ■

Example 3.4 Blood vessels, ureters, intestines, tracheas, and plant stems are examples of biological tubes. Even the left ventricle has been modeled as a thick-walled cylinder (Arts et al., 1979; Chadwick, 1982; Tozeren, 1983; Guccione et al., 1991; Taber, 1991b; Guccione et al., 1993; Taber et al., 1996). Thus, deformation of a tube, i.e., a hollow cylinder, is an important problem in biomechanics.

Consider **extension, inflation, and torsion of a hollow circular cylinder** (Fig. 3.7). For convenience, choose $(X^1, X^2, X^3) = (R, \Theta, Z)$ and (x^1, x^2, x^3) $= (r, \theta, z)$ as cylindrical polar coordinates for a point before and after deformation, respectively. In addition, assume that the deformation is described by the mapping

$$r = r(R), \qquad \theta = \Theta + \psi Z, \qquad z = \lambda Z, \tag{3.56}$$

where ψ is the angle of twist per unit undeformed length, and λ is the axial stretch ratio. Compute the tensors $\mathbf{F}$ and $\mathbf{E}$ in terms of both tensor and physical components.

Solution. For an arbitrary material element in the tube, the undeformed and deformed position vectors are

$$\begin{aligned} \mathbf{R} &= R\mathbf{e}_R + Z\mathbf{e}_Z \\ \mathbf{r} &= r\mathbf{e}_r + z\mathbf{e}_z \end{aligned} \tag{3.57}$$

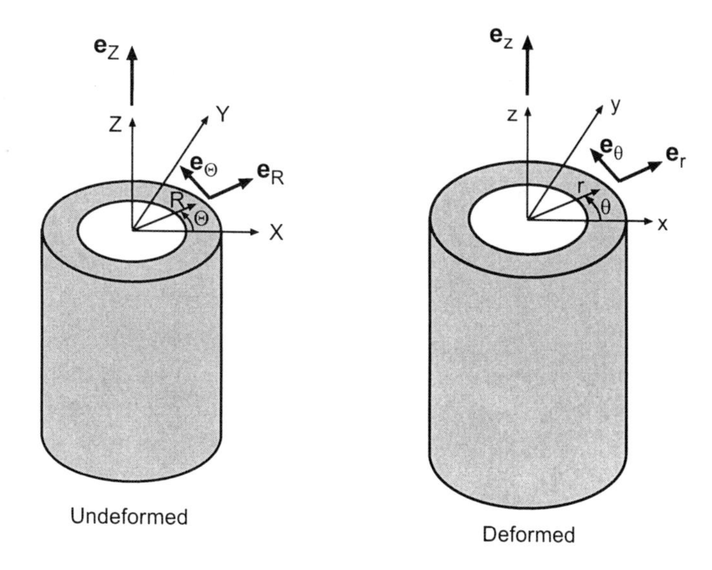

Fig. 3.7 Combined extension, inflation, and torsion of a cylinder.

where the geometry of Fig. 3.7 gives (with $\mathbf{e}_z = \mathbf{e}_Z = \text{constant}$)

$$\begin{aligned}
\mathbf{e}_R &= \mathbf{e}_x \cos\Theta + \mathbf{e}_y \sin\Theta \\
\mathbf{e}_\Theta &= -\mathbf{e}_x \sin\Theta + \mathbf{e}_y \cos\Theta \\
\mathbf{e}_r &= \mathbf{e}_x \cos\theta + \mathbf{e}_y \sin\theta \\
\mathbf{e}_\theta &= -\mathbf{e}_x \sin\theta + \mathbf{e}_y \cos\theta.
\end{aligned} \tag{3.58}$$

In these equations, $(\mathbf{e}_x, \mathbf{e}_y, \mathbf{e}_z)$ are Cartesian unit vectors fixed in the frame of reference, while $(\mathbf{e}_R, \mathbf{e}_\Theta, \mathbf{e}_Z)$ and $(\mathbf{e}_r, \mathbf{e}_\theta, \mathbf{e}_z)$ are polar unit vectors associated with the undeformed and deformed positions of the element, respectively (see Fig. 3.7). With Eqs. (3.56)–(3.58), Eqs. (3.10) and (3.11) give the base vectors

$$\begin{aligned}
\mathbf{G}_1 &= \mathbf{R}_{,R} = \mathbf{e}_R \\
\mathbf{G}_2 &= \mathbf{R}_{,\Theta} = R\mathbf{e}_\Theta \\
\mathbf{G}_3 &= \mathbf{R}_{,Z} = \mathbf{e}_Z
\end{aligned}$$

$$\begin{aligned} \mathbf{g}_1 &= \mathbf{r},_r = \mathbf{e}_r \\ \mathbf{g}_2 &= \mathbf{r},_\theta = r\mathbf{e}_\theta \\ \mathbf{g}_3 &= \mathbf{r},_z = \mathbf{e}_z \end{aligned}$$

$$\begin{aligned} \mathbf{g}_{1^*} &= \mathbf{r},_R = r'(R)\,\mathbf{e}_r \\ \mathbf{g}_{2^*} &= \mathbf{r},_\Theta = r\mathbf{e}_\theta \\ \mathbf{g}_{3^*} &= \mathbf{r},_Z = \psi r\mathbf{e}_\theta + \lambda\mathbf{e}_z. \end{aligned} \tag{3.59}$$

In terms of the convected base vectors $\mathbf{g}_I$, we can compute the deformation gradient tensor $\mathbf{F} = \mathbf{g}_I\mathbf{G}^I$ after determining the $\mathbf{G}^I$. Since the cylindrical polar coordinates (R, Θ, Z) are orthogonal, this is a relatively simple matter. In fact, we already have found the $\mathbf{G}^I$ in Example 2.3; Eqs. (2.23) give

$$\mathbf{G}^1 = \mathbf{e}_R, \qquad \mathbf{G}^2 = R^{-1}\mathbf{e}_\Theta, \qquad \mathbf{G}^3 = \mathbf{e}_Z. \tag{3.60}$$

For simplicity in this example, we write all of the (orthogonal) unit vectors with subscripts. Thus,

$$\begin{aligned} \mathbf{F} &= \mathbf{g}_{1^*}\mathbf{G}^1 + \mathbf{g}_{2^*}\mathbf{G}^2 + \mathbf{g}_{3^*}\mathbf{G}^3 \\ &= r'(R)\,\mathbf{e}_r\mathbf{e}_R + (r\mathbf{e}_\theta)(R^{-1}\mathbf{e}_\Theta) + (\psi r\mathbf{e}_\theta + \lambda\mathbf{e}_z)\mathbf{e}_Z \\ &= r'(R)\,\mathbf{e}_r\mathbf{e}_R + (r/R)\mathbf{e}_\theta\mathbf{e}_\Theta + \lambda\mathbf{e}_z\mathbf{e}_Z + \psi r\mathbf{e}_\theta\mathbf{e}_Z, \end{aligned} \tag{3.61}$$

and the matrix of components *relative to the basis* $\{\mathbf{e}_i\mathbf{e}_I\}$ is

$$\mathbf{F}_{(\mathbf{e}_i\mathbf{e}_I)} = \begin{bmatrix} r' & 0 & 0 \\ 0 & r/R & \psi r \\ 0 & 0 & \lambda \end{bmatrix}. \tag{3.62}$$

Note that, because the basis is composed of **unit dyads**, i.e., dyads of unit vectors, all of the components of $\mathbf{F}$ in these equations are nondimensional. In fact, as shown next, they also are the physical components of $\mathbf{F}$.

Relative to the natural basis, the tensor components of $\mathbf{F}$ are given by [see Eqs. $(3.15)_1$ and $(3.22)_1$]

$$\mathbf{F}_{(\mathbf{g}_i\mathbf{G}^I)} = [F^i_I] = [x^i,_I] = \begin{bmatrix} r,_R & r,_\Theta & r,_Z \\ \theta,_R & \theta,_\Theta & \theta,_Z \\ z,_R & z,_\Theta & z,_Z \end{bmatrix} = \begin{bmatrix} r' & 0 & 0 \\ 0 & 1 & \psi \\ 0 & 0 & \lambda \end{bmatrix} \tag{3.63}$$

where Eqs. (3.56) have been used. Adapting Eq. (2.125) gives the physical components

$$\boxed{\hat{F}^i_I = F^i_I \sqrt{\frac{g_{ii}}{G_{II}}} \qquad (i, I \text{ not summed})} \tag{3.64}$$

with Eqs. (3.59) yielding

$$\begin{aligned} [G_{IJ}] &= [\mathbf{G}_I \cdot \mathbf{G}_J] = \text{diag}\left[1, R^2, 1\right] \\ [g_{ij}] &= [\mathbf{g}_i \cdot \mathbf{g}_j] = \text{diag}\left[1, r^2, 1\right]. \end{aligned} \tag{3.65}$$

Combining Eqs. (3.63)–(3.65) and comparing with Eq. (3.62) confirms that

$$\mathbf{F}_{(\mathbf{e}_i \mathbf{e}_I)} = \left[\hat{F}^i_I\right]. \tag{3.66}$$

The deformation tensor components C_{IJ} and $(B^{-1})_{ij}$ now can be computed using Eqs. (3.31), (3.63), and (3.65). Alternatively, Eqs. $(3.31)_1$ and (3.59) give the C_{IJ} directly as

$$[C_{IJ}] = [g_{IJ}] = [\mathbf{g}_I \cdot \mathbf{g}_J] = \begin{bmatrix} r'^2 & 0 & 0 \\ 0 & r^2 & \psi r^2 \\ 0 & \psi r^2 & \lambda^2 + \psi^2 r^2 \end{bmatrix} \tag{3.67}$$

Equation $(3.30)_1$ then gives the Lagrangian strain components

$$[E_{IJ}] = \tfrac{1}{2}\left[C_{IJ} - G_{IJ}\right] = \frac{1}{2}\begin{bmatrix} r'^2 - 1 & 0 & 0 \\ 0 & r^2 - R^2 & \psi r^2 \\ 0 & \psi r^2 & \lambda^2 + \psi^2 r^2 - 1 \end{bmatrix} \tag{3.68}$$

where Eq. (3.65) has provided the G_{IJ}, and Eq. (3.38) gives the physical components

$$[\hat{E}_{IJ}] = \frac{1}{2}\begin{bmatrix} r'^2 - 1 & 0 & 0 \\ 0 & \dfrac{r^2}{R^2} - 1 & \dfrac{\psi r^2}{R} \\ 0 & \dfrac{\psi r^2}{R} & \lambda^2 + \psi^2 r^2 - 1 \end{bmatrix}. \tag{3.69}$$

These components should agree with those computed directly from Eq. $(3.27)_1$ using the physical components of $\mathbf{F}$ and $\mathbf{I}$. Using Eq. (3.61) and with $\mathbf{I}$ expressed in terms of unit dyads, we get

$$
\begin{aligned}
\mathbf{E} &= \tfrac{1}{2}(\mathbf{F}^T\cdot\mathbf{F} - \mathbf{I}) \\
&= \tfrac{1}{2}\{[r'\mathbf{e}_R\mathbf{e}_r + (r/R)\mathbf{e}_\Theta\mathbf{e}_\theta + \lambda\mathbf{e}_Z\mathbf{e}_z + \psi r\mathbf{e}_Z\mathbf{e}_\theta] \\
&\quad \cdot [r'\mathbf{e}_r\mathbf{e}_R + (r/R)\mathbf{e}_\theta\mathbf{e}_\Theta + \lambda\mathbf{e}_z\mathbf{e}_Z + \psi r\mathbf{e}_\theta\mathbf{e}_Z] \\
&\quad -(\mathbf{e}_R\mathbf{e}_R + \mathbf{e}_\Theta\mathbf{e}_\Theta + \mathbf{e}_Z\mathbf{e}_Z)\} \\
&= \tfrac{1}{2}\left[(r'^2-1)\mathbf{e}_R\mathbf{e}_R + (r^2/R^2-1)\mathbf{e}_\Theta\mathbf{e}_\Theta + (\psi r^2/R)\mathbf{e}_\Theta\mathbf{e}_Z \right. \\
&\quad \left. + (\psi r^2/R)\mathbf{e}_Z\mathbf{e}_\Theta + (\lambda^2 + \psi^2 r^2 - 1)\mathbf{e}_Z\mathbf{e}_Z\right] \\
&= \hat{E}_{IJ}\mathbf{e}^I\mathbf{e}^J.
\end{aligned} \tag{3.70}
$$

The last line follows from Eq. (3.41), since the $\mathbf{G}_I$ $(\mathbf{e}_I)$ are orthogonal vectors. Extracting the $\hat{E}_{IJ}$ from this expression shows that they do indeed agree with those of Eq. (3.69).

Finally, we note one last point. The matrix $[E_{IJ}]$ is always symmetric since $C_{IJ} = g_{IJ} = g_{JI}$. As shown by Eq. (3.69), the matrix of physical components also is symmetric in this problem. On the other hand, the matrix of mixed components of $\mathbf{E}$, given by $E^I_J = E_{JK}G^{KI}$, is

$$
[E^I_J] = \frac{1}{2}\begin{bmatrix} r'^2 - 1 & 0 & 0 \\ 0 & \dfrac{r^2}{R^2} - 1 & \dfrac{\psi r^2}{R^2} \\ 0 & \psi r^2 & \lambda^2 + \psi^2 r^2 - 1 \end{bmatrix} \tag{3.71}
$$

which is *not* symmetric. (Here, we have used the relation $G^{(II)} = 1/G_{(II)}$ for orthogonal coordinates.) This demonstrates that, even though $\mathbf{E} = \mathbf{E}^T$, $E^I_J \neq E^J_I$ in general [see Eqs. (2.40)]. ■

Example 3.5 If the tube of the previous example is elastic, it does not matter whether the cylinder is extended, inflated, or twisted first (or all three simultaneously). The final state of stress is the same in all cases, and so we did not consider the order of the imposed deformations. If dissipation occurs, however, the path the cylinder takes to its final configuration may affect the stresses, even though the final strain state is the same. Because viscoelastic effects often are important in biomechanics, it is useful to illustrate how such an analysis may proceed. In this example, therefore, consider the following two paths to the same final deformed configuration:

Path A:

1 Extension: $r_1 = R, \theta_1 = \Theta, z_1 = \lambda Z$
2 Inflation: $r_2 = r(r_1), \theta_2 = \theta_1, z_2 = z_1$
3 Torsion: $r = r_2, \theta = \theta_2 + \psi Z = \theta_2 + \psi z_2/\lambda, z = z_2$

Path B:

1 Torsion: $r_1 = R, \theta_1 = \Theta + \psi Z, z_1 = Z$
2 Extension: $r_2 = r_1, \theta_2 = \theta_1, z_2 = \lambda z_1$
3 Inflation: $r = r(r_2), \theta = \theta_2, z = z_2$

In both cases, the final location of an arbitrary element is given by Eqs. (3.56). Show that both paths produce the same total deformation gradient tensor (3.63).

Solution. For each step in Paths A and B, the following *relative* deformation gradients are computed [see Eq. (3.63)]:

$$
\begin{aligned}
\left[F^i_I\right]^{(A)}_1 &= \begin{bmatrix} r_{1,R} & r_{1,\Theta} & r_{1,Z} \\ \theta_{1,R} & \theta_{1,\Theta} & \theta_{1,Z} \\ z_{1,R} & z_{1,\Theta} & z_{1,Z} \end{bmatrix}^{(A)} = \begin{bmatrix} 1 & 0 & 0 \\ 0 & 1 & 0 \\ 0 & 0 & \lambda \end{bmatrix} \\
\left[F^i_I\right]^{(A)}_2 &= \begin{bmatrix} r_{2,r_1} & r_{2,\theta_1} & r_{2,z_1} \\ \theta_{2,r_1} & \theta_{2,\theta_1} & \theta_{2,z_1} \\ z_{2,r_1} & z_{2,\theta_1} & z_{2,z_1} \end{bmatrix}^{(A)} = \begin{bmatrix} r'(R) & 0 & 0 \\ 0 & 1 & 0 \\ 0 & 0 & 1 \end{bmatrix} \\
\left[F^i_I\right]^{(A)}_3 &= \begin{bmatrix} r_{,r_2} & r_{,\theta_2} & r_{,z_2} \\ \theta_{,r_2} & \theta_{,\theta_2} & \theta_{,z_2} \\ z_{,r_2} & z_{,\theta_2} & z_{,z_2} \end{bmatrix}^{(A)} = \begin{bmatrix} 1 & 0 & 0 \\ 0 & 1 & \psi/\lambda \\ 0 & 0 & 1 \end{bmatrix} \\
\left[F^i_I\right]^{(B)}_1 &= \begin{bmatrix} r_{1,R} & r_{1,\Theta} & r_{1,Z} \\ \theta_{1,R} & \theta_{1,\Theta} & \theta_{1,Z} \\ z_{1,R} & z_{1,\Theta} & z_{1,Z} \end{bmatrix}^{(B)} = \begin{bmatrix} 1 & 0 & 0 \\ 0 & 1 & \psi \\ 0 & 0 & 1 \end{bmatrix} \\
\left[F^i_I\right]^{(B)}_2 &= \begin{bmatrix} r_{2,r_1} & r_{2,\theta_1} & r_{2,z_1} \\ \theta_{2,r_1} & \theta_{2,\theta_1} & \theta_{2,z_1} \\ z_{2,r_1} & z_{2,\theta_1} & z_{2,z_1} \end{bmatrix}^{(B)} = \begin{bmatrix} 1 & 0 & 0 \\ 0 & 1 & 0 \\ 0 & 0 & \lambda \end{bmatrix} \\
\left[F^i_I\right]^{(B)}_3 &= \begin{bmatrix} r_{,r_2} & r_{,\theta_2} & r_{,z_2} \\ \theta_{,r_2} & \theta_{,\theta_2} & \theta_{,z_2} \\ z_{,r_2} & z_{,\theta_2} & z_{,z_2} \end{bmatrix}^{(B)} = \begin{bmatrix} r'(R) & 0 & 0 \\ 0 & 1 & 0 \\ 0 & 0 & 1 \end{bmatrix} \qquad (3.72)
\end{aligned}
$$

These matrices give the deformation gradients of each step relative to the geometry of the previous step.

With each of these matrices, we can compute a deformation gradient tensor $\mathbf{F}_k = (F^i_I)_k \mathbf{g}_i \mathbf{G}^I$, where k indicates a step in the deformation path. [Note that the $\mathbf{g}_i$ and $\mathbf{G}^I$ given by Eqs. (3.59) remain fixed.] Then, the successive deformations of a line element are given by

$$\begin{aligned} d\mathbf{r}_1 &= \mathbf{F}_1 \cdot d\mathbf{R} \\ d\mathbf{r}_2 &= \mathbf{F}_2 \cdot d\mathbf{r}_1 \\ d\mathbf{r} &= \mathbf{F}_3 \cdot d\mathbf{r}_2 \end{aligned}$$

and, therefore,

$$d\mathbf{r} = \mathbf{F}_3 \cdot \left[\mathbf{F}_2 \cdot (\mathbf{F}_1 \cdot d\mathbf{R}) \right].$$

This equation can be written

$$d\mathbf{r} = \mathbf{F} \cdot d\mathbf{R}$$

where

$$\mathbf{F} = \mathbf{F}_3 \cdot \mathbf{F}_2 \cdot \mathbf{F}_1 \tag{3.73}$$

is the total deformation gradient tensor. Alternatively, the chain rule gives the equivalent expression

$$[F^i_I] = \frac{\partial x^i}{\partial X^I} = \frac{\partial x^i}{\partial x_2^j} \frac{\partial x_2^j}{\partial x_1^k} \frac{\partial x_1^k}{\partial X_I} = [F^i_j]_3 [F^j_k]_2 [F^k_I]_1. \tag{3.74}$$

The reader may verify that substituting Eqs. (3.72) for either Path A or Path B reproduces Eq. (3.63) (see Problem **3–6**). This exercise demonstrates that the final deformation is independent of the deformation path; but again, if the material is inelastic, the stresses may depend on the path. ■

3.5 Strain-Displacement Relations

In this section, we write the strains in terms of displacements. Let $\mathbf{u}(X^I, t)$ be the displacement vector from the point P in B to its image p in b. In addition, the reference frames associated with B and b may be in relative motion, with the vector $\mathbf{b}(t)$ locating the position of reference frame o with respect to frame O (see Fig. 3.3, page 70). Then, the position vectors of P and p are related by

$$\mathbf{r}(X^I, t) + \mathbf{b}(t) = \mathbf{R}(X^I, t) + \mathbf{u}(X^I, t). \tag{3.75}$$

Note that, since it is independent of the point P, $\mathbf{b}$ does not depend on the coordinates X^I of P.

Substituting Eq. (3.75) into (3.21) and noting Eqs. (3.18) gives

$$\begin{aligned} \mathbf{F}^T &= \frac{\partial \mathbf{r}}{\partial \mathbf{R}} = \mathbf{I} + \frac{\partial \mathbf{u}}{\partial \mathbf{R}} = \mathbf{I} + \boldsymbol{\nabla}\mathbf{u} \\ \mathbf{F}^{-T} &= \frac{\partial \mathbf{R}}{\partial \mathbf{r}} = \mathbf{I} - \frac{\partial \mathbf{u}}{\partial \mathbf{r}} = \mathbf{I} - \overline{\boldsymbol{\nabla}}\mathbf{u} \end{aligned} \tag{3.76}$$

or, since $\mathbf{I}^T = \mathbf{I}$,

$$\begin{aligned} \mathbf{F} &= (\mathbf{F}^T)^T = \mathbf{I} + (\boldsymbol{\nabla}\mathbf{u})^T \\ \mathbf{F}^{-1} &= (\mathbf{F}^{-T})^T = \mathbf{I} - (\overline{\boldsymbol{\nabla}}\mathbf{u})^T. \end{aligned} \tag{3.77}$$

Inserting these equations into Eqs. (3.27) yields

$$\begin{aligned} 2\mathbf{E} &= \mathbf{F}^T\!\cdot\!\mathbf{F} - \mathbf{I} \\ &= [\mathbf{I} + \boldsymbol{\nabla}\mathbf{u}] \cdot [\mathbf{I} + (\boldsymbol{\nabla}\mathbf{u})^T] - \mathbf{I} \end{aligned}$$

$$\mathbf{E} = \tfrac{1}{2}[\boldsymbol{\nabla}\mathbf{u} + (\boldsymbol{\nabla}\mathbf{u})^T + (\boldsymbol{\nabla}\mathbf{u})\!\cdot\!(\boldsymbol{\nabla}\mathbf{u})^T] \tag{3.78}$$

$$\begin{aligned} 2\mathbf{e} &= \mathbf{I} - \mathbf{F}^{-T}\!\cdot\!\mathbf{F}^{-1} \\ &= \mathbf{I} - [\mathbf{I} - \overline{\boldsymbol{\nabla}}\mathbf{u}] \cdot [\mathbf{I} - (\overline{\boldsymbol{\nabla}}\mathbf{u})^T] \end{aligned}$$

$$\mathbf{e} = \tfrac{1}{2}[\overline{\boldsymbol{\nabla}}\mathbf{u} + (\overline{\boldsymbol{\nabla}}\mathbf{u})^T - (\overline{\boldsymbol{\nabla}}\mathbf{u})\!\cdot\!(\overline{\boldsymbol{\nabla}}\mathbf{u})^T]. \tag{3.79}$$

These **strain-displacement relations** illustrate the symmetry of the tensors $\mathbf{E}$ and $\mathbf{e}$. Moreover, they show that geometric nonlinearity comes from terms quadratic in the displacement gradients. This nonlinearity leads to substantial complexity when deformation is large; thus, when possible, it is advantageous to seek approximations to Eqs. (3.78) and (3.79), as will be discussed in Section 3.9.

We next derive component forms for the strain-displacement relations. Because the Lagrangian and the Eulerian strain tensors contain derivatives with respect to the undeformed and deformed bodies, respectively, the most convenient representations for these tensors are

$$\begin{aligned} \mathbf{E} &= E_{IJ}\,\mathbf{G}^I\mathbf{G}^J \\ \mathbf{e} &= e_{ij}\,\mathbf{g}^i\mathbf{g}^j = e_{I^*J^*}\,\mathbf{g}^I\mathbf{g}^J, \end{aligned} \tag{3.80}$$

and the displacement vector has the representations

$$\begin{aligned} \mathbf{u} &= u^I\mathbf{G}_I = u_I\mathbf{G}^I \\ &= u^i\mathbf{g}_i = u_i\mathbf{g}^i = u^{I^*}\mathbf{g}_I = u_{I^*}\mathbf{g}^I. \end{aligned} \tag{3.81}$$

The strain tensors depend on the displacement gradients given by

$$\begin{aligned} \boldsymbol{\nabla}\mathbf{u} &= \mathbf{G}^I\mathbf{u}_{,I} \\ \bar{\boldsymbol{\nabla}}\mathbf{u} &= \mathbf{g}^i\mathbf{u}_{,i} = \mathbf{g}^I\mathbf{u}_{,I} \end{aligned} \tag{3.82}$$

in which Eqs. (3.18) have provided the gradient operators. Now, substituting the covariant-component forms of Eqs. (3.81) and noting Eq. (2.141) yields

$$\begin{aligned} \boldsymbol{\nabla}\mathbf{u} &= \mathbf{G}^I(u_J\mathbf{G}^J)_{,I} = \mathbf{G}^I(u_J|_I\mathbf{G}^J) = u_J|_I\mathbf{G}^I\mathbf{G}^J \\ \bar{\boldsymbol{\nabla}}\mathbf{u} &= \mathbf{g}^i(u_j\mathbf{g}^j)_{,i} = \mathbf{g}^i(u_j||_i\mathbf{g}^j) = u_j||_i\mathbf{g}^i\mathbf{g}^j \\ &= \mathbf{g}^I(u_{J^*}\mathbf{g}^J)_{,I} = \mathbf{g}^I(u_{J^*}||_I\mathbf{g}^J) = u_{J^*}||_I\mathbf{g}^I\mathbf{g}^J \end{aligned} \tag{3.83}$$

where the single vertical lines and double vertical lines indicate covariant differentiation with respect to B and b, respectively. In particular, the three relations in Eqs. (3.83) involve differentiation of the base vectors $\mathbf{G}^I$, $\mathbf{g}^i$, and $\mathbf{g}^I$, respectively. Thus, Eq. $(2.142)_2$ gives

$$\boxed{\begin{aligned} u_I|_J &= u_{I,J} - u_K\Gamma^K_{IJ} \\ u_i||_j &= u_{i,j} - u_k\bar{\Gamma}^k_{ij} \\ u_{I^*}||_J &= u_{I^*,J} - u_{K^*}\bar{\Gamma}^K_{IJ} \end{aligned}} \tag{3.84}$$

where suitable modification of Eq. $(2.128)_2$ yields the Christoffel symbols

$$\boxed{\begin{aligned} \Gamma^K_{IJ} &= \mathbf{G}_{I,J}\cdot\mathbf{G}^K \\ \bar{\Gamma}^k_{ij} &= \mathbf{g}_{i,j}\cdot\mathbf{g}^k \\ \bar{\Gamma}^K_{IJ} &= \mathbf{g}_{I,J}\cdot\mathbf{g}^K, \end{aligned}} \tag{3.85}$$

with bar indicating a symbol defined in the deformed body b. In summary, $()|_I$ represents differentiation with respect to X^I and $\mathbf{G}_I$ in B, $()||_i$ represents differentiation with respect to x^i and $\mathbf{g}_i$ in b, and $()||_I$ represents differentiation with respect to X^I (convected) and $\mathbf{g}_I$ in b. In convected coordinates, $x^i = X^I$, $\mathbf{g}_i = \mathbf{g}_I$, and $u_i||_j = u_{I^*}||_J$; and the two forms for $\overline{\boldsymbol{\nabla}}\mathbf{u}$ in Eq. (3.83) are equivalent.

Getting back to the strain-displacement relations, we substitute Eqs. (3.83) into (3.78) to obtain

$$\begin{aligned} 2\mathbf{E} &= u_J|_I \mathbf{G}^I\mathbf{G}^J + u_J|_I \mathbf{G}^J\mathbf{G}^I + (u_J|_I \mathbf{G}^I\mathbf{G}^J)\cdot(u_L|_K \mathbf{G}^L\mathbf{G}^K) \\ &= u_J|_I \mathbf{G}^I\mathbf{G}^J + u_I|_J \mathbf{G}^I\mathbf{G}^J + u_J|_I\, u_L|_K G^{JL}\mathbf{G}^I\mathbf{G}^K \\ &= u_J|_I \mathbf{G}^I\mathbf{G}^J + u_I|_J \mathbf{G}^I\mathbf{G}^J + u_J|_I\, u^J|_K \mathbf{G}^I\mathbf{G}^K \\ &= (u_I|_J + u_J|_I + u_K|_I\, u^K|_J)\mathbf{G}^I\mathbf{G}^J \end{aligned} \tag{3.86}$$

and, similarly, Eq. (3.79) gives

$$\begin{aligned} 2\mathbf{e} &= u_j||_i \mathbf{g}^i\mathbf{g}^j + u_j||_i \mathbf{g}^j\mathbf{g}^i - (u_j||_i \mathbf{g}^i\mathbf{g}^j)\cdot(u_l||_k \mathbf{g}^l\mathbf{g}^k) \\ &= (u_i||_j + u_j||_i - u_k||_i\, u^k||_j)\mathbf{g}^i\mathbf{g}^j \end{aligned} \tag{3.87}$$

or

$$\begin{aligned} 2\mathbf{e} &= u_{J^*}||_I \mathbf{g}^I\mathbf{g}^J + u_{J^*}||_I \mathbf{g}^J\mathbf{g}^I - (u_{J^*}||_I \mathbf{g}^I\mathbf{g}^J)\cdot(u_{L^*}||_K \mathbf{g}^L\mathbf{g}^K) \\ &= (u_{I^*}||_J + u_{J^*}||_I - u_{K^*}||_I\, u^{K^*}||_J)\mathbf{g}^I\mathbf{g}^J. \end{aligned} \tag{3.88}$$

Note that G^{JL} in the second line of Eq. (3.86) was used to raise the subscript on u_L, and the indices J and K were interchanged in the last term of the last line. In Eqs. (3.87) and (3.88), g^{jl} and g^{JL} raised the subscripts on u_l and u_{L^*}, respectively. To show that g^{JL} is needed to raise the index of u_{L^*}, we consider the representations

$$\mathbf{u} = u_{I^*}\mathbf{g}^I = u^{I^*}\mathbf{g}_I.$$

Dotting with $\mathbf{g}^J$ yields

$$(u_{I^*}\mathbf{g}^I)\cdot\mathbf{g}^J = (u^{I^*}\mathbf{g}_I)\cdot\mathbf{g}^J$$

or

$$u_{I^*}g^{IJ} = u^{I^*}\delta_I^J = u^{J^*}, \tag{3.89}$$

which is the desired result.

Finally, comparing Eqs. (3.80) with (3.86)–(3.88) reveals that

$$\boxed{\begin{aligned} E_{IJ} &= \tfrac{1}{2}(u_I|_J + u_J|_I + u_K|_I\, u^K|_J) \\ e_{ij} &= \tfrac{1}{2}(u_i||_j + u_j||_i - u_k||_i\, u^k||_j) \\ &= \tfrac{1}{2}(u_{I^*}||_J + u_{J^*}||_I - u_{K^*}||_I\, u^{K^*}||_J). \end{aligned}} \quad (3.90)$$

These equations represent the scalar component forms of the strain-displacement relations.

3.6 Geometric Measures of Deformation

The deformation gradient, deformation, and strain tensors *characterize* the deformation of a body. In general, they have no direct physical interpretations like those of the stretch or stretch ratio. In practice, however, more physically meaningful measures often are useful. For example, regional deformation of a tissue or organ can be estimated by tracking the motions of markers (Waldman et al., 1985; Waldman et al., 1988; McCulloch et al., 1989; Villarreal et al., 1991; Hashima et al., 1993; Taber et al., 1994). This section explores the relations between strain and changes in length, angle, volume, and area.

3.6.1 *Stretch*

Let dS and ds be the lengths of a differential line element before and after deformation, respectively (see Fig. 3.4, page 74). The stretch ratio ds/dS is a function of the *specified* orientation of *either* the undeformed or the deformed element. In other words, we can compute a **Lagrangian stretch ratio** $\Lambda_{(\mathbf{N})}$ for an element that occupies a specified direction $\mathbf{N}$ in the undeformed body B or an **Eulerian stretch ratio** $\lambda_{(\mathbf{n})}$ for an element that occupies a specified direction $\mathbf{n}$ in the deformed body b, with $\mathbf{N}$ and $\mathbf{n}$ being unit vectors (Fig. 3.4). In general, $\mathbf{N}$ and $\mathbf{n}$ can be chosen arbitrarily in B and b, respectively. However, if $\mathbf{N}$ and $\mathbf{n}$ are bound to the same material line element, i.e., $\mathbf{N}$ is convected into $\mathbf{n}$ during deformation, then $\Lambda_{(\mathbf{N})} = \lambda_{(\mathbf{n})}$. On the other hand, if we choose $\mathbf{N} = \mathbf{n}$, then the material elements associated with each of these unit vectors differ if local rotation occurs, and so $\Lambda_{(\mathbf{N})} \neq \lambda_{(\mathbf{n})}$ in general.

In the Lagrangian and Eulerian descriptions, respectively, we can write

$$d\mathbf{R} = \mathbf{N}\,dS, \qquad d\mathbf{r} = \mathbf{n}\,ds \tag{3.91}$$

where $d\mathbf{r}$ is not necessarily the deformed image of $d\mathbf{R}$. Substitution into Eq. (3.19) yields

$$\mathbf{n}\,ds = \mathbf{F}\cdot\mathbf{N}\,dS \qquad \text{or} \qquad \mathbf{N}\,dS = \mathbf{F}^{-1}\cdot\mathbf{n}\,ds,$$

which gives the relations

$$\mathbf{n}\,\Lambda_{(\mathbf{N})} = \mathbf{F}\cdot\mathbf{N} = \mathbf{N}\cdot\mathbf{F}^T \tag{3.92}$$

$$\mathbf{N}\,\lambda_{(\mathbf{n})}^{-1} = \mathbf{F}^{-1}\cdot\mathbf{n} = \mathbf{n}\cdot\mathbf{F}^{-T}. \tag{3.93}$$

Because $\mathbf{N}$ and $\mathbf{n}$ are unit vectors, we can write

$$\begin{aligned}\Lambda_{(\mathbf{N})}^2 &= \left(\mathbf{n}\,\Lambda_{(\mathbf{N})}\right)\cdot\left(\mathbf{n}\,\Lambda_{(\mathbf{N})}\right) = \left(\mathbf{N}\cdot\mathbf{F}^T\right)\cdot\left(\mathbf{F}\cdot\mathbf{N}\right)\\ \lambda_{(\mathbf{n})}^{-2} &= \left(\mathbf{N}\,\lambda_{(\mathbf{n})}^{-1}\right)\cdot\left(\mathbf{N}\,\lambda_{(\mathbf{n})}^{-1}\right) = \left(\mathbf{n}\cdot\mathbf{F}^{-T}\right)\cdot\left(\mathbf{F}^{-1}\cdot\mathbf{n}\right),\end{aligned}$$

and using Eqs. (3.24) gives

$$\boxed{\begin{aligned}\Lambda_{(\mathbf{N})}^2 &= \mathbf{N}\cdot\mathbf{C}\cdot\mathbf{N}\\ \lambda_{(\mathbf{n})}^{-2} &= \mathbf{n}\cdot\mathbf{B}^{-1}\cdot\mathbf{n}.\end{aligned}} \tag{3.94}$$

In addition, analogous to the engineering strain (stretch) of Eq. (3.1), the Lagrangian and Eulerian **extension ratios** are defined by

$$\boxed{\begin{aligned}E_{(\mathbf{N})}^* &= \frac{ds - dS}{dS} = \Lambda_{(\mathbf{N})} - 1\\ e_{(\mathbf{n})}^* &= \frac{ds - dS}{ds} = 1 - \frac{1}{\lambda_{(\mathbf{n})}}.\end{aligned}} \tag{3.95}$$

Example 3.6 Suppose the components of the strain tensors $\mathbf{E}$ and $\mathbf{e}$ are known in the Cartesian coordinates $X^I = x^i$. Compute the extension ratios in terms of strain components for an element aligned with the X^1-axis in the undeformed body and for an element aligned with this axis in the deformed body.

Solution. In the first case, $\mathbf{N} = \mathbf{e}_1$, and in the second case, $\mathbf{n} = \mathbf{e}_1$. Thus, Eqs. (3.27) and (3.94) yield

$$\begin{aligned} \Lambda_{(1)}^2 &= \mathbf{e}_1 \cdot \mathbf{C} \cdot \mathbf{e}_1 = \mathbf{e}_1 \cdot (\mathbf{I} + 2\mathbf{E}) \cdot \mathbf{e}_1 = 1 + 2E_{11} \\ \lambda_{(1)}^{-2} &= \mathbf{e}_1 \cdot \mathbf{B}^{-1} \cdot \mathbf{e}_1 = \mathbf{e}_1 \cdot (\mathbf{I} - 2\mathbf{e}) \cdot \mathbf{e}_1 = 1 - 2e_{11}, \end{aligned} \tag{3.96}$$

which indicate that $\Lambda_{(1)} \neq \lambda_{(1)}$ in general. In this problem, the stretch ratios depend only on normal strains. However, if either $\mathbf{N}$ or $\mathbf{n}$ is not aligned with a coordinate axis, then shear strains would be involved. The extension ratios, from Eqs. (3.95), are

$$\begin{aligned} E_{(1)}^* &= \Lambda_{(1)} - 1 = (1 + 2E_{11})^{\frac{1}{2}} - 1 \\ e_{(1)}^* &= 1 - (\lambda_{(1)})^{-1} = 1 - (1 - 2e_{11})^{\frac{1}{2}}, \end{aligned} \tag{3.97}$$

and these equations give

$$\begin{aligned} E_{11} &= E_{(1)}^* + \tfrac{1}{2} E_{(1)}^{*2} \\ e_{11} &= e_{(1)}^* - \tfrac{1}{2} e_{(1)}^{*2}. \end{aligned} \tag{3.98}$$

These last relations show that if the extension ratios are small, then $E_{11} \cong E_{(1)}^*$ and $e_{11} \cong e_{(1)}^*$, i.e., the linear normal strain components have the usual physical interpretation of change in length divided by undeformed length. ■

Example 3.7 For the case of simple shear (see Example 3.3 and Fig. 3.6, page 82), compute the stretch ratios for elements that are parallel to the X^2-axis (a) before deformation and (b) after deformation.

Solution. (a) On setting $\mathbf{N} = \mathbf{e}_2$, Eqs. $(3.55)_1$ and $(3.94)_1$ give the stretch ratio

$$\Lambda_{(2)}^2 = \mathbf{e}_2 \cdot \mathbf{C} \cdot \mathbf{e}_2 = C_{22} = 1 + k^2.$$

Note that the geometry of Fig. 3.6 shows that the deformed length of side AB is $\sqrt{1 + k^2}$, in agreement with this result.

(b) Setting $\mathbf{n} = \mathbf{e}_2$ in Eq. $(3.94)_2$ gives

$$\lambda_{(2)}^{-2} = \mathbf{e}_2 \cdot \mathbf{B}^{-1} \cdot \mathbf{e}_2 = (B^{-1})_{22}.$$

Inverting the matrix of Eq. $(3.55)_2$ yields

$$[B_{ij}]^{-1} = \begin{bmatrix} 1 & -k & 0 \\ -k & 1 + k^2 & 0 \\ 0 & 0 & 1 \end{bmatrix},$$

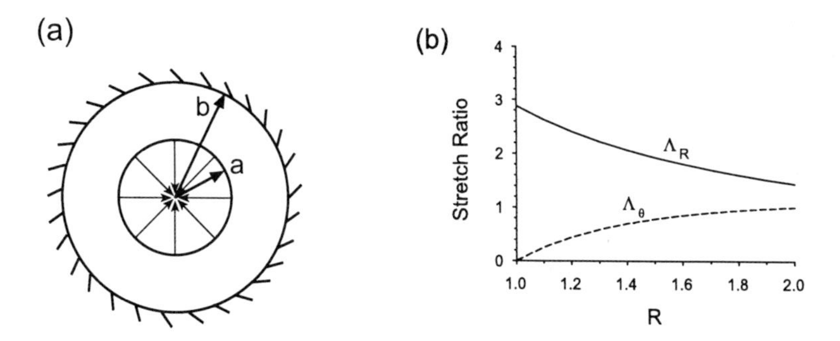

Fig. 3.8 (a) Model for wound healing: annular membrane with outer edge fixed, and inner edge moves to center. (b) Stretch ratios in model after healing. (inner edge: $R = 1$; outer edge: $R = 2$)

and so

$$\lambda_{(2)}^2 = (1 + k^2)^{-1}.$$

Again, we see that $\Lambda_{(2)} \neq \lambda_{(2)}$. For $k << 1$, however, $\Lambda_{(2)} \cong \lambda_{(2)} \cong 1$, i.e., to a first approximation, simple shear involves no extension in the linear case (relative to the coordinate axes). ■

Example 3.8 Wound healing is an important problem that involves mechanics. As a wound heals, the shape of its boundary affects the extent of scarring, possibly due to stresses generated during wound closure. Fibroblasts play a major role in this process, as these cells generate contractile forces and tractions that draw the edges of the wound together.

Consider a simple model for skin with a puncture wound. Before healing, the model consists of an annular membrane with an inner radius a and outer radius b (Fig. 3.8a). With the membrane fixed at $R = b$, the healing process moves the inner (cut) edge symmetrically inward toward the center. If a point originally a distance R from the center moves to the location

$$r(R) = b\,\frac{\log(R/a)}{\log(b/a)}, \tag{3.99}$$

compute the radial and circumferential stretch ratios.[2]

[2]The deformed geometry described by Eq. (3.99), chosen only for illustration, likely is not the actual solution to this problem. The true deformation field must be found by solving the full boundary value problem of nonlinear elasticity (see Chapter 6).

Solution. The geometry for this problem is similar to that of the tube in Example 3.4 (page 84). In two dimensions, the position vectors are

$$\mathbf{R} = R\,\mathbf{e}_R, \qquad \mathbf{r} = r\,\mathbf{e}_r. \tag{3.100}$$

Because there is no torsional deformation here, the unit vectors $\mathbf{e}_R = \mathbf{e}_r$ and $\mathbf{e}_\Theta = \mathbf{e}_\theta$ are given by Eqs. $(3.58)_{1,2}$.

With the gradient operator for cylindrical coordinates provided by Eq. (2.159), Eq. $(3.21)_1$ yields

$$\begin{aligned}\mathbf{F}^T &= \boldsymbol{\nabla}\mathbf{r} = \left(\mathbf{e}_R\frac{\partial}{\partial R} + \frac{\mathbf{e}_\Theta}{R}\frac{\partial}{\partial \Theta}\right)(r\,\mathbf{e}_r) \\ &= \frac{\partial r}{\partial R}\mathbf{e}_R\mathbf{e}_r + \frac{r}{R}\mathbf{e}_\Theta\frac{\partial \mathbf{e}_r}{\partial \Theta} \\ &= \frac{\partial r}{\partial R}\mathbf{e}_R\mathbf{e}_r + \frac{r}{R}\mathbf{e}_\Theta\mathbf{e}_\theta\end{aligned}$$

or

$$\mathbf{F} = \frac{\partial r}{\partial R}\mathbf{e}_r\mathbf{e}_R + \frac{r}{R}\mathbf{e}_\theta\mathbf{e}_\Theta. \tag{3.101}$$

To determine the stretch ratios, we compute the two-dimensional deformation tensor

$$\begin{aligned}\mathbf{C} &= \mathbf{F}^T\cdot\mathbf{F} \\ &= \left(\frac{\partial r}{\partial R}\right)^2\mathbf{e}_R\mathbf{e}_R + \left(\frac{r}{R}\right)^2\mathbf{e}_\Theta\mathbf{e}_\Theta,\end{aligned} \tag{3.102}$$

and Eq. $(3.94)_1$ gives the stretch ratios

$$\begin{aligned}\Lambda_R &= (\mathbf{e}_R\cdot\mathbf{C}\cdot\mathbf{e}_R)^{1/2} = \frac{\partial r}{\partial R} \\ \Lambda_\Theta &= (\mathbf{e}_\Theta\cdot\mathbf{C}\cdot\mathbf{e}_\Theta)^{1/2} = \frac{r}{R}.\end{aligned} \tag{3.103}$$

Comparison with Eq. (3.101) reveals that the stretch ratios in this problem are also the physical components of $\mathbf{F}$. It is relatively easy to show that this is true whenever $\mathbf{F}$ contains no off-diagonal (shear) terms. Finally, inserting Eq. (3.99) yields

$$\begin{aligned}\Lambda_R &= \frac{1}{\log(b/a)}\frac{b}{R} \\ \Lambda_\Theta &= \frac{1}{\log(b/a)}\frac{b}{R}\log\frac{R}{a}.\end{aligned} \tag{3.104}$$

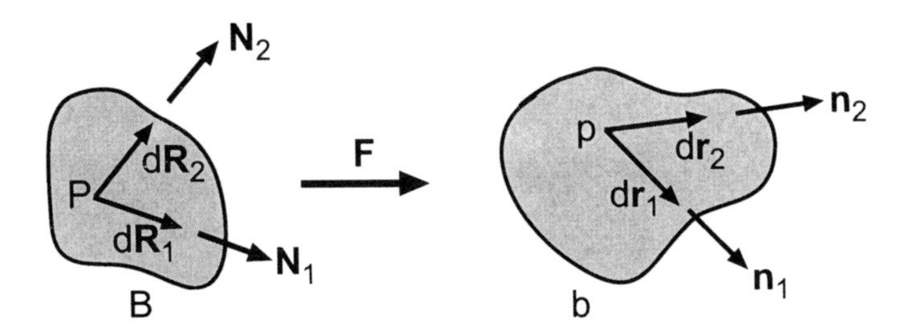

Fig. 3.9 Change in angle (shear) between two differential line elements.

Stretch ratio distributions for $a = 1$ and $b = 2$ are shown in Fig. 3.8b. Note that, for complete wound closure, the computed value $\Lambda_\Theta = 0$ at the center is not physically possible, as a length cannot shrink to a point. Hence, this deformation is possible only if the wound does not close completely. ■

3.6.2 *Shear*

Shear strains characterize the change in angle, due to deformation, between two line elements. The elements

$$\begin{aligned} d\mathbf{R}_1 &= \mathbf{N}_1\, dS_1 = \mathbf{F}^{-1}\cdot d\mathbf{r}_1 = d\mathbf{r}_1\cdot\mathbf{F}^{-T} \\ d\mathbf{R}_2 &= \mathbf{N}_2\, dS_2 = \mathbf{F}^{-1}\cdot d\mathbf{r}_2 = d\mathbf{r}_2\cdot\mathbf{F}^{-T} \end{aligned} \tag{3.105}$$

at a point P in B deform into the elements

$$\begin{aligned} d\mathbf{r}_1 &= \mathbf{n}_1\, ds_1 = \mathbf{F}\cdot d\mathbf{R}_1 = d\mathbf{R}_1\cdot\mathbf{F}^{T} \\ d\mathbf{r}_2 &= \mathbf{n}_2\, ds_2 = \mathbf{F}\cdot d\mathbf{R}_2 = d\mathbf{R}_2\cdot\mathbf{F}^{T} \end{aligned} \tag{3.106}$$

at the point p in b (Fig. 3.9), where the $\mathbf{N}_i$ and $\mathbf{n}_i$ are unit vectors and $\mathbf{F}$ is the local deformation gradient tensor [see Eqs. (3.19) and (3.20)].

The cosines of the angles between the undeformed and the deformed elements are

$$\begin{aligned}\cos(\mathbf{N}_1, \mathbf{N}_2) &= \mathbf{N}_1 \cdot \mathbf{N}_2 = \frac{d\mathbf{R}_1 \cdot d\mathbf{R}_2}{dS_1 dS_2} = \frac{(d\mathbf{r}_1 \cdot \mathbf{F}^{-T}) \cdot (\mathbf{F}^{-1} \cdot d\mathbf{r}_2)}{(d\mathbf{r}_1 \cdot \mathbf{B}^{-1} \cdot d\mathbf{r}_1)^{\frac{1}{2}} (d\mathbf{r}_2 \cdot \mathbf{B}^{-1} \cdot d\mathbf{r}_2)^{\frac{1}{2}}} \\ \cos(\mathbf{n}_1, \mathbf{n}_2) &= \mathbf{n}_1 \cdot \mathbf{n}_2 = \frac{d\mathbf{r}_1 \cdot d\mathbf{r}_2}{ds_1 ds_2} = \frac{(d\mathbf{R}_1 \cdot \mathbf{F}^{T}) \cdot (\mathbf{F} \cdot d\mathbf{R}_2)}{(d\mathbf{R}_1 \cdot \mathbf{C} \cdot d\mathbf{R}_1)^{\frac{1}{2}} (d\mathbf{R}_2 \cdot \mathbf{C} \cdot d\mathbf{R}_2)^{\frac{1}{2}}}\end{aligned} \tag{3.107}$$

in which Eqs. (3.24) have been used. With Eqs. (3.25), (3.94), (3.105), and (3.106), these relations can be written

$$\begin{aligned}\mathbf{N}_1 \cdot \mathbf{N}_2 &= \frac{\mathbf{n}_1 \cdot \mathbf{B}^{-1} \cdot \mathbf{n}_2}{(\mathbf{n}_1 \cdot \mathbf{B}^{-1} \cdot \mathbf{n}_1)^{\frac{1}{2}} (\mathbf{n}_2 \cdot \mathbf{B}^{-1} \cdot \mathbf{n}_2)^{\frac{1}{2}}} = \lambda_{(\mathbf{n}_1)} \lambda_{(\mathbf{n}_2)} (\mathbf{n}_1 \cdot \mathbf{B}^{-1} \cdot \mathbf{n}_2) \\ \mathbf{n}_1 \cdot \mathbf{n}_2 &= \frac{\mathbf{N}_1 \cdot \mathbf{C} \cdot \mathbf{N}_2}{(\mathbf{N}_1 \cdot \mathbf{C} \cdot \mathbf{N}_1)^{\frac{1}{2}} (\mathbf{N}_2 \cdot \mathbf{C} \cdot \mathbf{N}_2)^{\frac{1}{2}}} = \frac{\mathbf{N}_1 \cdot \mathbf{C} \cdot \mathbf{N}_2}{\Lambda_{(\mathbf{N}_1)} \Lambda_{(\mathbf{N}_2)}}.\end{aligned} \tag{3.108}$$

If $\mathbf{n}_1$ and $\mathbf{n}_2$ are given, the first equation provides $\mathbf{N}_1 \cdot \mathbf{N}_2$; if $\mathbf{N}_1$ and $\mathbf{N}_2$ are given, the second yields $\mathbf{n}_1 \cdot \mathbf{n}_2$. Of course, if there is no deformation, then $\mathbf{C} = \mathbf{B} = \mathbf{I}$ and both equations give $\mathbf{n}_1 \cdot \mathbf{n}_2 = \mathbf{N}_1 \cdot \mathbf{N}_2$. Finally, we define the **shears**

$$\Gamma_{(\mathbf{N}_1 \mathbf{N}_2)} = \gamma_{(\mathbf{n}_1 \mathbf{n}_2)} = \mathbf{n}_1 \cdot \mathbf{n}_2 - \mathbf{N}_1 \cdot \mathbf{N}_2, \tag{3.109}$$

and Eqs. (3.108) give

$$\begin{aligned}\Gamma_{(\mathbf{N}_1 \mathbf{N}_2)} &= \mathbf{N}_1 \cdot \left[\frac{\mathbf{C}}{\Lambda_{(\mathbf{N}_1)} \Lambda_{(\mathbf{N}_2)}} - \mathbf{I} \right] \cdot \mathbf{N}_2 \\ \gamma_{(\mathbf{n}_1 \mathbf{n}_2)} &= \mathbf{n}_1 \cdot \left[\mathbf{I} - \lambda_{(\mathbf{n}_1)} \lambda_{(\mathbf{n}_2)} \mathbf{B}^{-1} \right] \cdot \mathbf{n}_2.\end{aligned} \tag{3.110}$$

The shears provide measures of the change in angle between undeformed and deformed line elements.

Example 3.9 In Cartesian coordinates, the strain components E_{IJ} are known. Compute the shear in the $x^1 x^2$-plane in terms of the E_{IJ} for (a) an element that is rectangular in the undeformed body B, with edges parallel to the X^1 and X^2 axes; and (b) an element that is rectangular in the deformed body b, again with edges parallel to the X^1 and X^2 axes.

Solution. In the first case, $\mathbf{N}_1 = \mathbf{e}_1$ and $\mathbf{N}_2 = \mathbf{e}_2$, while in the second case, $\mathbf{n}_1 = \mathbf{e}_1$ and $\mathbf{n}_2 = \mathbf{e}_2$ ($\mathbf{e}_1 \cdot \mathbf{e}_2 = 0$). Thus, Eqs. (3.27), (3.96), and (3.110) yield

$$\begin{aligned} \Gamma_{(12)} &= \mathbf{e}_1 \cdot \left[\frac{\mathbf{I} + 2\mathbf{E}}{\Lambda_{(1)}\Lambda_{(2)}} - \mathbf{I} \right] \cdot \mathbf{e}_2 = \frac{2E_{12}}{[(1+2E_{11})(1+2E_{22})]^{\frac{1}{2}}} \\ \gamma_{(12)} &= \mathbf{e}_1 \cdot \left[\mathbf{I} - \lambda_{(1)}\lambda_{(2)}(\mathbf{I} - 2\mathbf{e}) \right] \cdot \mathbf{e}_2 = \frac{2e_{12}}{[(1-2e_{11})(1-2e_{22})]^{\frac{1}{2}}} \end{aligned} \tag{3.111}$$

for cases (a) and (b), respectively. Note that the shears depend on the shear strains, E_{12} and e_{12}, *and* on the normal strains. If the normal strains are small compared to unity, then

$$\Gamma_{(12)} \cong 2E_{12}, \qquad \gamma_{(12)} \cong 2e_{12},$$

which agree with the usual interpretation for the linear shear strain components. ■

Example 3.10 For simple shear (see Example 3.3 and Fig. 3.6, page 82), $\Gamma_{(12)}$ provides a measure of the change in angle between line elements originally aligned with the X^1 and X^2 axes. Determine $\Gamma_{(12)}$ in terms of k.

Solution. With $\mathbf{N}_1 = \mathbf{e}_1$ and $\mathbf{N}_2 = \mathbf{e}_2$, Eq. $(3.110)_1$ gives

$$\Gamma_{(12)} = \mathbf{e}_1 \cdot \left[\frac{\mathbf{C}}{\Lambda_{(1)}\Lambda_{(2)}} - \mathbf{I} \right] \cdot \mathbf{e}_2 = \frac{C_{12}}{\Lambda_{(1)}\Lambda_{(2)}} = \frac{C_{12}}{\sqrt{C_{11}C_{22}}}$$

where we have inserted $\Lambda_{(1)}^2 = \mathbf{e}_1 \cdot \mathbf{C} \cdot \mathbf{e}_1 = C_{11}$ and $\Lambda_{(2)}^2 = \mathbf{e}_2 \cdot \mathbf{C} \cdot \mathbf{e}_2 = C_{22}$. Using Eq. $(3.55)_1$, we get

$$\Gamma_{(12)} = \frac{k}{\sqrt{1+k^2}}.$$

A glance at the geometry of Fig. 3.6 reveals that $\Gamma_{(12)} = \sin \gamma$. ■

3.6.3 *Volume Change*

Consider a differential volume element with edges parallel to $d\mathbf{R}_{(1)}$, $d\mathbf{R}_{(2)}$, and $d\mathbf{R}_{(3)}$ that deforms into an element with edges parallel to $d\mathbf{r}_{(1)}$, $d\mathbf{r}_{(2)}$, and $d\mathbf{r}_{(3)}$, where

$$d\mathbf{R}_{(k)} = \mathbf{G}_I \, dX^I_{(k)}, \qquad d\mathbf{r}_{(k)} = \mathbf{g}_i \, dx^i_{(k)}. \tag{3.112}$$

Inserting these relations and Eq. $(3.22)_1$ into

$$d\mathbf{r}_{(k)} = \mathbf{F}\cdot d\mathbf{R}_{(k)}, \tag{3.113}$$

yields

$$\begin{aligned} \mathbf{g}_i \, dx^i_{(k)} &= (F^i_I \mathbf{g}_i \mathbf{G}^I)\cdot(\mathbf{G}_J \, dX^J_{(k)}) \\ &= F^i_I \mathbf{g}_i \delta^I_J \, dX^J_{(k)} \\ &= F^i_I \mathbf{g}_i \, dX^I_{(k)} \end{aligned}$$

or

$$dx^i_{(k)} = F^i_I \, dX^I_{(k)} = x^i{}_{,I} \; dX^I_{(k)}. \tag{3.114}$$

This relation, which involves Eq. (3.15), is an expression of the chain rule for differentiation.

The volumes of the parallelepiped before and after deformation, respectively, are

$$\begin{aligned} dV &= d\mathbf{R}_{(1)}\cdot(d\mathbf{R}_{(2)}\times d\mathbf{R}_{(3)}) \\ dv &= d\mathbf{r}_{(1)}\cdot(d\mathbf{r}_{(2)}\times d\mathbf{r}_{(3)}). \end{aligned} \tag{3.115}$$

Substituting Eqs. (3.112) and noting Eqs. $(2.96)_2$ and $(2.108)_1$ gives

$$\begin{aligned} dV &= \mathbf{G}_I\cdot(\mathbf{G}_J\times\mathbf{G}_K) \, dX^I_{(1)} dX^J_{(2)} dX^K_{(3)} \\ &= \sqrt{G}\, e_{IJK} \, dX^I_{(1)} dX^J_{(2)} dX^K_{(3)} \\ dv &= \mathbf{g}_i\cdot(\mathbf{g}_j\times\mathbf{g}_k) \, dx^i_{(1)} dx^j_{(2)} dx^k_{(3)} \\ &= \sqrt{g}\, e_{ijk} \, dx^i_{(1)} dx^j_{(2)} dx^k_{(3)} \end{aligned} \tag{3.116}$$

where e_{IJK} and e_{ijk} are the permutation symbols defined by Eq. (2.92), and

$$G = \det\left[G_{IJ}\right], \qquad g = \det\left[g_{ij}\right]. \tag{3.117}$$

Now, using Eqs. (3.114), $(2.91)_1$, and $(3.116)_1$ in order, we can write Eq. $(3.116)_2$ as

$$\begin{aligned} dv &= \sqrt{g}\,(e_{ijk}\,F^i_I F^j_J F^k_K)\,dX^I_{(1)} dX^J_{(2)} dX^K_{(3)} \\ &= \sqrt{g}\,(e_{IJK}\,j)\,dX^I_{(1)} dX^J_{(2)} dX^K_{(3)} \\ &= \sqrt{g}\,(dV/\sqrt{G})\,j \end{aligned}$$

where

$$\boxed{j \equiv \det\left[F^i_I\right].} \tag{3.118}$$

Finally, the **dilatation ratio** is defined to be

$$\boxed{J \equiv \frac{dv}{dV} = \sqrt{\frac{g}{G}}\,j,} \tag{3.119}$$

which is the ratio of the deformed to the undeformed volume of the element. An alternate form for J can be found by taking the determinant of Eq. (3.64) and comparing the result to (3.118) and (3.119) to obtain

$$\boxed{J = \det[\hat{F}^i_I]} \tag{3.120}$$

where the $\hat{F}^i_I$ are physical components of $\mathbf{F}$. (Note the distinction between j and J.)

Here, we mention two important special cases. First, in convected coordinates, since $F^i_I = x^i{}_{,I} = \delta^i_I\ (x^i = X^I)$, we have $j = \det\left[F^i_I\right] = 1$ and so $J = \sqrt{g/G}$. Second, if we choose $\mathbf{g}_i = \mathbf{G}_I$, then $g = G$, and hence $J = j = \det\left[F^i_I\right]$. Most authors use one or the other of these specializations of Eq. (3.119).

Before leaving our present discussion, we derive yet another representation for J. Equations $(2.126)_2$ and (3.120) suggest that $J = \det \mathbf{F}$, which turns out to be true. But since the F^i_I are two-point tensor components (referred to base vectors in two different reference frames), Eq. (2.125) shows that we must exercise caution in making this statement. In this regard, we write

$$\det \mathbf{C} = \det \mathbf{F}^T \cdot \mathbf{F} = \det \mathbf{F}^T \det \mathbf{F} = (\det \mathbf{F})^2. \tag{3.121}$$

In addition, Eqs. (2.71), (3.31), (3.117), and (3.118) yield

$$\begin{aligned}
\det \mathbf{C} &= \det\left[C^{I}_{\cdot J}\right] = \det\left[G^{IK}C_{KJ}\right] \\
&= \det\left[G^{IK}F^{i}_{K}F^{j}_{J}g_{ij}\right] \\
&= \det\left[G^{IK}\right]\det\left[F^{i}_{K}\right]\det\left[F^{j}_{J}\right]\det\left[g_{ij}\right] \\
&= (g/G)\, j^2,
\end{aligned}$$

and thus

$$\boxed{\det \mathbf{C} = J^2.} \tag{3.122}$$

In the manipulations above, we also have used the fact that $\det[G^{IK}G_{JK}] = \det[\delta^{I}_{J}] = 1$, and so $\det[G^{IK}] = 1/\det[G_{IK}] = 1/G$. Now, substituting Eq. (3.121) into (3.122) gives

$$\boxed{J = \det \mathbf{F}.} \tag{3.123}$$

At first glance, it appears that Eqs. (2.71) and (3.118) imply $j = \det \mathbf{F}$, which is generally inconsistent with Eqs. (3.119) and (3.123). Similar to the above discussion, the explanation is that j is expressed in terms of mixed components of $\mathbf{F}$ relative to both undeformed and deformed base vectors. The factor $\sqrt{g/G}$ in (3.119) links these bases. These subtle distinctions in computing the quantities j and J are easy to overlook when reading the literature.

Soft biological tissues usually are assumed to be incompressible. In this case, $J = 1$ ($dv = dV$) must be enforced as a material constraint, and the resulting deformation is called **isochoric**. As will be shown in Chapter 6, this assumption sometimes simplifies analyses.

The incompressibility assumption for soft tissues is based on the fact that most tissues are composed primarily of water by volume fraction. This assumption is reasonable if extracellular fluid is relatively immobile. In some tissues, including articular cartilage, however, significant fluid flow can occur as the tissue deforms. In this case, the solid skeleton should be considered compressible, and the problem should be analyzed using a biphasic theory such as poroelasticity or mixture theory (Armstrong et al., 1984; Holmes, 1986; Mow et al., 1986; Simon and Gaballa, 1988; Huyghe et al., 1992).

Example 3.11 Compute the dilatation ratios for the problems discussed in Examples 3.2–3.4, i.e., uniform extension, simple shear, and axisymmetric deformation of a tube.

Solution. For uniform extension, Eq. (3.44) expresses **F** in a Cartesian dyadic basis. Since the components are mixed and because the undeformed and deformed bases coincide, the definition for the determinant of a tensor (2.71) gives

$$J = \det \mathbf{F} = \begin{vmatrix} \lambda_1 & 0 & 0 \\ 0 & \lambda_2 & 0 \\ 0 & 0 & \lambda_3 \end{vmatrix} = \lambda_1 \lambda_2 \lambda_3.$$

(Here we note that the distinctions between covariant, contravariant, and mixed components vanish in Cartesian coordinates.) If a block of material is stretched by $\lambda_1 = \lambda$ in the X^1-direction and deforms freely in the other two directions, then symmetry demands that $\lambda_2 = \lambda_3$. If, in addition, the material is incompressible, then

$$J = \lambda \lambda_2^2 = 1$$

which gives $\lambda_2 = \lambda^{-1/2}$.

For simple shear, Eq. (3.54) gives

$$J = \det \mathbf{F} = \begin{vmatrix} 1 & k & 0 \\ 0 & 1 & 0 \\ 0 & 0 & 1 \end{vmatrix} = 1.$$

Hence, this deformation is isochoric for all values of the shear parameter k.

Lastly, combined extension, inflation, and torsion of a cylinder is treated in three ways. First, Eqs. (3.63) and (3.65) give

$$\begin{aligned} j &= \det \left[F^i_I\right] = \lambda r' \\ G &= \det \left[G_{IJ}\right] = R^2 \\ g &= \det \left[g_{ij}\right] = r^2, \end{aligned}$$

and so Eq. (3.119) yields

$$J = \sqrt{\frac{g}{G}}\, j = \lambda r' \frac{r}{R}. \tag{3.124}$$

Second, Eq. (3.122) gives

$$\begin{aligned} J^2 &= \det \mathbf{C} = \det \left[C^I_J\right] = \det \left[C_{JK} G^{KI}\right] \\ &= \det \left[C_{JK}\right] \det \left[G^{KI}\right] \\ &= \frac{1}{G} \det \left[C_{JK}\right]. \end{aligned}$$

With Eq. (3.67), this expression becomes

$$J^2 = \frac{1}{R^2} \begin{vmatrix} r'^2 & 0 & 0 \\ 0 & r^2 & \psi r^2 \\ 0 & \psi r^2 & \lambda^2 + \psi^2 r^2 \end{vmatrix} = \frac{\lambda^2 r'^2 r^2}{R^2}$$

which agrees with Eq. (3.124). Finally, we compute $J = \det \mathbf{F} = \det[\hat{F}^i_I]$. With Eq. (3.62) providing the physical components of $\mathbf{F}$, we have

$$J = \begin{vmatrix} r' & 0 & 0 \\ 0 & r/R & \psi r \\ 0 & 0 & \lambda \end{vmatrix} = \frac{\lambda r r'}{R}$$

which again agrees with the previous result. ■

3.6.4 *Area Change*

Consider a differential area element $d\mathbf{A}$ with edges parallel to $d\mathbf{R}_{(1)}$ and $d\mathbf{R}_{(2)}$ that deforms into an element $d\mathbf{a}$ with edges parallel to $d\mathbf{r}_{(1)}$ and $d\mathbf{r}_{(2)}$ (Fig. 3.10). Before and after deformation, respectively, we use elementary vector mechanics to write

$$\begin{aligned} d\mathbf{A} &= \mathbf{N}\, dA = d\mathbf{R}_{(1)} \times d\mathbf{R}_{(2)} \\ d\mathbf{a} &= \mathbf{n}\, da = d\mathbf{r}_{(1)} \times d\mathbf{r}_{(2)} \end{aligned} \tag{3.125}$$

where $\mathbf{N}$ and $\mathbf{n}$ are unit vectors normal to the elements with areas dA and da.

Substituting Eqs. (3.112) into (3.125) and noting Eq. $(2.96)_1$ yields

$$\begin{aligned} d\mathbf{A} &= (\mathbf{G}_I\, dX^I_{(1)}) \times (\mathbf{G}_J\, dX^J_{(2)}) = \mathbf{G}_I \times \mathbf{G}_J\, dX^I_{(1)} dX^J_{(2)} \\ &= \epsilon_{IJK}\, \mathbf{G}^K\, dX^I_{(1)} dX^J_{(2)} \end{aligned}$$

$$\begin{aligned} d\mathbf{a} &= (\mathbf{g}_i\, dx^i_{(1)}) \times (\mathbf{g}_j\, dx^j_{(2)}) = \mathbf{g}_i \times \mathbf{g}_j\, dx^i_{(1)} dx^j_{(2)} \\ &= \epsilon_{ijk}\, \mathbf{g}^k\, dx^i_{(1)} dx^j_{(2)}. \end{aligned} \tag{3.126}$$

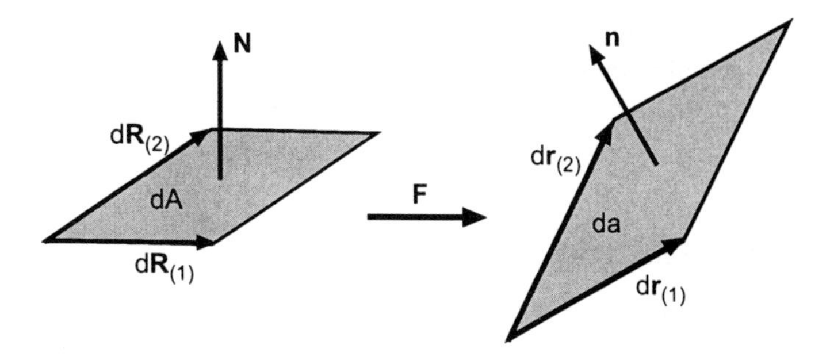

Fig. 3.10 Differential area element before and after deformation.

Next, using Eqs. $(2.108)_1$, $(3.16)_2$, and (3.114) to replace $\mathbf{g}^k$, $dx^i_{(1)}$, and $dx^i_{(2)}$ in $d\mathbf{a}$ gives

$$\begin{aligned} d\mathbf{a} &= \sqrt{g}\, e_{ijk}(F^k_K \mathbf{g}^K)(F^i_I\, dX^I_{(1)})(F^j_J dX^J_{(2)}) \\ &= \sqrt{g}\,(e_{ijk} F^i_I F^j_J F^k_K)\mathbf{g}^K\, dX^I_{(1)} dX^J_{(2)} \\ &= \sqrt{g}\,(e_{IJK}\, j)\mathbf{g}^K\, dX^I_{(1)} dX^J_{(2)}. \end{aligned} \tag{3.127}$$

In the last line, we have used the determinant formula $(2.91)_1$, with j defined by Eq. (3.118). Finally, this expression can be written in the form

$$\boxed{d\mathbf{a} = J\, d\mathbf{A}\cdot\mathbf{F}^{-1}} \tag{3.128}$$

as verified by the manipulations

$$\begin{aligned} d\mathbf{a} &= J(\epsilon_{IJK}\mathbf{G}^K\, dX^I_{(1)} dX^J_{(2)})\cdot(\mathbf{G}_L \mathbf{g}^L) \\ &= J(\sqrt{G}\, e_{IJK})\delta^K_L\, \mathbf{g}^L\, dX^I_{(1)} dX^J_{(2)} \\ &= j\sqrt{g}\, e_{IJK}\mathbf{g}^K\, dX^I_{(1)} dX^J_{(2)} \end{aligned}$$

which employ Eqs. $(3.126)_1$, $(3.22)_2$, $(2.108)_1$, and (3.119), in this order. This expression agrees with Eq. (3.127).

3.7 Principal Strains

At any point in a body, we always can find a unique set of orthogonal axes so that the shear strains relative to those axes vanish. These axes are called **principal axes of strain**, and the normal strain components relative to this system are called **principal strains**. In addition, the planes orthogonal to the principal axes are called **principal planes**. The principal axes (principal directions) and the corresponding principal strains at a point are determined by solving an eigenvalue problem.

3.7.1 *The Eigenvalue Problem*

Consider a line element that is oriented in the direction of the unit vector $\mathbf{N}$ in the undeformed body and $\mathbf{n}$ in the deformed body. In addition, let $\mathbf{T}$ and $\mathbf{t}$ be arbitrary unit vectors orthogonal to $\mathbf{N}$ and $\mathbf{n}$, respectively. The vectors $\mathbf{N}$ and $\mathbf{n}$ correspond to principal axes if

$$\begin{aligned} E_{(\mathbf{NN})} &= \mathbf{E}{:}\mathbf{NN} = \mathbf{N}{\cdot}\mathbf{E}{\cdot}\mathbf{N} \neq 0 \\ E_{(\mathbf{NT})} &= \mathbf{E}{:}\mathbf{NT} = \mathbf{N}{\cdot}\mathbf{E}{\cdot}\mathbf{T} = 0 \end{aligned} \tag{3.129}$$

and

$$\begin{aligned} e_{(\mathbf{nn})} &= \mathbf{e}{:}\mathbf{nn} = \mathbf{n}{\cdot}\mathbf{e}{\cdot}\mathbf{n} \neq 0 \\ e_{(\mathbf{nt})} &= \mathbf{e}{:}\mathbf{nt} = \mathbf{n}{\cdot}\mathbf{e}{\cdot}\mathbf{t} = 0 \end{aligned} \tag{3.130}$$

where $E_{(\mathbf{NN})}$ and $e_{(\mathbf{nn})}$ are principal Lagrangian and Eulerian strain components along $\mathbf{N}$ and $\mathbf{n}$, respectively, and $E_{(\mathbf{NT})}$ and $e_{(\mathbf{nt})}$ are shear strains.

Relative to the local orthogonal principal axes, $G_{IJ} = \delta_{IJ}$ and $g_{ij} = \delta_{ij}$, and so Eqs. (3.30) give

$$\begin{aligned} E_{IJ} &= \tfrac{1}{2}(C_{IJ} - \delta_{IJ}) \\ e_{ij} &= \tfrac{1}{2}[\delta_{ij} - (B^{-1})_{ij}]. \end{aligned} \tag{3.131}$$

These equations show that the shear components of $\mathbf{E}$ and $\mathbf{e}$ vanish when the corresponding components of $\mathbf{C}$ and $\mathbf{B}^{-1}$ vanish. In other words, the principal directions of $\mathbf{E}$ and $\mathbf{C}$ coincide, as do those of $\mathbf{e}$ and $\mathbf{B}^{-1}$. Thus, Eqs. (3.129) and (3.130) can be replaced by the conditions $\mathbf{N}{\cdot}\mathbf{C}{\cdot}\mathbf{N} \neq 0$, $\mathbf{N}{\cdot}\mathbf{C}{\cdot}\mathbf{T} = 0$ and $\mathbf{n}{\cdot}\mathbf{B}^{-1}{\cdot}\mathbf{n} \neq 0$, $\mathbf{n}{\cdot}\mathbf{B}^{-1}{\cdot}\mathbf{t} = 0$, which are satisfied if

$$\begin{aligned} \mathbf{N}{\cdot}\mathbf{C} &= \Lambda_{(\mathbf{N})}^2\,\mathbf{N} \\ \mathbf{n}{\cdot}\mathbf{B}^{-1} &= \lambda_{(\mathbf{n})}^{-2}\,\mathbf{n}. \end{aligned} \tag{3.132}$$

The stretch ratios on the right-hand sides of these relations follow from Eqs. (3.94). Rewriting Eqs. (3.132) yields the eigenvalue problems

$$\boxed{\begin{aligned}(\mathbf{C} - \Lambda^2_{(\mathbf{N})}\,\mathbf{I})\cdot\mathbf{N} &= \mathbf{0}\\ (\mathbf{B} - \lambda^2_{(\mathbf{n})}\,\mathbf{I})\cdot\mathbf{n} &= \mathbf{0}.\end{aligned}} \tag{3.133}$$

Solving these equations (see Section 2.5.3) provides the **principal stretch ratios** $\Lambda_{(\mathbf{N}_i)}$ and $\lambda_{(\mathbf{n}_i)}$ $(i = 1, 2, 3)$ and the corresponding principal axes $\mathbf{N}_i$ and $\mathbf{n}_i$. Equations $(3.5)_2$ and (3.7) then give the principal strains

$$\begin{aligned}E_{(\mathbf{N}_i)} &= \tfrac{1}{2}(\Lambda^2_{(\mathbf{N}_i)} - 1)\\ e_{(\mathbf{n}_i)} &= \tfrac{1}{2}(1 - \lambda^{-2}_{(\mathbf{n}_i)}).\end{aligned} \tag{3.134}$$

It can be shown that since $\mathbf{C}$ and $\mathbf{B}$ are symmetric tensors, their eigenvalues are real and their eigenvectors are mutually orthogonal. Also, as $\mathbf{n}$ is taken in the direction of the deformed element with undeformed orientation $\mathbf{N}$, $\lambda_{(\mathbf{n})} = \Lambda_{(\mathbf{N})}$, and so the eigenvalues of $\mathbf{C}$ and $\mathbf{B}$ are equal. Thus, the orthogonal principal axes $\mathbf{N}_i$ at the point P in the undeformed body translate and rotate into the orthogonal principal axes $\mathbf{n}_i$ at the point p in the deformed body. In addition, since the eigenvalues represent the deformation or strain components relative to the principal axes, we can write

$$\begin{aligned}\mathbf{C} &= \Lambda_1^2\,\mathbf{N}_1\mathbf{N}_1 + \Lambda_2^2\,\mathbf{N}_2\mathbf{N}_2 + \Lambda_3^2\,\mathbf{N}_3\mathbf{N}_3\\ \mathbf{B} &= \lambda_1^2\,\mathbf{n}_1\mathbf{n}_1 + \lambda_2^2\,\mathbf{n}_2\mathbf{n}_2 + \lambda_3^2\,\mathbf{n}_3\mathbf{n}_3\\ \mathbf{E} &= E_1\,\mathbf{N}_1\mathbf{N}_1 + E_2\,\mathbf{N}_2\mathbf{N}_2 + E_3\,\mathbf{N}_3\mathbf{N}_3\\ \mathbf{e} &= e_1\,\mathbf{n}_1\mathbf{n}_1 + e_2\,\mathbf{n}_2\mathbf{n}_2 + e_3\,\mathbf{n}_3\mathbf{n}_3\end{aligned} \tag{3.135}$$

where we have used the invariance properties of these tensors, along with the definitions $E_i \equiv E_{(\mathbf{N}_i)}$ and $e_i \equiv e_{(\mathbf{n}_i)}$.

3.7.2 *Strain Invariants*

Each of the deformation and strain tensors possesses principal invariants as defined by Eq. (2.77). Because $\mathbf{C}$ and $\mathbf{B}$ have the same eigenvalues, Eq. (2.76) indicates that they also have the same invariants, i.e.,

$$\boxed{\begin{aligned} I_1 &= \operatorname{tr}\mathbf{C} = \operatorname{tr}\mathbf{B} \\ I_2 &= \tfrac{1}{2}\left[(\operatorname{tr}\mathbf{C})^2 - \operatorname{tr}\mathbf{C}^2\right] = \tfrac{1}{2}\left[(\operatorname{tr}\mathbf{B})^2 - \operatorname{tr}\mathbf{B}^2\right] \\ I_3 &= \det\mathbf{C} = \det\mathbf{B}. \end{aligned}} \qquad (3.136)$$

In solid mechanics problems, the Lagrangian description often is more convenient than the Eulerian description. Thus, we now give component forms for the invariants of $\mathbf{C}$ in terms of the components of $\mathbf{C}$ and $\mathbf{E}$, where $\mathbf{C} = \mathbf{I} + 2\mathbf{E}$. Note that the invariants contain the trace and determinant, which Eqs. (2.66) and (2.71) define in terms of mixed tensor components. (Since $\mathbf{C}$ and $\mathbf{E}$ are symmetric tensors, $C^I_{\cdot J} = C_J^{\cdot I} \equiv C^I_J$ and $E^I_{\cdot J} = E_J^{\cdot I} \equiv E^I_J$.) Equations (3.30) and (3.31) give

$$\begin{aligned} \mathbf{C} &= C_{IJ}\mathbf{G}^I\mathbf{G}^J = g_{IJ}\mathbf{G}^I\mathbf{G}^J \\ \mathbf{E} &= E_{IJ}\mathbf{G}^I\mathbf{G}^J = \tfrac{1}{2}(C_{IJ} - G_{IJ})\mathbf{G}^I\mathbf{G}^J, \end{aligned}$$

and so the mixed components are

$$\begin{aligned} C^I_J &= \mathbf{G}^I\cdot\mathbf{C}\cdot\mathbf{G}_J = \mathbf{G}^I\cdot(g_{KL}\mathbf{G}^K\mathbf{G}^L)\cdot\mathbf{G}_J \\ &= g_{KL}\,G^{IK}\delta^L_J = g_{KJ}\,G^{IK} \end{aligned}$$

$$\begin{aligned} E^I_J &= \mathbf{G}^I\cdot\mathbf{E}\cdot\mathbf{G}_J = \tfrac{1}{2}(C_{KL} - G_{KL})\mathbf{G}^I\cdot(\mathbf{G}^K\mathbf{G}^L)\cdot\mathbf{G}_J \\ &= \tfrac{1}{2}(C_{KL} - G_{KL})G^{IK}\delta^L_J = \tfrac{1}{2}(C_{KJ} - G_{KJ})G^{IK} \\ &= = \tfrac{1}{2}(C^I_J - \delta^I_J) \end{aligned} \qquad (3.137)$$

in which Eqs. (2.85) and $(2.86)_1$ have been used. With Eq. (2.66), the first strain invariant becomes

$$\begin{aligned} I_1 &= \operatorname{tr}\left[C^I_J\right] = C^I_I = g_{KI}\,G^{IK} = g_{IJ}\,G^{IJ} \\ &= \operatorname{tr}\left[\delta^I_J + 2E^I_J\right] = \delta^I_I + 2E^I_I = 3 + 2E^I_I. \end{aligned} \qquad (3.138)$$

The second invariant requires

$$
\begin{aligned}
\mathbf{C}^2 &= \mathbf{C}\cdot\mathbf{C} = (C_{IJ}\mathbf{G}^I\mathbf{G}^J)\cdot(C_{KL}\mathbf{G}^K\mathbf{G}^L) \\
&= C_{IJ}C_{KL}G^{JK}\mathbf{G}^I\mathbf{G}^L \\
&= C_{IJ}C_L^J\mathbf{G}^I\mathbf{G}^L \\
&= C_{IK}C_J^K\mathbf{G}^I\mathbf{G}^J,
\end{aligned}
$$

and thus, by Eq. (2.65),

$$
\begin{aligned}
\operatorname{tr}\mathbf{C}^2 &= C_{IK}C_J^K\mathbf{G}^I\cdot\mathbf{G}^J = C_{IK}C_J^K G^{IJ} \\
&= C_K^J C_J^K = C_J^I C_I^J.
\end{aligned} \tag{3.139}
$$

Combining these relations yields

$$
\begin{aligned}
I_2 &= \tfrac{1}{2}(I_1^2 - C_J^I C_I^J) \\
&= \tfrac{1}{2}\left[I_1^2 - (g_{JK}G^{IK})(g_{IL}G^{JL})\right] \\
&= \tfrac{1}{2}\left[(3+2E_I^I)(3+2E_J^J) - (\delta_J^I + 2E_J^I)(\delta_I^J + 2E_I^J)\right] \\
&= 3 + 4E_I^I + 2(E_I^I E_J^J - E_J^I E_I^J).
\end{aligned} \tag{3.140}
$$

Finally, with Eqs. (2.71) and (3.121)–(3.123), we can write

$$
\begin{aligned}
I_3 &= \det\left[C_J^I\right] = \det\left[\delta_J^I + 2E_J^I\right] \\
&= (\det\mathbf{F})^2 \\
&= (g/G)j^2 = J^2.
\end{aligned} \tag{3.141}
$$

In summary, the strain invariants have the indicial forms

$$
\boxed{
\begin{aligned}
I_1 &= C_I^I = g_{IJ}G^{IJ} \\
&= 3 + 2E_I^I \\
I_2 &= \tfrac{1}{2}(C_I^I C_J^J - C_J^I C_I^J) \\
&= \tfrac{1}{2}(I_1^2 - g_{IL}g_{JK}G^{IK}G^{JL}) = g^{IJ}G_{IJ}I_3 \\
&= 3 + 4E_I^I + 2(E_I^I E_J^J - E_J^I E_I^J) \\
I_3 &= \det\left[C_J^I\right] = (\det\mathbf{F})^2 = (g/G)j^2 = J^2 \\
&= \det\left[\delta_J^I + 2E_J^I\right].
\end{aligned}
} \tag{3.142}
$$

By definition, these invariants have the same values when computed in any coordinate system. In principal coordinates, the shear strains vanish and the invariants become

$$\begin{aligned} I_1 &= 3 + 2(E_1 + E_2 + E_3) \\ I_2 &= 3 + 4(E_1 + E_2 + E_3) + 4(E_1 E_2 + E_2 E_3 + E_3 E_1) \\ I_3 &= (1 + 2E_1)(1 + 2E_2)(1 + 2E_3) \end{aligned} \tag{3.143}$$

where the E_i are principal Lagrangian strains.

Example 3.12 Compute the strain invariants for Examples 3.2–3.4, i.e., uniform extension, simple shear, and axisymmetric deformation of a tube.

Solution. For uniform extension, Eqs. $(3.47)_1$ and (3.142) yield (recall that $C^I_J = C_{IJ}$ in this case)

$$\begin{aligned} I_1 &= C^1_1 + C^2_2 + C^3_3 = \lambda_1^2 + \lambda_2^2 + \lambda_3^2 \\ I_2 &= \tfrac{1}{2}\left\{(C^1_1 + C^2_2 + C^3_3)^2 - \left[(C^1_1)^2 + (C^2_2)^2 + (C^3_3)^2\right]\right\} \\ &= C^1_1 C^2_2 + C^2_2 C^3_3 + C^3_3 C^1_1 \\ &= \lambda_1^2 \lambda_2^2 + \lambda_2^2 \lambda_3^2 + \lambda_3^2 \lambda_1^2 \\ I_3 &= \det\left[C^I_J\right] = \lambda_1^2 \lambda_2^2 \lambda_3^2. \end{aligned} \tag{3.144}$$

For simple shear, Eq. $(3.55)_1$ gives (again, $C^I_J = C_{IJ}$)

$$\begin{aligned} I_1 &= C^1_1 + C^2_2 + C^3_3 = 3 + k^2 \\ I_2 &= C^1_1 C^2_2 + C^2_2 C^3_3 + C^3_3 C^1_1 - (C^1_2)^2 = 3 + k^2 \\ I_3 &= \det\left[C^I_J\right] = 1. \end{aligned} \tag{3.145}$$

Finally, for combined extension, inflation, and torsion of a cylinder, Eqs. (3.71) give the mixed strain components E^I_J. Putting these into Eqs. (3.142) yields

$$\begin{aligned} I_1 &= 3 + 2(E^1_1 + E^2_2 + E^3_3) = r'^2 + \frac{r^2}{R^2} + \lambda^2 + \psi^2 r^2 \\ I_2 &= 3 + 4(E^1_1 + E^2_2 + E^3_3 + E^1_1 E^2_2 + E^2_2 E^3_3 + E^3_3 E^1_1 - E^2_3 E^3_2) \\ &= r'^2\left(\frac{r^2}{R^2} + \lambda^2 + \psi^2 r^2\right) + \frac{\lambda^2 r^2}{R^2} \\ I_3 &= \det\left[\delta^I_J + 2E^I_J\right] = \frac{\lambda^2 r'^2 r^2}{R^2}. \end{aligned} \tag{3.146}$$

■

3.8 Stretch and Rotation Tensors

As mentioned in Section 3.4, the deformation gradient tensor $\mathbf{F}$ includes information on both deformation and rigid-body rotation. In contrast, the deformation tensors, $\mathbf{C}$ and $\mathbf{B}$, and the strain tensors, $\mathbf{E}$ and $\mathbf{e}$, contain only deformation information. In this section, we obtain additional measures of deformation by separating the deformational and rotational parts of $\mathbf{F}$.

3.8.1 *Polar Decomposition*

The deformation gradient tensor can be decomposed into (1) a pure deformation followed by a pure rotation or (2) a pure rotation followed by a pure deformation. For small displacements and rotations, these motions are additive and commutative (see Section 3.9.1). For large displacements and rotations, this is not the case; rather, the rotation and deformation are multiplicative.

The **polar decomposition theorem** states that the deformation gradient tensor $\mathbf{F}$ can be decomposed uniquely into the two forms

$$\boxed{\mathbf{F} = \boldsymbol{\Theta}\cdot\mathbf{U} = \mathbf{V}\cdot\boldsymbol{\Theta}} \tag{3.147}$$

where $\mathbf{U}$ and $\mathbf{V}$ are positive definite symmetric tensors, and $\boldsymbol{\Theta}$ is a proper orthogonal tensor (since $\det\mathbf{F} = J > 0$, see Section 2.6.3). Since a proper orthogonal tensor produces a rigid-body rotation, $\boldsymbol{\Theta}$ is called the **rotation tensor**. Thus, all of the deformation is contained in $\mathbf{U}$ and $\mathbf{V}$, which are called the **right and left stretch tensors**, respectively. The relation $d\mathbf{r} = \mathbf{F}\cdot d\mathbf{R}$ shows that the first decomposition represents a stretch ($\mathbf{U}$) of $d\mathbf{R}$ followed by a rigid-body rotation ($\boldsymbol{\Theta}$) into $d\mathbf{r}$, whereas the second decomposition represents a rigid-body rotation ($\boldsymbol{\Theta}$) of $d\mathbf{R}$ followed by a stretch ($\mathbf{V}$) into $d\mathbf{r}$ (Fig. 3.11). Although the rotation $\boldsymbol{\Theta}$ is the same in both cases, $\mathbf{U} \neq \mathbf{V}$ in general.

In the following, we prove the polar decomposition theorem in two ways: first through a formal mathematical derivation and then via a relatively informal physical argument.

Mathematical Derivation

We begin our formal derivation by considering the deformation tensor $\mathbf{C}$ in principal coordinates. Relative to the principal axes $\mathbf{N}_i$, Eq. $(3.135)_1$ gives

$$\mathbf{C} = \Lambda_1^2\,\mathbf{N}_1\mathbf{N}_1 + \Lambda_2^2\,\mathbf{N}_2\mathbf{N}_2 + \Lambda_3^2\,\mathbf{N}_3\mathbf{N}_3 \tag{3.148}$$

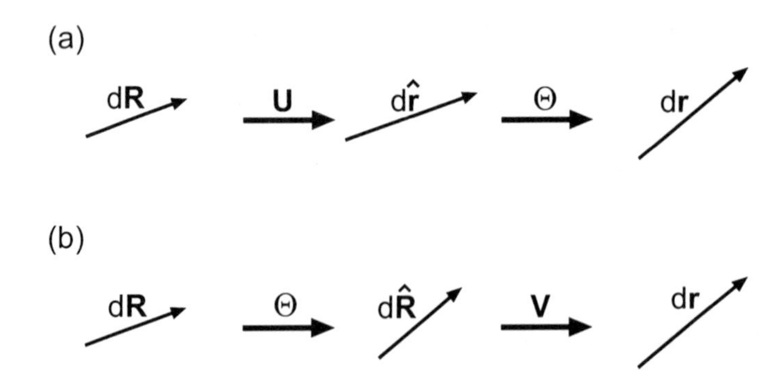

Fig. 3.11 Polar decomposition of a deformation. (a) Stretch $\mathbf{U}$ followed by rotation $\boldsymbol{\Theta}$. (b) Rotation $\boldsymbol{\Theta}$ followed by stretch $\mathbf{V}$.

where $\Lambda_i \equiv \Lambda_{(\mathbf{N}_i)}$. We already have seen that $\mathbf{C}$ is a symmetric tensor. Moreover, Eq. $(3.24)_1$ shows that it also is positive definite, i.e.,

$$\mathbf{C}{:}\mathbf{aa} = \mathbf{a}{\cdot}\mathbf{C}{\cdot}\mathbf{a} > 0 \tag{3.149}$$

for all $\mathbf{a} \neq \mathbf{0}$. (Here, $\mathbf{a} = d\mathbf{R}$ and $d\mathbf{R}{\cdot}\mathbf{C}{\cdot}d\mathbf{R} = ds^2 > 0$.) Next, we define the tensor

$$\mathbf{U} = \Lambda_1\,\mathbf{N}_1\mathbf{N}_1 + \Lambda_2\,\mathbf{N}_2\mathbf{N}_2 + \Lambda_3\,\mathbf{N}_3\mathbf{N}_3, \tag{3.150}$$

so that

$$\mathbf{C} = \mathbf{U}{\cdot}\mathbf{U} = \mathbf{U}^2. \tag{3.151}$$

Since $\mathbf{C}$ is symmetric and positive definite, so is $\mathbf{U}$. Now, if we can show that $\boldsymbol{\Theta}$ as defined by

$$\boldsymbol{\Theta} = \mathbf{F}{\cdot}\mathbf{U}^{-1}$$

is an orthogonal tensor, then the existence of the first decomposition of (3.147) will be proven. This equation and Table 2.3 (page 24) give

$$\begin{aligned} \boldsymbol{\Theta}^T{\cdot}\boldsymbol{\Theta} &= (\mathbf{F}{\cdot}\mathbf{U}^{-1})^T{\cdot}(\mathbf{F}{\cdot}\mathbf{U}^{-1}) \\ &= (\mathbf{U}^{-T}{\cdot}\mathbf{F}^T){\cdot}(\mathbf{F}{\cdot}\mathbf{U}^{-1}) \\ &= \mathbf{U}^{-1}{\cdot}(\mathbf{F}^T{\cdot}\mathbf{F}){\cdot}\mathbf{U}^{-1} \end{aligned}$$

since $\mathbf{U}^T = \mathbf{U}$. Substituting $\mathbf{F}^T \cdot \mathbf{F} = \mathbf{C} = \mathbf{U}^2$ yields

$$\begin{aligned} \boldsymbol{\Theta}^T \cdot \boldsymbol{\Theta} &= \mathbf{U}^{-1} \cdot (\mathbf{U} \cdot \mathbf{U}) \cdot \mathbf{U}^{-1} \\ &= (\mathbf{U}^{-1} \cdot \mathbf{U}) \cdot (\mathbf{U} \cdot \mathbf{U}^{-1}) \\ &= \mathbf{I}, \end{aligned}$$

which shows that $\boldsymbol{\Theta}$ is orthogonal under the definition of Eq. (2.112). It also can be shown that $\boldsymbol{\Theta}$ is a proper orthogonal tensor.

The existence of the second decomposition of Eq. (3.147) follows from the definition of a second-order tensor $\mathbf{V}$ in the form

$$\mathbf{V} = \boldsymbol{\Theta} \cdot \mathbf{U} \cdot \boldsymbol{\Theta}^T. \tag{3.152}$$

The manipulations

$$\begin{aligned} \mathbf{V}^T &= (\boldsymbol{\Theta} \cdot \mathbf{U} \cdot \boldsymbol{\Theta}^T)^T = (\boldsymbol{\Theta}^T)^T \cdot \mathbf{U}^T \cdot \boldsymbol{\Theta}^T \\ &= \boldsymbol{\Theta} \cdot \mathbf{U} \cdot \boldsymbol{\Theta}^T \\ &= \mathbf{V} \end{aligned}$$

show that $\mathbf{V}$ is a symmetric tensor, and the rigid-body rotations of $\mathbf{U}$ to produce $\mathbf{V}$ in (3.152) preserve its positive definite character. To obtain the second decomposition, we substitute for $\mathbf{U}$ from the first decomposition of (3.147) to get

$$\begin{aligned} \mathbf{V} &= \boldsymbol{\Theta} \cdot (\boldsymbol{\Theta}^{-1} \cdot \mathbf{F}) \cdot \boldsymbol{\Theta}^T \\ &= (\boldsymbol{\Theta} \cdot \boldsymbol{\Theta}^{-1}) \cdot \mathbf{F} \cdot \boldsymbol{\Theta}^T \\ &= \mathbf{I} \cdot \mathbf{F} \cdot \boldsymbol{\Theta}^{-1} \\ &= \mathbf{F} \cdot \boldsymbol{\Theta}^{-1} \end{aligned}$$

as orthogonality implies $\boldsymbol{\Theta}^T = \boldsymbol{\Theta}^{-1}$. This relation proves the existence of the second decomposition of (3.147).

To prove uniqueness of the first decomposition, we suppose that another decomposition exists such that

$$\mathbf{F} = \boldsymbol{\Theta}' \cdot \mathbf{U}'$$

where

$$\mathbf{U}' = \Lambda_1' \, \mathbf{N}_1 \mathbf{N}_1 + \Lambda_2' \, \mathbf{N}_2 \mathbf{N}_2 + \Lambda_3' \, \mathbf{N}_3 \mathbf{N}_3$$

and $\boldsymbol{\Theta}'$ is a rotation tensor. Then,

$$\begin{aligned}\mathbf{C} = \mathbf{F}^T \cdot \mathbf{F} &= (\boldsymbol{\Theta}' \cdot \mathbf{U}')^T \cdot (\boldsymbol{\Theta}' \cdot \mathbf{U}') \\ &= \mathbf{U}'^T \cdot (\boldsymbol{\Theta}'^T \cdot \boldsymbol{\Theta}') \cdot \mathbf{U}' \\ &= \mathbf{U}' \cdot \mathbf{U}' = \mathbf{U}'^2.\end{aligned}$$

Since the eigenvalues of $\mathbf{C}$ are unique, the Λ_i' must equal the Λ_i in Eq. (3.150). Thus, $\mathbf{U}' = \mathbf{U}$ which proves uniqueness. The uniqueness of $\boldsymbol{\Theta}$ and $\mathbf{V}$ follows from their definitions.

Finally, we note that substituting Eq. (3.147) into (3.25) yields

$$\begin{aligned}\mathbf{C} = \mathbf{F}^T \cdot \mathbf{F} &= (\boldsymbol{\Theta} \cdot \mathbf{U})^T \cdot (\boldsymbol{\Theta} \cdot \mathbf{U}) = (\mathbf{U}^T \cdot \boldsymbol{\Theta}^T) \cdot (\boldsymbol{\Theta} \cdot \mathbf{U}) \\ &= \mathbf{U}^T \cdot (\boldsymbol{\Theta}^T \cdot \boldsymbol{\Theta}) \cdot \mathbf{U} = \mathbf{U}^T \cdot \mathbf{I} \cdot \mathbf{U} \\ \mathbf{B} = \mathbf{F} \cdot \mathbf{F}^T &= (\mathbf{V} \cdot \boldsymbol{\Theta}) \cdot (\mathbf{V} \cdot \boldsymbol{\Theta})^T = (\mathbf{V} \cdot \boldsymbol{\Theta}) \cdot (\boldsymbol{\Theta}^T \cdot \mathbf{V}^T) \\ &= \mathbf{V} \cdot (\boldsymbol{\Theta} \cdot \boldsymbol{\Theta}^T) \cdot \mathbf{V}^T = \mathbf{V} \cdot \mathbf{I} \cdot \mathbf{V}^T\end{aligned}$$

where orthogonality of the rotation tensor has been used. Because $\mathbf{U}$ and $\mathbf{V}$ are symmetric tensors, we have

$$\boxed{\mathbf{C} = \mathbf{U}^2, \qquad \mathbf{B} = \mathbf{V}^2.} \tag{3.153}$$

The first relation again confirms Eq. (3.151).

Physical Derivation

Consider translation and rotation (without deformation) of the differential line element $d\mathbf{R}$ into the element $d\hat{\mathbf{R}}$ (Fig. 3.11b) according to

$$d\hat{\mathbf{R}} = \boldsymbol{\Theta} \cdot d\mathbf{R} + \mathbf{D} = d\mathbf{R} \cdot \boldsymbol{\Theta}^T + \mathbf{D} \tag{3.154}$$

where $\mathbf{D}$ is a translation vector and $\boldsymbol{\Theta}$ is the rotation tensor. For this motion, the deformation gradient tensor, given by Eqs. (2.154) and $(3.21)_1$, is

$$\hat{\mathbf{F}}^T = \frac{\partial \hat{\mathbf{R}}}{\partial \mathbf{R}} = \boldsymbol{\Theta}^T \qquad \rightarrow \qquad \hat{\mathbf{F}} = \boldsymbol{\Theta}. \tag{3.155}$$

Because Eq. (3.154) represents a rigid-body motion, the deformation tensors must equal the identity tensor. Hence,

$$\begin{aligned}\hat{\mathbf{C}} &= \hat{\mathbf{F}}^T \cdot \hat{\mathbf{F}} = \boldsymbol{\Theta}^T \cdot \boldsymbol{\Theta} = \mathbf{I} \\ \hat{\mathbf{B}} &= \hat{\mathbf{F}} \cdot \hat{\mathbf{F}}^T = \boldsymbol{\Theta} \cdot \boldsymbol{\Theta}^T = \mathbf{I}\end{aligned}$$

or

$$\boxed{\boldsymbol{\Theta}\cdot\boldsymbol{\Theta}^T = \boldsymbol{\Theta}^T\cdot\boldsymbol{\Theta} = \mathbf{I}} \tag{3.156}$$

which shows that $\boldsymbol{\Theta}$ does indeed satisfy the criteria for an orthogonal (rotation) tensor [see Eq. (2.112)]. Next, the rotated element $d\hat{\mathbf{R}}$ is deformed into the element $d\mathbf{r}$ (Fig. 3.11b) according to

$$\begin{aligned} d\mathbf{r} &= \mathbf{V}\cdot d\hat{\mathbf{R}} = \mathbf{V}\cdot(\boldsymbol{\Theta}\cdot d\mathbf{R} + \mathbf{D}) \\ &= (\mathbf{V}\cdot\boldsymbol{\Theta})\cdot d\mathbf{R} + \mathbf{V}\cdot\mathbf{D} \\ &= d\mathbf{R}\cdot(\mathbf{V}\cdot\boldsymbol{\Theta})^T + \mathbf{V}\cdot\mathbf{D}, \end{aligned} \tag{3.157}$$

and the deformation gradient tensor for the entire motion is

$$\mathbf{F}^T = \frac{\partial \mathbf{r}}{\partial \mathbf{R}} = (\mathbf{V}\cdot\boldsymbol{\Theta})^T \qquad \rightarrow \qquad \mathbf{F} = \mathbf{V}\cdot\boldsymbol{\Theta}. \tag{3.158}$$

Now, consider a deformation that first transforms the line element $d\mathbf{R}$ into $d\hat{\mathbf{r}}$ (Fig. 3.11a), without rotation, through

$$d\hat{\mathbf{r}} = \mathbf{U}\cdot d\mathbf{R}.$$

Then $d\hat{\mathbf{r}}$ translates and rotates into $d\mathbf{r}$ by

$$\begin{aligned} d\mathbf{r} &= \boldsymbol{\Theta}\cdot d\hat{\mathbf{r}} + \mathbf{D} \\ &= (\boldsymbol{\Theta}\cdot\mathbf{U})\cdot d\mathbf{R} + \mathbf{D} = d\mathbf{R}\cdot(\boldsymbol{\Theta}\cdot\mathbf{U})^T + \mathbf{D}, \end{aligned} \tag{3.159}$$

where we specify that the rotation tensor $\boldsymbol{\Theta}$ is the same one introduced above. Thus,

$$\mathbf{F}^T = \frac{\partial \mathbf{r}}{\partial \mathbf{R}} = (\boldsymbol{\Theta}\cdot\mathbf{U})^T \qquad \rightarrow \qquad \mathbf{F} = \boldsymbol{\Theta}\cdot\mathbf{U}. \tag{3.160}$$

Equations (3.158) and (3.160) are consistent with the polar decomposition theorem of Eq. (3.147).

In summary, the deformation gradient tensor $\mathbf{F}$ can be decomposed into a deformation $\mathbf{U}$ followed by a rotation $\boldsymbol{\Theta}$ or a rotation $\boldsymbol{\Theta}$ followed by a deformation $\mathbf{V}$. Although the rotation tensors are the same, $\mathbf{U} \neq \mathbf{V}$ in general. We next investigate some properties of the stretch tensors.

3.8.2 *Principal Stretch Ratios*

Consider again the mapping of the undeformed line element $d\mathbf{R}$ into the deformed element $d\mathbf{r}$ according to

$$d\mathbf{r} = \mathbf{F}\cdot d\mathbf{R}. \tag{3.161}$$

The representations

$$d\mathbf{R} = \mathbf{N}\, dS, \qquad d\mathbf{r} = \mathbf{n}\, ds \tag{3.162}$$

separate the motion into the rotation of the unit vector $\mathbf{N}$ into $\mathbf{n}$ and the stretching of the length dS into ds (see Fig. 3.4, page 74). Substituting Eqs. (3.162) into (3.161) and dividing by dS yields

$$\Lambda_{(\mathbf{N})}\, \mathbf{n} = \mathbf{F}\cdot\mathbf{N} \qquad \text{or} \qquad \lambda_{(\mathbf{n})}\, \mathbf{n} = \mathbf{F}\cdot\mathbf{N} \tag{3.163}$$

where $\Lambda_{(\mathbf{N})} = \lambda_{(\mathbf{n})} = ds/dS$ is the stretch ratio of the line element originally pointing in the direction of $\mathbf{N}$, as defined in Section 3.6.1.

Since $\mathbf{N}$ and $\mathbf{n}$ are unit vectors, $\mathbf{N}$ rotates into $\mathbf{n}$ by

$$\boxed{\mathbf{n} = \boldsymbol{\Theta}\cdot\mathbf{N}} \tag{3.164}$$

where $\boldsymbol{\Theta}$ is the rotation tensor, and the translation has been dropped. Direct substitution into this equation verifies the dyadic representation

$$\boldsymbol{\Theta} = \mathbf{n}\mathbf{N}. \tag{3.165}$$

Now, inserting Eqs. (3.147) and (3.164) into $(3.163)_1$ gives

$$\Lambda_{(\mathbf{N})}\, \boldsymbol{\Theta}\cdot\mathbf{N} = (\boldsymbol{\Theta}\cdot\mathbf{U})\cdot\mathbf{N}$$

or

$$\boldsymbol{\Theta}\cdot(\mathbf{U} - \Lambda_{(\mathbf{N})}\, \mathbf{I})\cdot\mathbf{N} = \mathbf{0}.$$

For any arbitrary rotation $\boldsymbol{\Theta}$, this relation provides the eigenvalue problem

$$\boxed{(\mathbf{U} - \Lambda_{(\mathbf{N})}\, \mathbf{I})\cdot\mathbf{N} = \mathbf{0}} \tag{3.166}$$

for the **principal stretch ratios** $\Lambda_{(\mathbf{N})}$ in the *unrotated* principal directions $\mathbf{N}$ at $\mathbf{R}$ in the undeformed body B. Recall, from Eq. (3.147), that the deformation tensor $\mathbf{U}$ operates on $d\mathbf{R}$ before the rotation due to $\boldsymbol{\Theta}$.

Similarly, substituting Eq. (3.147) into $(3.163)_2$ yields

$$\lambda_{(\mathrm{n})}\,\mathbf{n} = (\mathbf{V}\cdot\boldsymbol{\Theta})\cdot\mathbf{N} = \mathbf{V}\cdot(\boldsymbol{\Theta}\cdot\mathbf{N}) = \mathbf{V}\cdot\mathbf{n}$$

in which Eq. (3.164) has been used. This equation gives the eigenvalue problem

$$\boxed{(\mathbf{V} - \lambda_{(\mathrm{n})}\,\mathbf{I})\cdot\mathbf{n} = \mathbf{0}} \tag{3.167}$$

for the principal stretch ratios $\lambda_{(\mathrm{n})}$ in the *rotated* principal directions $\mathbf{n}$ at $\mathbf{r}$ in the deformed body b. This is consistent with the fact that $\mathbf{V}$ gives a deformation *after* the rotation due to $\boldsymbol{\Theta}$.

Here, we make several observations. First, comparing Eqs. (3.166) and (3.167) with Eqs. (3.133) suggests that the eigenvalues of the strain tensors $\mathbf{C}$ and $\mathbf{B}$ are equal to the squares of the eigenvalues (principal stretches) of the stretch tensors $\mathbf{U}$ and $\mathbf{V}$, respectively. Equations (3.148) and (3.150) also are consistent with this assertion. To show that this is indeed the case, we substitute Eqs. (3.153) into (3.133) to get

$$\begin{aligned} (\mathbf{U}^2 - \Lambda^2_{(\mathrm{N})}\,\mathbf{I})\cdot\mathbf{N} &= \mathbf{0} \\ (\mathbf{V}^2 - \lambda^2_{(\mathrm{n})}\,\mathbf{I})\cdot\mathbf{n} &= \mathbf{0}. \end{aligned}$$

These equations can be written

$$\begin{aligned} (\mathbf{U} + \Lambda_{(\mathrm{N})}\,\mathbf{I})\cdot(\mathbf{U} - \Lambda_{(\mathrm{N})}\,\mathbf{I})\cdot\mathbf{N} &= \mathbf{0} \\ (\mathbf{V} + \lambda_{(\mathrm{n})}\,\mathbf{I})\cdot(\mathbf{V} - \lambda_{(\mathrm{n})}\,\mathbf{I})\cdot\mathbf{n} &= \mathbf{0} \end{aligned}$$

which are satisfied by Eqs. (3.166) and (3.167). In addition, since the eigenvalues of $\mathbf{C}$ and $\mathbf{B}$ are equal, so are the eigenvalues of $\mathbf{U}$ and $\mathbf{V}$. Moreover, the principal directions of $\mathbf{C}$ and $\mathbf{B}$ correspond to those of $\mathbf{U}$ and $\mathbf{V}$, respectively. Finally, as shown by Eq. (3.164), the tensor $\boldsymbol{\Theta}$ rotates the principal directions of $\mathbf{U}$ (and $\mathbf{C}$) into those of $\mathbf{V}$ (and $\mathbf{B}$).

In summary, the tensors $\mathbf{C}$, $\mathbf{B}$, $\mathbf{U}$, and $\mathbf{V}$ are all symmetric and positive definite. The eigenvalues of $\mathbf{C}$ and $\mathbf{B}$ are $\Lambda^2_{(\mathbf{N})} = \lambda^2_{(\mathbf{n})}$, and those of $\mathbf{U}$ and $\mathbf{V}$ are $\Lambda_{(\mathbf{N})} = \lambda_{(\mathbf{n})}$, which are the principal stretch ratios. Moreover, $\mathbf{\Theta}$ rotates the equal and mutually orthogonal principal directions $\mathbf{N}_{(i)}$ of $\mathbf{C}$ and $\mathbf{U}$ at point P in the undeformed body into the equal and orthogonal principal directions $\mathbf{n}_{(i)}$ of $\mathbf{B}$ and $\mathbf{V}$ at the image p of P in the deformed body. Thus, we can extend Eqs. (3.164) and (3.165) to write

$$\boxed{\begin{aligned} \mathbf{n}_{(i)} &= \mathbf{\Theta}\cdot\mathbf{N}_{(i)} \\ \mathbf{\Theta} &= \sum_i \mathbf{n}_{(i)}\mathbf{N}_{(i)}. \end{aligned}} \tag{3.168}$$

Then we have, for example, $\mathbf{n}_{(1)} = \mathbf{\Theta}\cdot\mathbf{N}_{(1)} = (\mathbf{n}_{(1)}\mathbf{N}_{(1)} + \mathbf{n}_{(2)}\mathbf{N}_{(2)} + \mathbf{n}_{(3)}\mathbf{N}_{(3)}) \cdot\mathbf{N}_{(1)} = \mathbf{n}_{(1)}$, since $\mathbf{N}_{(1)}\cdot\mathbf{N}_{(2)} = \mathbf{N}_{(1)}\cdot\mathbf{N}_{(3)} = 0$.

Example 3.13 For simple shear of a cube (Example 3.3, page 81), determine the following:

(a) The eigenvalues and eigenvectors for the deformation tensors $\mathbf{C}$ and $\mathbf{B}$.
(b) The rotation tensor $\mathbf{\Theta}$ and the stretch tensors $\mathbf{U}$ and $\mathbf{V}$.

Solution. (a) The eigenvalue problems are given by Eqs. (3.133). For nontrivial solutions, we must have

$$\begin{aligned} \det(\mathbf{C} - \Lambda^2_{(\mathbf{N})}\,\mathbf{I}) &= 0 \\ \det(\mathbf{B} - \lambda^2_{(\mathbf{n})}\,\mathbf{I}) &= 0 \end{aligned} \tag{3.169}$$

or, noting Eq. (2.71),

$$\begin{aligned} \det\left[C^I_J - \Lambda^2_{(\mathbf{N})}\,\delta^I_J\right] &= 0 \\ \det\left[B^i_j - \lambda^2_{(\mathbf{n})}\,\delta^i_j\right] &= 0. \end{aligned} \tag{3.170}$$

Because $\mathbf{C}$ and $\mathbf{B}$ are symmetric tensors, we can write $C^I_{\cdot J} = C^{\cdot I}_J = C^I_J$ and $B^i_{\cdot j} = B^{\cdot i}_j = B^i_j$. Moreover, because the $\mathbf{G}_I$ and $\mathbf{g}_i$ are Cartesian (unit) base vectors in this problem (see Example 3.3), $C^I_J = C_{IJ}$ and $B^i_j = B_{ij}$. Thus,

Eqs. (3.55) and (3.170) give

$$\begin{vmatrix} 1-\Lambda_{(\mathbf{N})}^2 & k & 0 \\ k & 1+k^2-\Lambda_{(\mathbf{N})}^2 & 0 \\ 0 & 0 & 1-\Lambda_{(\mathbf{N})}^2 \end{vmatrix} = 0$$

$$\begin{vmatrix} 1+k^2-\lambda_{(\mathbf{n})}^2 & k & 0 \\ k & 1-\lambda_{(\mathbf{n})}^2 & 0 \\ 0 & 0 & 1-\lambda_{(\mathbf{n})}^2 \end{vmatrix} = 0. \qquad (3.171)$$

These characteristic equations provide the same eigenvalues, and so $\Lambda_{(\mathbf{N}_i)} = \lambda_{(\mathbf{n}_i)} \equiv \lambda_{(i)}$ for the *ith* eigenvalue. This is expected, because the $\lambda_{(i)}$ represent principal stretch ratios of an element that rotates from the direction $\mathbf{N}_{(i)}$ to the direction $\mathbf{n}_{(i)}$ during the deformation. We also note that expanding Eq. $(3.171)_1$ gives

$$-\Lambda_{(\mathbf{N})}^3 + I_1\,\Lambda_{(\mathbf{N})}^2 - I_2\,\Lambda_{(\mathbf{N})} + I_3 = 0$$

with the I_i given by Eq. (3.145). This agrees with the general characteristic equation (2.76).

Equations (3.171) yield the principal stretch ratios

$$\lambda_{(1,2)}^2 = 1 + \tfrac{1}{2}k^2 \pm k\sqrt{1+\tfrac{1}{4}k^2}, \qquad \lambda_{(3)} = 1 \qquad (3.172)$$

and Eqs. (3.55) and (3.133) provide the corresponding principal directions

$$\mathbf{N}_{(1,2)} = \frac{\mathbf{e}_1 + \left(\frac{1}{2}k \pm \sqrt{1+\frac{1}{4}k^2}\right)\mathbf{e}_2}{\sqrt{2+\frac{1}{2}k^2 \pm k\sqrt{1+\frac{1}{4}k^2}}}, \qquad \mathbf{N}_{(3)} = \mathbf{e}_3$$

$$\mathbf{n}_{(1,2)} = \frac{\mathbf{e}_1 + \left(-\frac{1}{2}k \pm \sqrt{1+\frac{1}{4}k^2}\right)\mathbf{e}_2}{\sqrt{2+\frac{1}{2}k^2 \mp k\sqrt{1+\frac{1}{4}k^2}}}, \qquad \mathbf{n}_{(3)} = \mathbf{e}_3. \qquad (3.173)$$

(The denominators make $\mathbf{N}_{(1,2)}$ and $\mathbf{n}_{(1,2)}$ unit vectors.) It is easy to show that the $\mathbf{N}_{(i)}$ are mutually orthogonal, as are the $\mathbf{n}_{(i)}$.

For the special case $k = 0.5$, these equations give $\lambda_{(1)} = 1.28$, $\lambda_{(2)} = 0.781$, $\mathbf{N}_1 = 0.615\mathbf{e}_1 + 0.788\mathbf{e}_2$, $\mathbf{N}_2 = 0.788\mathbf{e}_1 - 0.615\mathbf{e}_2$, $\mathbf{n}_1 = 0.788\mathbf{e}_1 + 0.615\mathbf{e}_2$, and $\mathbf{n}_2 = 0.615\mathbf{e}_1 - 0.788\mathbf{e}_2$. The eigenvectors (principal directions) are shown in Fig. 3.12.

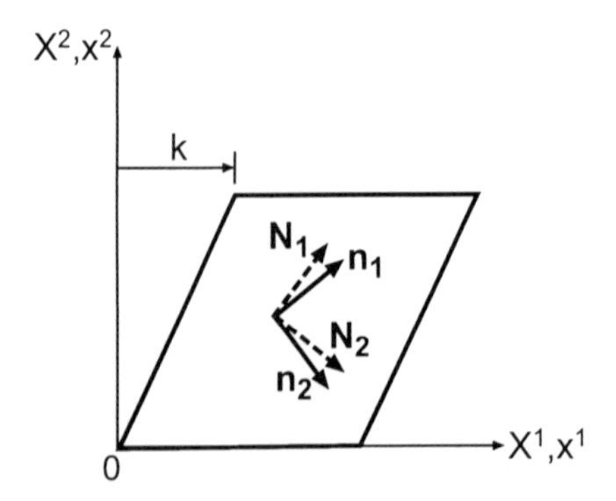

Fig. 3.12 Principal directions for simple shear of a unit cube ($k = 0.5$).

(b) Equations $(3.168)_2$ and (3.173) provide the rotation tensor $\mathbf{\Theta} = \mathbf{n}_{(1)}\mathbf{N}_{(1)} + \mathbf{n}_{(2)}\mathbf{N}_{(2)} + \mathbf{n}_{(3)}\mathbf{N}_{(3)}$, or

$$\mathbf{\Theta}_{(\mathbf{e}_i\mathbf{e}_j)} = \begin{bmatrix} (1+\frac{1}{4}k^2)^{-\frac{1}{2}} & \frac{1}{2}k(1+\frac{1}{4}k^2)^{-\frac{1}{2}} & 0 \\ -\frac{1}{2}k(1+\frac{1}{4}k^2)^{-\frac{1}{2}} & (1+\frac{1}{4}k^2)^{-\frac{1}{2}} & 0 \\ 0 & 0 & 1 \end{bmatrix} \tag{3.174}$$

which shows that $\mathbf{\Theta}$ is antisymmetric, i.e., $\mathbf{\Theta}^T = -\mathbf{\Theta}$. This tensor rotates the orthogonal triad $\mathbf{N}_{(i)}$ into $\mathbf{n}_{(i)}$ (see Problem **3–7**). The above matrix indicates that the rotation is about the X^3-axis.

Computing $\mathbf{U}$ and $\mathbf{V}$ from Eqs. (3.153) would be a difficult task. With $\mathbf{\Theta}$ known, however, Eqs. (3.147) yield

$$\boxed{\begin{aligned} \mathbf{U} &= \mathbf{\Theta}^{-1}\cdot\mathbf{F} = \mathbf{\Theta}^T\cdot\mathbf{F} \\ \mathbf{V} &= \mathbf{F}\cdot\mathbf{\Theta}^{-1} = \mathbf{F}\cdot\mathbf{\Theta}^T \end{aligned}} \tag{3.175}$$

as $\boldsymbol{\Theta}^{-1} = \boldsymbol{\Theta}^T$ due to orthogonality. Substituting Eqs. (3.54) and (3.174) gives

$$\begin{aligned}
\mathbf{U}_{(\mathbf{e}_i\mathbf{e}_j)} &= \begin{bmatrix} (1+\frac{1}{4}k^2)^{-\frac{1}{2}} & -\frac{1}{2}k(1+\frac{1}{4}k^2)^{-\frac{1}{2}} & 0 \\ \frac{1}{2}k(1+\frac{1}{4}k^2)^{-\frac{1}{2}} & (1+\frac{1}{4}k^2)^{-\frac{1}{2}} & 0 \\ 0 & 0 & 1 \end{bmatrix} \begin{bmatrix} 1 & k & 0 \\ 0 & 1 & 0 \\ 0 & 0 & 1 \end{bmatrix} \\
&= \begin{bmatrix} (1+\frac{1}{4}k^2)^{-\frac{1}{2}} & \frac{1}{2}k(1+\frac{1}{4}k^2)^{-\frac{1}{2}} & 0 \\ \frac{1}{2}k(1+\frac{1}{4}k^2)^{-\frac{1}{2}} & (1+\frac{1}{2}k^2)(1+\frac{1}{4}k^2)^{-\frac{1}{2}} & 0 \\ 0 & 0 & 1 \end{bmatrix} \qquad (3.176)
\end{aligned}$$

which is symmetric. The left stretch tensor $\mathbf{V}$ can be found similarly (see Problem **3–7**).

Finally, we mention some other calculations that can be carried out as checks. Equations (3.55) and (3.176) can be used to show that $\mathbf{C} = \mathbf{U}^2 = \mathbf{U}\cdot\mathbf{U}$. Also, since the $\mathbf{N}_{(i)}$ and $\mathbf{n}_{(i)}$ correspond to the directions of the principal stretch ratios, Eqs. (3.55), (3.94), and (3.173) should give

$$\lambda_{(i)}^2 = \mathbf{N}_{(i)}\cdot\mathbf{C}\cdot\mathbf{N}_{(i)} = \left[\mathbf{n}_{(i)}\cdot\mathbf{B}^{-1}\cdot\mathbf{n}_{(i)}\right]^{-1}.$$

Moreover, the shears should vanish in the principal directions, and Eqs. (3.110) should give $\Gamma_{(12)} = \Gamma_{(23)} = \Gamma_{(31)} = 0$ and $\gamma_{(12)} = \gamma_{(23)} = \gamma_{(31)} = 0$, which are true if $\mathbf{N}_{(i)}\cdot\mathbf{C}\cdot\mathbf{N}_{(j)} = \mathbf{n}_{(i)}\cdot\mathbf{B}^{-1}\cdot\mathbf{n}_{(j)} = 0$ for $i \neq j$. Verification of these relations is left to the interested, diligent, or skeptical reader (see Problem **3–7**). (Symbolic manipulation software is helpful here.) ■

3.9 Approximations

The geometric nonlinearity in the exact strain-displacement relations (3.78) can be a source of substantial analytical complication. Thus, it is advantageous to seek simplifying approximations when possible.

This section considers the following cases:

1. **Small displacement.** *Example:* Any problem in the linear theory of elasticity, e.g., bone.
2. **Small deformation, arbitrary displacement and rotation.** *Example:* The "elastica" problem, i.e., large bending of thin beams such as cilia or actin filaments.
3. **Small deformation, moderate displacement and rotation.** *Example:* Deflection a few times the thickness in a thin plate or shell, e.g., defor-

mation of the cornea by an indenter (tonometer) to measure intraocular pressure.

4 **Small rotation, arbitrary deformation.** *Example:* Concentrated load on a half space, e.g., a tumor pressing against an organ.

The approximations considered here are for general geometry. Of course, further simplification may be possible by taking advantage of symmetry, thinness, or other features of a particular problem. Note that we consider "deformation" equivalent to "strain."

The meanings of the terms "small" and "moderate" are not always obvious. Thus, we give general rules of thumb, based on the magnitudes of the strain and rotation tensors, $|\mathbf{E}| = (\mathbf{E}{:}\mathbf{E})^{1/2}$ and $|\boldsymbol{\Theta}| = (\boldsymbol{\Theta}{:}\boldsymbol{\Theta})^{1/2}$, which are analogous to the magnitude of a vector, $|\mathbf{a}| = (\mathbf{a}\cdot\mathbf{a})^{1/2}$. In addition, since displacements can be due to rigid-body motion, the magnitude of the "displacement" is actually based on the magnitude of the displacement *gradient*, $|\boldsymbol{\nabla}\mathbf{u}| = [(\boldsymbol{\nabla}\mathbf{u}){:}(\boldsymbol{\nabla}\mathbf{u})]^{1/2}$. With these definitions, "small" means $\lesssim 0.01$ and "moderate" means $\lesssim 0.2$.

3.9.1 *Small Displacement*

If displacements are small, then the convected coordinate system differs little from the undeformed system. Therefore, we can take $\mathbf{G}_i \cong \mathbf{g}_I = \mathbf{g}_i$ and $X^I \cong x^i$. Then, Eqs. (3.18) show that $\overline{\boldsymbol{\nabla}} \cong \boldsymbol{\nabla}$, and for $|\boldsymbol{\nabla}\mathbf{u}| << 1$, Eqs. (3.78) and (3.79) provide the **linear strain tensor**

$$\boxed{\mathbf{E}^* = \tfrac{1}{2}\left[(\boldsymbol{\nabla}\mathbf{u})^T + (\boldsymbol{\nabla}\mathbf{u})\right] \cong \mathbf{e}^*.} \tag{3.177}$$

In this case, the difference between the Lagrangian and Eulerian strain tensors becomes negligible. In addition, as shown in Appendix A, the **linear rotation tensor** is

$$\boxed{\boldsymbol{\Theta}^* = \tfrac{1}{2}\left[(\boldsymbol{\nabla}\mathbf{u})^T - (\boldsymbol{\nabla}\mathbf{u})\right]} \tag{3.178}$$

which represents the average rotation of all line elements passing through a point.

These relations show that $\mathbf{E}^* = \mathbf{E}^{*T}$, $\boldsymbol{\Theta}^* = -\boldsymbol{\Theta}^{*T}$, and

$$\begin{aligned} (\boldsymbol{\nabla}\mathbf{u})^T &= \mathbf{E}^* + \boldsymbol{\Theta}^* \\ \boldsymbol{\nabla}\mathbf{u} &= \mathbf{E}^* - \boldsymbol{\Theta}^*. \end{aligned} \tag{3.179}$$

Thus, Eq. $(3.76)_1$ gives

$$\boxed{\begin{aligned}\mathbf{F} &= \mathbf{I} + (\boldsymbol{\nabla}\mathbf{u})^T \cong \mathbf{I} + \mathbf{E}^* + \boldsymbol{\Theta}^* \\ \mathbf{F}^T &= \mathbf{I} + \boldsymbol{\nabla}\mathbf{u} \cong \mathbf{I} + \mathbf{E}^* - \boldsymbol{\Theta}^*.\end{aligned}} \tag{3.180}$$

Moreover, substitution into Eqs. (3.25) and (3.153) produces

$$\begin{aligned}\mathbf{U}^2 &= \mathbf{C} = \mathbf{F}^T\cdot\mathbf{F} = (\mathbf{I} + \mathbf{E}^* - \boldsymbol{\Theta}^*)\cdot(\mathbf{I} + \mathbf{E}^* + \boldsymbol{\Theta}^*) \\ &= \mathbf{I} + \mathbf{I}\cdot(\mathbf{E}^* + \boldsymbol{\Theta}^*) + \mathbf{I}\cdot(\mathbf{E}^* - \boldsymbol{\Theta}^*) + (\mathbf{E}^* - \boldsymbol{\Theta}^*)\cdot(\mathbf{E}^* + \boldsymbol{\Theta}^*) \\ &\cong \mathbf{I} + 2\mathbf{E}^*\end{aligned}$$

$$\mathbf{V}^2 = \mathbf{B} = \mathbf{F}\cdot\mathbf{F}^T \cong \mathbf{I} + 2\mathbf{E}^*$$

where squares and products of $\mathbf{E}^*$ and $\boldsymbol{\Theta}^*$ have been neglected. For $|\mathbf{E}^*| << 1$, therefore,

$$\boxed{\mathbf{U} \cong \mathbf{V} \cong \mathbf{I} + \mathbf{E}^*.} \tag{3.181}$$

Finally, Eq. $(3.147)_1$ gives the *finite* rotation tensor

$$\boldsymbol{\Theta} = \mathbf{F}\cdot\mathbf{U}^{-1} \cong (\mathbf{I} + \mathbf{E}^* + \boldsymbol{\Theta}^*)\cdot(\mathbf{I} - \mathbf{E}^*)$$

or

$$\boxed{\boldsymbol{\Theta} = \mathbf{I} + \boldsymbol{\Theta}^*} \tag{3.182}$$

where again higher-order terms have been dropped. Thus, Eq. $(3.180)_1$ can be written

$$\boxed{\mathbf{F} \cong \mathbf{E}^* + \boldsymbol{\Theta}} \tag{3.183}$$

which shows that the decomposition of $\mathbf{F}$ into a deformation and a rotation is additive and, therefore, commutative in the linear theory (see also Section A.2.2).

3.9.2 *Small Deformation*

So far, we have examined the two extremes of the kinematic theory, i.e., the full nonlinear theory and the linearized version. Developing intermediate theories is facilitated by expressing $\mathbf{E}$ in terms of $\mathbf{E}^*$ and $\boldsymbol{\Theta}^*$. (Here, we restrict attention to $\mathbf{E}$; the analysis for $\mathbf{e}$ is similar.) Substituting Eqs. (3.179) into (3.78) yields

$$\boxed{\mathbf{E} = \mathbf{E}^* + \tfrac{1}{2}(\mathbf{E}^* - \boldsymbol{\Theta}^*)\cdot(\mathbf{E}^* + \boldsymbol{\Theta}^*)} \tag{3.184}$$

which is an *exact* expression. This section considers the case when deformation is small, but no restrictions are made on the magnitudes of displacements or rotations.

"Small deformation" implies small extensions $E_{(I)}$ and shears $\Gamma_{(IJ)}$. (This is equivalent to stipulating "small principal stretches.") In this case, Eq. $(3.98)_1$ and similar relations show that $E_{(II)} \cong E_{(I)} << 1$, and then Eq. $(3.111)_1$ and similar relations show that $E_{IJ} \cong \frac{1}{2}\Gamma_{IJ}$ $(I \neq J)$. Similar to the linear theory, therefore,

$$\boxed{E_{IJ} \cong \begin{cases} E_{(I)}, & I = J \\ \tfrac{1}{2}\Gamma_{(IJ)}, & I \neq J. \end{cases}} \tag{3.185}$$

In other words, the normal strains correspond to stretches, and the shear strains correspond to angle changes. Moreover, since $|\mathbf{E}^*\cdot\mathbf{E}^*| << |\mathbf{E}^*|$, Eq. (3.184) gives the approximation

$$\boxed{\mathbf{E} \cong \mathbf{E}^* + \tfrac{1}{2}(\mathbf{E}^*\cdot\boldsymbol{\Theta}^* - \boldsymbol{\Theta}^*\cdot\mathbf{E}^* - \boldsymbol{\Theta}^*\cdot\boldsymbol{\Theta}^*).} \tag{3.186}$$

Finally, if the deformation is small, then

$$\det\left[\delta^I_J + 2E^I_J\right] = 1 + 2(E^1_1 + E^2_2 + E^3_3) + \cdots,$$

and Eq. $(3.142)_3$ gives

$$\boxed{J \cong 1 + E^I_I.} \tag{3.187}$$

3.9.3 *Small Deformation and Moderate Rotation*

By "moderate rotation," we mean that the rotations are small enough to be approximately additive, so that Eq. (3.182) holds. Then, if deformation is small, terms like $|\mathbf{E}^*\cdot\mathbf{\Theta}^*|$ can be ignored compared to $|\mathbf{E}^*|$, but $|\mathbf{\Theta}^*\cdot\mathbf{\Theta}^*|$ cannot. In this case, Eq. (3.184) reduces to

$$\boxed{\mathbf{E} \cong \mathbf{E}^* - \tfrac{1}{2}\mathbf{\Theta}^*\cdot\mathbf{\Theta}^* = \mathbf{E}^* + \tfrac{1}{2}\mathbf{\Theta}^*\cdot\mathbf{\Theta}^{*T}.} \tag{3.188}$$

3.9.4 *Small Rotation*

If rotations are small but deformation is arbitrary, then $|\mathbf{\Theta}^*\cdot\mathbf{\Theta}^*| << |\mathbf{E}^*|$ and Eq. (3.184) gives the approximation

$$\boxed{\mathbf{E} \cong \mathbf{E}^* + \tfrac{1}{2}(\mathbf{E}^*\cdot\mathbf{E}^* + \mathbf{E}^*\cdot\mathbf{\Theta}^* - \mathbf{\Theta}^*\cdot\mathbf{E}^*).} \tag{3.189}$$

3.10 Deformation Rates

The state of stress in an elastic body depends only on its final deformed configuration, regardless of how or how fast it got there. Soft tissues generally are viscoelastic, however, with viscous losses accompanying the deformation. In such materials, the final stress state depends on the entire history and rate of the deformation. Even for elastic bodies, moreover, rate formulations often are advantageous computationally, and we will use rate equations to develop constitutive relations for elastic materials in Chapter 5. This section presents measures for the rate of deformation at a point in a body relative to a given frame of reference. First, however, we discuss time differentiation of Lagrangian and Eulerian quantities.

3.10.1 *Time Rates of Change*

The mechanics of differentiating a physical quantity, such as density or temperature, with respect to time depends on whether the quantity is considered a Lagrangian or an Eulerian variable. In the Lagrangian description, a particle originally at the position $\mathbf{R}$ is followed through space and time, but the Eulerian description monitors time-dependent changes at a particular spatial position $\mathbf{r}$. In general, as a body deforms, different particles pass through $\mathbf{r}$. Since it is relatively easy in general to identify a reference configuration in a solid body, Lagrangian variables usually are used in solid mechanics problems. Researchers in fluid dynamics, however, usually prefer to work with Eulerian variables.

Consider a scalar quantity $\phi(\mathbf{R},t) = \phi(\mathbf{r},t)$. In general, the partial derivatives $\partial\phi(\mathbf{R},t)/\partial t$ and $\partial\phi(\mathbf{r},t)/\partial t$ differ. The former derivative is the time rate of change of ϕ following a particle ($\mathbf{R}$ = constant), while the latter is the rate of change of ϕ at a fixed point in space ($\mathbf{r}$ = constant). As a particle passes through the point located at $\mathbf{r}$, it instantaneously possesses the property $\phi(\mathbf{r},t)$.

Usually, we want to know how ϕ changes for a given particle. Since $\mathbf{R}$ is a constant, the Lagrangian description gives the time derivative

$$\frac{d\phi(\mathbf{R},t)}{dt} = \frac{\partial\phi(\mathbf{R},t)}{\partial t}. \tag{3.190}$$

Following the motion of a particle in the Eulerian description means that its position $\mathbf{r}$ changes with time. In this case, the chain rule yields

$$\frac{d\phi(\mathbf{r},t)}{dt} = \frac{\partial\phi(\mathbf{r},t)}{\partial t} + \frac{\partial\mathbf{r}}{\partial t}\cdot\frac{\partial\phi(\mathbf{r},t)}{\partial\mathbf{r}}$$

where Eq. (2.147) has been used in taking the derivative with respect to the vector $\mathbf{r}$. Since $\mathbf{r} = \mathbf{r}(\mathbf{R},t)$, the velocity of a particle is

$$\boxed{\mathbf{v} = \frac{d\mathbf{r}}{dt} = \frac{\partial\mathbf{r}}{\partial t},} \tag{3.191}$$

and Eq. $(3.18)_2$ gives $\partial/\partial\mathbf{r} = \overline{\boldsymbol{\nabla}}$. Thus,

$$\frac{d\phi(\mathbf{r},t)}{dt} = \frac{\partial\phi(\mathbf{r},t)}{\partial t} + \mathbf{v}\cdot\overline{\boldsymbol{\nabla}}\phi(\mathbf{r},t) \tag{3.192}$$

in which the first term represents the time rate of change of ϕ at a fixed position $\mathbf{r}$, and the second term gives the rate of change as the particle is convected through space.

The physical meaning of the convection term in (3.192) is illustrated by considering the following. Suppose ϕ has a different value but is constant at each point in space. Then, $\partial\phi/\partial t = 0$ at each point, but a particle traveling through space takes on different local values of ϕ as it passes through each point, i.e., $d\phi/dt = \mathbf{v}\cdot\overline{\boldsymbol{\nabla}}\phi \neq 0$.

With the definition

$$\boxed{\frac{d}{dt} \equiv \frac{\partial}{\partial t} + \mathbf{v}\cdot\overline{\boldsymbol{\nabla}},} \tag{3.193}$$

$d\phi/dt$ becomes either Eq. (3.190) or Eq. (3.192), depending on whether $\phi = \phi(\mathbf{R}, t)$ or $\phi = \phi(\mathbf{r}, t)$ [since $\overline{\boldsymbol{\nabla}}\phi(\mathbf{R}, t) = 0$]. In component form, with $\mathbf{v} = v^i\mathbf{g}_i$ and $\overline{\boldsymbol{\nabla}} = \mathbf{g}^i\partial/\partial x^i$, Eq. (3.193) becomes

$$\boxed{\frac{d}{dt} = \frac{\partial}{\partial t} + v^i(\ )_{,i}\,.} \tag{3.194}$$

Consider now the time derivative of a vector. We compute, for example, the acceleration of a particle as

$$\mathbf{a} = \frac{d\mathbf{v}}{dt}. \tag{3.195}$$

For the Lagrangian and Eulerian description, respectively, Eq. (3.193) gives

$$\boxed{\begin{aligned} \mathbf{a} &= \frac{d\mathbf{v}(\mathbf{R}, t)}{dt} = \frac{\partial \mathbf{v}}{\partial t} \\ \mathbf{a} &= \frac{d\mathbf{v}(\mathbf{r}, t)}{dt} = \frac{\partial \mathbf{v}}{\partial t} + \mathbf{v}\cdot\overline{\boldsymbol{\nabla}}\mathbf{v}. \end{aligned}} \tag{3.196}$$

Vector and tensor components are written most conveniently in terms of the undeformed basis $\{\mathbf{G}_I\}$ (coordinates X^I) for Lagrangian quantities and in terms of the deformed basis $\{\mathbf{g}_i\}$ (coordinates x^i) for Eulerian quantities. Both of these coordinate systems are fixed in the reference frame and so are independent of time. In contrast, since the base vectors $\mathbf{g}_I$ move with the deformation, they are time-dependent. Thus, we take

$$\begin{aligned} \mathbf{v}(\mathbf{R}, t) &= v^I(X^I, t)\,\mathbf{G}_I, \qquad & \mathbf{v}(\mathbf{r}, t) &= v^i(x^i, t)\,\mathbf{g}_i \\ \mathbf{a}(\mathbf{R}, t) &= a^I(X^I, t)\,\mathbf{G}_I, \qquad & \mathbf{a}(\mathbf{r}, t) &= a^i(x^i, t)\,\mathbf{g}_i. \end{aligned} \tag{3.197}$$

Substituting these expressions into Eq. $(3.196)_1$ gives simply

$$a^I = \frac{\partial v^I}{\partial t}, \tag{3.198}$$

but finding a^i requires a little more work. First, we note that Eq. (2.161) gives

$$\overline{\boldsymbol{\nabla}}\mathbf{v} = v^i|_j\, \mathbf{g}^j \mathbf{g}_i,$$

and inserting this relation and Eq. (3.197) into $(3.196)_2$ yields

$$\begin{aligned} \mathbf{a} &= \frac{\partial}{\partial t}(v^i \mathbf{g}_i) + (v^k \mathbf{g}_k)\cdot(v^i|_j \mathbf{g}^j \mathbf{g}_i) \\ &= \frac{\partial v^i}{\partial t}\mathbf{g}_i + v^k v^i|_j \delta^j_k \mathbf{g}_i \\ &= \left(\frac{\partial v^i}{\partial t} + v^j v^i|_j\right)\mathbf{g}_i. \end{aligned}$$

Thus, we can write

$$\boxed{\mathbf{a} = \frac{Dv^i}{Dt}\mathbf{g}_i} \tag{3.199}$$

where

$$\boxed{\frac{D}{Dt} \equiv \frac{\partial}{\partial t} + v^j(\ )|_j} \tag{3.200}$$

is called the **material derivative** or the **convected derivative** . As Eqs. (3.195) and (3.199) indicate, the material derivatives of v^j are the contravariant components of $d\mathbf{v}/dt$ with respect to the deformed basis. Finally, Eqs. (3.197) and (3.199) give

$$a^i = \frac{Dv^i}{Dt}. \tag{3.201}$$

Comparing Eqs. (3.194) and (3.200) reveals that the operators d/dt and D/Dt are not quite the same. The former is to be used when differentiating vectors and tensors, whereas the latter is for differentiating vector or tensor components. They are equivalent, however, in Cartesian coordinates or when the argument is a scalar function. In these cases, there is no distinction between covariant differentiation and partial differentiation.

Example 3.14 At $t = 0$, a particle is located at the Cartesian coordinates X_I, and for $t \geq 0$, it occupies the Cartesian coordinates

$$\begin{aligned} x_1 &= X_1(1 + at^2 X_2) \\ x_2 &= X_2 \\ x_3 &= X_3 \end{aligned} \tag{3.202}$$

where a is a constant. (For convenience, only subscripts are used here.) Compute the displacement, velocity, and acceleration of the particle.

Solution. With the position vectors being

$$\begin{aligned} t = 0: &\qquad \mathbf{R} = X_1\mathbf{e}_1 + X_2\mathbf{e}_2 + X_3\mathbf{e}_3 \\ t \geq 0: &\qquad \mathbf{r} = x_1\mathbf{e}_1 + x_2\mathbf{e}_2 + x_3\mathbf{e}_3, \end{aligned} \tag{3.203}$$

the displacement vector is

$$\mathbf{u} = \mathbf{r} - \mathbf{R} = at^2 X_1 X_2\, \mathbf{e}_1. \tag{3.204}$$

In terms of the material coordinates X_I, the velocity vector is

$$\mathbf{v} = \frac{d\mathbf{r}}{dt} = \frac{d\mathbf{u}}{dt} = 2atX_1X_2\,\mathbf{e}_1. \tag{3.205}$$

The velocity can be expressed in terms of the spatial coordinates x_i by solving Eqs. (3.202) for the X_I and substituting into (3.205) to obtain

$$\mathbf{v} = \frac{2atx_1x_2}{1 + at^2x_2}\,\mathbf{e}_1. \tag{3.206}$$

Finally, we compute the acceleration of the particle by differentiating both the Lagrangian representation (3.205) and the Eulerian representation (3.206) for the velocity. We should obtain the same answer in both cases. In the first case, Eq. $(3.196)_1$ gives

$$\mathbf{a} = \frac{d\mathbf{v}}{dt} = \frac{\partial \mathbf{v}}{\partial t} = 2aX_1X_2\,\mathbf{e}_1. \tag{3.207}$$

In the second case, Eq. $(3.196)_2$ gives

$$\mathbf{a} = \frac{d\mathbf{v}}{dt} = \frac{\partial \mathbf{v}}{\partial t} + \mathbf{v}\cdot\overline{\boldsymbol{\nabla}}\mathbf{v}$$

where Eq. (3.206) provides

$$\begin{aligned}\frac{\partial \mathbf{v}}{\partial t} &= 2ax_1x_2\left[\frac{1}{1+at^2x_2} - \frac{2at^2x_2}{(1+at^2x_2)^2}\right]\mathbf{e}_1 \\ \mathbf{v}\cdot\overline{\boldsymbol{\nabla}}\mathbf{v} &= \frac{2atx_1x_2}{1+at^2x_2}\frac{\partial}{\partial x_1}\left[\frac{2atx_1x_2}{1+at^2x_2}\right]\mathbf{e}_1 = \frac{4a^2t^2x_1x_2^2}{(1+at^2x_2)^2}\mathbf{e}_1\end{aligned}$$

since $\overline{\boldsymbol{\nabla}} = \mathbf{e}_1\partial/\partial x_1 + \mathbf{e}_2\partial/\partial x_2 + \mathbf{e}_3\partial/\partial x_3$. Combining these relations yields

$$\mathbf{a} = \frac{2ax_1x_2}{1+at^2x_2}\mathbf{e}_1, \tag{3.208}$$

and substituting Eqs. (3.202) for x_1 and x_2 shows that this relation is equivalent to (3.207). ■

3.10.2 *Rate-of-Deformation and Spin Tensors*

We now examine the deformation rate in the *current* configuration, i.e., using the Eulerian description. We begin by taking the time derivative d/dt, denoted by a superposed dot, of the deformed element $d\mathbf{r} = \mathbf{F}\cdot d\mathbf{R}$ to get the differential velocity vector

$$d\mathbf{v} \equiv d\dot{\mathbf{r}} = \dot{\mathbf{F}}\cdot d\mathbf{R} = \dot{\mathbf{F}}\cdot(\mathbf{F}^{-1}\cdot d\mathbf{r})$$

since $d\mathbf{R}$ is constant. This expression can be written

$$d\mathbf{v} = \mathbf{L}\cdot d\mathbf{r} = d\mathbf{r}\cdot\mathbf{L}^T \tag{3.209}$$

where

$$\boxed{\mathbf{L} \equiv \dot{\mathbf{F}}\cdot\mathbf{F}^{-1}.} \tag{3.210}$$

In addition, with Eqs. $(2.154)_1$ and $(3.18)_2$, Eq. (3.209) gives

$$\boxed{\mathbf{L}^T = \frac{\partial \mathbf{v}}{\partial \mathbf{r}} = \overline{\boldsymbol{\nabla}}\mathbf{v}.} \tag{3.211}$$

The tensor $\mathbf{L}$ is called the **velocity gradient tensor**.

In terms of components with respect to the deformed body, the velocity vector is given by

$$\mathbf{v} = v^i\mathbf{g}_i = v_i\mathbf{g}^i = v^{I^*}\mathbf{g}_I = v_{I^*}\mathbf{g}^I, \tag{3.212}$$

and Eqs. (2.141) and $(3.18)_2$ yield, for example,

$$\mathbf{L}^T = \overline{\boldsymbol{\nabla}}\mathbf{v} = \mathbf{g}^j \mathbf{v}_{,j} = \mathbf{g}^j (v^i\|_j \mathbf{g}_i) = v^i\|_j \mathbf{g}^j \mathbf{g}_i$$

where the covariant derivative is defined by a relation similar to Eq. $(3.84)_2$. This and similar manipulations yield

$$\boxed{\begin{aligned} \mathbf{L} &= v^i\|_j \mathbf{g}_i \mathbf{g}^j = v_i\|_j \mathbf{g}^i \mathbf{g}^j \\ &= v^{I^*}\|_J \mathbf{g}_I \mathbf{g}^J = v_{I^*}\|_J \mathbf{g}^I \mathbf{g}^J. \end{aligned}} \tag{3.213}$$

The **rate-of-deformation tensor** (or stretch rate tensor) and the **spin tensor** (or vorticity tensor) are defined to be the symmetric and antisymmetric components of $\mathbf{L}$, respectively, i.e. [see Eqs. (2.69)]

$$\boxed{\begin{aligned} \mathbf{D} &\equiv \tfrac{1}{2}(\mathbf{L} + \mathbf{L}^T) = \mathbf{D}^T \\ \boldsymbol{\Omega} &\equiv \tfrac{1}{2}(\mathbf{L} - \mathbf{L}^T) = -\boldsymbol{\Omega}^T. \end{aligned}} \tag{3.214}$$

Substituting Eq. (3.211) gives

$$\begin{aligned} \mathbf{D} &\equiv \tfrac{1}{2}\left[(\overline{\boldsymbol{\nabla}}\mathbf{v})^T + (\overline{\boldsymbol{\nabla}}\mathbf{v})\right] \\ \boldsymbol{\Omega} &\equiv \tfrac{1}{2}\left[(\overline{\boldsymbol{\nabla}}\mathbf{v})^T - (\overline{\boldsymbol{\nabla}}\mathbf{v})\right] \end{aligned} \tag{3.215}$$

which resemble the *linear* strain and rotation tensors of Eqs. (3.177) and (3.178). In fact, in the linear case with $\overline{\boldsymbol{\nabla}} \cong \boldsymbol{\nabla}$ and $\mathbf{v} \cong \dot{\mathbf{u}}$, we see that

$$\mathbf{D} \cong \dot{\mathbf{E}}^*, \qquad \boldsymbol{\Omega} \cong \dot{\boldsymbol{\Theta}}^*. \tag{3.216}$$

There is, however, an important distinction. Whereas Eqs. (3.177) and (3.178) are approximate relations when displacement gradients are small, Eqs. (3.215) are *exact* for arbitrarily large velocity gradients. Also, adding Eqs. (3.214) gives

$$\boxed{\mathbf{L} = \mathbf{D} + \boldsymbol{\Omega}} \tag{3.217}$$

which agrees with a time derivative of Eq. $(3.180)_1$, with $\dot{\mathbf{F}} \cong (\boldsymbol{\nabla}\dot{\mathbf{u}})^T = (\boldsymbol{\nabla}\mathbf{v})^T = \mathbf{L}$ in the linear case.

Consider now the time derivative of $ds^2 = d\mathbf{r}\cdot d\mathbf{r}$ given by

$$\frac{d}{dt}(ds^2) = d\dot{\mathbf{r}}\cdot d\mathbf{r} + d\mathbf{r}\cdot d\dot{\mathbf{r}} = 2\,d\mathbf{r}\cdot d\mathbf{v}$$
$$= 2\,d\mathbf{r}\cdot\mathbf{L}\cdot d\mathbf{r} \qquad (3.218)$$

in which Eq. (3.209) has been substituted. Inserting Eq. (3.217) yields

$$\frac{d}{dt}(ds^2) = 2\,d\mathbf{r}\cdot\mathbf{D}\cdot d\mathbf{r} \qquad (3.219)$$

since

$$\begin{aligned} 2\,d\mathbf{r}\cdot\boldsymbol{\Omega}\cdot d\mathbf{r} &= d\mathbf{r}\cdot(\mathbf{L} - \mathbf{L}^T)\cdot d\mathbf{r} \\ &= d\mathbf{r}\cdot(\mathbf{L}\cdot d\mathbf{r}) - (d\mathbf{r}\cdot\mathbf{L}^T)\cdot d\mathbf{r} \\ &= d\mathbf{r}\cdot d\mathbf{v} - d\mathbf{v}\cdot d\mathbf{r} \\ &= 0 \end{aligned}$$

where Eqs. (3.209) and $(3.214)_2$ have been used. Thus, $\mathbf{D}$ characterizes the rate of change of ds^2. Next, substituting Eq. (3.19) into (3.219) gives

$$\begin{aligned} \frac{d}{dt}(ds^2) &= 2\,(d\mathbf{R}\cdot\mathbf{F}^T)\cdot\mathbf{D}\cdot(\mathbf{F}\cdot d\mathbf{R}) \\ &= 2\,d\mathbf{R}\cdot(\mathbf{F}^T\cdot\mathbf{D}\cdot\mathbf{F})\cdot d\mathbf{R}, \end{aligned}$$

and differentiating Eq. (3.26) yields

$$\frac{d}{dt}(ds^2 - dS^2) = \frac{d}{dt}(ds^2) = 2\,d\mathbf{R}\cdot\dot{\mathbf{E}}\cdot d\mathbf{R}.$$

Comparing these relations defines the **Lagrangian strain-rate tensor** as

$$\boxed{\dot{\mathbf{E}} = \mathbf{F}^T\cdot\mathbf{D}\cdot\mathbf{F}.} \qquad (3.220)$$

Similarly, Eqs. $(3.26)_2$ and (3.209) give

$$\begin{aligned} \frac{d}{dt}(ds^2 - dS^2) &= \frac{d}{dt}(ds^2) = \frac{d}{dt}(2d\mathbf{r}\cdot\mathbf{e}\cdot d\mathbf{r}) \\ &= 2\,(d\dot{\mathbf{r}}\cdot\mathbf{e}\cdot d\mathbf{r} + d\mathbf{r}\cdot\dot{\mathbf{e}}\cdot d\mathbf{r} + d\mathbf{r}\cdot\mathbf{e}\cdot d\dot{\mathbf{r}}) \\ &= 2\,(d\mathbf{r}\cdot\mathbf{L}^T\cdot\mathbf{e}\cdot d\mathbf{r} + d\mathbf{r}\cdot\dot{\mathbf{e}}\cdot d\mathbf{r} + d\mathbf{r}\cdot\mathbf{e}\cdot\mathbf{L}\cdot d\mathbf{r}) \\ &= 2\,d\mathbf{r}\cdot(\mathbf{L}^T\cdot\mathbf{e} + \dot{\mathbf{e}} + \mathbf{e}\cdot\mathbf{L})\cdot d\mathbf{r}, \end{aligned}$$

and comparison with Eq. (3.219) gives the **Eulerian strain-rate tensor**

$$\boxed{\dot{\mathbf{e}} = \mathbf{D} - \mathbf{e}\cdot\mathbf{L} - \mathbf{L}^T\cdot\mathbf{e}.} \tag{3.221}$$

Note that $\dot{\mathbf{e}} = d\mathbf{e}/dt$, with d/dt defined by Eq. (3.193).

Example 3.15 Of course, deformation rate tensors are particularly useful in rate-dependent problems, e.g., if the body is viscoelastic, plastic, or poroelastic. For a cylinder undergoing combined extension, inflation, and torsion (Example 3.4, page 84), derive (a) the velocity vector $\mathbf{v}$ in terms of natural components, (b) the velocity gradient tensor $\mathbf{L}$, and (c) $\dot{\mathbf{F}}$ by differentiating Eq. (3.61) directly and also using $\dot{\mathbf{F}} = \mathbf{L}\cdot\mathbf{F}$ from Eq. (3.210).

Solution. (a) Adding time dependency, we write Eqs. (3.56) as

$$r = r(R,t), \qquad \theta = \Theta + \psi(t)Z, \qquad z = \lambda(t)Z, \tag{3.222}$$

which can be inverted to give

$$R = R(r,t), \qquad \Theta = \theta - \psi z/\lambda, \qquad Z = z/\lambda. \tag{3.223}$$

Taking time derivatives of Eqs. (3.222) yields

$$\begin{aligned} \dot{r} &= \dot{r}(R(r,t),t) = \dot{r}(r,t) \\ \dot{\theta} &= \dot{\psi}(t)\, z/\lambda(t) = \dot{\theta}(z,t) \\ \dot{z} &= \dot{\lambda}(t)\, z/\lambda(t) = \dot{z}(z,t) \end{aligned} \tag{3.224}$$

to be used in an Eulerian formulation.

Now, with the deformed position vector given by $\mathbf{r} = r\mathbf{e}_r + z\mathbf{e}_z$, the velocity of a material element is

$$\mathbf{v} = \dot{\mathbf{r}} = \dot{r}\mathbf{e}_r + r\dot{\mathbf{e}}_r + \dot{z}\mathbf{e}_z$$

where the time derivatives of the unit vectors $\mathbf{e}_r$ and $\mathbf{e}_\theta$, found from Eqs. (3.58), are ($\mathbf{e}_z$ =constant)

$$\boxed{\dot{\mathbf{e}}_r = \dot{\theta}\mathbf{e}_\theta, \qquad \dot{\mathbf{e}}_\theta = -\dot{\theta}\mathbf{e}_r.} \tag{3.225}$$

Combining these relations gives

$$\mathbf{v} = \dot{r}\mathbf{e}_r + \dot{\theta}r\,\mathbf{e}_\theta + \dot{z}\,\mathbf{e}_z = v^i\mathbf{g}_i \tag{3.226}$$

in which the $\mathbf{g}^i$ are given by Eqs. (3.59). Hence, we have

$$v^1 = \dot{r}, \qquad v^2 = \dot{\theta}, \qquad v^3 = \dot{z} \tag{3.227}$$

(b) The velocity gradient tensor can be computed by inserting Eq. (3.226) into (3.211) and taking derivatives of the components and base vectors directly (see Problem **3–20**). For a change of pace, however, we use Eq. (3.213) in the form

$$\mathbf{L} = v^i||_j \, \mathbf{g}_i\mathbf{g}^j \tag{3.228}$$

where modifying Eq. $(2.142)_1$ gives

$$v^i||_j = v^i{}_{,j} + v^k \bar{\Gamma}^i_{jk} \tag{3.229}$$

with the $\bar{\Gamma}^i_{jk}$ defined by Eq. $(3.85)_2$. Since the base vectors $\mathbf{g}_i$ correspond to the cylindrical polar coordinate system (r, θ, z), Eqs. (2.139) provide the only nonzero Christoffel symbols:

$$\bar{\Gamma}^1_{22} = -r, \qquad \bar{\Gamma}^2_{12} = \bar{\Gamma}^2_{21} = r^{-1}. \tag{3.230}$$

Inserting these relations and Eqs. (3.227) into (3.229) yields

$$\mathbf{L}_{(\mathbf{g}_i\mathbf{g}^j)} = \left[v^i||_j\right] = \begin{bmatrix} \dot{r}_{,r} & -r\dot{\theta} & 0 \\ \dfrac{\dot{\theta}}{r} & \dfrac{\dot{r}}{r} & \dot{\theta}_{,z} \\ 0 & 0 & \dot{z}_{,z} \end{bmatrix}. \tag{3.231}$$

From this equation, with care taken to keep the base vectors in order, $\mathbf{D}$ and $\mathbf{\Omega}$ can be computed from Eqs. (3.214).

(c) In the first approach, differentiating Eq. (3.61) gives ($\mathbf{e}_R$, $\mathbf{e}_\Theta$, and $\mathbf{e}_Z = \mathbf{e}_z$ are constants with respect to time)

$$\begin{aligned} \dot{\mathbf{F}} \;=\; & \dot{r}_{,R}\,\mathbf{e}_r\mathbf{e}_R + r_{,R}\,\dot{\mathbf{e}}_r\mathbf{e}_R + \frac{\dot{r}}{R}\mathbf{e}_\theta\mathbf{e}_\Theta + \frac{r}{R}\dot{\mathbf{e}}_\theta\mathbf{e}_\Theta \\ & + (\dot{\psi}r + \psi\dot{r})\mathbf{e}_\theta\mathbf{e}_Z + \psi r\dot{\mathbf{e}}_\theta\mathbf{e}_Z + \dot{\lambda}\mathbf{e}_z\mathbf{e}_Z \end{aligned}$$

and substituting Eqs. (3.225) produces

$$\dot{\mathbf{F}}_{(\mathbf{e}_i\mathbf{e}_I)} = \begin{bmatrix} \dot{r}_{,R} & -\dfrac{r\dot{\theta}}{R} & -\psi r\dot{\theta} \\ r_{,R}\,\dot{\theta} & \dfrac{\dot{r}}{R} & (\dot{\psi}r + \psi\dot{r}) \\ 0 & 0 & \dot{\lambda} \end{bmatrix}. \tag{3.232}$$

In the second approach, substituting Eqs. (3.22)$_1$ and (3.228) yields

$$\begin{aligned}\dot{\mathbf{F}} &= \mathbf{L}\cdot\mathbf{F} = (v^i||_j\, \mathbf{g}_i\mathbf{g}^j)\cdot(F^k_K\, \mathbf{g}_k\mathbf{G}^K) \\ &= v^i||_j F^k_K \mathbf{g}_i \delta^j_k \mathbf{G}^K \\ &= v^i||_j F^j_K \mathbf{g}_i \mathbf{G}^K.\end{aligned}$$

Thus, Eqs. (3.63) and (3.231) give

$$\begin{aligned}\dot{\mathbf{F}}_{(\mathbf{g}_i\mathbf{G}^K)} &= \left[v^i||_j\right]\left[F^j_K\right] \\ &= \begin{bmatrix} \dot{r}_{,r} & -r\dot{\theta} & 0 \\ \dfrac{\dot{\theta}}{r} & \dfrac{\dot{r}}{r} & \dot{\theta}_{,z} \\ 0 & 0 & \dot{z}_{,z} \end{bmatrix} \begin{bmatrix} r_{,R} & 0 & 0 \\ 0 & 1 & \psi \\ 0 & 0 & \lambda \end{bmatrix} \\ &= \begin{bmatrix} \dot{r}_{,r}\, r_{,R} & -r\dot{\theta} & -r\psi\dot{\theta} \\ \dfrac{\dot{\theta} r_{,R}}{r} & \dfrac{\dot{r}}{r} & \dfrac{\dot{r}}{r}\psi + \lambda\dot{\theta}_{,z} \\ 0 & 0 & \lambda\dot{z}_{,z} \end{bmatrix}. \end{aligned} \tag{3.233}$$

Showing that Eqs. (3.232) and (3.233) are identical requires some work. They first need to be expressed in terms of the same basis. One way to do this is to write the latter equation in the form

$$\dot{\mathbf{F}} = \begin{bmatrix} \dot{F}^1_1\mathbf{g}_1\mathbf{G}^1 & \dot{F}^1_2\mathbf{g}_1\mathbf{G}^2 & \dot{F}^1_3\mathbf{g}_1\mathbf{G}^3 \\ \dot{F}^2_1\mathbf{g}_2\mathbf{G}^1 & \dot{F}^2_2\mathbf{g}_2\mathbf{G}^2 & \dot{F}^2_3\mathbf{g}_2\mathbf{G}^3 \\ \dot{F}^3_1\mathbf{g}_3\mathbf{G}^1 & \dot{F}^3_2\mathbf{g}_3\mathbf{G}^2 & \dot{F}^3_3\mathbf{g}_3\mathbf{G}^3 \end{bmatrix}.$$

Then, on substituting Eqs. (3.59) and (3.60), Eq. (3.233) becomes

$$\dot{\mathbf{F}}_{(\mathbf{e}_i\mathbf{e}_I)} = \begin{bmatrix} r_{,R}\,\dot{r}_{,r} & -\dfrac{r}{R}\dot{\theta} & -r\psi\dot{\theta} \\ r_{,R}\,\dot{\theta} & \dfrac{\dot{r}}{R} & \dot{r}\psi + \lambda r\dot{\theta}_{,z} \\ 0 & 0 & \lambda\dot{z}_{,z} \end{bmatrix}. \tag{3.234}$$

With the exception of $\dot{F}^1_1$, $\dot{F}^2_3$, and $\dot{F}^3_3$, this expression is the same as Eq. (3.232).

These other terms can be shown to be equivalent by noting that

$$
\begin{aligned}
r_{,R}\,\dot{r}_{,r} &= \frac{\partial \dot{r}}{\partial r}\frac{\partial r}{\partial R} = \frac{\partial \dot{r}}{\partial R} = \dot{r}_{,R} \\
\lambda r\,\dot{\theta}_{,z} &= \lambda r\left(\frac{\dot{\psi}}{\lambda}\right) = r\dot{\psi} \\
\lambda \dot{z}_{,z} &= \lambda\left(\frac{\dot{\lambda}}{\lambda}\right) = \dot{\lambda}
\end{aligned}
\tag{3.235}
$$

where the chain rule and Eqs. (3.224) have been used. Putting these relations into (3.234) reproduces Eq. (3.232). ■

3.11 Compatibility Conditions

In Eqs. (3.90), six independent components of the strain tensor $\mathbf{E}$ (or $\mathbf{e}$) are related to three components of the displacement vector $\mathbf{u}$. If $\mathbf{u}$ is given, then all of the strain components can be computed directly. However, if $\mathbf{E}$ is prescribed, the strain-displacement equations provide a system of six differential equations for only three unknown displacement components. In this case, as in the linear theory of elasticity, the strain components must satisfy some additional **compatibility conditions** in order to ensure a single-valued continuous displacement field.

In the linear theory, the compatibility conditions can be found by eliminating the displacement components from the strain-displacement relations (see Section A.2.4). In the nonlinear theory, this procedure is virtually hopeless. An alternate derivation uses a theorem of Riemann (Eringen, 1962), which states that *for a symmetric tensor* $\mathbf{T}$ *to be a metric tensor for a Euclidean space, it is necessary and sufficient that* $\mathbf{T}$ *be nonsingular and positive definite and that the Riemann-Christoffel tensor* $R^{(\mathbf{T})\,m}_{\cdot ijk}$ *formed from it vanish identically.* We noted previously [see below Eq. (2.168)] that this condition implies that the order of covariant differentiation is immaterial. Since $\mathbf{C}$ is a positive definite tensor, this theorem and Eq. (2.168) give

$$
\boxed{R^{(\mathbf{C})\,M}_{\cdot IJK} = \Gamma^{(\mathbf{C})\,M}_{IK,J} - \Gamma^{(\mathbf{C})\,M}_{IJ,K} + \Gamma^{(\mathbf{C})\,M}_{LJ}\Gamma^{(\mathbf{C})\,L}_{IK} - \Gamma^{(\mathbf{C})\,M}_{LK}\Gamma^{(\mathbf{C})\,L}_{IJ} = 0}
\tag{3.236}
$$

where, from Eqs. (2.133) and (2.135),

$$
\begin{aligned}
2\Gamma^{(\mathbf{C})}_{IJK} &= C_{JK},_I + C_{KI},_J - C_{IJ},_K \\
\Gamma^{(\mathbf{C})\,L}_{IJ} &= C^{KL}\Gamma_{IJK}.
\end{aligned}
\tag{3.237}
$$

If the C_{IJ} satisfy Eq. (3.236), then they also represent the components of the deformed metric tensor, as indicated by Eq. $(3.31)_1$. Thus, Eq. (3.236) provides $3^4 = 81$ compatibility conditions. Fortunately, using $C_{IJ} = C_{JI}$ and $\Gamma^K_{IJ} = \Gamma^K_{JI}$ reduces the number of independent and nonidentically vanishing equations to six, as in the linear theory [see Eqs. (A.86)]. Similar equations can be found for E_{IJ} ($C_{IJ} = \delta_{IJ} + 2E_{IJ}$) and e_{ij} (in terms of B_{ij}).

3.12 Problems

3–1 A body is in a state of *plane strain* relative to the xy-plane, i.e., $E_{zx} = E_{zy} = E_{zz} = 0$. Assume that all components of the strain tensor $\mathbf{E}$ are known relative to the Cartesian axes (x, y, z). If another set of axes $(\bar{x}, \bar{y}, \bar{z})$ are defined by rotating the (x, y, z) axes by an angle θ about the z-axis, use the general formula $E_{ij} = \mathbf{g}_i \cdot \mathbf{E} \cdot \mathbf{g}_j$ to compute the strain components relative to the rotated axes.

3–2 In two-dimensional polar coordinates (R, Θ), the displacement vector can be written

$$\mathbf{u}(R, \Theta) = u_R \mathbf{e}_R + u_\Theta \mathbf{e}_\Theta$$

and the gradient operator has the form

$$\boldsymbol{\nabla} = \mathbf{e}_R \frac{\partial}{\partial R} + \frac{\mathbf{e}_\Theta}{R} \frac{\partial}{\partial \Theta}.$$

Using Eq. (3.78), express the Lagrangian strain components E_{RR}, $E_{\Theta\Theta}$, $E_{R\Theta}$ and $E_{\Theta R}$ in terms of the displacements u_R and u_Θ.

3–3 Using the relation $\mathbf{F}^{-1} \cdot \mathbf{F} = \mathbf{I} = \mathbf{G}_I \mathbf{G}^I$, show that $\mathbf{F}^{-1} = F^I_i \mathbf{G}_I \mathbf{g}^i$.

3–4 The points in an undeformed body are described by the Cartesian coordinates (X_1, X_2, X_3). During deformation, the points move to the coordinates

$$x_1 = X_1 + 2X_2X_3, \qquad x_2 = X_2 + 2X_3^2, \qquad x_3 = X_3.$$

(a) Compute the base vectors $\mathbf{G}_I$, $\mathbf{G}^I$, $\mathbf{g}_I$, and $\mathbf{g}^I$. Show that $\mathbf{g}_I \cdot \mathbf{g}^J = \delta^J_I$.

(b) Use Eq. $(3.15)_1$ to compute the deformation gradients F^i_I and show that $\mathbf{F} = F^i_I \mathbf{g}_i \mathbf{G}^I = \mathbf{g}_I \mathbf{G}^I$.

(c) Compute the covariant components C_{IJ} of $\mathbf{C}$ using each of the relations $\mathbf{C} = \mathbf{F}^T \cdot \mathbf{F}$, $C_{IJ} = g_{IJ}$, and $C_{IJ} = F^i_I F^j_J g_{ij}$.

(d) Compute the stretch ratio for a fiber that originally is located at the coordinates (-2,1,-1) and points in the direction of the vector $\mathbf{e}_1 + 2\mathbf{e}_3$.

(e) Compute the dilatation ratio using the expressions $J = j\sqrt{\frac{g}{G}}$ and $J = \det \mathbf{F}$.

(f) Write the physical components of $\mathbf{E}$.

3–5 In material Cartesian coordinates, the displacement field in a body is

$$\mathbf{u} = X_2 X_3 \mathbf{e}_1 + X_3^2 \mathbf{e}_2.$$

Compute the strain tensors $\mathbf{E}$ and $\mathbf{e}$.

3–6 In Example 3.5 (page 88), two deformation paths were considered for extension, inflation, and torsion of a circular tube. Show that both paths lead to the same total $\mathbf{F}$ as given by Eq. (3.61).

3–7 Consider the problem of simple shear of a cube (Example 3.13, page 120).

(a) Show that the rotation tensor of Eq. (3.174) transforms the eigenvectors $\mathbf{N}_{(i)}$ of Eqs. (3.173) into the vectors $\mathbf{n}_{(i)}$.

(b) Compute the left stretch tensor $\mathbf{V}$.

(c) Show that the shears are zero relative to the principal axes of stretch.

3–8 For "small deformation," use Eq. (3.186) to write the approximate strain-displacement relations in Cartesian coordinates.

3–9 In Cartesian coordinates, a deformation is defined by

$$\begin{aligned} x_1 &= X_1 \cos\theta + X_2 \sin\theta \\ x_2 &= -X_1 \sin\theta + X_2 \cos\theta \\ x_3 &= X_3. \end{aligned}$$

Compute $\mathbf{C}$, $\mathbf{E}$, $\mathbf{U}$, and $\boldsymbol{\Theta}$. Explain the result.

3–10 Consider 2-D deformation of a body from configuration B to b, as defined by the relations

$$\begin{aligned} x_1 &= 0.1X_1(1+2X_1+X_2) \\ x_2 &= 0.2X_2(1+X_2), \end{aligned}$$

where X_i and x_i are Cartesian coordinates of a particle in B and b, respectively. An infinitesimal line element $\mathbf{a}_0 = 2\mathbf{e}_1 + 3\mathbf{e}_2$ located at the point $(-1, 2)$ in B deforms into the line element $\mathbf{a}$ in b. Determine $\mathbf{a}$.

3–11 Relative to Cartesian coordinates, the state of strain at a point in a body is given by

$$\mathbf{E} = \begin{bmatrix} 0.5 & 0.3 & 0 \\ 0.3 & 0.4 & -0.1 \\ 0 & -0.1 & 0.2 \end{bmatrix}.$$

Determine the change in angle between two lines, emanating from the point, that are parallel to the vectors $2\mathbf{e}_1 + 2\mathbf{e}_2 + \mathbf{e}_3$ and $3\mathbf{e}_1 - 6\mathbf{e}_3$ in the undeformed body.

3–12 In Cartesian spatial coordinates (x_1, x_2, x_3), the displacement field in a body is

$$\mathbf{u} = k\left[x_1^2\mathbf{e}_x + x_2x_3\mathbf{e}_y + (2x_1x_3 + x_1^2)\mathbf{e}_z\right]$$

where k is a constant. Determine the Lagrangian extensional strain of a fiber that lies in the direction $\mathbf{e}_x + \mathbf{e}_y + \mathbf{e}_z$ at the point $(1, 2, 3)$ in the deformed body.

3–13 A cylindrical tube undergoes the deformation given by

$$\begin{aligned} r &= R \\ \theta &= \Theta + \phi(R) \\ z &= Z + w(R) \end{aligned}$$

where (R, Θ, Z) and (r, θ, z) are polar coordinates of a point in the tube before and after deformation, respectively, and ϕ and w are scalar functions of R.

(a) Describe the deformation in words for the cases (i) $\phi = 0$ and (ii) $w = 0$.

(b) Compute $\mathbf{F}$, $\mathbf{C}$, and $\mathbf{E}$ using dyadic methods.

(c) Compute $E_{\theta\theta}$, E_{RZ}, and $e_{\theta z}$ using Eqs. (3.90).

3–14 In Cartesian coordinates, the deformation of a thin rectangular sheet is given by

$$\mathbf{r} = (\lambda_1 X_1 + k_1 X_2)\mathbf{e}_1 + (k_2 X_1 + \lambda_2 X_2)\mathbf{e}_2 + \lambda_3 X_3 \mathbf{e}_3.$$

(a) Compute the tensors $\mathbf{F}$, $\mathbf{C}$, $\mathbf{E}$, $\mathbf{U}$, and $\boldsymbol{\Theta}$. Show that $\boldsymbol{\Theta} \cdot \boldsymbol{\Theta}^T = \mathbf{I}$.

(b) For $\lambda_1 = 1.1$, $\lambda_2 = 1.25$, $k_1 = 0.15$, and $k_2 = -0.2$, determine the principal values and directions of $\mathbf{E}$. Verify that the principal directions are mutually orthogonal.

(c) Compute the strain invariants using Eqs. (3.142) and show that they are consistent with the characteristic equation from part (b).

3–15 The deformation gradient tensor at a point in a body is

$$\mathbf{F} = 0.1(4\mathbf{e}_1\mathbf{e}_1 - \mathbf{e}_1\mathbf{e}_2 + 2\mathbf{e}_2\mathbf{e}_1 + 3\mathbf{e}_2\mathbf{e}_2 + \mathbf{e}_3\mathbf{e}_3)$$

in which the $\mathbf{e}_i$ are Cartesian base vectors.

(a) Compute $\mathbf{F}$, $\mathbf{B}$, $\mathbf{C}$, $\mathbf{U}$, $\mathbf{V}$, and $\boldsymbol{\Theta}$.

(b) Compute the strain tensors $\mathbf{E}$ and $\mathbf{e}$.

(c) Determine the eigenvalues and eigenvectors of $\mathbf{B}$ and $\mathbf{C}$.

(d) Show the following:

i. $\boldsymbol{\Theta} = \sum_{i=1}^{3} \mathbf{n}_{(i)}\mathbf{N}_{(i)}$, where $\mathbf{n}_{(i)}$ and $\mathbf{N}_{(i)}$ are the eigenvectors of $\mathbf{B}$ and $\mathbf{C}$, respectively.

ii. $\mathbf{n}_{(i)} = \boldsymbol{\Theta} \cdot \mathbf{N}_{(i)}$ for $i = 1, 2, 3$.

3–16 For the deformation of Problem **3–14**, consider the vectors

$$\begin{aligned} \mathbf{a} &= \mathbf{e}_1 \cos\alpha + \mathbf{e}_2 \sin\alpha \\ \mathbf{b} &= \mathbf{e}_1 \cos\beta + \mathbf{e}_2 \sin\beta \end{aligned}$$

(a) Compute the stretch ratio for a line element that lies in the direction of $\mathbf{a}$ in the undeformed sheet.

(b) Compute the stretch ratio for a line element that lies in the direction of $\mathbf{a}$ in the deformed sheet.

(c) Compute the change in angle due to deformation between line elements that lie in the directions of $\mathbf{a}$ and $\mathbf{b}$ in the undeformed sheet.

(d) Compute the ratio of the deformed to the undeformed area of an element that is normal to the direction of $\mathbf{a}$ in the undeformed sheet.

3–17 The deformation of a body is described by the relations

$$
\begin{aligned}
x_1 &= X_1(1 + at^3 X_2 + btX_3 e^t) \\
x_2 &= X_2 \\
x_3 &= X_3
\end{aligned}
$$

where the X_i and x_i are material and spatial Cartesian coordinates, respectively, a and b are constants, and t is time.

(a) Compute the velocity field in terms of material coordinates and in terms of spatial coordinates.
(b) At $t = 0$, a point is located at the coordinates (1,2,4). For $a = 5$, determine the location of this point when $t = 2$. In addition, compute its velocity at this time using both the Lagrangian and Eulerian forms from part (a).
(c) Compute the acceleration field using both the Lagrangian and Eulerian forms of Dv_i/Dt. Show that the two formulas are equivalent.
(d) Compute the rate-of-deformation tensor $\mathbf{D}$ in terms of spatial coordinates.

3–18 In Cartesian coordinates, the motion of a particle is described by $\mathbf{r} = (X_1 + X_2 t)\mathbf{e}_1 + (X_2 + X_2^2 t^2)\mathbf{e}_2 + X_3\mathbf{e}_3$. In terms of the spatial coordinates x_1 and x_2, the temperature distribution is $\theta = (x_1^2 + x_2^2)t$. For a particle that is located at $(2, -3, 1)$ at $t = 0$, compute $d\theta/dt$ at $t = 1.5$.

3–19 In Cartesian coordinates, the deformation of a body is described by the equations

$$
\begin{aligned}
x_1 &= X_1 + 0.2X_2 t^2 \\
x_2 &= X_2 + 0.2X_1 t^2 \\
x_3 &= X_3.
\end{aligned}
$$

(a) Determine the velocity components at $t = 1.5$ of the particle that occupied the point $(2, 3, 4)$ when $t = 1.0$.
(b) Calculate the acceleration components of the same particle when $t = 2$.

3–20 Derive Eq. (3.231) by substituting (3.226) into (3.211).

3–21 The velocity field in a body is

$$
\mathbf{v} = \frac{x_1\mathbf{e}_1 + x_2\mathbf{e}_2 + x_3\mathbf{e}_3}{\alpha + t}
$$

where the x_i are Cartesian coordinates and α is a constant. Compute the acceleration field and explain the result.

3–22 In Cartesian coordinates, the deformation of a unit cube is given by

$$\begin{aligned} x_1 &= X_1 + kX_2^2 t \\ x_2 &= X_2 \\ x_3 &= X_3. \end{aligned}$$

(a) Sketch the deformed cube in the x_1x_2-plane for a time $t = t_0$.

(b) Compute the rate-of-deformation tensor $\mathbf{D}$.

(c) Because $J = \det \mathbf{F} = J(\mathbf{F})$, we can write $\dot{J} = \dot{\mathbf{F}}{:}\partial J/\partial \mathbf{F}$. Using this expression, the last formula in Table 2.5 (page 33), the relation $\mathbf{L} = \dot{\mathbf{F}} \cdot \mathbf{F}^{-1}$, and the identity $\mathbf{I} : \mathbf{A} = \operatorname{tr} \mathbf{A}$ for any second-order tensor $\mathbf{A}$, show that the dilatation rate is given in terms of the velocity $\mathbf{v}$ by

$$\dot{J} = J\overline{\boldsymbol{\nabla}} \cdot \mathbf{v}.$$

Compute $\dot{J}$ for this problem.

Chapter 4

Analysis of Stress

In general terms, stress is defined as force per unit area. But which area: the undeformed area or some other reference area? For small deformation, the distinction is immaterial, since changes in area are negligible. For large deformation, however, we must distinguish between the undeformed and deformed geometries. In this case, the only physically meaningful definition for stress is force per unit *deformed* area, which is called **true stress**. But the deformed geometry of a solid body is not known *a priori*, often is difficult to measure, and may be time-dependent. Experimentalists, therefore, find it useful to define a **Lagrangian stress** or **engineering stress** as the force per unit *undeformed* (reference) area. In addition, the governing equations sometimes have simpler forms or may be easier to solve when written in terms of yet another type of stress (to be defined later). We call any stress that is not the true stress a **pseudostress**, since it has no physical significance. Pseudostresses always must be converted into true stress for physical interpretation. This chapter introduces various types of stress tensor and discusses the relations between them.

4.1 Body and Contact Forces

External forces on an object consist of **body forces**, which act at a distance, and **contact forces**, which act directly on surfaces. In an Eulerian formulation, we denote the body force per unit deformed volume by $\mathbf{f}(\mathbf{r},t)$; in a Lagrangian formulation, the body force per unit undeformed volume is $\mathbf{f}_0(\mathbf{R},t)$. Then, with dV and dv being the respective undeformed and deformed volumes of a material element, we have

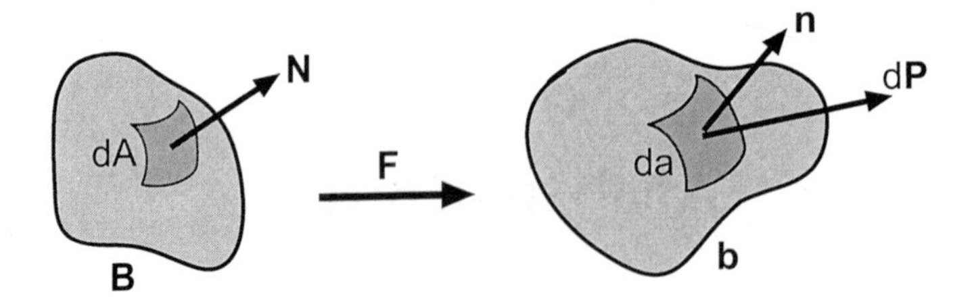

Fig. 4.1 Deformation of area element $d\mathbf{A}$ into area element $d\mathbf{a}$. The contact force $d\mathbf{P}$ acts on $d\mathbf{a}$.

$$\mathrm{f}_0 = J\,\mathrm{f} \tag{4.1}$$

where $J = dv/dV$.

Contact forces act either on the external surfaces of a body, including the surfaces of cavities, or on the fictitious internal surface enclosing a volume element. Internal contact forces are exerted on the element by the material adjacent to its surface. The intensity of a contact force, i.e., the force per unit area, is represented by the **traction vector**. The following paragraphs define three types of traction vector, which will be used later to define three types of stress tensor.

Consider an area element $d\mathbf{A}$ in an undeformed body B that, under the action of applied loads, deforms into the area element $d\mathbf{a}$ in the deformed body b (Fig. 4.1). As in Eqs. (3.125), we can write

$$\begin{aligned} d\mathbf{A} &= \mathbf{N}\,dA \\ d\mathbf{a} &= \mathbf{n}\,da \end{aligned} \tag{4.2}$$

where dA and da are the physical areas of the elements, and $\mathbf{N}$ and $\mathbf{n}$ are unit normals to the elements. By convention, the normal to a surface points outward from the material enclosed by the surface. If $d\mathbf{P}$ is a contact force acting on the deformed area $d\mathbf{a}$, then the **true traction (stress) vector** is defined as

$$\mathbf{T}^{(\mathbf{n})}(\mathbf{r},t) = \frac{d\mathbf{P}}{da} \tag{4.3}$$

in the limit as $da \to 0$. In general, $\mathbf{T}^{(\mathbf{n})}$ varies from point to point in a body. Moreover, as the orientation of $d\mathbf{a}(\mathbf{n})$ changes at a point, so does $d\mathbf{P}$ and therefore $\mathbf{T}^{(\mathbf{n})}$.

Next, two **pseudotraction (pseudostress) vectors** are defined. The first, which

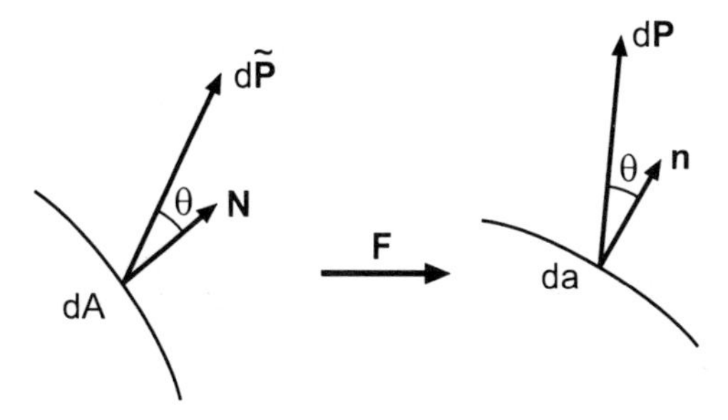

Fig. 4.2 Force vector $d\mathbf{P}$ acting on deformed area $d\mathbf{a}$ and pseudoforce vector $d\tilde{\mathbf{P}}$ acting on undeformed area $d\mathbf{A}$.

is an engineering-type traction, is simply the force per unit undeformed area, i.e.,

$$\mathbf{T}^{(\mathbf{N})}(\mathbf{R},t) = \frac{d\mathbf{P}}{dA}. \tag{4.4}$$

Note that, even though $\mathbf{T}^{(\mathbf{N})}$ is referred to the undeformed area $d\mathbf{A}$, $d\mathbf{P}$ still acts on the deformed area $d\mathbf{a}$ (Fig. 4.1). If $d\mathbf{A} \cong d\mathbf{a}$, then $\mathbf{T}^{(\mathbf{N})} \cong \mathbf{T}^{(\mathbf{n})}$, an approximation used in the linear theory (see Section A.3.1).

The second pseudotraction vector is defined in terms of a fictitious (pseudo) force $d\tilde{\mathbf{P}}$ acting on $d\mathbf{A}$ as defined by the relation

$$d\mathbf{P} = \mathbf{F}\cdot d\tilde{\mathbf{P}} = d\tilde{\mathbf{P}} \cdot \mathbf{F}^T. \tag{4.5}$$

This transformation is analogous to the deformation of the undeformed length element $d\mathbf{R}$ into the deformed element $d\mathbf{r}$, i.e., $d\mathbf{r} = \mathbf{F}\cdot d\mathbf{R}$. Thus, the orientation of $d\tilde{\mathbf{P}}$ relative to $d\mathbf{A}$ is the same as the orientation of $d\mathbf{P}$ relative to $d\mathbf{a}$ (Fig. 4.2). In terms of this pseudoforce, the second pseudotraction vector is

$$\tilde{\mathbf{T}}^{(\mathbf{N})}(\mathbf{R},t) = \frac{d\tilde{\mathbf{P}}}{dA}, \tag{4.6}$$

and combining Eqs. (4.3)–(4.6) yields

$$\boxed{d\mathbf{P} = \mathbf{T}^{(\mathbf{n})}da = \mathbf{T}^{(\mathbf{N})}dA = \tilde{\mathbf{T}}^{(\mathbf{N})}\cdot\mathbf{F}^T dA.} \tag{4.7}$$

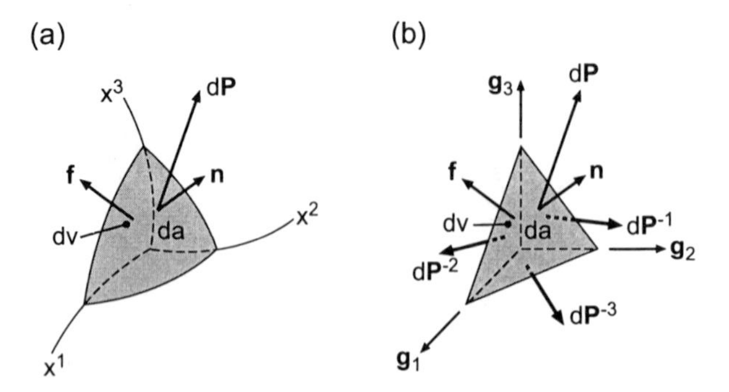

Fig. 4.3 Differential volume element. (a) Element carved out by general curvilinear coordinate surfaces. (b) Element in limit as volume approaches zero.

4.2 Stress Tensors

Using the traction vectors of Eq. (4.7), we will define three stress tensors: the Cauchy stress tensor and the first and second Piola-Kirchhoff stress tensors. The relations between these tensors are discussed in this section, while the relations between their various components are derived in Section 4.3.

4.2.1 *Cauchy Stress Tensor*

When subjected to applied forces and moments, an object and all of its component parts must obey Newton's laws of motion. Consider a tetrahedron at a point p in a deformed body that consists of three faces carved out by the surfaces of a general curvilinear coordinate system and an arbitrary fourth face (Fig. 4.3a). In the limit as the dimensions of the tetrahedral element shrink to zero, its edges become approximately straight and its faces planar (Fig. 4.3b).

The coordinate system that we use to define the element is immaterial. For convenience, the present derivation is based on a system corresponding to the base vectors $\mathbf{g}_i$. The three coordinate faces of the element then correspond to the surfaces x^i = constant (Fig. 4.3).

Let $d\hat{a}_i$ and $\mathbf{n}^i$ denote the physical area and unit normal of the element face x^i= constant (the x^i-face), and let da and $\mathbf{n}$ denote the area and normal of the arbitrary (oblique) face (Fig. 4.3b). The outward-directed normal to the x^3-face, for example, is $\mathbf{n}^3 = -\mathbf{g}_1 \times \mathbf{g}_2/|\mathbf{g}_1 \times \mathbf{g}_2| = -\mathbf{g}^3/\sqrt{\mathbf{g}^3 \cdot \mathbf{g}^3} = -\mathbf{g}^3/\sqrt{g^{33}}$ by

Eq. (2.17), and in general

$$\mathbf{n}^i = -\frac{\mathbf{g}^i}{\sqrt{g^{(ii)}}}. \tag{4.8}$$

Thus, the vector representations of the four faces of the tetrahedron can be written $\mathbf{n}\,da$ and $\mathbf{n}^{(i)}d\hat{a}_{(i)} = -\mathbf{g}^{(i)}d\hat{a}_{(i)}/\sqrt{g^{(ii)}}$ $(i = 1, 2, 3)$.

To relate these areas, we note that they form a closed surface, and it can be shown that the vectorial sum of the areas must vanish, i.e.,

$$\mathbf{n}\,da + \mathbf{n}^i d\hat{a}_i = \mathbf{0}. \tag{4.9}$$

This can be seen by considering the limiting case where the element collapses into a triangle in the plane x^3 = constant (see Fig. 4.3). In this case, $d\hat{a}_1 = d\hat{a}_2 = 0$ and Eq. (4.9) gives $\mathbf{n}\,da = -\mathbf{n}^i d\hat{a}_i = -\mathbf{n}^3 d\hat{a}_3$. Inspection shows that $\mathbf{n} = -\mathbf{n}^3$, and so $da = d\hat{a}_3$ as expected. Thus, Eqs. (4.8) and (4.9) yield

$$\mathbf{n}\,da = -\mathbf{n}^i d\hat{a}_i = \mathbf{g}^i d\hat{a}_i/\sqrt{g^{(ii)}}. \tag{4.10}$$

Setting

$$d\mathbf{a} = \mathbf{n}\,da = \mathbf{g}^i da_i \tag{4.11}$$

and comparing with Eq. (4.10) shows that the *covariant* components of the area vector $d\mathbf{a}$ are related to the *physical* areas of the x^i-faces by

$$da_i = \frac{d\hat{a}_i}{\sqrt{g^{(ii)}}}. \tag{4.12}$$

Now, let $d\mathbf{P}$ and $d\mathbf{P}^{-i}$ be the contact forces acting on the element faces with unit normals $\mathbf{n}$ and $\mathbf{n}^i$ respectively (Fig. 4.3b). The element also is subjected to a body force $\mathbf{f}$ per unit deformed volume. Then, Newton's second law of motion for the element gives

$$d\mathbf{P} + d\mathbf{P}^{-1} + d\mathbf{P}^{-2} + d\mathbf{P}^{-3} + \mathbf{f}\,dv = (\rho\,dv)\mathbf{a} \tag{4.13}$$

where ρ is the mass density and dv the volume of the element, and $\mathbf{a}$ is the acceleration of its center of mass. The minus sign in the superscript of $d\mathbf{P}^{-i}$ indicates that the force acts on a negative face of the element, i.e., the normal to the face makes an acute angle with the $-x^i$-direction (Fig. 4.3). The face of the element adjacent to this surface is positive, with a normal making an acute angle with the

$+x^i$-direction, and so the contact force on this adjacent area is denoted $d\mathbf{P}^i$.[1] By Newton's third law of action-reaction, $d\mathbf{P}^{-i} = -d\mathbf{P}^i$ and Eq. (4.13) becomes

$$d\mathbf{P} - d\mathbf{P}^1 - d\mathbf{P}^2 - d\mathbf{P}^3 + \mathbf{f}\,dv = \rho\mathbf{a}\,dv. \tag{4.14}$$

The true traction vector acting on the area da is defined by Eq. (4.3), and similarly

$$\hat{\mathbf{T}}^i = \frac{d\mathbf{P}^{(i)}}{d\hat{a}_{(i)}} \tag{4.15}$$

is the true traction vector acting on the x^i-face. Substituting Eqs. (4.3) and (4.15) into (4.14) yields

$$\mathbf{T}^{(\mathbf{n})}da - \hat{\mathbf{T}}^i\,d\hat{a}_i + \mathbf{f}\,dv = \rho\mathbf{a}\,dv \tag{4.16}$$

in which the summation convention applies to the index i.

Next, we examine the limit as the tetrahedral element shrinks to a point. Dividing Eq. (4.16) through by da and letting da and dv approach zero gives

$$\mathbf{T}^{(\mathbf{n})} = \hat{\mathbf{T}}^i\,\frac{d\hat{a}_i}{da} \tag{4.17}$$

since the linear dimension dv/da of the element approaches zero. Note that, in this limit, all four faces of the tetrahedron pass through the same point. The area ratio $d\hat{a}_i/da$ is obtained by taking the dot product of Eq. (4.10) with $\mathbf{g}_j$ to get

$$\mathbf{n}\,da\cdot\mathbf{g}_j = \mathbf{g}^i\cdot\mathbf{g}_j\frac{d\hat{a}_i}{\sqrt{g^{(ii)}}} = \delta^i_j\frac{d\hat{a}_i}{\sqrt{g^{(ii)}}} = \frac{d\hat{a}_j}{\sqrt{g^{(jj)}}},$$

which gives

$$\frac{d\hat{a}_i}{da} = (\mathbf{n}\cdot\mathbf{g}_i)\sqrt{g^{(ii)}}. \tag{4.18}$$

Combining this relation with Eq. (4.17) yields

$$\mathbf{T}^{(\mathbf{n})} = (\mathbf{n}\cdot\mathbf{g}_i)\sqrt{g^{(ii)}}\,\hat{\mathbf{T}}^i. \tag{4.19}$$

With the definition of yet another pseudotraction vector

$$\boxed{\mathbf{T}^i \equiv \sqrt{g^{(ii)}}\,\hat{\mathbf{T}}^i,} \tag{4.20}$$

[1] In nonorthogonal coordinates, the normals to the element faces may not point in the coordinate directions (hence the acute angle reference).

Eq. (4.19) becomes

$$\boxed{\mathbf{T}^{(\mathbf{n})} = (\mathbf{n} \cdot \mathbf{g}_i)\, \mathbf{T}^i = n_i \mathbf{T}^i} \tag{4.21}$$

where the n_i are the covariant components of $\mathbf{n}$. This equation shows that if the traction vectors $\mathbf{T}^i$ are known across three different planes passing through a point, then the traction across any plane through that point can be computed.

Next, we express the pseudotraction vector of Eq. (4.20) in the form

$$\mathbf{T}^i = \sigma^{ij} \mathbf{g}_j \tag{4.22}$$

where the σ^{ij} are the contravariant components of the vector $\mathbf{T}^i$. Substituting this relation into Eq. (4.21) yields the **Cauchy stress formula**

$$\boxed{\mathbf{T}^{(\mathbf{n})} = \mathbf{n} \cdot \boldsymbol{\sigma}} \tag{4.23}$$

where

$$\boxed{\boldsymbol{\sigma} = \sigma^{ij}\, \mathbf{g}_i \mathbf{g}_j.} \tag{4.24}$$

Because $\mathbf{T}^{(\mathbf{n})}$ is a true traction vector, $\boldsymbol{\sigma}$ is called the **true stress tensor** or the **Cauchy stress tensor**. Of course, the Cauchy stress components with respect to any basis can be obtained by double-dotting $\boldsymbol{\sigma}$ with the appropriate dyad.

To compute physical components of stress, we recall that the true traction vector $\hat{\mathbf{T}}^i$ defined by Eq. (4.15) has the physical meaning of force per unit deformed (actual) area. Thus, the physical stress components $\hat{\sigma}^{ij}$ are defined by

$$\hat{\mathbf{T}}^i = \hat{\sigma}^{ij} \mathbf{e}_j \tag{4.25}$$

where $\mathbf{e}_j$ is the unit vector along $\mathbf{g}_j$. Inserting Eqs. (4.22) and (4.25) into (4.20) gives (since $\mathbf{e}_j = \mathbf{g}_j / \sqrt{g_{(jj)}}$)

$$\sigma^{ij} \mathbf{g}_j = \sqrt{g^{(ii)}}\, \hat{\sigma}^{ij} \mathbf{e}_j = \sqrt{g^{(ii)}}\, \hat{\sigma}^{ij} \left(\frac{\mathbf{g}_j}{\sqrt{g_{(jj)}}} \right),$$

which implies

$$\boxed{\hat{\sigma}^{ij} = \sqrt{\frac{g_{(jj)}}{g^{(ii)}}}\, \sigma^{ij}.} \tag{4.26}$$

4.2.2 *Pseudostress Tensors*

Since it is referred to the deformed body, the Cauchy stress tensor is an appropriate stress measure to use in an Eulerian formulation. In a Lagrangian formulation, on the other hand, there often are advantages to working with a pseudostress tensor referred to the undeformed, or a reference, configuration. Thus, we now define two commonly used pseudostress tensors based on the pseudotraction vectors $\mathbf{T}^{(\mathrm{N})}$ and $\tilde{\mathbf{T}}^{(\mathrm{N})}$ of Eqs. (4.4) and (4.6).

In analogy with Eq. (4.23), the **first Piola-Kirchhoff stress tensor** t and the **second Piola-Kirchhoff stress tensor** s are defined through the relations

$$\boxed{\begin{aligned} \mathbf{T}^{(\mathrm{N})} &= \mathbf{N}\cdot\mathrm{t} \\ \tilde{\mathbf{T}}^{(\mathrm{N})} &= \mathbf{N}\cdot\mathrm{s} \end{aligned}} \tag{4.27}$$

where $\mathbf{N}$ is the unit normal to the undeformed area element $d\mathbf{A}$ that deforms into the area element $d\mathrm{a}$ with normal $\mathbf{n}$ (Fig. 4.1). Substituting Eqs. (4.23) and (4.27) into (4.7) yields

$$d\mathbf{P} = \mathbf{n}\cdot\boldsymbol{\sigma}\; da = \mathbf{N}\cdot\mathrm{t}\; dA = \mathbf{N}\cdot\mathrm{s}\cdot\mathbf{F}^T\; dA, \tag{4.28}$$

and using Eqs. (4.2) gives

$$\boxed{d\mathbf{P} = d\mathrm{a}\cdot\boldsymbol{\sigma} = d\mathbf{A}\cdot\mathrm{t} = d\mathbf{A}\cdot\mathrm{s}\cdot\mathbf{F}^T.} \tag{4.29}$$

The undeformed and deformed areas are related by Eq. (3.128), which gives

$$d\mathrm{a}\cdot\boldsymbol{\sigma} = (J\; d\mathbf{A}\cdot\mathbf{F}^{-1})\cdot\boldsymbol{\sigma},$$

and so Eq. (4.29) implies

$$J\,\mathbf{F}^{-1}\cdot\boldsymbol{\sigma} = \mathrm{t} = \mathrm{s}\cdot\mathbf{F}^T$$

or

$$\boxed{\boldsymbol{\sigma} = J^{-1}\mathbf{F}\cdot\mathrm{t} = J^{-1}\mathbf{F}\cdot\mathrm{s}\cdot\mathbf{F}^T.} \tag{4.30}$$

This relation links the three stress tensors through the deformation gradient tensor. Solving for the pseudostress tensors yields

$$\boxed{\begin{aligned} \mathrm{t} &= J\,\mathbf{F}^{-1}\cdot\boldsymbol{\sigma} = \mathrm{s}\cdot\mathbf{F}^T \\ \mathrm{s} &= J\,\mathbf{F}^{-1}\cdot\boldsymbol{\sigma}\cdot\mathbf{F}^{-T} = \mathrm{t}\cdot\mathbf{F}^{-T}. \end{aligned}} \tag{4.31}$$

Which stress tensor to use in formulating a given mechanics problem is a matter of convenience and personal preference. Each has both advantages and disadvantages, which will become clear later. Later, we will show that $\boldsymbol{\sigma}$ is a symmetric tensor (Section 4.6.2), i.e., $\boldsymbol{\sigma} = \boldsymbol{\sigma}^T$, and so Eq. $(4.31)_2$ gives

$$\begin{aligned} \mathrm{s}^T &= J(\mathbf{F}^{-1}\cdot\boldsymbol{\sigma}\cdot\mathbf{F}^{-T})^T = J(\mathbf{F}^{-T})^T\cdot(\mathbf{F}^{-1}\cdot\boldsymbol{\sigma})^T \\ &= J\mathbf{F}^{-1}\cdot(\boldsymbol{\sigma}^T\cdot\mathbf{F}^{-T}) = J\mathbf{F}^{-1}\cdot\boldsymbol{\sigma}\cdot\mathbf{F}^{-T} \\ &= \mathrm{s}. \end{aligned}$$

Thus, s also is a symmetric tensor. However, because $\mathbf{F}$ is generally not symmetric, Eq. $(4.31)_1$ shows that t is not symmetric in general. For solid mechanics problems involving large deformations, therefore, the second Piola-Kirchhoff stress tensor s often is the tensor of choice for two primary reasons: (1) It is referred to the known undeformed configuration; and (2) it is symmetric. Once s is determined, the other stress tensors can be computed from Eqs. (4.30) and (4.31). But again, we emphasize that only the physical components of the Cauchy stress tensor provide physically meaningful results.

Atluri (1984) discusses several other stress tensors that are useful in certain situations. Equations (4.31), for example, suggest defining another stress tensor $\boldsymbol{\tau} = J\boldsymbol{\sigma}$, which is called simply the **Kirchhoff stress tensor**, but we do not use $\boldsymbol{\tau}$ in this book. For small displacement, all of these tensors are essentially equal to the linear stress tensor described in Appendix A. Indeed, Eq. (4.30) shows that $\boldsymbol{\sigma} \cong \mathrm{t} \cong \mathrm{s}$ when $J \to 1$ and $\mathbf{F} \to \mathbf{I}$. We note, however, that since $\mathbf{F}$ contains both deformation and rigid-body rotation, stipulating small strain alone is not a sufficient condition for this equivalency.

4.3 Relations Between Stress Components

The components of stress can be a source of confusion in learning nonlinear elasticity. Different authors use different notations, different coordinate bases, and different derivations. Often a certain set of stress components is referred to as a "stress tensor," suggesting that other types of components of the actual stress tensor do not exist. Working with the stress tensors in dyadic form alleviates much of this confusion.

Like any tensor, the stress tensors can be expressed in terms of components relative to any dyadic basis, with the corresponding set of components defining the state of stress at a point in a body. Our choice of basis includes dyads composed of the natural base vectors $\mathbf{G}_I$, $\mathbf{g}_i$, or $\mathbf{g}_I$ and their reciprocals, with some stress components offering certain advantages over others. Here, only the most commonly used stress components are considered, which happen to be contravariant components of the three stress tensors. When desired, covariant and mixed components can be computed by lowering superscripts in the usual manner.

Because the Cauchy stress tensor $\boldsymbol{\sigma}$ is referred to the deformed configuration, it seems natural to represent this tensor in terms of the base vectors $\mathbf{g}_i$ or $\mathbf{g}_I$ in the deformed body. Likewise, since $\mathbf{t}$ and $\mathbf{s}$ are referred to the undeformed body, expressing these tensors in terms of the undeformed base vectors $\mathbf{G}_I$ is warranted. Moreover, Eq. (4.30) indicates that $\mathbf{t}$ falls between $\boldsymbol{\sigma}$ and $\mathbf{s}$ in some respects. Thus, components of $\mathbf{t}$ relative to mixed dyads of undeformed and deformed base vectors also are useful.

With these considerations in mind, we examine stress components given by the following representations:

$$\boxed{\begin{aligned} \boldsymbol{\sigma} &= \sigma^{ij}\mathbf{g}_i\mathbf{g}_j = \sigma^{I^*J^*}\mathbf{g}_I\mathbf{g}_J \\ \mathbf{t} &= t^{IJ}\mathbf{G}_I\mathbf{G}_J = \mathbf{t}^{Ij}\mathbf{G}_I\mathbf{g}_j = \mathbf{t}^{IJ^*}\mathbf{G}_I\mathbf{g}_J \\ \mathbf{s} &= s^{IJ}\mathbf{G}_I\mathbf{G}_J. \end{aligned}} \tag{4.32}$$

Table 4.1 lists the notations used for stress components in two other popular texts. Relations between the various stress components can be found by substituting Eqs. (3.22) and (4.32) into (4.30). For example, we can write

$$\begin{aligned} \boldsymbol{\sigma} &= \sigma^{ij}\mathbf{g}_i\mathbf{g}_j = J^{-1}\mathbf{F}\cdot\mathbf{t} \\ &= J^{-1}\left(F^i_J\mathbf{g}_i\mathbf{G}^J\right)\cdot\left(t^{Ij}\mathbf{G}_I\mathbf{g}_j\right) \\ &= J^{-1}F^i_J t^{Ij}\delta^J_I\mathbf{g}_i\mathbf{g}_j \\ &= J^{-1}F^i_I t^{Ij}\mathbf{g}_i\mathbf{g}_j \end{aligned} \tag{4.33}$$

which implies

$$\sigma^{ij} = J^{-1}F^i_I\, t^{Ij}.$$

Table 4.1 Stress components of other authors

Current	Green & Zerna (1968)	Eringen (1962)
σ^{ij}		t^{ij}
$\sigma^{I^*J^*}$	τ^{ij}	
t^{IJ}	t^{ij}	
s^{IJ}	s^{ij}	T^{IJ}
σ^{I^*J}	π^{ij}	
t^{Ij}		T^{Ij}

Similarly,

$$\begin{aligned}\boldsymbol{\sigma} &= \sigma^{ij}\mathbf{g}_i\mathbf{g}_j = J^{-1}\mathbf{F}\cdot\mathbf{s}\cdot\mathbf{F}^T \\ &= J^{-1}\left(F^i_K\mathbf{g}_i\mathbf{G}^K\right)\cdot\left(s^{IJ}\mathbf{G}_I\mathbf{G}_J\right)\cdot\left(F^j_L\mathbf{G}^L\mathbf{g}_j\right) \\ &= J^{-1}F^i_KF^j_Ls^{IJ}\delta^K_I\delta^L_J\mathbf{g}_i\mathbf{g}_j \\ &= J^{-1}F^i_IF^j_Js^{IJ}\mathbf{g}_i\mathbf{g}_j \end{aligned} \tag{4.34}$$

gives

$$\sigma^{ij} = J^{-1}F^i_IF^j_J\,s^{IJ}.$$

Combining these expressions yields

$$\boxed{\sigma^{ij} = J^{-1}F^i_I\,t^{Ij} = J^{-1}F^i_IF^j_J\,s^{IJ},} \tag{4.35}$$

which relates the stress components used by Eringen (1962; 1980) (see Table 4.1).

Relations for the convected stress components $\sigma^{I^*J^*}$ can be found by setting $x^i = X^I$ in Eq. (4.35), but it is not clear where to place asterisks. One way to clarify matters is to substitute Eq. (4.35) into the dyadic representations $(4.32)_1$. This procedure gives, for example,

$$\begin{aligned}\boldsymbol{\sigma} &= \sigma^{ij}\mathbf{g}_i\mathbf{g}_j = J^{-1}F^i_I\,t^{Ij}\mathbf{g}_i\mathbf{g}_j \\ &= J^{-1}F^I_K\,t^{KJ^*}\mathbf{g}_I\mathbf{g}_J\end{aligned}$$

where the dummy index I has been changed to K, i and j have been set to I and J ($x^i = X^I$), and the asterisk indicates that the superscript J is attached to the base vector $\mathbf{g}_J$. Since $F^I_K = \partial X^I/\partial X^K = \delta^I_K$ by Eq. (3.15), the above expression can be written

$$\begin{aligned}\boldsymbol{\sigma} = \sigma^{I^*J^*}\mathbf{g}_I\mathbf{g}_J &= J^{-1}\delta^I_K t^{KJ^*}\mathbf{g}_I\mathbf{g}_J \\ &= J^{-1}t^{IJ^*}\mathbf{g}_I\mathbf{g}_J\end{aligned}$$

which implies

$$\sigma^{I^*J^*} = J^{-1}\,t^{IJ^*}.$$

Perhaps a more straightforward (and safer) method is to substitute $\mathbf{F} = \mathbf{g}_K\mathbf{G}^K$ from Eq. $(3.22)_1$, along with Eqs. (4.32), into (4.30) to get

$$\begin{aligned}\boldsymbol{\sigma} = \sigma^{I^*J^*}\mathbf{g}_I\mathbf{g}_J &= J^{-1}\mathbf{F}\cdot\mathbf{t} \\ &= J^{-1}\left(\mathbf{g}_K\mathbf{G}^K\right)\cdot\left(t^{IJ^*}\mathbf{G}_I\mathbf{g}_J\right) \\ &= J^{-1}t^{IJ^*}\delta^K_I\,\mathbf{g}_K\mathbf{g}_J \\ &= J^{-1}t^{IJ^*}\mathbf{g}_I\mathbf{g}_J\end{aligned}$$

which agrees with the expression derived above. Similarly,

$$\begin{aligned}\boldsymbol{\sigma} = \sigma^{I^*J^*}\mathbf{g}_I\mathbf{g}_J &= J^{-1}\mathbf{F}\cdot\mathbf{s}\cdot\mathbf{F}^T \\ &= J^{-1}\left(\mathbf{g}_K\mathbf{G}^K\right)\cdot\left(s^{IJ}\mathbf{G}_I\mathbf{G}_J\right)\cdot\left(\mathbf{G}^L\mathbf{g}_L\right) \\ &= J^{-1}s^{IJ}\delta^K_I\,\delta^L_J\mathbf{g}_K\mathbf{g}_L \\ &= J^{-1}s^{IJ}\mathbf{g}_I\mathbf{g}_J\end{aligned}$$

gives

$$\sigma^{I^*J^*} = J^{-1}s^{IJ}.$$

Combining these relations yields

$$\boxed{\sigma^{I^*J^*} = J^{-1}\,t^{IJ^*} = J^{-1}\,s^{IJ},} \tag{4.36}$$

which indicates that $t^{IJ^*} = s^{IJ}$. Again, working with dyadics is recommended to avoid confusion.

Comparing Eqs. (4.35) and (4.36) reveals that the relations between stress components simplify when convected base vectors are used. Most of the finite elasticity (and biomechanics) literature, however, uses expressions of the form (4.35), sacrificing the added complexity for the convenience of describing the deformed configuration in terms of an independent coordinate system (x^i). Furthermore, we see that calling stress components "tensors" can be misleading; σ^{ij}

certainly is quite different from $\sigma^{I^*J^*}$ in general. In fact, for an incompressible material ($J = 1$), Eq. (4.36) gives $\sigma^{I^*J^*} = t^{IJ^*} = s^{IJ}$, suggesting that all of these stress "tensors" are the same. Clearly, this is not true for the *actual* stress tensors $\boldsymbol{\sigma}$, $\mathbf{t}$, and $\mathbf{s}$.

Finally, the relation between s^{IJ} and t^{IJ} is derived. Substituting Eqs. $(3.22)_1$ and $(4.32)_3$ into $(4.31)_1$ yields

$$\begin{aligned}
\mathbf{t} &= \mathbf{s}\cdot\mathbf{F}^T \\
&= \left(s^{IJ}\mathbf{G}_I\mathbf{G}_J\right)\cdot\left(\mathbf{G}^K\mathbf{g}_K\right) \\
&= s^{IJ}\delta_J^K\mathbf{G}_I\mathbf{g}_K \\
&= s^{IJ}\mathbf{G}_I\mathbf{g}_{J,}
\end{aligned}$$

and extracting the components gives

$$\begin{aligned}
t^{IJ} &= \mathbf{G}^I\cdot\mathbf{t}\cdot\mathbf{G}^J \\
&= \mathbf{G}^I\cdot\left(s^{KL}\mathbf{G}_K\mathbf{g}_L\right)\cdot\mathbf{G}^J \\
&= s^{KL}\delta_K^I\mathbf{g}_L\cdot\mathbf{G}^J \\
&= s^{IL}\mathbf{g}_L\cdot\mathbf{G}^J
\end{aligned}$$

or

$$t^{IJ} = s^{IK}\left(\mathbf{g}_K\cdot\mathbf{G}^J\right). \tag{4.37}$$

This expression can be written in another form by noting that $\mathbf{r} = \mathbf{R} + \mathbf{u}$, and so Eqs. (3.10) and (3.11) give

$$\begin{aligned}
\mathbf{g}_I = \mathbf{r}_{,I} &= \mathbf{R}_{,I} + \mathbf{u}_{,I} = \mathbf{G}_I + \mathbf{u}_{,I} \\
&= \mathbf{G}_I + \left(u^K\mathbf{G}_K\right)_{,I} \\
&= \mathbf{G}_I + u^K|_I\mathbf{G}_K
\end{aligned}$$

where the covariant derivative comes from Eq. (2.141). Thus,

$$\begin{aligned}
\mathbf{g}_I\cdot\mathbf{G}^J &= \left(\mathbf{G}_I + u^K|_I\mathbf{G}_K\right)\cdot\mathbf{G}^J \\
&= \delta_I^J + u^K|_I\delta_K^J \\
&= \delta_I^J + u^J|_I
\end{aligned}$$

and Eq. (4.37) becomes

$$\boxed{t^{IJ} = s^{IK}(\delta_K^J + u^J|_K).} \tag{4.38}$$

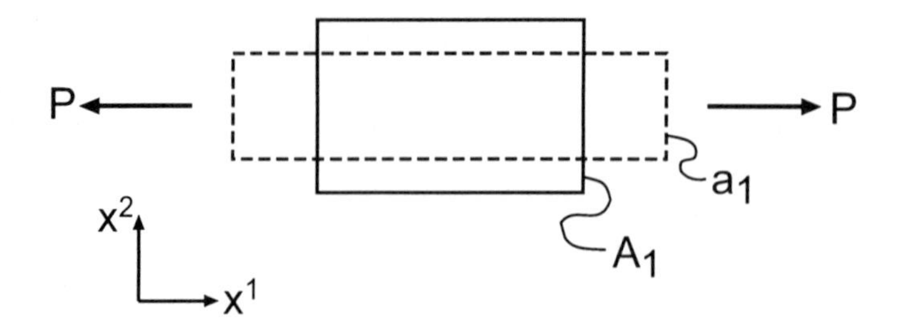

Fig. 4.4 Extension of a rectangular block.

Example 4.1 Consider uniform extension of a rectangular block of tissue due to an applied load P in the Cartesian x^1-direction (Fig. 4.4). The face on which the load acts has cross-sectional areas A_1 and a_1 before and after deformation, respectively, and the stretch ratios in the coordinate directions are λ_1, λ_2, and λ_3. If the load is distributed uniformly over the cross section, compute the following:

(a) The relation between the Cauchy (true) stress σ^{11} and the first Piola-Kirchhoff (engineering) stress t^{11} in terms of the stretch ratios. Do this first using direct geometrical analysis and then using Eq. (4.35).
(b) The relation between the three stress components σ^{11}, t^{11}, and s^{11} in terms of λ_1 only, if the block is composed of incompressible material.

Solution. (a) By definition, the Cauchy stress is

$$\sigma^{11} = P/a_1, \tag{4.39}$$

and the first Piola-Kirchhoff stress is

$$t^{11} = P/A_1. \tag{4.40}$$

According to the geometry,

$$a_1/A_1 = \lambda_2\lambda_3$$

where λ_2 and λ_3 are stretch ratios. Combining these equations gives

$$\sigma^{11} = \frac{A_1}{a_1}t^{11} = \frac{t^{11}}{\lambda_2\lambda_3}. \tag{4.41}$$

Next, we derive this relation from Eq. (4.35). For uniform extension, Eq. (3.45) gives

$$[F^i_I] = \text{diag}\,[\lambda_1, \lambda_2, \lambda_3], \tag{4.42}$$

and so Eq. (4.35) yields

$$\sigma^{11} = J^{-1} F^1_1 t^{11} = \frac{1}{\lambda_1 \lambda_2 \lambda_3} \cdot \lambda_1 t^{11} = \frac{t^{11}}{\lambda_2 \lambda_3}$$

which agrees with (4.41).

(b) Equation (4.35) gives the second Piola-Kirchhoff stress

$$\begin{aligned} s^{11} &= J(F^1_1)^{-2} \sigma^{11} = \frac{\lambda_1 \lambda_2 \lambda_3}{\lambda_1^2}\, \sigma^{11} \\ &= \frac{\lambda_2 \lambda_3}{\lambda_1}\, \sigma^{11}. \end{aligned} \tag{4.43}$$

If the block is incompressible, then

$$\det[F^i_I] = \lambda_1 \lambda_2 \lambda_3 = 1.$$

Thus, $\lambda_2 \lambda_3 = \lambda_1^{-1}$ and Eqs. (4.41) and (4.43) give

$$\sigma^{11} = \lambda_1 t^{11} = \lambda_1^2 s^{11}. \tag{4.44}$$

■

Example 4.2 An **aneurysm** is a local bulge in the wall of a pressurized shell-like structure. Regional weakening of the wall due to injury or disease can lead to aneurysms in the heart and arteries. In the left ventricle, for example, an aneurysm may develop after a myocardial infarction. Typically, an infarction is caused by blockage of a coronary artery that cuts off blood flow to the myocardium downstream of the blockage, resulting in a region of dead muscle. Because the affected muscle no longer contracts, it may bulge outward during systole. As the infarct heals, the dead muscle is replaced by relatively stiff scar tissue. If the infarct is small, the bulging may disappear during healing. In large transmural infarcts, however, the bulge may transform into a permanent aneurysm with potentially life-threatening consequences. Although it has not been proven, abnormally high wall stress may play a role in aneurysm formation.

Consider a model for a relatively soft aneurysm consisting of a thick-walled hemispherical shell that is fixed around its base (Fig. 4.5). When subjected to an

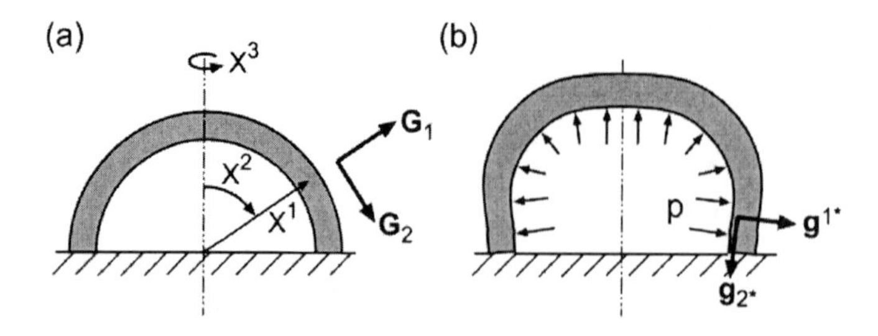

Fig. 4.5 Inflation of a hemispherical dome. (a) Undeformed configuration. (b) Deformed configuration.

internal pressure p, the shell undergoes large deformation. The boundary conditions for this problem consist of zero displacement and rotation at the fixed base, zero traction on the outer surface, and a specified pressure on the inner surface. Write the boundary condition for the inner surface.

Solution. In setting up the pressure boundary condition, it is important to note that the pressure exerts a force per unit area normal to the *deformed* surface (Fig. 4.5b). Thus, the force on an element da of the deformed inner surface is given by

$$d\mathbf{P} = -(p\,da)\mathbf{n} \tag{4.45}$$

where $\mathbf{n}$ is the outward-directed unit normal to da. The boundary condition can be written in terms of any of the stress tensors. But since the pressure follows the surface, it is convenient to express the condition in terms of the convected Cauchy stress components $\sigma^{I^*J^*}$, which then can be transformed if desired. Equations (4.28) and (4.45) give the boundary condition

$$\text{On } a: \qquad \mathbf{n}\cdot\boldsymbol{\sigma} = -p\,\mathbf{n}, \tag{4.46}$$

which next is written in terms of the $\sigma^{I^*J^*}$.

Suppose a point in the undeformed shell is located at the spherical polar coordinates X^I, where X^1, X^2, and X^3 are the radial, meridional, and circumferential coordinates, respectively (Fig. 4.5a). During the deformation, the undeformed base vectors $\mathbf{G}_I$ are convected to the base vectors $\mathbf{g}_I$, with $\mathbf{g}_{2^*}$ following the changing contour of the shell meridian (Fig. 4.5b).

The normal to the inner surface $\mathbf{n}$ points along the vector $-\mathbf{g}^{1^*}$, since it is normal to both $\mathbf{g}_{2^*}$ and $\mathbf{g}_{3^*}$ (Fig. 4.5b). Thus, $\mathbf{n} = -\mathbf{g}^{1^*}/\sqrt{g^{1^*1^*}}$ and Eq. (4.46) gives

$$\left(\frac{-\mathbf{g}^{1^*}}{\sqrt{g^{1^*1^*}}}\right)\cdot\left(\sigma^{I^*J^*}\mathbf{g}_I\mathbf{g}_J\right)=-p\left(\frac{-\mathbf{g}^{1^*}}{\sqrt{g^{1^*1^*}}}\right)$$

or

$$\sigma^{I^*J^*}\delta_I^{1^*}\mathbf{g}_J=\sigma^{1^*J^*}\mathbf{g}_J=-p\,\mathbf{g}^{1^*}.$$

Dotting both sides of this expression with $\mathbf{g}^K$ yields the boundary condition

$$\text{On } a:\quad \sigma^{1^*K^*}=-p\,g^{1^*K}\qquad\qquad (K=1,2,3)\tag{4.47}$$

which provides conditions on $\sigma^{1^*1^*}$, $\sigma^{1^*2^*}$, and $\sigma^{1^*3^*}$, i.e., on the components of the pseudotraction vector $\mathbf{T}^1$ acting on the inner surface [see Eq. (4.22)]. ■

4.4 Physical Interpretation of Stress Components

Recall that the stress components are the components of the traction vector, which depends on the orientation of the plane on which it acts. Consider, for example, the pseudotraction vector

$$\mathbf{T}^i=\sigma^{ij}\mathbf{g}_j,$$

which acts on an area element with a normal oriented in the direction of $\mathbf{g}^i$.[2] This equation shows that the first index of σ^{ij} corresponds to the plane (orthogonal to $\mathbf{g}^i$) on which the stress acts, and the second index corresponds to the direction of action ($\mathbf{g}_j$). Interpreting other types of stress components follows similar reasoning.

Consider now the dyadic representations

$$\boldsymbol{\sigma}=\sigma^{ij}\mathbf{g}_i\mathbf{g}_j=\sigma_{ij}\mathbf{g}^i\mathbf{g}^j=\sigma^i_{\cdot j}\mathbf{g}_i\mathbf{g}^j=\sigma_i^{\cdot j}\mathbf{g}^i\mathbf{g}_j.\tag{4.48}$$

The relation involving σ^{ij} indicates that the direction of the surface normal ($\mathbf{g}^i$) is given by raising the index of the first base vector of the dyadic, and the stress acts in the direction of the second base vector. In fact, for all components of the stress dyadic, changing the position of the index of the first base vector (up or down) gives the direction of the surface normal. Thus, σ_{ij} acts on a surface normal to $\mathbf{g}_i$ and points along $\mathbf{g}^j$, and so on.

[2]As shown in Fig. 4.3 (page 148), $d\mathbf{P}^{-i}$ acts on the element face with a normal in the direction of $-\mathbf{g}^i$, and so $d\mathbf{P}^i$ acts on a face with a normal in the direction of $\mathbf{g}^i$.

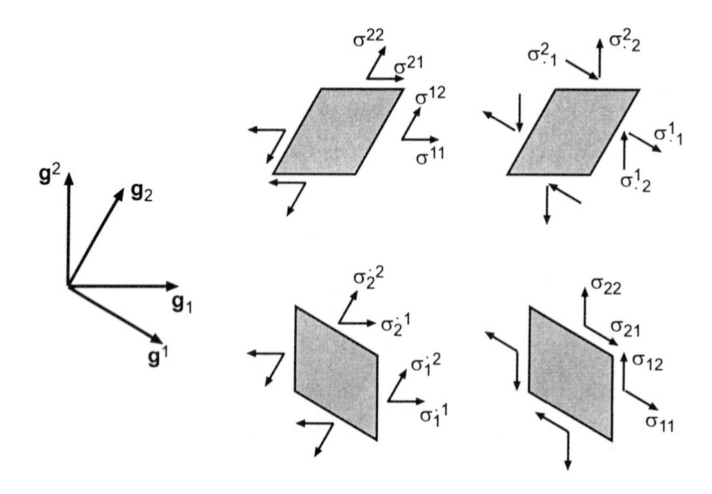

Fig. 4.6 Stress components on two-dimensional differential elements.

All four sets of stress components are illustrated for two-dimensional elements in the x^1x^2-plane (Fig. 4.6). In each case, the shape of the element is dictated by the orientations of the $\mathbf{g}_i$ for σ^{ij} and $\sigma^{i}_{\cdot j}$ and by the $\mathbf{g}^i$ for σ_{ij} and $\sigma_{i}^{\cdot j}$, i.e., the faces of the element are perpendicular to the $\mathbf{g}^i$ and $\mathbf{g}_i$, respectively. The figure suggests that the σ^{ij} may be the easiest to visualize, since the faces of the element align with the coordinate curves, and the stresses point along the coordinate directions. This is one reason we focused on contravariant stress components in the previous section.

Similar interpretations apply to components relative to the $\{\mathbf{G}_I\}$ and $\{\mathbf{g}_I\}$ bases. These alternative base vectors simply carve out differently shaped material elements. Being referred to areas and directions that are convected with the deformation, the Cauchy stress components $\sigma^{I^*J^*}$ are advantageous in visualizing the mechanics. Of course, only the physical components

$$\hat{\sigma}^{I^*J^*} = \sqrt{\frac{g_{(JJ)}}{g^{(II)}}}\,\sigma^{I^*J^*}, \tag{4.49}$$

given by modifying Eq. (4.26), have a true physical meaning.

4.5 Principal Stresses

In general, the traction vector acting on an area element in a deformed body does not align with the normal to the area. If these vectors do align, then no shear stresses act across the area. At any point in a deformed body, it is always possible to find a unique set of three mutually orthogonal surfaces on which the shear stress components vanish. These surfaces are called **principal planes**, the normals to these surfaces define the **principal axes of stress**, and the normal stress components acting on these surfaces are called **principal stresses**.

Consider a plane with unit normal $\mathbf{n}$ passing through a point p in a deformed body. If the traction vector $\mathbf{T}^{(\mathbf{n})}$ points in the direction of $\mathbf{n}$, then we can write

$$\mathbf{T}^{(\mathbf{n})} = \sigma \mathbf{n} \tag{4.50}$$

where σ is the normal component of stress on the plane. Because $\boldsymbol{\sigma}$ is a symmetric tensor, $\mathbf{n} \cdot \boldsymbol{\sigma} = \boldsymbol{\sigma}^T \cdot \mathbf{n} = \boldsymbol{\sigma} \cdot \mathbf{n}$, and so substituting Eq. (4.23) into (4.50) gives the eigenvalue problem

$$\boxed{(\boldsymbol{\sigma} - \sigma \mathbf{I}) \cdot \mathbf{n} = \mathbf{0}.} \tag{4.51}$$

Solving this equation provides the principal Cauchy stresses σ_i and the corresponding principal directions $\mathbf{n}_i$ $(i = 1, 2, 3)$. Since $\boldsymbol{\sigma}$ is symmetric, the eigenvalues are real and the eigenvectors are mutually orthogonal.

Principal stresses, therefore, are determined by solving an eigenvalue problem similar to that used to find the principal strains. The principal axes of stress and strain, however, do not always coincide. In general anisotropic materials, for example, normal stresses alone can induce shear strains.

4.6 Equations of Motion

Every particle of a solid body, as well as the entire body itself, must obey Newton's laws of motion. Section 4.2 investigated the consequences of these laws for a volume element that shrinks to a point. That analysis led to the definition of the Cauchy stress tensor. Here, we use Newton's second law in the form of the principles of linear and angular momentum to write the global equations of motion for a body. Manipulating these relations then provides the local equations of motion for an infinitesimal element.

4.6.1 *Principle of Linear Momentum*

Consider a deformed body of mass density $\rho(\mathbf{r}, t)$ that is subjected to a traction $\mathbf{T}^{(\mathbf{n})}(\mathbf{r}, t)$ acting over its surface area a and a body force $\mathbf{f}(\mathbf{r}, t)$ acting over its volume v (Fig. 4.7). Applying the law of conservation of linear momentum to the entire body gives

$$\int_a \mathbf{T}^{(\mathbf{n})}\, da + \int_v \mathbf{f}\; dv = \frac{d}{dt}\int_v \mathbf{v}\; \rho dv \tag{4.52}$$

where $\mathbf{v}(\mathbf{r}, t)$ is the velocity of the mass center of the volume element dv. This relation represents the Eulerian form of the global equation of motion for the body, with the various quantities being functions of the deformed position $\mathbf{r}$.

Substituting Eq. (4.23) into the area integral of (4.52) and applying the divergence theorem (2.191) yields

$$\int_a \mathbf{T}^{(\mathbf{n})}\, da = \int_a \mathbf{n} \cdot \boldsymbol{\sigma}\; da = \int_v \bar{\boldsymbol{\nabla}} \cdot \boldsymbol{\sigma}\; dv \tag{4.53}$$

where $\bar{\boldsymbol{\nabla}}$ is the gradient operator in the deformed body b, as defined by Eq. $(3.18)_2$. In addition, carrying out the differentiation on the right-hand-side of Eq. (4.52) gives

$$\begin{aligned} \frac{d}{dt}\int_v \mathbf{v}\; \rho dv &= \int_v \frac{d}{dt}(\mathbf{v}\; \rho dv) \\ &= \int_v \left[\frac{d\mathbf{v}}{dt}\rho dv + \mathbf{v}\frac{d}{dt}(\rho dv)\right] \\ &= \int_v \mathbf{a}\; \rho dv \end{aligned} \tag{4.54}$$

in which $d(\rho\, dv)/dt = 0$ by conservation of mass and $\mathbf{a} = d\mathbf{v}/dt$ is the acceleration of the element dv. With these expressions, Eq. (4.52) can be written

$$\int_v (\bar{\boldsymbol{\nabla}} \cdot \boldsymbol{\sigma} + \mathbf{f} - \rho\mathbf{a})\; dv = \mathbf{0}. \tag{4.55}$$

Finally, because it must hold for an arbitrary volume in the deformed body, this relation implies the local equation of motion

$$\boxed{\bar{\boldsymbol{\nabla}} \cdot \boldsymbol{\sigma} + \mathbf{f} = \rho\mathbf{a}} \tag{4.56}$$

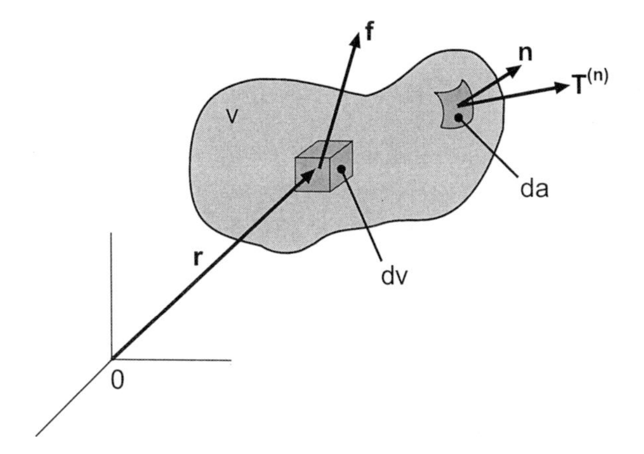

Fig. 4.7 Forces acting on deformed body.

where the acceleration $\mathbf{a}(\mathbf{r}, t)$ is provided by Eq. (3.196)$_2$. If inertia effects can be neglected, then we can set $\mathbf{a} = \mathbf{0}$, and Eq. (4.56) becomes an equilibrium equation.

4.6.2 *Principle of Angular Momentum*

Equation (4.52) governs the linear motion of a body. The rotational motion must satisfy the law of conservation of angular momentum

$$\int_a \mathbf{r} \times \mathbf{T}^{(\mathbf{n})}\, da + \int_v \mathbf{r} \times \mathbf{f}\ dv = \frac{d}{dt} \int_v (\mathbf{r} \times \mathbf{v})\, \rho dv. \tag{4.57}$$

To set the stage for converting the surface integral of Eq. (4.57) into a volume integral, we first note that the Cauchy stress tensor can be written in the form

$$\boldsymbol{\sigma} = \sigma^{ij} \mathbf{g}_i \mathbf{g}_j = \mathbf{g}_i (\sigma^{ij} \mathbf{g}_j) = \mathbf{g}_i \mathbf{T}^i$$

where Eq. (4.22) has been used. This relation and Eq. (4.23) give

$$\begin{aligned} \mathbf{r} \times \mathbf{T}^{(\mathbf{n})} &= \mathbf{r} \times (\mathbf{n} \cdot \boldsymbol{\sigma}) = \mathbf{r} \times (\mathbf{n} \cdot \mathbf{g}_i \mathbf{T}^i) = (\mathbf{n} \cdot \mathbf{g}_i)\, \mathbf{r} \times \mathbf{T}^i \\ &= \mathbf{n} \cdot (\mathbf{g}_i \mathbf{r} \times \mathbf{T}^i) \end{aligned}$$

in which $\mathbf{n} \cdot \mathbf{g}_i$ could be moved because it is a scalar. With this equation and the divergence theorem (2.191), the first term in Eq. (4.57) becomes

$$\int_a \mathbf{r} \times \mathbf{T}^{(\mathbf{n})}\, da = \int_a \mathbf{n} \cdot (\mathbf{g}_i \mathbf{r} \times \mathbf{T}^i)\, da = \int_v \bar{\boldsymbol{\nabla}} \cdot (\mathbf{g}_i \mathbf{r} \times \mathbf{T}^i)\, dv. \tag{4.58}$$

Next, differentiating the integrand, setting $\mathbf{r}_{,i} = \mathbf{g}_i$, and using $(2.128)_2$ yields

$$\begin{aligned}
\bar{\boldsymbol{\nabla}} \cdot (\mathbf{g}_i \mathbf{r} \times \mathbf{T}^i) &= \left(\mathbf{g}^k \frac{\partial}{\partial x^k}\right) \cdot (\mathbf{g}_i \mathbf{r} \times \mathbf{T}^i) \\
&= \mathbf{g}^k \cdot (\mathbf{g}_{i,k}\, \mathbf{r} \times \mathbf{T}^i + \mathbf{g}_i \mathbf{r}_{,k} \times \mathbf{T}^i + \mathbf{g}_i \mathbf{r} \times \mathbf{T}^i{}_{,k}) \\
&= \Gamma^k_{ik}\, \mathbf{r} \times \mathbf{T}^i + \delta^k_i \mathbf{g}_k \times \mathbf{T}^i + \delta^k_i \mathbf{r} \times \mathbf{T}^i{}_{,k} \\
&= \Gamma^k_{ik}\, \mathbf{r} \times \mathbf{T}^i + \mathbf{g}_i \times \mathbf{T}^i + \mathbf{r} \times \mathbf{T}^i{}_{,i},
\end{aligned}$$

which can be simplified by noting that

$$\begin{aligned}
\bar{\boldsymbol{\nabla}} \cdot \boldsymbol{\sigma} &= \left(\mathbf{g}^k \frac{\partial}{\partial x^k}\right) \cdot (\mathbf{g}_i \mathbf{T}^i) \\
&= \mathbf{g}^k \cdot (\mathbf{g}_{i,k}\, \mathbf{T}^i + \mathbf{g}_i T^i{}_{,k}) \\
&= \Gamma^k_{ik}\, \mathbf{T}^i + \delta^k_i \mathbf{T}^i{}_{,k} \\
&= \Gamma^k_{ik}\, \mathbf{T}^i + \mathbf{T}^i{}_{,i}.
\end{aligned}$$

Combining these expressions shows that Eq. (4.58) can be written

$$\int_a \mathbf{r} \times \mathbf{T}^{(\mathbf{n})}\, da = \int_v [\mathbf{g}_i \times \mathbf{T}^i + \mathbf{r} \times (\bar{\boldsymbol{\nabla}} \cdot \boldsymbol{\sigma})]\, dv. \tag{4.59}$$

In addition, the right-hand-side of Eq. (4.57) is

$$\begin{aligned}
\frac{d}{dt} \int_v (\mathbf{r} \times \mathbf{v})\, \rho\, dv &= \int_v \frac{d}{dt}[(\mathbf{r} \times \mathbf{v})\, \rho\, dv] \\
&= \int_v \left[\rho\, dv\, \frac{d}{dt}(\mathbf{r} \times \mathbf{v}) + (\mathbf{r} \times \mathbf{v})\, \frac{d}{dt}(\rho\, dv)\right] \\
&= \int_v \left[\rho\, dv\, (\dot{\mathbf{r}} \times \mathbf{v} + \mathbf{r} \times \dot{\mathbf{v}}) + (\mathbf{r} \times \mathbf{v})\, \frac{d}{dt}(\rho\, dv)\right] \\
&= \int_v (\mathbf{r} \times \mathbf{a})\, \rho\, dv.
\end{aligned} \tag{4.60}$$

Here, we have noted that $\dot{\mathbf{r}} \times \mathbf{v} = \mathbf{0}$ since $\dot{\mathbf{r}} = \mathbf{v}$, and $d(\rho\, dv)/dt = 0$ by conservation of mass.

Substituting Eqs. (4.59) and (4.60) into (4.57) and rearranging yields

$$\int_v [\mathbf{g}_i \times \mathbf{T}^i + \mathbf{r} \times (\bar{\boldsymbol{\nabla}} \cdot \boldsymbol{\sigma} + \mathbf{f} - \rho \mathbf{a})]\, dv = \mathbf{0}. \tag{4.61}$$

The term in parentheses vanishes by the equation of motion (4.56). Thus, because this relation must hold for an arbitrary volume, the principle of angular momentum reduces to

$$\mathbf{g}_i \times \mathbf{T}^i = \mathbf{0}. \tag{4.62}$$

The consequence of this result is revealed by substituting Eq. (4.22) and using $(2.96)_1$ to obtain

$$\mathbf{g}_i \times \mathbf{T}^i = \mathbf{g}_i \times \sigma^{ij}\mathbf{g}_j = \epsilon_{ijk}\sigma^{ij}\mathbf{g}^k = \mathbf{0}$$

which can be written

$$\epsilon_{ijk}\sigma^{ij}\mathbf{g}^k = \mathbf{0} \qquad \text{or} \qquad \epsilon_{jik}\sigma^{ji}\mathbf{g}^k = \mathbf{0}.$$

Finally, since $\epsilon_{jik} = -\epsilon_{ijk}$, adding these expressions yields

$$\epsilon_{ijk}(\sigma^{ij} - \sigma^{ji})\mathbf{g}^k = \mathbf{0} \tag{4.63}$$

which implies that $\sigma^{ij} = \sigma^{ji}$, i.e., *the Cauchy stress tensor is symmetric* ($\boldsymbol{\sigma} = \boldsymbol{\sigma}^T$). In Section 4.2.2, we used this fact to show that the second Piola-Kirchhoff stress tensor s also is symmetric.

4.6.3 *Lagrangian Form of Equation of Motion*

For small deformation, the gradient operator $\bar{\nabla}$ in the equation of motion (4.56) can be replaced by ∇, so that derivatives are taken with respect to the undeformed coordinates. When the deformation is large, however, derivatives must be taken with respect to the deformed coordinates, which are not known *a priori*. Because this can complicate matters considerably, it is useful to express the equations of motion for arbitrarily large deformation in a Lagrangian, or material, form that involves ∇ rather than $\bar{\nabla}$. A similar concern motivated the introduction of the pseudostress tensors in Section 4.2.2.

A relatively painless way to derive the appropriate equation is to rewrite the principle of linear momentum (4.52) in terms of Lagrangian quantities. For a body that originally occupied a volume V enclosed by a surface area A, we have

$$\int_A \mathbf{T}^{(\mathrm{N})}\, dA + \int_V \mathrm{f}_0\, dV = \frac{d}{dt}\int_V \mathbf{v}\,\rho_0 dV \tag{4.64}$$

where the surface traction $\mathbf{T}^{(\mathrm{N})}(\mathbf{R}, t)$ and body force $\mathrm{f}_0(\mathbf{R}, t)$ are defined by Eqs. (4.4) and (4.1), respectively, and $\rho_0(\mathbf{R}, t)$ is the mass density of the undeformed volume element dV. Note that, although the integrals are taken over the undeformed body, the forces actually are applied to the deformed body.

From here, the derivation follows that given in Section 4.6.1. First, with Eqs. $(4.27)_1$, and (2.191), the surface integral of (4.64) becomes

$$\int_A \mathbf{T}^{(\mathbf{N})}\, dA = \int_A \mathbf{N} \cdot \mathbf{t}\, dA = \int_V \boldsymbol{\nabla} \cdot \mathbf{t}\, dV. \tag{4.65}$$

Second, as Eq. (4.54) shows,

$$\frac{d}{dt} \int_V \mathbf{v}\, \rho_0 dV = \int_V \mathbf{a}\, \rho_0 dV \tag{4.66}$$

since $\rho_0\, dV = \rho\, dv$ is the constant mass of an element. Substituting these relations into (4.64) yields

$$\int_V (\boldsymbol{\nabla} \cdot \mathbf{t} + \mathbf{f}_0 - \rho_0 \mathbf{a})\, dV = 0, \tag{4.67}$$

which implies the local equation of motion

$$\boxed{\boldsymbol{\nabla} \cdot \mathbf{t} + \mathbf{f}_0 = \rho_0 \mathbf{a}.} \tag{4.68}$$

Although Eqs. (4.56) and (4.68) appear similar, there are significant differences. Since the derivatives are computed relative to the known undeformed coordinates, Eq. (4.68) is easier to solve in principle. Moreover, with $\mathbf{v} = \mathbf{v}(\mathbf{R}, t)$, the acceleration is given by Eq. $(3.196)_1$, rather than $(3.196)_2$. In contrast to $\boldsymbol{\sigma}$, however, the first Piola-Kirchhoff stress tensor $\mathbf{t}$ is generally not symmetric. Thus, working with the symmetric second Piola-Kirchhoff stress tensor $\mathbf{s}$ often is more convenient, and putting Eq. $(4.31)_1$ into (4.68) gives

$$\boxed{\boldsymbol{\nabla} \cdot (\mathbf{s} \cdot \mathbf{F}^T) + \mathbf{f}_0 = \rho_0 \mathbf{a}.} \tag{4.69}$$

But this equation is more complicated than Eq. (4.68). As is becoming apparent, all large-deformation formulations possess both inherent advantages and disadvantages. While the stress and strain measures of choice vary, the pseudostress tensor $\mathbf{s}$ and the Lagrangian strain tensor $\mathbf{E}$ are used often in finite elasticity problems. Besides their symmetry and being referred to the undeformed configuration, these tensors lead to a convenient form of the material constitutive relations (see Chapter 5).

Example 4.3 As we have just seen, deriving the Lagrangian form of the equation of motion from a direct application of the principle of linear momentum is straightforward. Deriving Eq. (4.68) from Eq. (4.56), or vice versa, is a more complicated task. Show that these two forms of the equations of motion are indeed equivalent.

Solution. Beginning with Eq. (4.68), we substitute Eqs. (4.1) and $(4.31)_1$ and use the relation

$$J = \frac{dv}{dV} = \frac{\rho_0}{\rho} \tag{4.70}$$

to obtain

$$J^{-1}\boldsymbol{\nabla} \cdot (J\,\mathbf{F}^{-1} \cdot \boldsymbol{\sigma}) + \mathbf{f} = \rho\mathbf{a}. \tag{4.71}$$

A glance at Eq. (4.56) reveals that our task boils down to showing that the first term in (4.71) is equivalent to $\bar{\boldsymbol{\nabla}} \cdot \boldsymbol{\sigma}$. But this is easier said than done. With $\boldsymbol{\nabla}$ given by Eq. $(3.18)_1$, expanding this term gives

$$\begin{aligned} J^{-1}\boldsymbol{\nabla} \cdot (J\,\mathbf{F}^{-1} \cdot \boldsymbol{\sigma}) &= J^{-1}\left(\mathbf{G}^I \frac{\partial}{\partial X^I}\right) \cdot (J\,\mathbf{F}^{-1} \cdot \boldsymbol{\sigma}) \\ &= J^{-1}\mathbf{G}^I \cdot (J_{,I}\,\mathbf{F}^{-1} \cdot \boldsymbol{\sigma} + J\,\mathbf{F}^{-1}_{,I} \cdot \boldsymbol{\sigma} + J\,\mathbf{F}^{-1} \cdot \boldsymbol{\sigma}_{,I}). \end{aligned} \tag{4.72}$$

In the following, we examine separately each term of this equation.

To compute $J_{,I} = \partial J/\partial X^I$ in the first term, we use Eq. (3.119), which gives $J = j\sqrt{g/G}$. Then, working in convected coordinates for convenience, we can set $j = 1$ (see Section 3.6.3) to obtain

$$\begin{aligned} J_{,I} = \left(\sqrt{\frac{g}{G}}\right)_{,I} &= \frac{\sqrt{G}(\sqrt{g})_{,I} - \sqrt{g}(\sqrt{G})_{,I}}{G} \\ &= \sqrt{\frac{g}{G}}\left[\frac{(\sqrt{g})_{,I}}{\sqrt{g}} - \frac{(\sqrt{G})_{,I}}{\sqrt{G}}\right]. \end{aligned} \tag{4.73}$$

Specializing Eq. (2.140) gives

$$\begin{aligned} (\sqrt{g})_{,I} &= \sqrt{g}\,\bar{\Gamma}^J_{JI} \\ (\sqrt{G})_{,I} &= \sqrt{G}\,\Gamma^J_{JI} \end{aligned} \tag{4.74}$$

where the Christoffel symbols are defined by Eqs. (3.85), and so Eq. (4.73) can be written

$$\boxed{J^{-1}J_{,I} = \bar{\Gamma}^{J}_{JI} - \Gamma^{J}_{JI}.} \tag{4.75}$$

With the help of Eq. $(3.22)_2$, the first term on the right-hand-side of Eq. (4.72) now becomes

$$\begin{aligned} J^{-1}\mathbf{G}^I \cdot (J_{,I}\, \mathbf{F}^{-1} \cdot \boldsymbol{\sigma}) &= J^{-1}J_{,I}\, \mathbf{G}^I \cdot (\mathbf{G}_K \mathbf{g}^K) \cdot \boldsymbol{\sigma} \\ &= J^{-1}J_{,I}\, \delta^I_K \mathbf{g}^K \cdot \boldsymbol{\sigma} \\ &= (\bar{\Gamma}^J_{JI} - \Gamma^J_{JI})\, \mathbf{g}^I \cdot \boldsymbol{\sigma}. \end{aligned} \tag{4.76}$$

The second term on the right-hand-side of Eq. (4.72) requires

$$\begin{aligned} \mathbf{F}^{-1}_{,I} &= (\mathbf{G}_K \mathbf{g}^K)_{,I} = \mathbf{G}_{K,I}\, \mathbf{g}^K + \mathbf{G}_K \mathbf{g}^K{}_{,I} \\ &= (\Gamma^J_{KI} \mathbf{G}_J)\mathbf{g}^K - \mathbf{G}_K(\bar{\Gamma}^K_{JI}\mathbf{g}^J) \end{aligned}$$

in which Eqs. (2.127) and (2.130) have provided the derivatives of the base vectors. Inserting this expression into the second term of (4.72) gives

$$\begin{aligned} J^{-1}\mathbf{G}^I \cdot (J\,\mathbf{F}^{-1}{}_{,I} \cdot \boldsymbol{\sigma}) &= \mathbf{G}^I \cdot (\Gamma^J_{KI}\mathbf{G}_J \mathbf{g}^K - \bar{\Gamma}^K_{JI}\mathbf{G}_K \mathbf{g}^J) \cdot \boldsymbol{\sigma} \\ &= (\Gamma^J_{KI}\delta^I_J \mathbf{g}^K - \bar{\Gamma}^K_{JI}\delta^I_K \mathbf{g}^J) \cdot \boldsymbol{\sigma} \\ &= (\Gamma^I_{KI}\mathbf{g}^K - \bar{\Gamma}^I_{JI}\mathbf{g}^J) \cdot \boldsymbol{\sigma} \\ &= (\Gamma^J_{IJ} - \bar{\Gamma}^J_{IJ})\mathbf{g}^I \cdot \boldsymbol{\sigma} \end{aligned} \tag{4.77}$$

where dummy indices have been redefined.

Next, using Eq. $(3.18)_2$ in the last term of (4.72) yields

$$\begin{aligned} J^{-1}\mathbf{G}^I \cdot (J\,\mathbf{F}^{-1} \cdot \boldsymbol{\sigma}_{,I}) &= \mathbf{G}^I \cdot (\mathbf{G}_K \mathbf{g}^K) \cdot \boldsymbol{\sigma}_{,I} \\ &= \delta^I_K \mathbf{g}^K \cdot \boldsymbol{\sigma}_{,I} = \mathbf{g}^I \cdot \boldsymbol{\sigma}_{,I} \\ &= \left(\mathbf{g}^I \frac{\partial}{\partial X^I}\right) \cdot \boldsymbol{\sigma} \\ &= \bar{\boldsymbol{\nabla}} \cdot \boldsymbol{\sigma}. \end{aligned} \tag{4.78}$$

Now, substituting Eqs. (4.76)–(4.78) into (4.72) gives (since $\Gamma^J_{IJ} = \Gamma^J_{JI}$)

$$J^{-1}\boldsymbol{\nabla} \cdot (J\,\mathbf{F}^{-1} \cdot \boldsymbol{\sigma}) = \bar{\boldsymbol{\nabla}} \cdot \boldsymbol{\sigma}, \tag{4.79}$$

which proves (finally!) that Eq. (4.71), or Eq. (4.68), is equivalent to Eq. (4.56). ■

4.6.4 *Component Forms of Equations of Motion*

In terms of the three stress tensors, the equations of motion (4.56), (4.68), and (4.69) are

$$\begin{aligned} \bar{\nabla} \cdot \sigma + \mathbf{f} &= \rho \mathbf{a} \\ \nabla \cdot \mathbf{t} + \mathbf{f}_0 &= \rho_0 \mathbf{a} \\ \nabla \cdot (\mathbf{s} \cdot \mathbf{F}^T) + \mathbf{f}_0 &= \rho_0 \mathbf{a}. \end{aligned} \tag{4.80}$$

When dyadic notation is used, obtaining the component forms of these equations is straightforward. Here, we write the equations in terms of the following sets of components:

$$\begin{aligned} \sigma &= \sigma^{ij} \mathbf{g}_i \mathbf{g}_j = \sigma^{I^* J^*} \mathbf{g}_I \mathbf{g}_J \\ \mathbf{f} &= f^j \mathbf{g}_j = f^{J^*} \mathbf{g}_J \\ \mathbf{a} &= a^j \mathbf{g}_j = a^{J^*} \mathbf{g}_J \end{aligned} \tag{4.81}$$

$$\begin{aligned} \mathbf{t} &= t^{IJ} \mathbf{G}_I \mathbf{G}_J \\ \mathbf{f}_0 &= f_0^J \mathbf{G}_J \\ \mathbf{a} &= a^J \mathbf{G}_J \end{aligned} \tag{4.82}$$

$$\begin{aligned} \mathbf{s} &= s^{IJ} \mathbf{G}_I \mathbf{G}_J \\ \mathbf{f}_0 &= f_0^J \mathbf{G}_J = f_0^{J^*} \mathbf{g}_J \\ \mathbf{a} &= a^J \mathbf{G}_J = a^{J^*} \mathbf{g}_J. \end{aligned} \tag{4.83}$$

The derivation requires the gradient operators

$$\begin{aligned} \bar{\nabla} &= \mathbf{g}^i \frac{\partial}{\partial x^i} = \mathbf{g}^I \frac{\partial}{\partial X^I} \\ \nabla &= \mathbf{G}^I \frac{\partial}{\partial X^I}, \end{aligned} \tag{4.84}$$

as provided by Eqs. (3.18) and the Christoffel symbols

$$\begin{aligned} \bar{\Gamma}_{ij}^k &= \mathbf{g}_{i,j} \cdot \mathbf{g}^k \\ \bar{\Gamma}_{IJ}^K &= \mathbf{g}_{I,J} \cdot \mathbf{g}^K \\ \Gamma_{IJ}^K &= \mathbf{G}_{I,J} \cdot \mathbf{G}^K \end{aligned} \tag{4.85}$$

which are given by Eqs. (3.85).

Consider first the divergences in Eqs. (4.80). For the first of these expressions, Eqs. $(4.81)_1$ and $(4.84)_1$ give the two forms

$$\begin{aligned}\bar{\nabla}\cdot\boldsymbol{\sigma} &= \left(\mathbf{g}^k\frac{\partial}{\partial x^k}\right)\cdot\boldsymbol{\sigma} = \mathbf{g}^k\cdot(\sigma^{ij}\mathbf{g}_i\mathbf{g}_j)_{,k} \\ &= \left(\mathbf{g}^K\frac{\partial}{\partial X^K}\right)\cdot\boldsymbol{\sigma} = \mathbf{g}^K\cdot(\sigma^{I^*J^*}\mathbf{g}_I\mathbf{g}_J)_{,K}\end{aligned}$$

and noting Eq. (2.144) yields

$$\begin{aligned}\bar{\nabla}\cdot\boldsymbol{\sigma} &= \mathbf{g}^k\cdot(\sigma^{ij}|_k\,\mathbf{g}_i\mathbf{g}_j) = \sigma^{ij}|_k\,\delta_i^k\mathbf{g}_j = \sigma^{ij}|_i\,\mathbf{g}_j \\ &= \mathbf{g}^K\cdot(\sigma^{I^*J^*}|_K\,\mathbf{g}_I\mathbf{g}_J) = \sigma^{I^*J^*}|_K\,\delta_I^K\,\mathbf{g}_J \\ &= \sigma^{I^*J^*}|_I\,\mathbf{g}_J. \end{aligned} \tag{4.86}$$

In these equations the vertical bar indicates covariant differentiation with respect to the coordinates in the deformed body. Adapting Eq. $(2.145)_2$ gives

$$\boxed{\begin{aligned}\sigma^{ij}|_i &= \sigma^{ij}{}_{,i} + \sigma^{kj}\bar{\Gamma}^i_{ik} + \sigma^{ik}\bar{\Gamma}^j_{ik} \\ \sigma^{I^*J^*}|_I &= \sigma^{I^*J^*}{}_{,I} + \sigma^{K^*J^*}\bar{\Gamma}^I_{IK} + \sigma^{I^*K^*}\bar{\Gamma}^J_{IK}.\end{aligned}} \tag{4.87}$$

Similarly, for the second of (4.80), Eqs. (2.144), $(4.82)_1$, and $(4.84)_2$ give

$$\begin{aligned}\nabla\cdot\mathbf{t} &= \left(\mathbf{G}^K\frac{\partial}{\partial X^K}\right)\cdot\mathbf{t} = \mathbf{G}^K\cdot(t^{IJ}\mathbf{G}_I\mathbf{G}_J)_{,K} \\ &= \mathbf{G}^K\cdot(t^{IJ}||_K\mathbf{G}_I\mathbf{G}_J) = t^{IJ}||_K\delta_I^K\mathbf{G}_J \\ &= t^{IJ}||_I\mathbf{G}_J\end{aligned} \tag{4.88}$$

where

$$\boxed{t^{IJ}||_I = t^{IJ}{}_{,I} + t^{KJ}\Gamma^I_{IK} + t^{IK}\Gamma^J_{IK}.} \tag{4.89}$$

The double vertical bar denotes covariant differentiation with respect to the coordinates in the undeformed body.

The last of Eqs. (4.80) requires a little more work. With the relation $\mathbf{F} = \mathbf{g}_I\mathbf{G}^I$,

Eqs. $(4.83)_1$ and $(4.84)_2$ give

$$\begin{aligned}
\boldsymbol{\nabla}\cdot(\mathbf{s}\cdot\mathbf{F}^T) &= \boldsymbol{\nabla}\cdot(s^{IJ}\mathbf{G}_I\mathbf{G}_J\cdot\mathbf{G}^K\mathbf{g}_K)\\
&= \boldsymbol{\nabla}\cdot(s^{IJ}\mathbf{G}_I\delta_J^K\mathbf{g}_K)\\
&= \left(\mathbf{G}^K\frac{\partial}{\partial X^K}\right)\cdot(s^{IJ}\mathbf{G}_I\mathbf{g}_J)\\
&= \mathbf{G}^K\cdot(s^{IJ},_K\mathbf{G}_I\mathbf{g}_J + s^{IJ}\mathbf{G}_{I},_K\mathbf{g}_J + s^{IJ}\mathbf{G}_I\mathbf{g}_{J},_K)\\
&= s^{IJ},_K\delta_I^K\mathbf{g}_J + s^{IJ}\mathbf{G}^K\cdot\mathbf{G}_{I},_K\mathbf{g}_J + s^{IJ}\delta_I^K\mathbf{g}_{J},_K\\
&= s^{IJ},_I\mathbf{g}_J + s^{IJ}\mathbf{G}^K\cdot\mathbf{G}_{I},_K\mathbf{g}_J + s^{IJ}\mathbf{g}_{J},_I\,.
\end{aligned}$$

Simplifying this expression using Eqs. (2.127) and $(4.85)_{2,3}$ yields

$$\boldsymbol{\nabla}\cdot(\mathbf{s}\cdot\mathbf{F}^T) = s^{IJ}|||_I\mathbf{g}_J \tag{4.90}$$

where

$$\boxed{s^{IJ}|||_I = s^{IJ},_I + s^{IJ}\Gamma_{IK}^K + s^{IK}\bar{\Gamma}_{IK}^J.} \tag{4.91}$$

Here, the triple vertical bar indicates a mixed covariant derivative with respect to both the undeformed and the deformed body.

Substituting all of these relations into Eqs. (4.80) yields the equations of motion

$$\boxed{\begin{aligned}
\sigma^{ij}|_i + f^j &= \rho a^j\\
\sigma^{I^*J^*}|_I + f^{J^*} &= \rho a^{J^*}\\
t^{IJ}||_I + f_0^J &= \rho_0 a^J\\
s^{IJ}|||_I + f_0^{J^*} &= \rho_0 a^{J^*}.
\end{aligned}} \tag{4.92}$$

An alternate equation involving s^{IJ} can be found by substituting Eq. (4.38) into the third of these relations to get

$$\boxed{[s^{IK}(\delta_K^J + u^J|_K)]|||_I + f_0^J = \rho_0 a^J.} \tag{4.93}$$

Example 4.4 Consider a cylindrical polar coordinate system with unit base vectors $\{\mathbf{e}_r, \mathbf{e}_\theta, \mathbf{e}_z\}$ (Fig. 4.8).

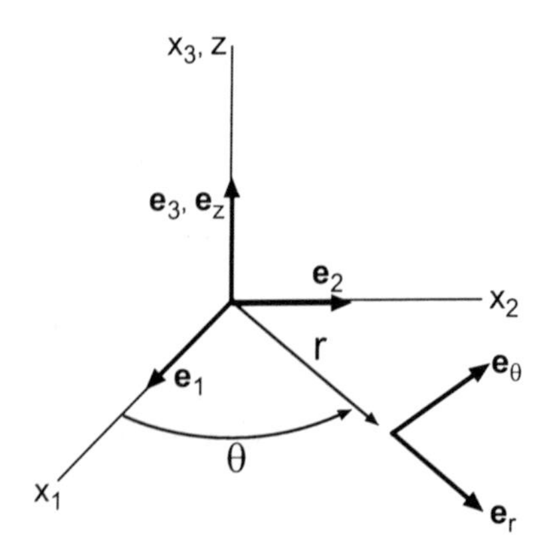

Fig. 4.8 Cylindrical polar coordinate system.

(a) Derive the equations of motion for large deformation in terms of the tensor components σ^{ij}.
(b) Show that the stress components relative to $\{\mathbf{e}_r, \mathbf{e}_\theta, \mathbf{e}_z\}$ are the physical components of the Cauchy stress tensor.
(c) Show that, when displacements are small, the equations derived in (a) are equivalent to those of the linear theory of elasticity (see Example A.7, page 364).

Solution. (a) The tensor-component form of the equation of motion, provided by Eqs. $(4.87)_1$ and $(4.92)_1$, is

$$\sigma^{ij}{}_{,i} + \sigma^{kj}\bar{\Gamma}^i_{ik} + \sigma^{ik}\bar{\Gamma}^j_{ik} + f^j = \rho a^j. \tag{4.94}$$

With $(x^1, x^2, x^3) = (r, \theta, z)$, the only nonzero Christoffel symbols in cylindrical polar coordinates are

$$\bar{\Gamma}^1_{22} = -r, \quad \bar{\Gamma}^2_{12} = \bar{\Gamma}^2_{21} = r^{-1} \tag{4.95}$$

as given by Eqs. (2.139). Hence, expanding Eq. (4.94) yields the relations

$$\begin{aligned}
\sigma^{11}{}_{,1} + \sigma^{21}{}_{,2} + \sigma^{31}{}_{,3} + \frac{1}{r}\sigma^{11} - r\sigma^{22} + f^1 &= \rho a^1 \\
\sigma^{12}{}_{,1} + \sigma^{22}{}_{,2} + \sigma^{32}{}_{,3} + \frac{3}{r}\sigma^{12} + f^2 &= \rho a^2 \\
\sigma^{13}{}_{,1} + \sigma^{23}{}_{,2} + \sigma^{33}{}_{,3} + \frac{1}{r}\sigma^{13} + f^3 &= \rho a^3.
\end{aligned} \tag{4.96}$$

(b) Relative to the base vectors $\{\mathbf{e}_r, \mathbf{e}_\theta, \mathbf{e}_z\}$, the Cauchy stress tensor can be written

$$\begin{aligned}\boldsymbol{\sigma} &= \sigma_{rr}\mathbf{e}_r\mathbf{e}_r + \sigma_{r\theta}\mathbf{e}_r\mathbf{e}_\theta + \sigma_{rz}\mathbf{e}_r\mathbf{e}_z \\ &+\sigma_{\theta r}\mathbf{e}_\theta\mathbf{e}_r + \sigma_{\theta\theta}\mathbf{e}_\theta\mathbf{e}_\theta + \sigma_{\theta z}\mathbf{e}_\theta\mathbf{e}_z \\ &+\sigma_{zr}\mathbf{e}_z\mathbf{e}_r + \sigma_{z\theta}\mathbf{e}_z\mathbf{e}_\theta + \sigma_{zz}\mathbf{e}_z\mathbf{e}_z. \end{aligned} \tag{4.97}$$

Consider now the tensor components σ^{ij}. Because cylindrical polar coordinates are orthogonal, we can set $g^{(ii)} = 1/g_{(ii)}$ and Eq. (4.26) gives the physical stress components

$$\hat{\sigma}^{ij} = \sqrt{g_{(ii)}g_{(jj)}}\,\sigma^{ij}. \tag{4.98}$$

Thus, the Cauchy stress tensor can be written in the form

$$\boldsymbol{\sigma} = \sigma^{ij}\mathbf{g}_i\mathbf{g}_j = \hat{\sigma}^{ij}\mathbf{e}_i\mathbf{e}_j \tag{4.99}$$

where the unit vectors are defined by $\mathbf{e}_i = \mathbf{g}_i/\sqrt{g_{(ii)}}$. Comparing this expression with Eq. (4.97) reveals that $\sigma_{rr}, \sigma_{\theta\theta}$, etc., are physical components of stress.

(c) To show that Eqs. (4.96) are consistent with Eqs. (A.119), we introduce physical components into the former. With the components of the metric tensor being [see Eqs. (2.137)]

$$[g_{ij}] = \text{diag}\,[1, r^2, 1],$$

Eq. (4.98) gives

$$\begin{aligned}\boldsymbol{\sigma}_{(\mathbf{e}_i\mathbf{e}_j)} &= \begin{bmatrix} \hat{\sigma}^{11} & \hat{\sigma}^{12} & \hat{\sigma}^{13} \\ \hat{\sigma}^{21} & \hat{\sigma}^{22} & \hat{\sigma}^{23} \\ \hat{\sigma}^{31} & \hat{\sigma}^{32} & \hat{\sigma}^{33} \end{bmatrix} = \begin{bmatrix} \hat{\sigma}^{rr} & \hat{\sigma}^{r\theta} & \hat{\sigma}^{rz} \\ \hat{\sigma}^{\theta r} & \hat{\sigma}^{\theta\theta} & \hat{\sigma}^{\theta z} \\ \hat{\sigma}^{zr} & \hat{\sigma}^{z\theta} & \hat{\sigma}^{zz} \end{bmatrix} \\ &= \begin{bmatrix} \sigma^{11} & r\sigma^{12} & \sigma^{13} \\ r\sigma^{21} & r^2\sigma^{22} & r\sigma^{23} \\ \sigma^{31} & r\sigma^{32} & \sigma^{33} \end{bmatrix}. \end{aligned} \tag{4.100}$$

In addition, the physical components of the body force and acceleration are defined by

$$\begin{aligned}\mathbf{f} &= f^i\mathbf{g}_i = \hat{f}^r\mathbf{e}_r + \hat{f}^\theta\mathbf{e}_\theta + \hat{f}^z\mathbf{e}_z \\ \mathbf{a} &= a^i\mathbf{g}_i = \hat{a}^r\mathbf{e}_r + \hat{a}^\theta\mathbf{e}_\theta + \hat{a}^z\mathbf{e}_z. \end{aligned} \tag{4.101}$$

Since Eqs. (2.136) give

$$\mathbf{g}_1 = \mathbf{e}_r, \qquad \mathbf{g}_2 = r\mathbf{e}_\theta, \qquad \mathbf{g}_3 = \mathbf{e}_z,$$

we have

$$\begin{aligned} \hat{f}^r &= f^1, \qquad \hat{f}^\theta = rf^2, \qquad \hat{f}^z = f^3 \\ \hat{a}^r &= a^1, \qquad \hat{a}^\theta = ra^2, \qquad \hat{a}^z = a^3. \end{aligned} \tag{4.102}$$

Finally, substituting Eqs. (4.100) and (4.102) into (4.96) yields

$$\begin{aligned} \frac{\partial \hat{\sigma}^{rr}}{\partial r} + \frac{1}{r}\frac{\partial \hat{\sigma}^{\theta r}}{\partial \theta} + \frac{\partial \hat{\sigma}^{zr}}{\partial z} + \frac{\hat{\sigma}^{rr} - \hat{\sigma}^{\theta\theta}}{r} + \hat{f}^r &= \rho \hat{a}^r \\ \frac{\partial \hat{\sigma}^{r\theta}}{\partial r} + \frac{1}{r}\frac{\partial \hat{\sigma}^{\theta\theta}}{\partial \theta} + \frac{\partial \hat{\sigma}^{z\theta}}{\partial z} + \frac{2\hat{\sigma}^{r\theta}}{r} + \hat{f}^\theta &= \rho \hat{a}^\theta \\ \frac{\partial \hat{\sigma}^{rz}}{\partial r} + \frac{1}{r}\frac{\partial \hat{\sigma}^{\theta z}}{\partial \theta} + \frac{\partial \hat{\sigma}^{zz}}{\partial z} + \frac{\hat{\sigma}^{rz}}{r} + \hat{f}^z &= \rho \hat{a}^z \end{aligned} \tag{4.103}$$

which agree with Eqs. (A.119). ■

4.7 Problems

4–1 Relative to Cartesian coordinates (x, y, z), the Cauchy stress tensor at a point is given by the matrix

$$[\sigma_{ij}] = \begin{bmatrix} 40 & 30 & -10 \\ 30 & 60 & 25 \\ -10 & 25 & 80 \end{bmatrix}.$$

(a) Determine the true traction vector across a plane defined by the relation

$$F = 2x + 2y + z - C = 0$$

where C is a constant. *Hint:* A vector normal to a surface with equation $F = 0$ is given by $\mathbf{a} = \nabla F$.

(b) The traction vector can be written in the form $\mathbf{T}^{(\mathbf{n})} = \sigma_{nn}\mathbf{n} + \boldsymbol{\sigma}_{ns}\mathbf{s}$, where $\mathbf{n}$ and $\mathbf{s}$ are unit vectors normal and tangent to the plane F (n not summed). Determine the stress components σ_{nn} and $\boldsymbol{\sigma}_{ns}$.

(c) Determine the principal Cauchy stresses and the directions in which they act.

4–2 Direction cosines of a unit vector $\mathbf{n}$ relative to Cartesian axes (x, y, z) are given by $\ell = \mathbf{n} \cdot \mathbf{e}_x$, $m = \mathbf{n} \cdot \mathbf{e}_y$, and $n = \mathbf{n} \cdot \mathbf{e}_z$. Suppose (ℓ, m, n) are known for the unit normal at a point on the boundary of a body. At this point, moreover, the true surface traction vector has a magnitude $\sqrt{3}a$ and is oriented at the same angle relative to all three positive coordinate axes. If $\sigma_{yy} = \sigma_{xz} = \sigma_{yz} = 0$, determine the remaining Cauchy stress components.

4–3 The Cauchy stress tensor at a point in a body is

$$\boldsymbol{\sigma} = a\,\mathbf{e}_1\mathbf{e}_1 + b\,\mathbf{e}_2\mathbf{e}_2 + c\,\mathbf{e}_3\mathbf{e}_3 + d(\mathbf{e}_1\mathbf{e}_3 + \mathbf{e}_3\mathbf{e}_1) + e(\mathbf{e}_2\mathbf{e}_3 + \mathbf{e}_3\mathbf{e}_2)$$

with respect to a Cartesian coordinate system x_i.

(a) Determine the unit normal vector of a plane parallel to the x_3-axis on which the true traction vector is tangent to the plane.

(b) Consider a coordinate system $\bar{x}_i$ that is obtained by rotating the x_i system through an angle θ about the x_1-axis. Compute the stress components relative to the rotated system.

4–4 Relative to spherical polar coordinates (r, θ, ϕ), the physical components of the Cauchy stress tensor at a point in a body are

$$\boldsymbol{\sigma}_{(\mathbf{e}_r, \mathbf{e}_\theta, \mathbf{e}_\phi)} = \begin{bmatrix} \hat{\sigma}^{rr} & 0 & 0 \\ 0 & 10 & 20 \\ 0 & 20 & \hat{\sigma}^{\phi\phi} \end{bmatrix}.$$

If the principal stresses at this point are $\sigma_1 = 10$, $\sigma_2 = -10$, and $\sigma_3 = 30$, find $\hat{\sigma}^{rr}$ and $\hat{\sigma}^{\phi\phi}$.

4–5 Relative to Cartesian coordinates x_i, the stress distribution in a body is

$$[\sigma_{ij}] = \begin{bmatrix} x_1 + x_2 & \sigma_{12} & 0 \\ \sigma_{12} & x_1 - 2x_2 & 0 \\ 0 & 0 & x_2 \end{bmatrix}.$$

Assume that the body is in equilibrium with no body forces. If the true stress vector across the plane $x_1 = 2$ is

$$\mathbf{T}^{(\mathbf{n})} = (1 + x_2)\mathbf{e}_1 + (3 - x_2)\mathbf{e}_2,$$

find $\sigma_{12}(x_1, x_2)$.

4–6 Relative to Cartesian coordinates (x, y, z) the stress distribution in a body is

$$\boldsymbol{\sigma}_{(\mathbf{e}_i\mathbf{e}_j)} = \begin{bmatrix} xy & -y & 0 \\ -y & x-y & 0 \\ 0 & 0 & y^3+2z \end{bmatrix}.$$

(a) Write the natural and the physical components of $\boldsymbol{\sigma}$ relative to the cylindrical coordinates (r, θ, z).

(b) For the given stress distribution, under what conditions will the body be in equilibrium?

4–7 In cylindrical coordinates, the first Piola-Kirchhoff stress tensor at the point $(r, \theta, z) = (5, \pi/3, -2)$ is

$$\mathbf{t} = 2\mathbf{g}_r\mathbf{g}_r + 3\mathbf{g}_\theta\mathbf{g}_\theta - \mathbf{g}_\theta\mathbf{g}_z - (2/3)\mathbf{g}_z\mathbf{g}_\theta,$$

and the deformation gradient tensor is

$$\mathbf{F} = \mathbf{g}_r\mathbf{g}_r + 2\mathbf{g}_\theta\mathbf{g}_\theta + 3\mathbf{g}_z\mathbf{g}_z - \mathbf{g}_\theta\mathbf{g}_z.$$

(a) Find the following cylindrical polar components of the Cauchy stress tensor: σ^{ij}, $\sigma^{i}_{\cdot j}$, $\sigma_{j}^{\cdot i}$.

(b) Find the tensor components of part (a) in spherical polar coordinates.

4–8 A circular cylinder of inner radius a and outer radius b is loaded by a uniform circumferential Cauchy shear stress τ_0 on its outer surface.

(a) Determine the Cauchy shear stress τ_i that must be applied to the inner surface to maintain equilibrium. Write the stress dyadic at any point in the cylinder in terms of the cylindrical polar coordinate basis $\{\mathbf{e}_r, \mathbf{e}_\theta, \mathbf{e}_z\}$.

(b) Find the Cartesian components of Cauchy stress at any point in the cylinder.

4–9 The deformation from configuration B to configuration b is defined by the relations

$$\begin{aligned} x_1 &= X_1 + 0.2X_2X_3 \\ x_2 &= X_2 + 0.2X_3^2 \\ x_3 &= X_3 \end{aligned}$$

where X_i and x_i are Cartesian coordinates. In addition, the second Piola-Kirchhoff stress tensor in b is given by

$$\mathbf{s} = 2X_1\mathbf{e}_1\mathbf{e}_1 - X_1X_2\mathbf{e}_1\mathbf{e}_2 - X_1X_2\mathbf{e}_2\mathbf{e}_1 + (3X_1 + X_2)\mathbf{e}_2\mathbf{e}_2$$

where the $\mathbf{e}_i$ are unit vectors along the x_i.

(a) Determine the Cauchy stress tensor $\boldsymbol{\sigma}$ in terms of the basis $\mathbf{e}_i\mathbf{e}_j$ for the point in b that was located at $(-1, 2, 1)$ in B.
(b) Determine the stress components $\sigma^{I^*J^*}$ and $\sigma_{I}^{\cdot J^*}$ at the same point.
(c) Determine the true traction vector $\mathbf{T}^{(\mathbf{n})}$ on a plane that passes through this point and is normal to the vector $\mathbf{e}_1 + \mathbf{e}_2$ in b.
(d) Determine the pseudotraction vector $\mathbf{T}^{(\mathbf{N})}$ on a plane that passes through this point and is normal to the vector $\mathbf{e}_1 + \mathbf{e}_2$ in B.

4–10 At a point p in a body, the deformation gradient tensor is

$$\mathbf{F} = 4\mathbf{e}_1\mathbf{e}_1 - \mathbf{e}_1\mathbf{e}_2 + 2\mathbf{e}_2\mathbf{e}_1 + 3\mathbf{e}_2\mathbf{e}_2 + \mathbf{e}_3\mathbf{e}_3$$

and the Cauchy stress tensor is

$$\boldsymbol{\sigma} = -2\mathbf{e}_1\mathbf{e}_1 + 3\mathbf{e}_1\mathbf{e}_2 - 3\mathbf{e}_2\mathbf{e}_1 + 4\mathbf{e}_2\mathbf{e}_2 + \mathbf{e}_3\mathbf{e}_3$$

in which the $\mathbf{e}_i$ are Cartesian base vectors.

(a) Compute the Piola-Kirchhoff stress tensors $\mathbf{t}$ and $\mathbf{s}$.
(b) Consider a plane in the deformed body that contains p and is normal to the vector $\mathbf{e}_1 - 3\mathbf{e}_2$. Determine the traction vector $\mathbf{T}^{(\mathbf{n})}$ across this plane.
(c) Compute the contravariant and the covariant components of $\boldsymbol{\sigma}$ relative to the basis defined by the vectors

$$\begin{aligned} \mathbf{g}_1 &= \mathbf{e}_1 - 3\mathbf{e}_2 \\ \mathbf{g}_2 &= 2\mathbf{e}_1 - \mathbf{e}_2 \\ \mathbf{g}_3 &= \mathbf{e}_1 + \mathbf{e}_2 + \mathbf{e}_3. \end{aligned}$$

Sketch the components on an element defined by the $\mathbf{g}_i$.

(d) Compute the physical stress components relative to the basis of part (c).

4–11 In terms of the parabolic cylindrical coordinates x^i, Cartesian coordinates z^i are given by

$$\begin{aligned} z^1 &= a(x^2 - x^1) \\ z^2 &= 2a(x^1 x^2)^{1/2} \\ z^3 &= x^3. \end{aligned}$$

If the Cauchy stress tensor is

$$\boldsymbol{\sigma} = x^1 x^2 \mathbf{g}_1\mathbf{g}_1 - 6x^2\mathbf{g}_2\mathbf{g}_2 + 3(x^1 - x^2)(\mathbf{g}_1\mathbf{g}_2 + \mathbf{g}_2\mathbf{g}_1),$$

where the $\mathbf{g}_i$ are parabolic base vectors, write $\boldsymbol{\sigma}$ in the form $\boldsymbol{\sigma} = \sigma^i_{\cdot j}\mathbf{g}_i\mathbf{g}^j$.

4–12 Write the stress components t^{IJ} in terms of s^{I^*J} and σ^{iJ} in terms of s^{IJ^*}.

4–13 Consider a problem in which both the undeformed and deformed configurations are described in terms of spherical polar coordinates (R, Θ, Φ) and (r, θ, ϕ), respectively.

(a) Using Eqs. (4.80), derive the differential equations of motion in terms of σ^{ij}, t^{IJ}, and s^{IJ}. Repeat the derivation using Eqs. (4.92).
(b) Write the equations of motion in terms of the $\hat{\sigma}^{ij}$.

4–14 Let the undeformed and deformed positions of a point in a circular cylinder be described by the cylindrical coordinates (R, Θ, Z) and (r, θ, z), respectively. A surface traction $\mathbf{T}^{(\mathbf{n})} = -p\mathbf{n} + \tau\mathbf{s}$ acts on its curved surface, where $p(\Theta)$ is a pressure, $\tau(\Theta)$ is a shear stress, and $\mathbf{n}(\Theta)$ and $\mathrm{s}(\Theta)$ are the unit normal and circumferential tangent to the deformed surface. Due to this loading, a point in the cylinder originally at the position $\mathbf{R} = R\mathbf{e}_R + Z\mathbf{e}_Z$ moves to the location $\mathbf{r} = R(1 + b\cos^2\Theta)\mathbf{e}_R + Z\mathbf{e}_Z$, where b is a constant.

If the cylinder initially has a radius a, write the boundary condition on the curved surface in terms of (a) Cauchy stress components σ^{IJ} and $\sigma^{I^*J^*}$; and (b) cylindrical second Piola-Kirchhoff stress components s^{IJ} and $s^{I^*J^*}$.

4–15 Consider a solid circular cylinder of deformed radius a. In Cartesian coordinates, the Cauchy stress tensor is

$$\boldsymbol{\sigma} = x^2\mathbf{e}_x\mathbf{e}_x + y^2\mathbf{e}_y\mathbf{e}_y + xy(\mathbf{e}_x\mathbf{e}_y + \mathbf{e}_y\mathbf{e}_x).$$

Determine the body forces and the surface tractions that must be applied to the cylinder to maintain equilibrium. Write the result in terms of stress components relative to the spatial cylindrical polar coordinates (r, θ, z).

Chapter 5
Constitutive Relations

The analyses of deformation and stress in the previous chapters are valid for any solid body that can be represented as a continuum, regardless of the type of material that comprises the body. Deformation and stress, however, are linked by constitutive relations that must be found experimentally for each type of material. In biomechanics, determining constitutive relations is complicated by material nonlinearity, complex geometry, the composite nature of biological tissues, and the influence of a wide range of environmental variables. In general, even the functional forms of these relations are unknown, but thermodynamic considerations place important restrictions on their form. This chapter considers some of these restrictions and discusses some strategies for determining constitutive relations for soft tissues.

5.1 Thermodynamics of Deformation

This section considers the first and second laws of thermodynamics for a **thermomechanical continuum** in which thermal and mechanical effects dominate the behavior, with electrical, chemical, and other effects ignored. Note that the present treatment is not exhaustive, and the reader is expected to be familiar with basic concepts such as entropy.

5.1.1 *First Law of Thermodynamics*

For a thermomechanical continuum, the principle of conservation of energy can be written in the form

$$\dot{K} + \dot{U} = P + Q \tag{5.1}$$

where K is the kinetic energy, U is the internal energy, P is the mechanical power input, and Q is the rate of heat input. The left-hand-side of this equation represents the time-rate of change of the total energy in the system, and the right-hand-side is the rate at which energy is added to the system. The various quantities in this equation are written below in both Eulerian and Lagrangian forms.

Consider a body of mass density $\rho_0(\mathbf{R})$ and volume V in the reference configuration. After deformation, these quantities become $\rho(\mathbf{r}, t)$ and v, respectively. The total kinetic energy for the body is

$$K = \tfrac{1}{2} \int_v \rho \, \mathbf{v} \cdot \mathbf{v} \, dv = \tfrac{1}{2} \int_V \rho_0 \, \mathbf{v} \cdot \mathbf{v} \, dV \tag{5.2}$$

where $\mathbf{v}(\mathbf{R}, t) = \mathbf{v}(\mathbf{r}, t)$ is the velocity vector for a point that moves from the undeformed location $\mathbf{R}$ to the deformed location $\mathbf{r}$. In addition, the total internal energy contained in the body is

$$U = \int_v \rho u \, dv = \int_V \rho_0 u \, dV \tag{5.3}$$

where u is the internal energy per unit mass. Here, the internal energy consists of thermal energy and strain energy.

The power input is the rate of work done on the body by applied loads, which include surface and body forces. In terms of the body force vectors in Eq. (4.1) and the surface traction vectors defined by Eqs. (4.3) and (4.4), the total power input is

$$\begin{aligned} P &= \int_a \mathbf{T}^{(\mathbf{n})} \cdot \mathbf{v} \, da + \int_v \mathbf{f} \cdot \mathbf{v} \, dv \\ &= \int_A \mathbf{T}^{(\mathbf{N})} \cdot \mathbf{v} \, dA + \int_V \mathbf{f}_0 \cdot \mathbf{v} \, dV \end{aligned} \tag{5.4}$$

in which A and a are the surface areas of the body in the undeformed and deformed configurations, respectively.

Finally, the body gains thermal energy by heat flowing across its surface and heat generated by internal sources. Let $\mathbf{q}$ and $\mathbf{q}_0$ be the outward-directed heat flux vectors per unit deformed and undeformed surface area, respectively, and let r be the rate of heat production per unit mass due to internal sources. Then, the rate of heat added to the body is

$$\begin{aligned} Q &= -\int_a \mathbf{q}\cdot\mathbf{n}\,da + \int_v \rho r\,dv \\ &= -\int_A \mathbf{q}_0\cdot\mathbf{N}\,dA + \int_V \rho_0 r\,dV \end{aligned} \tag{5.5}$$

where $\mathbf{n}$ and $\mathbf{N}$ are the respective (outward-directed) unit normals to the deformed and undeformed surface.

The various terms in Eq. (5.1) are next expressed in alternate forms so that they can be combined. First, Eq. (5.2) gives

$$\begin{aligned} \dot{K} &= \tfrac{1}{2}\int_v \left[\frac{d}{dt}(\rho\,dv)\,\mathbf{v}\cdot\mathbf{v} + \frac{d}{dt}(\mathbf{v}\cdot\mathbf{v})\rho\,dv\right] \\ &= \tfrac{1}{2}\int_v (\dot{\mathbf{v}}\cdot\mathbf{v} + \mathbf{v}\cdot\dot{\mathbf{v}})\rho\,dv = \int_v \mathbf{v}\cdot\dot{\mathbf{v}}\,\rho\,dv \\ &= \int_v \mathbf{v}\cdot(\bar{\boldsymbol{\nabla}}\cdot\boldsymbol{\sigma} + \mathbf{f})\,dv \end{aligned} \tag{5.6}$$

in which mass conservation gives $d(\rho\,dv)/dt = 0$, and substituting for $\rho\dot{\mathbf{v}} = \rho\mathbf{a}$ from the equation of motion (4.56) produces the third line. Similar manipulations with the Lagrangian form for K in Eq. (5.2) yield

$$\dot{K} = \int_V \mathbf{v}\cdot(\boldsymbol{\nabla}\cdot\mathbf{t} + \mathbf{f}_0)\,dV \tag{5.7}$$

in which Eq. (4.68) has been used.

Next, we recast the terms $\mathbf{v}\cdot(\bar{\boldsymbol{\nabla}}\cdot\boldsymbol{\sigma})$ and $\mathbf{v}\cdot(\boldsymbol{\nabla}\cdot\mathbf{t})$ in the last two equations. The complexities of the following operations make it convenient to work in Cartesian components and then convert the results back to direct notation. With all indices written temporarily as subscripts, we have

$$\begin{aligned} \mathbf{v}\cdot(\bar{\boldsymbol{\nabla}}\cdot\boldsymbol{\sigma}) &= \mathbf{v}\cdot\left(\mathbf{e}_k\frac{\partial}{\partial x_k}\cdot\sigma_{ij}\mathbf{e}_i\mathbf{e}_j\right) \\ &= \mathbf{v}\cdot(\sigma_{ij},_k\,\delta_{ki}\mathbf{e}_j) = \mathbf{v}\cdot(\sigma_{ij},_i\,\mathbf{e}_j) \\ &= (v_k\mathbf{e}_k)\cdot(\sigma_{ij},_i\,\mathbf{e}_j) = v_k\sigma_{ij},_i\,\delta_{kj} \\ &= v_j\sigma_{ij},_i \\ &= (v_j\sigma_{ij}),_i - v_j,_i\,\sigma_{ij}. \end{aligned} \tag{5.8}$$

To convert this expression back to direct notation, we note the following:

$$
\begin{aligned}
\bar{\boldsymbol{\nabla}}\cdot(\boldsymbol{\sigma}\cdot\mathbf{v}) &= \bar{\boldsymbol{\nabla}}\cdot(\sigma_{ij}\mathbf{e}_i\mathbf{e}_j\cdot v_k\mathbf{e}_k) = \bar{\boldsymbol{\nabla}}\cdot(\sigma_{ij}v_k\delta_{jk}\mathbf{e}_i) \\
&= \left(\mathbf{e}_k\frac{\partial}{\partial x_k}\right)\cdot(\sigma_{ij}v_j\mathbf{e}_i) = (\sigma_{ij}v_j)_{,k}\,\delta_{ki} = (\sigma_{ij}v_j)_{,i} \\
\boldsymbol{\sigma}:\bar{\boldsymbol{\nabla}}\mathbf{v} &= (\sigma_{ij}\mathbf{e}_i\mathbf{e}_j):\left(\mathbf{e}_k\frac{\partial}{\partial x_k}v_l\mathbf{e}_l\right) = \sigma_{ij}v_{l,k}\,(\mathbf{e}_i\cdot\mathbf{e}_k)(\mathbf{e}_j\cdot\mathbf{e}_l) \\
&= \sigma_{ij}v_{l,k}\,\delta_{ik}\delta_{jl} = \sigma_{ij}v_{j,i}\,.
\end{aligned}
$$

Thus, Eq. (5.8) and similar manipulations for $\mathbf{v}\cdot(\boldsymbol{\nabla}\cdot\mathbf{t})$ yield

$$
\begin{aligned}
\mathbf{v}\cdot(\bar{\boldsymbol{\nabla}}\cdot\boldsymbol{\sigma}) &= \bar{\boldsymbol{\nabla}}\cdot(\boldsymbol{\sigma}\cdot\mathbf{v}) - \boldsymbol{\sigma}:\bar{\boldsymbol{\nabla}}\mathbf{v} \\
\mathbf{v}\cdot(\boldsymbol{\nabla}\cdot\mathbf{t}) &= \boldsymbol{\nabla}\cdot(\mathbf{t}\cdot\mathbf{v}) - \mathbf{t}:\boldsymbol{\nabla}\mathbf{v}.
\end{aligned} \tag{5.9}
$$

In addition, Eqs. (3.215) and $(3.76)_1$ give $\bar{\boldsymbol{\nabla}}\mathbf{v} = \mathbf{D} - \boldsymbol{\Omega}$ and $\boldsymbol{\nabla}\mathbf{u} = \mathbf{F}^T - \mathbf{I}$, where $\mathbf{D}$ is the symmetric rate-of-deformation tensor and $\boldsymbol{\Omega}$ is the antisymmetric spin tensor. Thus, the last terms of Eqs. (5.9) can be written

$$
\begin{aligned}
\boldsymbol{\sigma}:\bar{\boldsymbol{\nabla}}\mathbf{v} &= \boldsymbol{\sigma}:(\mathbf{D}-\boldsymbol{\Omega}) = \boldsymbol{\sigma}:\mathbf{D} \\
\mathbf{t}:\boldsymbol{\nabla}\mathbf{v} &= \mathbf{t}:\boldsymbol{\nabla}\dot{\mathbf{u}} = \mathbf{t}:\dot{\mathbf{F}}^T,
\end{aligned} \tag{5.10}
$$

where $\boldsymbol{\sigma}:\boldsymbol{\Omega} = 0$ since the double-dot product of a symmetric tensor ($\boldsymbol{\sigma}$) and an antisymmetric tensor ($\boldsymbol{\Omega}$) is zero. Now, substituting Eqs. (5.9) and (5.10) into (5.6) and (5.7) yields

$$
\begin{aligned}
\dot{K} &= \int_v [\bar{\boldsymbol{\nabla}}\cdot(\boldsymbol{\sigma}\cdot\mathbf{v}) - \boldsymbol{\sigma}:\mathbf{D} + \mathbf{f}\cdot\mathbf{v}]\,dv \\
&= \int_V [\boldsymbol{\nabla}\cdot(\mathbf{t}\cdot\mathbf{v}) - \mathbf{t}:\dot{\mathbf{F}}^T + \mathbf{f}_0\cdot\mathbf{v}]\,dV.
\end{aligned} \tag{5.11}
$$

The Eulerian form of the second term in Eq. (5.1), given by differentiating Eq. (5.3), is

$$
\begin{aligned}
\dot{U} &= \int_v \left[\frac{d}{dt}(\rho\,dv)\,u + \dot{u}\rho\,dv\right] \\
&= \int_v \dot{u}\rho\,dv.
\end{aligned} \tag{5.12}
$$

Similarly, the Lagrangian form is

$$
\dot{U} = \int_V \dot{u}\rho_0\,dV. \tag{5.13}
$$

Next, the surface integrals in Eqs. (5.4) and (5.5) are converted into volume integrals. Substituting Eqs. (4.23) and $(4.27)_1$ into (5.4) and using the divergence theorem (2.191) yields

$$\begin{aligned} P &= \int_a \mathbf{n}\cdot\boldsymbol{\sigma}\cdot\mathbf{v}\,da + \int_v \mathbf{f}\cdot\mathbf{v}\,dv \\ &= \int_v [\bar{\boldsymbol{\nabla}}\cdot(\boldsymbol{\sigma}\cdot\mathbf{v}) + \mathbf{f}\cdot\mathbf{v}]\,dv \end{aligned} \tag{5.14}$$

in terms of Eulerian quantities and

$$\begin{aligned} P &= \int_A \mathbf{N}\cdot\mathbf{t}\cdot\mathbf{v}\,dA + \int_V \mathbf{f}_0\cdot\mathbf{v}\,dV \\ &= \int_V [\boldsymbol{\nabla}\cdot(\mathbf{t}\cdot\mathbf{v}) + \mathbf{f}_0\cdot\mathbf{v}]\,dV \end{aligned} \tag{5.15}$$

in terms of Lagrangian quantities. Similarly, applying the divergence theorem to Eq. (5.5) gives

$$\begin{aligned} Q &= \int_v (\rho r - \bar{\boldsymbol{\nabla}}\cdot\mathbf{q})\,dv \\ &= \int_V (\rho_0 r - \boldsymbol{\nabla}\cdot\mathbf{q}_0)\,dV. \end{aligned} \tag{5.16}$$

Finally, inserting Eqs. (5.11)–(5.16) into (5.1) and simplifying yields

$$\begin{aligned} \int_v (\rho\dot{u} - \boldsymbol{\sigma}:\mathbf{D} - \rho r + \bar{\boldsymbol{\nabla}}\cdot\mathbf{q})\,dv &= 0 \\ \int_V (\rho_0\dot{u} - \mathbf{t}:\dot{\mathbf{F}}^T - \rho_0 r + \boldsymbol{\nabla}\cdot\mathbf{q}_0)\,dV &= 0 \end{aligned} \tag{5.17}$$

which are the global forms of the first law of thermodynamics (conservation of energy). For an arbitrary volume element that deforms from dV into dv, these relations imply

$$\boxed{\begin{aligned} \rho\dot{u} - \boldsymbol{\sigma}:\mathbf{D} - \rho r + \bar{\boldsymbol{\nabla}}\cdot\mathbf{q} &= 0 \\ \rho_0\dot{u} - \mathbf{t}:\dot{\mathbf{F}}^T - \rho_0 r + \boldsymbol{\nabla}\cdot\mathbf{q}_0 &= 0, \end{aligned}} \tag{5.18}$$

which are the local forms of the first law. According to these equations, the rate of increase in internal energy of a volume element ($\rho\dot{u}$ or $\rho_0\dot{u}$) is equal to the sum of the rate of work done by the stresses on the element, i.e., the **stress power** ($\boldsymbol{\sigma}:\mathbf{D}$

or $\mathbf{t} : \dot{\mathbf{F}}^T$), the rate of internal heat production (ρr or $\rho_0 r$), and the rate of heat flow into the element ($-\bar{\boldsymbol{\nabla}} \cdot \mathbf{q}$ or $-\boldsymbol{\nabla} \cdot \mathbf{q}_0$).

The stress power appears in Eqs. (5.18) in terms of the Cauchy and first Piola-Kirchhoff stress tensors. We now derive the relation between these two forms for the stress power. First, since $\boldsymbol{\sigma} : \boldsymbol{\Omega} = 0$, we can write

$$\boldsymbol{\sigma} : \mathbf{D} = \boldsymbol{\sigma} : (\mathbf{D} + \boldsymbol{\Omega}) = \boldsymbol{\sigma} : \mathbf{L} = \boldsymbol{\sigma} : (\dot{\mathbf{F}} \cdot \mathbf{F}^{-1}) \tag{5.19}$$

where Eqs. (3.217) and (3.210) have been substituted. Next, applying formulas from Table 2.2 (page 23) with $\mathbf{T} = \boldsymbol{\sigma}$, $\mathbf{U} = \dot{\mathbf{F}}$, and $\mathbf{V} = \mathbf{F}^{-1}$ gives

$$\begin{aligned}
\boldsymbol{\sigma} : \mathbf{D} &= \dot{\mathbf{F}} : (\boldsymbol{\sigma} \cdot \mathbf{F}^{-T}) \\
&= (\boldsymbol{\sigma} \cdot \mathbf{F}^{-T}) : \dot{\mathbf{F}} = (\boldsymbol{\sigma} \cdot \mathbf{F}^{-T})^T : \dot{\mathbf{F}}^T \\
&= (\mathbf{F}^{-1} \cdot \boldsymbol{\sigma}^T) : \dot{\mathbf{F}}^T = (\mathbf{F}^{-1} \cdot \boldsymbol{\sigma}) : \dot{\mathbf{F}}^T
\end{aligned}$$

where the third line uses the symmetry property of the Cauchy stress tensor. Finally, substituting Eq. $(4.31)_1$ gives the result

$$\boldsymbol{\sigma} : \mathbf{D} = J^{-1} \mathbf{t} : \dot{\mathbf{F}}^T. \tag{5.20}$$

A third form for the stress power can be written in terms of the second Piola-Kirchhoff stress tensor $\mathbf{s}$. Substituting Eq. (3.220) for $\mathbf{D}$ into (5.20) yields

$$\mathbf{t} : \dot{\mathbf{F}}^T = J \boldsymbol{\sigma} : \mathbf{D} = J \boldsymbol{\sigma} : (\mathbf{F}^{-T} \cdot \dot{\mathbf{E}} \cdot \mathbf{F}^{-1}).$$

This expression is transformed by setting $\mathbf{T} = \mathbf{F}^{-1}$, $\mathbf{U}^T = \mathbf{F}^{-T} \cdot \dot{\mathbf{E}}$, and $\mathbf{V} = \boldsymbol{\sigma}$ in the next-to-last formula of Table 2.2 to get

$$\begin{aligned}
\mathbf{t} : \dot{\mathbf{F}}^T &= J\, (\mathbf{F}^{-T} \cdot \dot{\mathbf{E}})^T : (\mathbf{F}^{-1} \cdot \boldsymbol{\sigma}^T) \\
&= J\, (\mathbf{F}^{-1} \cdot \boldsymbol{\sigma}^T) : (\dot{\mathbf{E}}^T \cdot \mathbf{F}^{-1}) \\
&= J\, (\mathbf{F}^{-1} \cdot \boldsymbol{\sigma}) : (\dot{\mathbf{E}} \cdot \mathbf{F}^{-1})
\end{aligned}$$

since $\boldsymbol{\sigma} = \boldsymbol{\sigma}^T$ and $\mathbf{E} = \mathbf{E}^T$. Now, on setting $\mathbf{T} = \dot{\mathbf{E}}$, $\mathbf{U} = \mathbf{F}^{-1} \cdot \boldsymbol{\sigma}$, and $\mathbf{V}^T = \mathbf{F}^{-1}$, the next-to-last line in Table 2.2 gives

$$\begin{aligned}
\mathbf{t} : \dot{\mathbf{F}}^T &= J\, \dot{\mathbf{E}} : (\mathbf{F}^{-1} \cdot \boldsymbol{\sigma} \cdot \mathbf{F}^{-T}) \\
&= \dot{\mathbf{E}} : \mathbf{s} = \mathbf{s} : \dot{\mathbf{E}},
\end{aligned} \tag{5.21}$$

in which Eq. $(4.31)_2$ has been used. Thus, Eqs. (5.20) and (5.21) give

$$\boxed{J\boldsymbol{\sigma} : \mathbf{D} = \mathbf{t} : \dot{\mathbf{F}}^T = \mathbf{s} : \dot{\mathbf{E}},} \tag{5.22}$$

and so an alternative Lagrangian form for the first law of thermodynamics $(5.18)_2$ is

$$\boxed{\rho_0 \dot{u} - \mathbf{s} : \dot{\mathbf{E}} - \rho_0 r + \boldsymbol{\nabla} \cdot \mathbf{q}_0 = 0.} \tag{5.23}$$

Example 5.1 Show by direct manipulation that the two equations in (5.18) are equivalent.

Solution. Multiplying the second relation by J^{-1} gives

$$\rho \dot{u} - \boldsymbol{\sigma} : \mathbf{D} - \rho r + J^{-1} \boldsymbol{\nabla} \cdot \mathbf{q}_0 = 0$$

where $\rho_0 = \rho J$ and Eq. (5.22) have been used. All that is left to do is to show that $J^{-1} \boldsymbol{\nabla} \cdot \mathbf{q}_0 = \bar{\boldsymbol{\nabla}} \cdot \mathbf{q}$. To do this, we first note that $\mathbf{q}_0$ is the heat flux per unit undeformed area dA. Since the area vector $d\mathbf{A}$ deforms into $d\mathbf{a}$, the total amount of heat per unit time passing across the area element is

$$\mathbf{q}_0 \cdot d\mathbf{A} = \mathbf{q} \cdot d\mathbf{a}. \tag{5.24}$$

Next, substituting Eq. (3.128) gives

$$\begin{aligned} \mathbf{q}_0 \cdot d\mathbf{A} &= \mathbf{q} \cdot (J\, d\mathbf{A} \cdot \mathbf{F}^{-1}) \\ &= J\, \mathbf{q} \cdot (\mathbf{F}^{-T} \cdot d\mathbf{A}) \\ &= J\, (\mathbf{q} \cdot \mathbf{F}^{-T}) \cdot d\mathbf{A}, \end{aligned}$$

and so

$$\mathbf{q}_0 = J\, \mathbf{q} \cdot \mathbf{F}^{-T} = J\, \mathbf{F}^{-1} \cdot \mathbf{q}. \tag{5.25}$$

Thus, we have

$$J^{-1} \boldsymbol{\nabla} \cdot \mathbf{q}_0 = J^{-1} \boldsymbol{\nabla} \cdot (J\, \mathbf{F}^{-1} \cdot \mathbf{q}).$$

Finally, replacing $\boldsymbol{\sigma}$ by $\mathbf{q}_0$ in Eq. (4.79) shows that this expression becomes

$$J^{-1} \boldsymbol{\nabla} \cdot \mathbf{q}_0 = \bar{\boldsymbol{\nabla}} \cdot \mathbf{q}, \tag{5.26}$$

which is the desired result. ■

5.1.2 *Second Law of Thermodynamics*

The first law of thermodynamics states that work can be converted into heat and *vice versa*, so long as the total energy contained in a closed system remains constant. However, dissipated heat energy, such as that due to friction or viscosity, cannot be converted into work. The second law of thermodynamics restricts the direction of energy conversion processes. This principle is based on the concept of **entropy**, which is a measure of the disorder in a system due to an increase in heat energy. During a **reversible process**, the total entropy in a system is conserved, but during an **irreversible process**, a system gains entropy due to heat input. Friction is an example of an irreversible process.

In general terms, the specific entropy η (entropy per unit mass) is defined through the inequality

$$\Delta\eta = \eta_2 - \eta_1 \geq \int_1^2 \frac{d\bar{Q}}{T} \tag{5.27}$$

between two thermodynamic states, where $\bar{Q}$ is the heat input per unit mass and T is the absolute temperature. Equality holds for a reversible process, whereas the inequality indicates that entropy is generated during an irreversible process. Equation (5.27) defines a change in entropy as a measure of the change in energy dissipation with respect to temperature.

For a solid body, the second law of thermodynamics states that the time-rate of change of the total entropy in the body is greater than or equal to the sum of the influx of entropy through the surface of the body and the entropy generated by internal heat sources. In mathematical terms, using Eq. (5.5) to generalize Eq. (5.27) gives the second law in Eulerian form as

$$\frac{d}{dt}\int_v \rho\eta\, dv \geq -\int_a \frac{\mathbf{q}}{T}\cdot\mathbf{n}\, da + \int_v \frac{\rho r}{T}\, dv \tag{5.28}$$

and in Lagrangian form as

$$\frac{d}{dt}\int_V \rho_0\eta\, dV \geq -\int_A \frac{\mathbf{q}_0}{T}\cdot\mathbf{N}\, dA + \int_V \frac{\rho_0 r}{T}\, dV. \tag{5.29}$$

The local forms of the second law follow from applying the divergence theorem (2.191) to the last terms of the above equations to obtain

$$\begin{aligned}\int_v \left[\rho\dot{\eta} - \frac{\rho r}{T} + \bar{\nabla}\cdot\left(\frac{\mathbf{q}}{T}\right)\right] dv &\geq 0 \\ \int_V \left[\rho_0\dot{\eta} - \frac{\rho_0 r}{T} + \nabla\cdot\left(\frac{\mathbf{q}_0}{T}\right)\right] dV &\geq 0.\end{aligned} \tag{5.30}$$

These integral relations imply the local equations

$$\boxed{\begin{aligned}\dot{\eta} - \frac{r}{T} + \frac{1}{\rho}\bar{\nabla}\cdot\left(\frac{\mathbf{q}}{T}\right) &\geq 0 \\ \dot{\eta} - \frac{r}{T} + \frac{1}{\rho_0}\nabla\cdot\left(\frac{\mathbf{q}_0}{T}\right) &\geq 0\end{aligned}} \tag{5.31}$$

which are the Eulerian and Lagrangian forms of the **Clausius-Duhem inequality**. These equations can be written in alternate forms by expanding the terms involving the gradient operators. To do this, we use the defining relations (3.18) to get

$$\begin{aligned}\bar{\nabla}\cdot\left(\frac{\mathbf{q}}{T}\right) &= \left(\mathbf{g}_i\,\frac{\partial}{\partial x^i}\right)\cdot\left(\frac{\mathbf{q}}{T}\right) \\ &= \mathbf{g}_i\cdot\left(\frac{\mathbf{q}_{,i}T - \mathbf{q}_i T_{,i}}{T^2}\right) \\ &= \frac{1}{T}\,\mathbf{g}_i\cdot\mathbf{q}_{,i} - \frac{T_{,i}}{T^2}\,\mathbf{g}_i\cdot\mathbf{q} \\ &= \frac{1}{T}\left(\mathbf{g}_i\,\frac{\partial}{\partial x^i}\right)\cdot\mathbf{q} - \frac{1}{T^2}\,\mathbf{q}\cdot\left(\mathbf{g}_i\,\frac{\partial T}{\partial x^i}\right) \\ &= \frac{1}{T}\,\bar{\nabla}\cdot\mathbf{q} - \frac{1}{T^2}\,\mathbf{q}\cdot\bar{\nabla}T\end{aligned}$$

with a similar expression for $\nabla\cdot(\mathbf{q}_0/T)$. Thus, Eqs. (5.31) become

$$\boxed{\begin{aligned}\dot{\eta} - \frac{r}{T} + \frac{1}{\rho T}\,\bar{\nabla}\cdot\mathbf{q} - \frac{1}{\rho T^2}\,\mathbf{q}\cdot\bar{\nabla}T &\geq 0 \\ \dot{\eta} - \frac{r}{T} + \frac{1}{\rho_0 T}\,\nabla\cdot\mathbf{q}_0 - \frac{1}{\rho_0 T^2}\,\mathbf{q}_0\cdot\nabla T &\geq 0.\end{aligned}} \tag{5.32}$$

Since heat does not flow naturally from a cooler toward a warmer part of a body, it is always true that

$$\mathbf{q} \cdot \bar{\boldsymbol{\nabla}} T \leq 0 \qquad \text{and} \qquad \mathbf{q}_0 \cdot \boldsymbol{\nabla} T \leq 0,$$

i.e., the temperature gradient and the heat flux have opposite signs unless $\bar{\boldsymbol{\nabla}} T = 0$, in which case $\mathbf{q} = \mathbf{0}$. These terms represent the entropy introduced into the body by heat conduction. This observation and Eqs. (5.32) imply that the second law of thermodynamics also can be written

$$\begin{aligned} \dot{\eta} - \frac{r}{T} + \frac{1}{\rho T} \bar{\boldsymbol{\nabla}} \cdot \mathbf{q} &\geq 0 \\ \dot{\eta} - \frac{r}{T} + \frac{1}{\rho_0 T} \boldsymbol{\nabla} \cdot \mathbf{q}_0 &\geq 0, \end{aligned} \tag{5.33}$$

where equality holds for a reversible system.

5.2 Fundamental Constitutive Principles

The possible forms for constitutive equations are restricted by some fundamental postulates based on physical and intuitive arguments. The constitutive principles for a thermomechanical material include the following:

1. **Coordinate Invariance.** Constitutive equations must be independent of the coordinate system used to describe the motion of a body.
2. **Determinism**. The values of the dependent constitutive variables at a point in a body are determined by the histories of all points in the body.
3. **Local Action**. The values of the dependent constitutive variables at a point p in a body are not affected significantly by the values of the independent variables outside an arbitrarily small neighborhood of p.
4. **Equipresence**. An independent variable that appears in one constitutive equation for a material must be present in all constitutive equations for the material, unless its presence violates some other fundamental principle.
5. **Material Frame Indifference (Material Objectivity)**. Constitutive equations must be form invariant under rigid motions of the spatial (observer) reference frame.

6 **Physical Admissibility**. Constitutive equations must be consistent with the fundamental physical balance laws (mass, momentum, energy) and the law of entropy (second law of thermodynamics).
7 **Material Symmetry**. Constitutive equations must be form invariant under certain rigid transformations of material frames, depending on the symmetries inherent in the material.

The remainder of this section uses the first six of these principles, in the listed order, to develop general constitutive relations for a thermoelastic material. Consideration of the principle of material symmetry is deferred to Section 5.3, which examines specific forms for an elastic material in which thermal effects are ignored.

5.2.1 *Principle of Coordinate Invariance*

A deforming body is oblivious to any coordinate system used to follow its motion. Thus, like the other basic equations of mechanics, constitutive equations should be developed in tensor form. Doing this ensures satisfaction of the principle of coordinate invariance.

5.2.2 *Principle of Determinism*

In a general material, the stress distribution at a given time depends not only on the instantaneous deformation and temperature, but also on the entire history of the deformation and temperature up until this time. Deforming a viscoelastic material, for example, involves viscous dissipation of energy, which cannot be recovered. Since the amount of energy lost depends on the path taken to a particular configuration, the stresses depend on the entire deformation history.

During the deformation of an ideal *thermoelastic* material, no energy is dissipated. Thus, the state of stress at time t depends only on the deformation and temperature at t. In an ideal *elastic* material, temperature effects are neglected, and so the stress depends only on the instantaneous deformation.

5.2.3 *Principle of Local Action*

To determine the theoretical implications of the principle of local action, we consider the motion $\mathbf{r}(\mathbf{R},t)$ and temperature $T(\mathbf{R},t)$ at a point p in a body. In addition, the motion and temperature of an arbitrary point within a small neighborhood of p are given by $\mathbf{r}(\bar{\mathbf{R}},t)$ and $T(\bar{\mathbf{R}},t)$. Here, $\mathbf{R}$ and $\bar{\mathbf{R}}$ are the position vectors to the two nearby points in the undeformed body. For a sufficiently smooth deformation in the vicinity of p, Taylor series expansions yield

$$\begin{aligned} \mathbf{r}(\bar{\mathbf{R}},t) &= \mathbf{r}(\mathbf{R},t) + (\bar{\mathbf{R}} - \mathbf{R})\cdot\frac{\partial \mathbf{r}}{\partial \mathbf{R}} + \cdots \\ T(\bar{\mathbf{R}},t) &= T(\mathbf{R},t) + (\bar{\mathbf{R}} - \mathbf{R})\cdot\frac{\partial T}{\partial \mathbf{R}} + \cdots . \end{aligned} \tag{5.34}$$

Through the principle of local action, these equations indicate that the motion and temperature at a point in a thermoelastic body can be determined from the values of these quantities and their spatial derivatives at a nearby point.

In this book, we consider only **simple materials**, for which derivatives higher than the first are ignored in Eqs. (5.34). Then, these equations and the principle of determinism suggest that the constitutive equations for a thermoelastic material can be written in the form

$$\begin{aligned} \boldsymbol{\sigma} &= \boldsymbol{\sigma}(\mathbf{R}, \mathbf{F}, T, \boldsymbol{\nabla} T) \\ \mathbf{q} &= \mathbf{q}(\mathbf{R}, \mathbf{F}, T, \boldsymbol{\nabla} T) \\ u &= u(\mathbf{R}, \mathbf{F}, T, \boldsymbol{\nabla} T) \\ \eta &= \eta(\mathbf{R}, \mathbf{F}, T, \boldsymbol{\nabla} T) \end{aligned} \tag{5.35}$$

since $\partial/\partial\mathbf{R} = \boldsymbol{\nabla}$ and $\boldsymbol{\nabla}\mathbf{r} = \mathbf{F}^T$. The dependency on $\mathbf{R}$ allows for an inhomogeneous material. In an initially homogeneous thermoelastic body, therefore, the dependent constitutive variables at a point depend only on the instantaneous values of the deformation gradient tensor, the temperature, and the temperature gradient at that point.

5.2.4 *Principle of Equipresence*

The constitutive relations (5.35) satisfy the principle of equipresence, since the same independent variables appear in each equation. The usefulness of this principle can be appreciated when other effects are to be included. For example, if chemical effects are important, then chemical potentials can be added to the list of dependencies for each constitutive variable.

In some instances, it is useful to use entropy as an independent variable, rather than temperature. Then, by the principle of equipresence, Eqs. (5.35) can be replaced by the alternate constitutive relations

$$\begin{aligned} \boldsymbol{\sigma} &= \boldsymbol{\sigma}(\mathbf{R}, \mathbf{F}, \eta, \boldsymbol{\nabla} \eta) \\ \mathbf{q} &= \mathbf{q}(\mathbf{R}, \mathbf{F}, \eta, \boldsymbol{\nabla} \eta) \\ u &= u(\mathbf{R}, \mathbf{F}, \eta, \boldsymbol{\nabla} \eta) \\ T &= T(\mathbf{R}, \mathbf{F}, \eta, \boldsymbol{\nabla} \eta). \end{aligned} \tag{5.36}$$

5.2.5 *Principle of Material Frame Indifference*

The fundamental idea behind the principle of material frame indifference is that, if relativistic effects are ignored, the events occurring at a point in a body must be independent of the motion of the observer. Because internal forces, lengths, and temperature are frame indifferent (objective) quantities (see Section 2.1), the form of the constitutive equations must be invariant under rigid motions of the reference frame. Here, ignoring shifts in the time frame, we derive a mathematical formulation of this principle.

As discussed in Section 2.1, a change in coordinates is not the same as a change in reference frame. There, it was pointed out that the form of Newton's second law of motion $\mathbf{f} = m\mathbf{a}$ is invariant under a change of coordinates in a given frame of reference, but it is not invariant under a general change of frame. The reason for this is that acceleration is not frame indifferent (see below). Writing an equation in tensor form ensures invariance under a *time-independent* change in coordinates, but ensuring invariance under a *time-dependent* change between frames in relative motion requires that the tensors satisfy certain transformation relations. (Note that a scalar point function, such as temperature, is inherently frame indifferent.) These transformation relations are derived in the following paragraphs.

Position Vectors

Consider two reference frames A and A^* that are in relative motion (Fig. 5.1). A set of Cartesian coordinate axes is fixed in each frame, with observers stationed at the respective origins O and O^*. It is important to note that these observers are "glued" in position so that they move with their respective frames. Thus, a point p fixed in frame A appears stationary to the observer at O, but to the observer at O^*, p appears to be moving. Thus, position is *not* frame indifferent. Obviously, the same is true for velocity and acceleration.

Although position is not an objective quantity, the position vectors $\mathbf{r}$ and $\mathbf{r}^*$ of p with respect to O and O^*, respectively, are related (Fig. 5.1). Let the translation and rotation of frame A relative to frame A^* be described by the displacement vector $\mathbf{b}(t)$ and the rotation tensor $\mathbf{Q}(t)$, with the latter satisfying the conditions

$$\mathbf{Q}^T \cdot \mathbf{Q} = \mathbf{Q} \cdot \mathbf{Q}^T = \mathbf{I}, \qquad \det\, \mathbf{Q} = 1 \tag{5.37}$$

for a proper orthogonal tensor (see Section 2.6.3). Then, while $\mathbf{r}$ is constant rela-

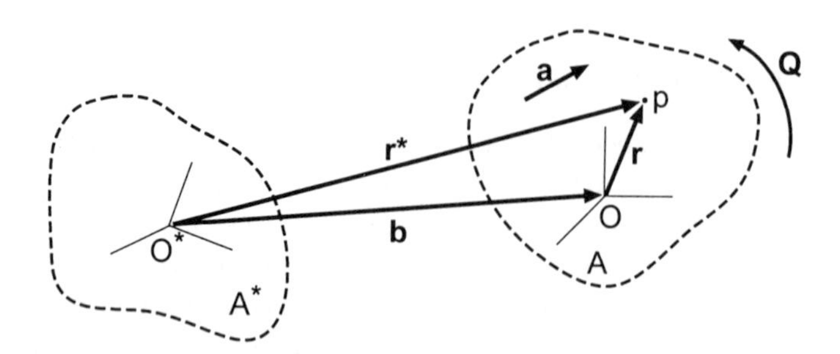

Fig. 5.1 Two reference frames (A and A^*) in relative motion. The point p and the vector $\mathbf{a}$ are fixed in frame A, and $\mathbf{r}$ and $\mathbf{r}^*$ are position vectors of p relative to coordinate systems fixed in A and A^*, respectively.

tive to the observer in A, the observer in A^* sees $\mathbf{r}$ as the rotating vector $\mathbf{Q}(t) \cdot \mathbf{r}$. Thus, as seen from O^*, the position of p is

$$\mathbf{r}^* = \mathbf{Q} \cdot \mathbf{r} + \mathbf{b}. \tag{5.38}$$

Note that this equation does not seem to agree with our global view (relative to the "fixed" stars) of the geometry in Fig. 5.1, which suggests that $\mathbf{r}^* = \mathbf{r} + \mathbf{b}$. The reason for this seeming contradiction is that Eq. (5.38) expresses a relation between position vectors as they *appear* to observers in relative motion. Although not frame indifferent, motions that satisfy Eq. (5.38) are called **equivalent motions**, as they represent the motion of the same point as seen by observers in two different reference frames.

Vectors

Equation (5.38) links the two distinct position vectors $\mathbf{r}$ and $\mathbf{r}^*$, which locate the same point in space. We now examine how the observers in frames A and A^* view a single vector $\mathbf{a}$ that is fixed in A (Fig. 5.1). Obviously, $\mathbf{a}$ does not appear to move relative to the observer in A, just as a tree does not appear to move to a person standing on the surface of the earth. But relative to the observer in A^*, the vector $\mathbf{a}$ translates and rotates. Translation does not change a vector, but rotation *does*. Thus, even though $\mathbf{a}$ is not changing in frame A, the relative motion of frame A^* makes $\mathbf{a}$ *appear* to be changing according to the relation

$$\mathbf{a}^* = \mathbf{Q} \cdot \mathbf{a} \tag{5.39}$$

as viewed by the observer at O^*. This transformation ensures that the vector $\mathbf{a}$ is frame indifferent.

In other words, if the observers at O and O^* simultaneously measure the length and orientation of the vector $\mathbf{a}$, the vectors constructed from these measurements must satisfy Eq. (5.39) at each instant of time. Thus, although $\mathbf{a}$ appears to be changing to the observer in A^*, it actually is constant relative to the observer (or a body) moving with frame A.

Example 5.2 Using the transformation (5.39), show that (a) the length of a vector and the angle between two vectors fixed in frame A are objective quantities; and (b) that velocity is *not* an objective quantity.

Solution. (a) Clearly, lengths and angles are scalar invariants under changes in the frame of reference. As measured in A, the squared length of $\mathbf{a}$ is $ds^2 = \mathbf{a} \cdot \mathbf{a}$. As measured in A^*, the squared length of $\mathbf{a}$ is

$$\begin{aligned} ds^{*2} &= \mathbf{a}^* \cdot \mathbf{a}^* = (\mathbf{Q} \cdot \mathbf{a}) \cdot (\mathbf{Q} \cdot \mathbf{a}) \\ &= (\mathbf{a} \cdot \mathbf{Q}^T) \cdot (\mathbf{Q} \cdot \mathbf{a}) = \mathbf{a} \cdot (\mathbf{Q}^T \cdot \mathbf{Q}) \cdot \mathbf{a} \\ &= \mathbf{a} \cdot \mathbf{I} \cdot \mathbf{a} = \mathbf{a} \cdot \mathbf{a} \\ &= ds^2 \end{aligned} \tag{5.40}$$

where Eqs. (5.37) and (5.39) have been used. Thus, as expected, the length of $\mathbf{a}$ is independent of the motion of the observer.

Next, consider the angle between two vectors $\mathbf{a}$ and $\mathbf{b}$ fixed in frame A. As seen from A^*, Eq. (5.39) gives

$$\mathbf{a}^* = \mathbf{Q} \cdot \mathbf{a}, \qquad \mathbf{b}^* = \mathbf{Q} \cdot \mathbf{b}, \tag{5.41}$$

and so

$$\begin{aligned} \mathbf{a}^* \cdot \mathbf{b}^* &= (\mathbf{Q} \cdot \mathbf{a}) \cdot (\mathbf{Q} \cdot \mathbf{b}) = (\mathbf{a} \cdot \mathbf{Q}^T) \cdot (\mathbf{Q} \cdot \mathbf{b}) \\ &= \mathbf{a} \cdot (\mathbf{Q}^T \cdot \mathbf{Q}) \cdot \mathbf{b} = \mathbf{a} \cdot \mathbf{I} \cdot \mathbf{b} \\ &= \mathbf{a} \cdot \mathbf{b}. \end{aligned} \tag{5.42}$$

Thus, $\cos(\mathbf{a}, \mathbf{b}) = \cos(\mathbf{a}^*, \mathbf{b}^*)$, which implies that angles are invariant under changes of reference frame.

(b) If the velocity of a point in frame A is $\dot{\mathbf{r}}$, then Eq. (5.38) shows that the

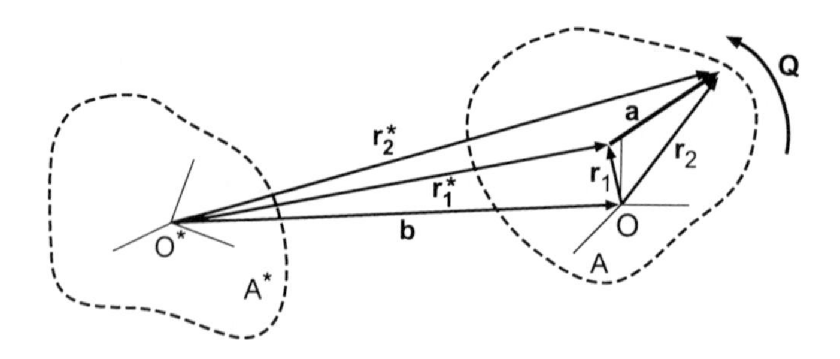

Fig. 5.2 Position vectors to the ends of a vector $\mathbf{a}$ fixed in reference frame A, which is in motion relative to frame A^*.

velocity of the point as seen from A^* is

$$\begin{aligned} \mathbf{v}^* &= \dot{\mathbf{r}}^* = \mathbf{Q} \cdot \dot{\mathbf{r}} + \dot{\mathbf{Q}} \cdot \mathbf{r} + \dot{\mathbf{b}} \\ &= \mathbf{Q} \cdot \mathbf{v} + \dot{\mathbf{Q}} \cdot \mathbf{r} + \dot{\mathbf{b}}. \end{aligned} \tag{5.43}$$

Clearly, $|\mathbf{v}^*| \neq |\mathbf{v}|$ and a similar calculation shows that acceleration also is not frame indifferent. ■

Example 5.3 Use Eq. (5.38) to derive Eq. (5.39).

Solution. Consider the position vectors to the ends of the vector $\mathbf{a}$, which is fixed in frame A (Fig. 5.2). Relative to the observers in A and A^*, respectively, the geometry gives

$$\mathbf{a} = \mathbf{r}_2 - \mathbf{r}_1, \qquad \mathbf{a}^* = \mathbf{r}_2^* - \mathbf{r}_1^* \tag{5.44}$$

where Eq. (5.38) gives

$$\begin{aligned} \mathbf{r}_1^* &= \mathbf{Q} \cdot \mathbf{r}_1 + \mathbf{b} \\ \mathbf{r}_2^* &= \mathbf{Q} \cdot \mathbf{r}_2 + \mathbf{b}. \end{aligned} \tag{5.45}$$

Combining these relations yields

$$\mathbf{a}^* = \mathbf{r}_2^* - \mathbf{r}_1^* = \mathbf{Q} \cdot (\mathbf{r}_2 - \mathbf{r}_1) = \mathbf{Q} \cdot \mathbf{a},$$

which is Eq. (5.39). ■

Tensors

The objectivity condition for a second-order tensor $\mathbf{T}$ follows from its defining relation (2.30). Suppose the observer in frame A witnesses the transformation

$$\mathbf{b} = \mathbf{T} \cdot \mathbf{a} \tag{5.46}$$

between the two vectors $\mathbf{a}$ and $\mathbf{b}$ fixed in A. Frame indifference requires that this transformation have the same form relative to the observer in A^*, i.e.,

$$\mathbf{b}^* = \mathbf{T}^* \cdot \mathbf{a}^* \tag{5.47}$$

where $\mathbf{a}^*$ and $\mathbf{b}^*$ satisfy Eqs. (5.41), and $\mathbf{T}^*$ is the tensor $\mathbf{T}$ as seen from A^*.

Inserting Eq. (5.46) into $(5.41)_2$ yields

$$\begin{aligned} \mathbf{b}^* &= \mathbf{Q} \cdot \mathbf{b} = \mathbf{Q} \cdot (\mathbf{T} \cdot \mathbf{a}) = \mathbf{Q} \cdot \mathbf{T} \cdot (\mathbf{Q}^T \cdot \mathbf{a}^*) \\ &= (\mathbf{Q} \cdot \mathbf{T} \cdot \mathbf{Q}^T) \cdot \mathbf{a}^* \end{aligned}$$

in which the substitution $\mathbf{a} = \mathbf{Q}^T \cdot \mathbf{a}^*$ follows from Eqs. $(5.41)_1$ and (5.37). Comparing this expression with Eq. (5.47) reveals that

$$\mathbf{T}^* = \mathbf{Q} \cdot \mathbf{T} \cdot \mathbf{Q}^T \tag{5.48}$$

is the appropriate transformation relation.

Example 5.4 Suppose the governing equation for a physical system has the form

$$\mathbf{a} = \mathbf{A} \cdot \mathbf{B} \cdot \mathbf{b} + \mathbf{C} \cdot \mathbf{c} \tag{5.49}$$

relative to an observer in frame A, where $\mathbf{a}$, $\mathbf{b}$, and $\mathbf{c}$ are vectors and $\mathbf{A}$, $\mathbf{B}$, and $\mathbf{C}$ are second-order tensors. According to the principle of material frame indifference, this equation must be seen as

$$\mathbf{a}^* = \mathbf{A}^* \cdot \mathbf{B}^* \cdot \mathbf{b}^* + \mathbf{C}^* \cdot \mathbf{c}^* \tag{5.50}$$

to an observer in frame A^*. Show that the vector and tensor transformation relations ensure that Eqs. (5.49) and (5.50) are equivalent.

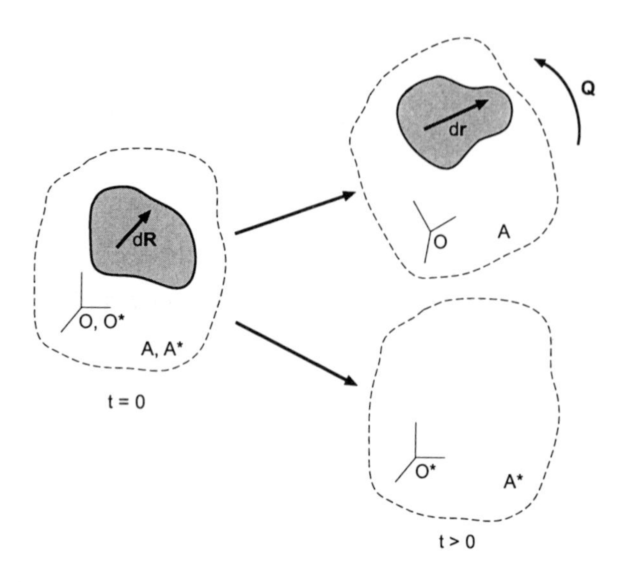

Fig. 5.3 Deformation of a body in reference frame A as seen by observers stationed in frames A and A^*, which are in relative motion. At $t = 0$, these frames coincide.

Solution. With Eqs. (5.37), (5.39), and (5.48), Eq. (5.50) takes the form

$$\begin{aligned}
\mathbf{Q}\cdot\mathbf{a} &= (\mathbf{Q}\cdot\mathbf{A}\cdot\mathbf{Q}^T)\cdot(\mathbf{Q}\cdot\mathbf{B}\cdot\mathbf{Q}^T)\cdot(\mathbf{Q}\cdot\mathbf{b}) \\
&\quad +(\mathbf{Q}\cdot\mathbf{C}\cdot\mathbf{Q}^T)\cdot(\mathbf{Q}\cdot\mathbf{c}) \\
&= \mathbf{Q}\cdot\mathbf{A}\cdot(\mathbf{Q}^T\cdot\mathbf{Q})\cdot\mathbf{B}\cdot(\mathbf{Q}^T\cdot\mathbf{Q})\cdot\mathbf{b} \\
&\quad +\mathbf{Q}\cdot\mathbf{C}\cdot(\mathbf{Q}^T\cdot\mathbf{Q})\cdot\mathbf{c} \\
&= \mathbf{Q}\cdot\mathbf{A}\cdot\mathbf{I}\cdot\mathbf{B}\cdot\mathbf{I}\cdot\mathbf{b}+\mathbf{Q}\cdot\mathbf{C}\cdot\mathbf{I}\cdot\mathbf{c} \\
&= \mathbf{Q}\cdot(\mathbf{A}\cdot\mathbf{B}\cdot\mathbf{b}+\mathbf{C}\cdot\mathbf{c}).
\end{aligned}$$

For arbitrary $\mathbf{Q}$, this relation implies Eq. (5.49), confirming its objective character. ■

Deformation Gradient Tensor

The deformation gradient tensor $\mathbf{F}$ contains information about the deformation and rigid-body rotation of a material element in a body. Deformation is an objective quantity, since it describes changes in lengths and angles, and, *relative to a given reference configuration*, rotation also is frame indifferent. Thus, $\mathbf{F}$ is an objective measure. While it may seem that, being a second-order tensor, $\mathbf{F}$

should transform according to Eq. (5.48) under a change of frame, the need for a reference configuration actually alters the transformation relation.

To determine the proper transformation, we consider two reference frames A and A^* that coincide at $t = 0$, when the body is undeformed (Fig. 5.3). At this instant, the observers stationed in both frames view the undeformed line element $d\mathbf{R}$ in the same way. For $t > 0$, the two frames move apart, with the deformed body following the rigid-body motion of frame A (Fig. 5.3). The vector $d\mathbf{r}$, the deformed image of $d\mathbf{R}$, then appears differently to the observers in A and A^*, with Eq. (5.39) giving

$$d\mathbf{r}^* = \mathbf{Q}\cdot d\mathbf{r}. \tag{5.51}$$

Material frame indifference demands that the forms of the defining relations for $\mathbf{F}$ be the same in all reference frames. Thus, Eq. (3.19) gives

$$\begin{aligned} d\mathbf{r} &= \mathbf{F}\cdot d\mathbf{R} \\ d\mathbf{r}^* &= \mathbf{F}^*\cdot d\mathbf{R} \end{aligned} \tag{5.52}$$

as seen by the observers in A and A^*, respectively. Substituting Eq. ($5.52)_1$ into (5.51) yields

$$d\mathbf{r}^* = \mathbf{Q}\cdot(\mathbf{F}\cdot d\mathbf{R}) = (\mathbf{Q}\cdot\mathbf{F})\cdot d\mathbf{R},$$

and comparison with $(5.52)_2$ shows that

$$\mathbf{F}^* = \mathbf{Q}\cdot\mathbf{F}. \tag{5.53}$$

This relation, which has the form of Eq. (5.39), indicates that the deformation gradient tensor transforms like a vector under a change of reference frame.

Summary

Under a change of reference frame, the following transformations apply:

$$\begin{array}{rrcl} \textbf{Motion:} & \mathbf{r}^* & = & \mathbf{Q}\cdot\mathbf{r}+\mathbf{b} \\ \textbf{Scalar:} & \phi^* & = & \phi \\ \textbf{Vector:} & \mathbf{a}^* & = & \mathbf{Q}\cdot\mathbf{a} \\ \textbf{Second-Order Tensor:} & \mathbf{T}^* & = & \mathbf{Q}\cdot\mathbf{T}\cdot\mathbf{Q}^T \\ \textbf{Deformation Gradient Tensor:} & \mathbf{F}^* & = & \mathbf{Q}\cdot\mathbf{F} \end{array} \tag{5.54}$$

As discussed above, position vectors are never frame indifferent, but scalars always are, and objective vectors and tensors must transform according to the above relations. Within a single frame of reference, equations are coordinate invariant if they are written in tensor form. They also are frame indifferent if all of their variables transform according to Eqs. $(5.54)_{2\text{-}5}$ under a change of frame.

Example 5.5 Show that the rate-of-deformation tensor $\mathbf{D}$ is an objective quantity.

Solution. Recall that this tensor is defined by

$$\mathbf{D} = \tfrac{1}{2}[\bar{\nabla}\mathbf{v} + (\bar{\nabla}\mathbf{v})^T], \tag{5.55}$$

where $\mathbf{v}$ is the particle velocity and $\bar{\nabla}$ is the gradient operator with respect to the deformed body. If $\mathbf{D}$ is the rate-of-deformation tensor observed in frame A, then

$$\mathbf{D}^* = \tfrac{1}{2}[\bar{\nabla}^*\mathbf{v}^* + (\bar{\nabla}^*\mathbf{v}^*)^T] \tag{5.56}$$

is the same tensor as seen by an observer in frame A^*.

Like $\mathbf{F}$, $\mathbf{D}$ is a measure of local deformation. Unlike $\mathbf{F}$, however, $\mathbf{D}$ depends only on the configuration of the deformed body. Frame indifference, therefore, requires that $\mathbf{D}$ and $\mathbf{D}^*$ satisfy the relation

$$\mathbf{D}^* = \mathbf{Q} \cdot \mathbf{D} \cdot \mathbf{Q}^T \tag{5.57}$$

as given by Eq. $(5.54)_4$. To show that this expression is valid, we need the transformation for $\bar{\nabla}$ under a change of frame. Consider the length element $d\mathbf{r}$ in frame A, which is seen as

$$d\mathbf{r}^* = \mathbf{Q}\cdot d\mathbf{r}$$

in A^*. Since $\mathbf{Q}^T = \mathbf{Q}^{-1}$, this relation also can be written in the form

$$d\mathbf{r} = \mathbf{Q}^T\cdot d\mathbf{r}^* = d\mathbf{r}^*\cdot\mathbf{Q}, \tag{5.58}$$

and Eqs. $(2.154)_1$ and $(3.18)_2$ show that the length elements are related by

$$d\mathbf{r} = d\mathbf{r}^*\cdot\frac{\partial\mathbf{r}}{\partial\mathbf{r}^*} = d\mathbf{r}^*\cdot\bar{\nabla}^*\mathbf{r} \tag{5.59}$$

where $\bar{\nabla}^* \equiv \partial/\partial\mathbf{r}^*$. Comparing Eqs. (5.58) and (5.59) yields the relation

$$\mathbf{Q} = \bar{\nabla}^*\mathbf{r}. \tag{5.60}$$

Now, the transformation

$$\boxed{\bar{\boldsymbol{\nabla}}^* = \mathbf{Q} \cdot \bar{\boldsymbol{\nabla}}} \tag{5.61}$$

can be confirmed by the manipulations

$$\begin{aligned} \bar{\boldsymbol{\nabla}}^* \mathbf{r} &= \mathbf{Q} \cdot \bar{\boldsymbol{\nabla}} \mathbf{r} \\ &= \mathbf{Q} \cdot \frac{\partial \mathbf{r}}{\partial \mathbf{r}} = \mathbf{Q} \cdot \mathbf{I} = \mathbf{Q}, \end{aligned}$$

which agrees with Eq. (5.60). Equation (5.61) shows that the gradient operator transforms as a vector under a change of frame.

Turning now to the rate-of-deformation tensor, we first differentiate Eq. $(5.54)_1$ to get

$$\begin{aligned} \mathbf{v}^* &= \dot{\mathbf{r}}^* = \dot{\mathbf{Q}} \cdot \mathbf{r} + \mathbf{Q} \cdot \dot{\mathbf{r}} + \dot{\mathbf{b}} \\ &= \mathbf{r} \cdot \dot{\mathbf{Q}}^T + \dot{\mathbf{r}} \cdot \mathbf{Q}^T + \dot{\mathbf{b}}. \end{aligned} \tag{5.62}$$

Because $\dot{\mathbf{r}} = \mathbf{v}$ and $\mathbf{b}$ and $\mathbf{Q}$ are functions only of time, we have

$$\begin{aligned} \bar{\boldsymbol{\nabla}}^* \mathbf{v}^* &= (\bar{\boldsymbol{\nabla}}^* \mathbf{r}) \cdot \dot{\mathbf{Q}}^T + (\bar{\boldsymbol{\nabla}}^* \mathbf{v}) \cdot \mathbf{Q}^T \\ &= \mathbf{Q} \cdot \dot{\mathbf{Q}}^T + \mathbf{Q} \cdot (\bar{\boldsymbol{\nabla}} \mathbf{v}) \cdot \mathbf{Q}^T \end{aligned} \tag{5.63}$$

where Eqs. (5.60) and (5.61) have been used in the last line Thus,

$$(\bar{\boldsymbol{\nabla}}^* \mathbf{v}^*)^T = \dot{\mathbf{Q}} \cdot \mathbf{Q}^T + \mathbf{Q} \cdot (\bar{\boldsymbol{\nabla}} \mathbf{v})^T \cdot \mathbf{Q}^T \tag{5.64}$$

since $(\mathbf{Q} \cdot \mathbf{T} \cdot \mathbf{Q}^T)^T = (\mathbf{Q}^T)^T \cdot (\mathbf{Q} \cdot \mathbf{T})^T = \mathbf{Q} \cdot \mathbf{T}^T \cdot \mathbf{Q}^T$. Next, substituting these equations into (5.56) gives

$$\mathbf{D}^* = \tfrac{1}{2}[\mathbf{Q} \cdot \dot{\mathbf{Q}}^T + \dot{\mathbf{Q}} \cdot \mathbf{Q}^T + \mathbf{Q} \cdot (\bar{\boldsymbol{\nabla}} \mathbf{v} + (\bar{\boldsymbol{\nabla}} \mathbf{v})^T) \cdot \mathbf{Q}^T]. \tag{5.65}$$

Finally, we note that

$$\mathbf{Q} \cdot \dot{\mathbf{Q}}^T + \dot{\mathbf{Q}} \cdot \mathbf{Q}^T = \frac{d}{dt}(\mathbf{Q} \cdot \mathbf{Q}^T) = \frac{d}{dt}(\mathbf{I}) = \mathbf{0}.$$

Hence, with $\mathbf{D}$ given by Eq. (5.55), (5.65) agrees with (5.57). Thus, although velocity is not an objective quantity, the rate-of-deformation tensor $\mathbf{D}$ is frame indifferent. ■

Implications for Constitutive Equations

Contact forces within a body are assumed to be frame indifferent, as are the differential areas on which they act. Thus, the Cauchy stress tensor is an objective quantity. Since deformation and temperature also are objective, it follows that the constitutive relations (5.35) for a thermoelastic material are frame indifferent.

Consider a simple material with a constitutive relation of the form[1]

$$\boldsymbol{\sigma} = \mathbf{g}_{(\sigma)}(\mathbf{F}) \tag{5.66}$$

where the tensor function $\mathbf{g}_{(\sigma)}$ is called the **response function**. Here, we assume that the response function is determined through experiments conducted in frame A. Then, if the functional form of the constitutive equation is independent of the motion of the observer, Eq. (5.66) must appear as

$$\boldsymbol{\sigma}^* = \mathbf{g}_{(\sigma)}(\mathbf{F}^*) \tag{5.67}$$

to an observer in frame A^*, where $\boldsymbol{\sigma}^*$ and $\mathbf{F}^*$ are related to $\boldsymbol{\sigma}$ and $\mathbf{F}$ through the appropriate objectivity relations. Since the Cauchy stress tensor depends only on the forces and geometry of the deformed body, $\boldsymbol{\sigma}$ transforms according to Eq. $(5.54)_4$ under a change of frame.[2] Hence, using Eqs. $(5.54)_{4\text{-}5}$ in (5.67) yields

$$\mathbf{Q} \cdot \boldsymbol{\sigma} \cdot \mathbf{Q}^T = \mathbf{g}_{(\sigma)}(\mathbf{Q} \cdot \mathbf{F}). \tag{5.68}$$

This equation must hold for an arbitrary rotation $\mathbf{Q}$. Convenient forms for the constitutive relation are obtained by choosing $\mathbf{Q} = \boldsymbol{\Theta}^T$, where $\boldsymbol{\Theta}$ is the rotation tensor defined by the polar decomposition (3.147). In this case, (5.68) becomes

$$\boldsymbol{\Theta}^T \cdot \boldsymbol{\sigma} \cdot \boldsymbol{\Theta} = \mathbf{g}_{(\sigma)}(\boldsymbol{\Theta}^T \cdot \mathbf{F}). \tag{5.69}$$

Moreover, since $\boldsymbol{\Theta}^T = \boldsymbol{\Theta}^{-1}$, Eq. (3.147) gives $\mathbf{U} = \boldsymbol{\Theta}^T \cdot \mathbf{F}$ and Eq. (5.69) yields

$$\boldsymbol{\sigma} = \boldsymbol{\Theta} \cdot \mathbf{g}_{(\sigma)}(\mathbf{U}) \cdot \boldsymbol{\Theta}^T,$$

[1] For convenience, the other independent variables of Eq. $(5.35)_1$ are not listed explicitly.

[2] Equation $(5.54)_4$ may not apply to the Piola-Kirchhoff stress tensors $\mathbf{t}$ and $\mathbf{s}$, since these tensors depend on reference configurations in addition to that of the deformed body.

which separates the contributions to the stress tensor of pure deformation $\mathbf{U}$ and rotation $\boldsymbol{\Theta}$. Other forms follow by noting Eqs. (3.27) and (3.153), which give $\mathbf{U}^2 = \mathbf{C} = \mathbf{I}+2\mathbf{E}$. Hence, Eq. (5.66) and the principle of material frame indifference provide constitutive equations in the forms

$$\boxed{\begin{aligned} \boldsymbol{\sigma} &= \boldsymbol{\Theta} \cdot \mathbf{g}_{(\sigma)}(\mathbf{U}) \cdot \boldsymbol{\Theta}^T \\ \text{or} \quad \boldsymbol{\sigma} &= \boldsymbol{\Theta} \cdot \mathbf{f}_{(\sigma)}(\mathbf{C}) \cdot \boldsymbol{\Theta}^T \\ \text{or} \quad \boldsymbol{\sigma} &= \boldsymbol{\Theta} \cdot \mathbf{h}_{(\sigma)}(\mathbf{E}) \cdot \boldsymbol{\Theta}^T. \end{aligned}} \tag{5.70}$$

To write the constitutive relations in terms of the Piola-Kirchhoff stress tensors, we use Eq. (4.30):

$$\boldsymbol{\sigma} = J^{-1}\,\mathbf{F} \cdot \mathbf{t} = J^{-1}\,\mathbf{F} \cdot \mathbf{s} \cdot \mathbf{F}^T. \tag{5.71}$$

Substituting Eq. $(5.70)_1$ gives

$$J^{-1}\,\mathbf{F} \cdot \mathbf{t} = \boldsymbol{\Theta} \cdot \mathbf{g}_{(\sigma)}(\mathbf{U}) \cdot \boldsymbol{\Theta}^T$$

or

$$J^{-1}\,(\boldsymbol{\Theta}^T \!\cdot\! \mathbf{F}) \cdot \mathbf{t} = \mathbf{g}_{(\sigma)}(\mathbf{U}) \cdot \boldsymbol{\Theta}^T.$$

Since $\boldsymbol{\Theta}^T \!\cdot\! \mathbf{F} = \mathbf{U}$ and Eqs. (3.123) and (5.37) give $J = \det \mathbf{F} = \det(\boldsymbol{\Theta} \cdot \mathbf{U}) = \det \boldsymbol{\Theta} \det \mathbf{U} = \det \mathbf{U}$, we have

$$\mathbf{t} = \mathbf{g}_{(t)}(\mathbf{U}) \cdot \boldsymbol{\Theta}^T$$

where

$$\mathbf{g}_{(t)}(\mathbf{U}) = (\det \mathbf{U})\,\mathbf{U}^{-1} \!\cdot\! \mathbf{g}_{(\sigma)}(\mathbf{U}).$$

Thus, in terms of the first Piola-Kirchhoff stress tensor, the constitutive relations take the forms

$$\boxed{\begin{aligned} \mathbf{t} &= \mathbf{g}_{(t)}(\mathbf{U}) \cdot \boldsymbol{\Theta}^T \\ \text{or} \quad \mathbf{t} &= \mathbf{f}_{(t)}(\mathbf{C}) \cdot \boldsymbol{\Theta}^T \\ \text{or} \quad \mathbf{t} &= \mathbf{h}_{(t)}(\mathbf{E}) \cdot \boldsymbol{\Theta}^T. \end{aligned}} \tag{5.72}$$

For the second Piola-Kirchhoff stress tensor, Eqs. (5.71) and $(5.72)_1$ give

$$\mathbf{s} = \mathbf{t} \cdot \mathbf{F}^{-T} = \mathbf{g}_{(t)}(\mathbf{U}) \cdot \boldsymbol{\Theta}^T \cdot \mathbf{F}^{-T},$$

and the symmetry property of s allows us to write

$$\begin{aligned}
\mathbf{s} &= \mathbf{s}^T = (\boldsymbol{\Theta}^T \cdot \mathbf{F}^{-T})^T \cdot [\mathbf{g}_{(t)}(\mathbf{U})]^T \\
&= (\mathbf{F}^{-1} \cdot \boldsymbol{\Theta}) \cdot [\mathbf{g}_{(t)}(\mathbf{U})]^T \\
&= \mathbf{U}^{-1} \cdot [\mathbf{g}_{(t)}(\mathbf{U})]^T \equiv \mathbf{g}_{(s)}(\mathbf{U}).
\end{aligned}$$

The last expression follows from the equation $\mathbf{U}^{-1} = \mathbf{F}^{-1} \cdot \boldsymbol{\Theta}$, as verified by the manipulations

$$\begin{aligned}
\mathbf{U} \cdot \mathbf{U}^{-1} &= (\boldsymbol{\Theta}^{-1} \cdot \mathbf{F}) \cdot (\mathbf{F}^{-1} \cdot \boldsymbol{\Theta}) \\
&= \boldsymbol{\Theta}^{-1} \cdot \mathbf{I} \cdot \boldsymbol{\Theta} = \mathbf{I}
\end{aligned}$$

where Eq. (3.147) has been used. Thus, the constitutive equations also can be written in the forms

$$\boxed{\begin{aligned}
& \mathbf{s} = \mathbf{g}_{(s)}(\mathbf{U}) \\
\text{or}\quad & \mathbf{s} = \mathbf{f}_{(s)}(\mathbf{C}) \\
\text{or}\quad & \mathbf{s} = \mathbf{h}_{(s)}(\mathbf{E}).
\end{aligned}} \tag{5.73}$$

These relations, written in terms of the second Piola-Kirchhoff stress tensor, are particularly convenient since the rotation does not appear explicitly. For small rotation ($\boldsymbol{\Theta} \cong \mathbf{I}$), however, Eqs. (5.70), (5.72), and (5.73) all reduce to the same form.

5.2.6 *Principle of Physical Admissibility*

The principle of physical admissibility stipulates that constitutive relations must not conflict with any of the basic laws of continuum mechanics. For example, we already have seen that the principle of angular momentum leads to the conclusion that the Cauchy stress tensor is symmetric (Section 4.6.2). Thus, the constitutive equations must not violate this symmetry. Here, we examine the consequences of the first and second laws of thermodynamics on the form of the constitutive equations.

The present development focuses on a Lagrangian formulation, with the second Piola-Kirchhoff stress tensor $\mathbf{s}$ and the Lagrangian strain tensor $\mathbf{E}$ taken as basic variables. Thus, the constitutive equations (5.35) and (5.36) are replaced by the alternate relations

$$\begin{aligned} \mathbf{s} &= \mathbf{s}(\mathbf{R}, \mathbf{E}, T, \nabla T) \\ \mathbf{q}_0 &= \mathbf{q}_0(\mathbf{R}, \mathbf{E}, T, \nabla T) \\ u &= u(\mathbf{R}, \mathbf{E}, T, \nabla T) \\ \eta &= \eta(\mathbf{R}, \mathbf{E}, T, \nabla T) \end{aligned} \tag{5.74}$$

and

$$\begin{aligned} \mathbf{s} &= \mathbf{s}(\mathbf{R}, \mathbf{E}, \eta, \nabla \eta) \\ \mathbf{q}_0 &= \mathbf{q}_0(\mathbf{R}, \mathbf{E}, \eta, \nabla \eta) \\ u &= u(\mathbf{R}, \mathbf{E}, \eta, \nabla \eta) \\ T &= T(\mathbf{R}, \mathbf{E}, \eta, \nabla \eta). \end{aligned} \tag{5.75}$$

The state of stress in a thermoelastic body depends only on the instantaneous deformation and temperature (or entropy) fields. Since no energy is dissipated during loading, the deformation is a reversible process governed by the thermodynamic relations

$$\boxed{\begin{aligned} \rho_0 \dot{u} - \mathbf{s} : \dot{\mathbf{E}} - \rho_0 r + \nabla \cdot \mathbf{q}_0 &= 0 \\ \dot{\eta} - \frac{r}{T} + \frac{1}{\rho_0 T} \nabla \cdot \mathbf{q}_0 &= 0 \end{aligned}} \tag{5.76}$$

as given by Eqs. (5.23) and $(5.33)_2$. Eliminating r between these two equations yields

$$\boxed{\rho_0 (T\dot{\eta} - \dot{u}) + \mathbf{s} : \dot{\mathbf{E}} = 0.} \tag{5.77}$$

In the following, the general constitutive relations (5.74) and (5.75) are substituted into this equation to examine the cases for which (1) temperature is an independent variable and (2) entropy is an independent variable.

Case 1: When temperature is taken as an independent variable, it is convenient to introduce the **free-energy function**

$$\boxed{\psi = u - T\eta} \tag{5.78}$$

as a replacement for the internal energy u. Then, Eq. $(5.74)_3$ is replaced by

$$\psi = \psi(\mathbf{R}, \mathbf{E}, T, \nabla T), \tag{5.79}$$

and using Eq. (5.78) to eliminate u in (5.77) gives

$$\boxed{-\rho_0(\dot{\psi} + \eta\dot{T}) + \mathbf{s} : \dot{\mathbf{E}} = 0.} \tag{5.80}$$

Next, substituting Eq. (5.79) into (5.80) and using the chain rule yields

$$\left(\mathbf{s} - \rho_0 \frac{\partial \psi}{\partial \mathbf{E}}\right) : \dot{\mathbf{E}} - \rho_0 \left(\eta + \frac{\partial \psi}{\partial T}\right) \dot{T} - \rho_0 \frac{\partial \psi}{\partial \nabla T} \cdot \nabla \dot{T} = 0.$$

For independent strain and temperature distributions, this equation implies the relations

$$\begin{aligned} \mathbf{s} &= \rho_0 \frac{\partial \psi}{\partial \mathbf{E}} \\ \eta &= -\frac{\partial \psi}{\partial T} \\ \frac{\partial \psi}{\partial \nabla T} &= \mathbf{0}. \end{aligned} \tag{5.81}$$

Thus, stress and entropy can be computed directly from ψ, and the last equation shows that ψ is independent of the temperature gradient. The first two of (5.81) represent the constitutive relations for a thermoelastic material with temperature taken as an independent variable.

If the deformation is **isothermal** ($\dot{T} = 0$), then $\psi = \psi(\mathbf{R}, \mathbf{E})$ and the material behaves elastically. In this case, we define the **strain-energy density function**

$$W(\mathbf{R}, \mathbf{E}) = \rho_0 \psi \tag{5.82}$$

per unit undeformed volume. Since the density ρ_0 of the undeformed body is independent of the deformation, Eq. $(5.81)_1$ gives the constitutive equation

$$\mathbf{s} = \frac{\partial W}{\partial \mathbf{E}}. \tag{5.83}$$

Case 2: If entropy is taken as an independent variable, then the internal energy is retained in the formulation. Substituting Eq. $(5.75)_3$ into (5.77) and using the chain rule gives

$$\left(\mathrm{s} - \rho_0 \frac{\partial u}{\partial \mathbf{E}}\right) : \dot{\mathbf{E}} + \rho_0 \left(T - \frac{\partial u}{\partial \eta}\right) \dot{\eta} - \rho_0 \frac{\partial u}{\partial \boldsymbol{\nabla} \eta} \cdot \boldsymbol{\nabla} \dot{\eta} = 0.$$

For independent strain and entropy fields, this equation yields

$$\begin{aligned} \mathrm{s} &= \rho_0 \frac{\partial u}{\partial \mathbf{E}} \\ T &= \frac{\partial u}{\partial \eta} \\ \frac{\partial u}{\partial \boldsymbol{\nabla} \eta} &= \mathbf{0}. \end{aligned} \tag{5.84}$$

Thus, stress and temperature can be computed directly from u, and the last equation shows that u is independent of the entropy gradient. The first two of (5.84) represent the constitutive relations for a thermoelastic material with entropy taken as an independent variable.

If the deformation is **isentropic** ($\dot{\eta} = 0$), then $u = u(\mathbf{R}, \mathbf{E})$ and the material behaves elastically. In this case, we take

$$W(\mathbf{R}, \mathbf{E}) = \rho_0 u, \tag{5.85}$$

and Eq. $(5.84)_1$ takes the form of Eq. (5.83).

A material possessing a constitutive relation of the form (5.83) is called a **hyperelastic material**. In other words, the mechanical properties of a hyperelastic material are characterized completely by a scalar strain-energy density function W. During an isothermal deformation, W is associated with the free energy per unit undeformed volume. During an isentropic deformation, W corresponds to the internal energy per unit undeformed volume.

The constitutive relation (5.83) can be written readily in terms of the Cauchy and first Piola-Kirchhoff stress tensors. Substitution into (4.30) and $(4.31)_1$ yields

$$\begin{aligned} \boldsymbol{\sigma} &= J^{-1}\, \mathbf{F} \cdot \frac{\partial W}{\partial \mathbf{E}} \cdot \mathbf{F}^T \\ \mathrm{t} &= \frac{\partial W}{\partial \mathbf{E}} \cdot \mathbf{F}^T. \end{aligned} \tag{5.86}$$

An alternate form for t can be found by noting that, via Eq. (5.22), the stress power $\mathrm{s} : \dot{\mathbf{E}}$ in Eq. (5.77) can be replaced by $\mathrm{t} : \dot{\mathbf{F}}^T$. Moreover, we replace s by

$\mathbf{t}$ and $\mathbf{E}$ by $\mathbf{F}^T$ in Eqs. (5.74) and (5.75). Then, manipulations similar to those discussed above lead to

$$\mathbf{t} = \frac{\partial W}{\partial \mathbf{F}^T}. \tag{5.87}$$

In summary, the constitutive equations for a hyperelastic material can be written in the following forms:

$$\boxed{\begin{aligned} \boldsymbol{\sigma} &= J^{-1}\,\mathbf{F}\cdot\frac{\partial W}{\partial \mathbf{E}}\cdot\mathbf{F}^T \\ \mathbf{t} &= \frac{\partial W}{\partial \mathbf{E}}\cdot\mathbf{F}^T = \frac{\partial W}{\partial \mathbf{F}^T} \\ \mathbf{s} &= \frac{\partial W}{\partial \mathbf{E}}. \end{aligned}} \tag{5.88}$$

Note that these relations are consistent with $\boldsymbol{\sigma}$ and $\mathbf{s}$ being symmetric tensors (since $\mathbf{E} = \mathbf{E}^T$), while $\mathbf{t}$ is not symmetric in general.

Scalar forms of these equations can be found by direct substitution of Eqs. (3.22), (3.29), and (4.32). Thus, Eq. $(5.88)_1$ gives

$$\begin{aligned} \sigma^{kl}\mathbf{g}_k\mathbf{g}_l &= J^{-1}(F^k_K\mathbf{g}_k\mathbf{G}^K)\cdot\left(\frac{\partial W}{\partial E_{IJ}}\,\mathbf{G}_I\mathbf{G}_J\right)\cdot(F^l_L\mathbf{G}^L\mathbf{g}_l) \\ &= J^{-1}F^k_K F^l_L\,\frac{\partial W}{\partial E_{IJ}}\,\delta^K_I\,\delta^L_J\mathbf{g}_k\mathbf{g}_l \\ &= J^{-1}F^k_I F^l_J\,\frac{\partial W}{\partial E_{IJ}}\,\mathbf{g}_k\mathbf{g}_l \end{aligned} \tag{5.89}$$

or

$$\begin{aligned} \sigma^{I^*J^*}\mathbf{g}_I\mathbf{g}_J &= J^{-1}(\mathbf{g}_K\mathbf{G}^K)\cdot\left(\frac{\partial W}{\partial E_{IJ}}\,\mathbf{G}_I\mathbf{G}_J\right)\cdot(\mathbf{G}^L\mathbf{g}_L) \\ &= J^{-1}\,\frac{\partial W}{\partial E_{IJ}}\,\delta^K_I\,\delta^L_J\mathbf{g}_K\mathbf{g}_L \\ &= J^{-1}\,\frac{\partial W}{\partial E_{IJ}}\,\mathbf{g}_I\mathbf{g}_J; \end{aligned} \tag{5.90}$$

Eq. $(5.88)_2$ gives

$$\begin{aligned} t^{Ik}\mathbf{G}_I\mathbf{g}_k &= \left(\frac{\partial W}{\partial E_{IJ}}\mathbf{G}_I\mathbf{G}_J\right)\cdot(F^k_K\mathbf{G}^K\mathbf{g}_k) \\ &= \frac{\partial W}{\partial E_{IJ}}\, F^k_K\delta^K_J\,\mathbf{G}_I\mathbf{g}_k \\ &= \frac{\partial W}{\partial E_{IJ}}\, F^k_J\,\mathbf{G}_I\mathbf{g}_k \end{aligned} \tag{5.91}$$

or

$$t^I_{\cdot j}\mathbf{G}_I\mathbf{g}^j = \frac{\partial W}{\partial(F^j_I\,\mathbf{G}^I\mathbf{g}_j)} = \frac{\partial W}{\partial F^j_I}\,\mathbf{G}_I\mathbf{g}^j; \tag{5.92}$$

and Eq. $(5.88)_3$ gives

$$s^{IJ}\,\mathbf{G}_I\mathbf{G}_J = \frac{\partial W}{\partial E_{IJ}}\,\mathbf{G}_I\mathbf{G}_J. \tag{5.93}$$

From these equations, the component forms of the constitutive relations for a hyperelastic material can be written as

$$\boxed{\begin{aligned} \sigma^{ij} &= J^{-1}F^i_I F^j_J\,\frac{\partial W}{\partial E_{IJ}} \\ \sigma^{I^*J^*} &= J^{-1}\,\frac{\partial W}{\partial E_{IJ}} \\ t^{Ij} &= F^j_J\,\frac{\partial W}{\partial E_{IJ}} \\ t^I_{\cdot j} &= \frac{\partial W}{\partial F^j_I} \\ s^{IJ} &= \frac{\partial W}{\partial E_{IJ}}. \end{aligned}} \tag{5.94}$$

Other forms can be found similarly.

Example 5.6 Derive Eq. $(5.94)_3$ directly from $(5.94)_4$.

Solution. To do this, we need the relation

$$E_{IJ} = \tfrac{1}{2}(F^i_I F^j_J g_{ij} - G_{IJ}) \tag{5.95}$$

as given by Eqs. $(3.30)_1$ and $(3.31)_1$. Then, with the chain rule and $W = W(E_{IJ})$, Eq. $(5.94)_4$ can be written

$$\begin{aligned} t^I_{\cdot j} &= \frac{\partial W}{\partial F^j_I} = \frac{\partial W}{\partial E_{KL}} \frac{\partial E_{KL}}{\partial F^j_I} \\ &= \frac{\partial W}{\partial E_{KL}} \frac{\partial}{\partial F^j_I} \left(\tfrac{1}{2} F^m_K F^n_L g_{mn}\right) \\ &= \frac{1}{2} \frac{\partial W}{\partial E_{KL}} g_{mn} \left(\frac{\partial F^m_K}{\partial F^j_I} F^n_L + F^m_K \frac{\partial F^n_L}{\partial F^j_I} \right) \\ &= \frac{1}{2} \frac{\partial W}{\partial E_{KL}} g_{mn} \left(\delta^m_j \delta^I_K F^n_L + F^m_K \delta^n_j \delta^I_L \right). \end{aligned}$$

Contractions over the Kronecker deltas convert this expression into

$$t^I_{\cdot j} = \frac{1}{2} \left(\frac{\partial W}{\partial E_{IL}} g_{jn} F^n_L + \frac{\partial W}{\partial E_{KI}} g_{mj} F^m_K \right),$$

and renaming indices and noting the symmetry of E_{IJ} and g_{ij} yields

$$t^I_{\cdot j} = \frac{\partial W}{\partial E_{IJ}} g_{jn} F^n_J.$$

Finally, using contravariant components of the metric tensor to raise a subscript gives

$$\begin{aligned} t^{Ik} &= g^{jk} t^I_{\cdot j} = g^{jk} \frac{\partial W}{\partial E_{IJ}} g_{jn} F^n_J \\ &= \delta^k_n \frac{\partial W}{\partial E_{IJ}} F^n_J \\ &= \frac{\partial W}{\partial E_{IJ}} F^k_J \end{aligned}$$

where Eq. (2.85) has been used. This relation agrees with Eq. $(5.94)_3$. ■

It is important to note that the strain components in the strain-energy function must be treated independently in evaluating Eqs. (5.94). In other words, we cannot set $E_{JI} = E_{IJ}$ in the expression for W before carrying out the differentiation $\partial W / \partial E_{IJ}$.

Consider, for example, a material characterized by the strain-energy density function

$$W = C(E_{12}^2 + E_{21}^2) \tag{5.96}$$

with C being a material constant. Equation $(5.94)_5$ gives

$$\begin{aligned} s^{12} &= \frac{\partial W}{\partial E_{12}} = 2CE_{12} \\ s^{21} &= \frac{\partial W}{\partial E_{21}} = 2CE_{21}, \end{aligned}$$

and setting $E_{21} = E_{12}$ gives $s^{21} = s^{12}$ as it should. However, if we set $E_{21} = E_{12}$ *a priori* to obtain

$$W = 2CE_{12}^2, \tag{5.97}$$

then we get

$$\begin{aligned} s^{12} &= \frac{\partial W}{\partial E_{12}} = 4CE_{12} \\ s^{21} &= \frac{\partial W}{\partial E_{21}} = 0, \end{aligned}$$

which is not correct. Some authors, therefore, replace Eq. $(5.94)_5$ by the equivalent expression

$$s^{IJ} = \frac{1}{2}\left(\frac{\partial W}{\partial E_{IJ}} + \frac{\partial W}{\partial E_{JI}}\right), \tag{5.98}$$

which gives the correct result using either (5.96) or (5.97).

5.3 Strain-Energy Density Function

Thus far, we have explored the implications of six of the seven principles listed in Section 5.2. The seventh, the principle of material symmetry, is used herein to deduce the form of the strain-energy density function for orthotropic, transversely isotropic, and isotropic hyperelastic materials.[3] These types of material symmetries often are used to characterize soft tissues.

This section deals only with the general manner in which W must depend on the strain components for each type of material, and some specific functional

[3]Theorems for tensor invariants (Spencer, 1984; Zheng, 1994) can be used to develop constitutive equations in perhaps a more mathematically elegant fashion, but this is beyond the scope of this book.

forms are listed. Section 5.7 discusses strategies for determining material parameters. It is convenient in this section to work in local Cartesian coordinates, and so superscript notation is not used.

5.3.1 *Orthotropic Material*

A material is orthotropic if each point of the unstressed material possesses three mutually orthogonal planes of symmetry. The symmetry is due to the microstructure, which may consist of aligned, orthogonally oriented fibers embedded in an isotropic matrix. Skin is but one example. The Cartesian axes (Z_1, Z_2, Z_3) are chosen to lie in the undeformed body along the intersections of the local planes of symmetry, making these planes the local coordinate planes. We call the Z_I axes **principal material axes**. In general, the orientation of these axes can vary from point to point in a body, but the principle of local action ensures that we need only consider a representative point in developing constitutive relations.

Due to the material symmetry, the constitutive relations must be independent of a reflection of any of the principal material axes. In other words, the constitutive equations must be form-invariant if the Z_I axes are replaced by the $\bar{Z}_I = -Z_I$ axes. It then follows that the strain-energy function must be independent of this coordinate transformation.

Let ε_{IJ} and $\bar{\varepsilon}_{IJ}$ be the components of the Lagrangian strain tensor with respect to the Z_I and $\bar{Z}_I$ coordinates, respectively. Then, we assume that the strain-energy density function has the functional dependencies[4]

$$\begin{aligned} W &= W(\varepsilon_{11}, \varepsilon_{22}, \varepsilon_{33}, \varepsilon_{12}, \varepsilon_{23}, \varepsilon_{31}) \\ W &= W(\bar{\varepsilon}_{11}, \bar{\varepsilon}_{22}, \bar{\varepsilon}_{33}, \bar{\varepsilon}_{12}, \bar{\varepsilon}_{23}, \bar{\varepsilon}_{31}) \end{aligned} \tag{5.99}$$

where Eq. (3.33) gives

$$\begin{aligned} \varepsilon_{IJ} &= \frac{1}{2}\left(\frac{\partial z_k}{\partial Z_I}\frac{\partial z_k}{\partial Z_J} - \delta_{IJ}\right) \\ \bar{\varepsilon}_{IJ} &= \frac{1}{2}\left(\frac{\partial z_k}{\partial \bar{Z}_I}\frac{\partial z_k}{\partial \bar{Z}_J} - \delta_{IJ}\right). \end{aligned} \tag{5.100}$$

Here, the z_i are Cartesian coordinates of the deformed image of the point located at Z_I (or $\bar{Z}_I$) in the undeformed body. Note that the z_i coordinate system is independent of both Z_I and $\bar{Z}_I$. With these definitions, we consider sequentially

[4]For convenience, we set $\varepsilon_{JI} = \varepsilon_{IJ}$ *a priori* and use Eq. (5.98) to compute the stress components.

the following symmetry transformations:

$$\begin{aligned}
\mathbf{T}_1: \quad & \bar{Z}_1 = -Z_1, && \bar{Z}_2 = Z_2, && \bar{Z}_3 = Z_3 \\
& \bar{\varepsilon}_{11} = \varepsilon_{11}, && \bar{\varepsilon}_{22} = \varepsilon_{22}, && \bar{\varepsilon}_{33} = \varepsilon_{33} \\
& \bar{\varepsilon}_{12} = -\varepsilon_{12}, && \bar{\varepsilon}_{23} = \varepsilon_{23}, && \bar{\varepsilon}_{31} = -\varepsilon_{31} \\
\mathbf{T}_2: \quad & \bar{Z}_1 = Z_1, && \bar{Z}_2 = -Z_2, && \bar{Z}_3 = Z_3 \\
& \bar{\varepsilon}_{11} = \varepsilon_{11}, && \bar{\varepsilon}_{22} = \varepsilon_{22}, && \bar{\varepsilon}_{33} = \varepsilon_{33} \\
& \bar{\varepsilon}_{12} = -\varepsilon_{12}, && \bar{\varepsilon}_{23} = -\varepsilon_{23}, && \bar{\varepsilon}_{31} = \varepsilon_{31} \\
\mathbf{T}_3: \quad & \bar{Z}_1 = Z_1, && \bar{Z}_2 = Z_2, && \bar{Z}_3 = -Z_3 \\
& \bar{\varepsilon}_{11} = \varepsilon_{11}, && \bar{\varepsilon}_{22} = \varepsilon_{22}, && \bar{\varepsilon}_{33} = \varepsilon_{33} \\
& \bar{\varepsilon}_{12} = \varepsilon_{12}, && \bar{\varepsilon}_{23} = -\varepsilon_{23}, && \bar{\varepsilon}_{31} = -\varepsilon_{31}.
\end{aligned} \tag{5.101}$$

Under the transformation $\mathbf{T}_1$, Eq. $(5.99)_2$ becomes

$$W = W(\varepsilon_{11}, \varepsilon_{22}, \varepsilon_{33}, -\varepsilon_{12}, \varepsilon_{23}, -\varepsilon_{31}),$$

and consistency with $(5.99)_1$ requires that W take the form

$$W = W(\varepsilon_{11}, \varepsilon_{22}, \varepsilon_{33}, \varepsilon_{23}, \varepsilon_{12}^2, \varepsilon_{31}^2, \varepsilon_{12}\varepsilon_{31}). \tag{5.102}$$

Next, applying $\mathbf{T}_2$ to this relation gives

$$W = W(\varepsilon_{11}, \varepsilon_{22}, \varepsilon_{33}, -\varepsilon_{23}, \varepsilon_{12}^2, \varepsilon_{31}^2, -\varepsilon_{12}\varepsilon_{31}),$$

and consistency with (5.102) requires

$$W = W(\varepsilon_{11}, \varepsilon_{22}, \varepsilon_{33}, \varepsilon_{12}^2, \varepsilon_{23}^2, \varepsilon_{31}^2, \varepsilon_{12}\varepsilon_{23}\varepsilon_{31}). \tag{5.103}$$

Furthermore, Eq. $(3.142)_3$ gives the third strain invariant

$$I_3 = \det(\delta_{IJ} + 2\varepsilon_{IJ})$$

in Cartesian coordinates. Expanding the determinant reveals that $\varepsilon_{12}\varepsilon_{23}\varepsilon_{31}$ can be expressed in terms of I_3 and the other arguments listed in Eq. (5.103), which therefore can be written in the form

$$\boxed{W = W(\varepsilon_{11}, \varepsilon_{22}, \varepsilon_{33}, \varepsilon_{12}^2, \varepsilon_{23}^2, \varepsilon_{31}^2, I_3).} \tag{5.104}$$

Applying $\mathbf{T}_3$ to this relation results in no further modification. Due to their high water content, soft tissues often are assumed to be incompressible. In this case, the strain invariant $I_3 = J^2 = 1$ is deleted from the list of arguments in Eq. (5.104).

As a specific example, consider blood vessels. To a first approximation, arteries can be treated as orthotropic circular cylinders, with the principal material directions defined by the cylindrical polar coordinates (R, Θ, Z). Experiments indicate that arterial tissue is nearly incompressible ($I_3 = 1$) and stiffens with increasing strain. Thus, one possible form for the strain-energy function is (Humphrey, 2002)

$$\begin{aligned} W &= C(e^Q - 1) \\ Q &= a_1\varepsilon_{RR}^2 + a_2\varepsilon_{\Theta\Theta}^2 + a_3\varepsilon_{ZZ}^2 + 2a_4\varepsilon_{RR}\varepsilon_{\Theta\Theta} + 2a_5\varepsilon_{\Theta\Theta}\varepsilon_{ZZ} \\ &\quad + 2a_6\varepsilon_{ZZ}\varepsilon_{RR} + a_7(\varepsilon_{R\Theta}^2 + \varepsilon_{\Theta R}^2) + a_8(\varepsilon_{\Theta Z}^2 + \varepsilon_{Z\Theta}^2) + a_9(\varepsilon_{ZR}^2 + \varepsilon_{RZ}^2) \end{aligned} \tag{5.105}$$

where C and the a_i are material constants, and the ε_{IJ} are physical components of strain relative to the polar system.[5] This form for W is consistent with Eq. (5.104). Once W is known, the stress-strain relations can be obtained using Eq. (5.98).

5.3.2 *Transversely Isotropic Material*

In an undeformed transversely isotropic material, each point possesses a single axis of symmetry, taken here as the Z_3-axis. In the Z_1Z_2-plane, the material is isotropic, i.e., the material properties in this plane are the same in all directions, while the properties in the Z_3 direction are different. This type of material may be composed, for example, of fibers (oriented in the Z_3 direction) embedded in an isotropic matrix. Skeletal muscle often is treated as a transversely isotropic material.

Since the Z_3-axis is an axis of symmetry, W must be invariant under the transformation $\mathbf{T}_3$ of Eq. (5.101). Moreover, since the properties are isotropic in the Z_1Z_2-plane, W also must be invariant for any rotation θ of the Z_1 and Z_2 axes about the Z_3-axis. The appropriate transformation for this case is

$$\mathbf{T}_4: \quad \begin{aligned} \bar{Z}_1 &= Z_1\cos\theta + Z_2\sin\theta \\ \bar{Z}_2 &= -Z_1\sin\theta + Z_2\cos\theta \\ \bar{Z}_3 &= Z_3 \end{aligned}$$

[5]In the neighborhood of a point, polar coordinate axes approximate Cartesian axes. The basis for writing W in terms of physical strain components is discussed in Section 5.4.4.

$$
\begin{aligned}
\bar{\varepsilon}_{11} &= \varepsilon_{11}\cos^2\theta + \varepsilon_{22}\sin^2\theta + 2\varepsilon_{12}\sin\theta\cos\theta \\
\bar{\varepsilon}_{22} &= \varepsilon_{11}\sin^2\theta + \varepsilon_{22}\cos^2\theta - 2\varepsilon_{12}\sin\theta\cos\theta \\
\bar{\varepsilon}_{33} &= \varepsilon_{33} \\
\bar{\varepsilon}_{12} &= (\varepsilon_{22} - \varepsilon_{11})\sin\theta\cos\theta + \varepsilon_{12}(\cos^2\theta - \sin^2\theta) \\
\bar{\varepsilon}_{23} &= \varepsilon_{23}\cos\theta - \varepsilon_{31}\sin\theta \\
\bar{\varepsilon}_{31} &= \varepsilon_{23}\sin\theta + \varepsilon_{31}\cos\theta.
\end{aligned} \tag{5.106}
$$

These relations represent the standard equations for strain transformation under a rotation about the Z_3-axis [see Eqs. (A.53) in Appendix A].

For the transformation $\mathbf{T}_4$, our main task is to find combinations of the strain components that are independent of the rotation θ. It is relatively easy to deduce that Eqs. (5.106) satisfy

$$
\begin{aligned}
\bar{\varepsilon}_{11} + \bar{\varepsilon}_{22} &= \varepsilon_{11} + \varepsilon_{22} \\
\bar{\varepsilon}_{31}^2 + \bar{\varepsilon}_{23}^2 &= \varepsilon_{31}^2 + \varepsilon_{23}^2 \\
\bar{\varepsilon}_{33} &= \varepsilon_{33},
\end{aligned} \tag{5.107}
$$

but other combinations are more difficult to find. Fortunately, Green and Adkins (1970) have done the job for us. Direct substitution of Eqs. (5.106) verifies the relations

$$
\begin{aligned}
\bar{E}_1^2 + \bar{E}_2^2 &= E_1^2 + E_2^2 \\
\bar{F}_1^2 + \bar{F}_2^2 &= F_1^2 + F_2^2 \\
\bar{E}_1\bar{F}_1 + \bar{E}_2\bar{F}_2 &= E_1F_1 + E_2F_2 \\
\bar{E}_1\bar{F}_2 - \bar{E}_2\bar{F}_1 &= E_1F_2 - E_2F_1
\end{aligned} \tag{5.108}
$$

where

$$
\begin{aligned}
E_1 &= \varepsilon_{11} - \varepsilon_{22} \\
E_2 &= 2\varepsilon_{12} \\
F_1 &= \varepsilon_{31}^2 - \varepsilon_{23}^2 \\
F_2 &= 2\varepsilon_{31}\varepsilon_{23},
\end{aligned} \tag{5.109}
$$

and $\bar{E}_1$, $\bar{E}_2$, $\bar{F}_1$, and $\bar{F}_2$ are given by placing bars over all quantities in Eqs. (5.109).

For W independent of rotation about the Z^3-axis, therefore, Eqs. (5.107) and

(5.108) indicate that the strain-energy function has the form

$$\begin{aligned} W &= W(\varepsilon_{11}+\varepsilon_{22}, \varepsilon_{31}^2+\varepsilon_{23}^2, \varepsilon_{33}, E_1^2+E_2^2, F_1^2+F_2^2, \\ &\qquad E_1F_1+E_2F_2, E_1F_2-E_2F_1), \end{aligned} \tag{5.110}$$

which also satisfies the transformation $\mathbf{T}_3$ of (5.101). Further manipulations show that this expression can be written more compactly. First, Eqs. (5.109) give

$$\begin{aligned} E_1^2+E_2^2 &= (\varepsilon_{11}+\varepsilon_{22})^2 - 4(\varepsilon_{11}\varepsilon_{22}-\varepsilon_{12}^2) \\ F_1^2+F_2^2 &= (\varepsilon_{31}^2+\varepsilon_{23}^2)^2 \\ E_1F_1+E_2F_2 &= 2\det[\varepsilon_{IJ}] + (\varepsilon_{11}+\varepsilon_{22})(\varepsilon_{31}^2+\varepsilon_{23}^2) \\ &\quad -2\varepsilon_{33}(\varepsilon_{11}\varepsilon_{22}-\varepsilon_{12}^2) \end{aligned}$$

$$(E_1F_2-E_2F_1)^2 = \begin{vmatrix} E_1^2+E_2^2 & E_1F_1+E_2F_2 \\ E_1F_1+E_2F_2 & F_1^2+F_2^2 \end{vmatrix},$$

and comparison with the terms in (5.110) reveals that W also can be written as

$$W = W(\varepsilon_{11}+\varepsilon_{22}, \varepsilon_{31}^2+\varepsilon_{23}^2, \varepsilon_{33}, \varepsilon_{11}\varepsilon_{22}-\varepsilon_{12}^2, \det[\varepsilon_{IJ}]). \tag{5.111}$$

Now, comparing these terms with those in the strain invariants I_1, I_2, and I_3 of Eq. (3.142) shows that the strain-energy function for a transversely isotropic material can be simplified to[6]

$$\boxed{W = W(I_1, I_2, I_3, I_4, I_5)} \tag{5.112}$$

where

$$\boxed{\begin{aligned} I_4 &= \varepsilon_{33} \\ I_5 &= \varepsilon_{31}^2+\varepsilon_{23}^2. \end{aligned}} \tag{5.113}$$

In direct form, it is easy to verify that these invariants can be written

$$\boxed{\begin{aligned} I_4 &= \mathbf{e}_3\cdot\mathbf{E}\cdot\mathbf{e}_3 \\ I_5 &= \mathbf{e}_3\cdot\mathbf{E}^2\cdot\mathbf{e}_3 - I_4^2 \end{aligned}} \tag{5.114}$$

[6]It is important to note here that any combination of invariants also is invariant.

where $\mathbf{e}_3$ is the unit vector in the fiber direction.

Heart muscle consists of layers that contain aligned fibers embedded in an isotropic matrix. Many researchers, therefore, treat each layer as transversely isotropic. For passive heart muscle, possible forms for the strain-energy density function include

$$W = b_1\, e^{b_2(I_1-3)} + b_3(\lambda_f - 1)^m \tag{5.115}$$

and

$$W = \sum_{i=0}^{n}\sum_{j=0}^{n} c_{ij}(I_1 - 3)^i(\lambda_f - 1)^j \tag{5.116}$$

where λ_f is the fiber stretch ratio ($\lambda_f^2 = 1 + 2\varepsilon_{33} = 1 + 2I_4$), the b_i and c_{ij} are material constants, and m is an even integer. Equation (5.115) separates the effects of the isotropic matrix (first term) from those of the fibers (second term). In contrast, Eq. (5.116) includes coupling between the matrix and the fibers.

5.3.3 *Isotropic Material*

In the undeformed configuration, the mechanical properties of an isotropic material are independent of direction.[7] Thus, interchanging any two coordinate axes should not affect the form of W. Setting $\bar{Z}_1 = Z_2$, $\bar{Z}_2 = Z_3$, and $\bar{Z}_3 = Z_1$ does not affect the three invariants defined by Eq. (3.142), but I_4 and I_5 of Eq. (5.113) change, i.e., they are no longer invariants. Thus, Eq. (5.112) reduces to

$$\boxed{W = W(I_1, I_2, I_3)} \tag{5.117}$$

for an isotropic material.

Although most soft tissues are not isotropic, some undifferentiated tissues in the early embryo, can be treated as approximately isotropic and incompressible ($I_3 = 1$), with mechanical properties similar to those of rubber. In this case, the strain-energy density function reduces to the form

$$W = W(I_1, I_2). \tag{5.118}$$

[7]Note that, after an initially isotropic, nonlinear material is deformed, its tangent moduli change. Thus, if the deformation is not isotropic, the deformed material itself becomes anisotropic. For this reason, material symmetries are defined in the unstressed material.

A commonly used strain-energy function for rubber-like materials is the Mooney-Rivlin form (Mooney, 1940; Rivlin, 1947)

$$W = b_1(I_1 - 3) + b_2(I_2 - 3), \tag{5.119}$$

where b_1 and b_2 are material constants.[8] If $b_2 = 0$, this expression reduces to the so-called neo-Hookean form. More general forms that contain Eq. (5.119) as special cases include (Rivlin and Saunders, 1951)

$$W = b_1(I_1 - 3) + f(I_2 - 3), \tag{5.120}$$

with f being a general function, and (Rivlin, 1956)

$$W = \sum_{i=0}^{n} \sum_{j=0}^{n} c_{ij}(I_1 - 3)^i (I_2 - 3)^j. \tag{5.121}$$

Note the similarity of Eqs. (5.116) and (5.121). Another form used for W is (Ogden, 1997)

$$W = \sum_{i=1}^{n} a_n(\lambda_1^{b_n} + \lambda_2^{b_n} + \lambda_3^{b_n} - 3) \tag{5.122}$$

where the λ_i are principal stretch ratios. This last expression takes advantage of the fact that the strain invariants always can be written in terms of principal strains. A disadvantage is that, except for some geometrically symmetric problems, the eigenvalue problem for the principal strains must be solved as part of the solution procedure.

If the water in a tissue is mobile, such as in articular cartilage, then the effects of bulk material compressibility should not be ignored. For compressible materials, W must include the effects of material volume change through the strain invariant I_3. One possible form for W, given by generalizing Eq. (5.121), is

$$W = \sum_{i=0}^{n} \sum_{j=0}^{n} \sum_{k=0}^{n} c_{ijk}(I_1 - 3)^i (I_2 - 3)^j (I_3 - 1)^k \tag{5.123}$$

where the c_{ijk} are constants. Another form was proposed by Blatz and Ko (1962), who studied the strain-energy density function for foam rubbers. Using theoretical

[8]The 3s in Eq. (5.119) are required to make $W = 0$ when all strains are zero. These constants do not affect the stresses, which are given by differentiating W.

and experimental arguments, they suggested the function

$$
\begin{aligned}
W &= \frac{\mu\alpha}{2}\left[I_1 - 3 + \frac{1-2\nu}{\nu}\left(I_3^{-\nu/(1-2\nu)} - 1\right)\right] \\
&\quad + \frac{\mu(1-\alpha)}{2}\left[I_2/I_3 - 3 + \frac{1-2\nu}{\nu}\left(I_3^{\nu/(1-2\nu)} - 1\right)\right]
\end{aligned}
\tag{5.124}
$$

where μ, ν, and α $(0 \le \alpha \le 1)$ are constants. For small strain, μ becomes the shear modulus and ν becomes Poisson's ratio. (These parameters have no physical meaning for large strain.) In addition, for $I_3 = 1$, W takes the Mooney-Rivlin form (5.119) for an incompressible rubber-like material. Finally, setting $\alpha = 0$ and $\nu = 0.25$ yields results in good agreement with some experimental data for foam rubber (Blatz and Ko, 1962). In this case, Eq. (5.124) becomes

$$
W = (\mu/2)(I_2/I_3 + 2I_3^{1/2} - 5). \tag{5.125}
$$

Most soft tissues stiffen with increasing strain levels. This behavior is captured, for example, by a function of the form

$$
W = C(e^Q - 1) \tag{5.126}
$$

where Q is given by the right-hand side of Eq. (5.119) or (5.124) for an incompressible or compressible tissue, respectively.

5.4 Stress-Strain Relations

In the previous section, the strain-energy density function was expressed in terms of Lagrangian strain components referred to a local set of Cartesian axes oriented along the principal material directions. Equation (5.98) can be used to obtain the second Piola-Kirchhoff stress components relative to these material axes. The resulting stress-strain relations then can be expressed in terms of stress and strain components relative to any desired curvilinear coordinate system through appropriate tensor transformations.

This procedure is recommended, since it promotes understanding of the fundamentals and allows for relatively straightforward adaptation of an analysis to general types of materials. Some authors [e.g., Green and Zerna (1968) and Green and Adkins (1970)], however, prefer working with more specialized equations. Indeed, this approach has advantages in some problems. Thus, we now derive stress-strain relations for hyperelastic materials possessing the symmetries considered in the previous section.

First, we introduce some notation. At a given point P in the undeformed body, three coordinate systems are defined (Fig. 5.4): (1) Cartesian coordinates Z^I with base vectors $\mathbf{e}_I$ along the principal material axes; (2) orthogonal curvilinear coordinates $X^{I'}$ with base vectors $\mathbf{G}_{I'}$ parallel to the $\mathbf{e}_I$ at P; and (3) general curvilinear coordinates X^I with base vectors $\mathbf{G}_I$. This last coordinate system is chosen for convenience in formulating a particular problem. Relative to these systems, the Lagrangian strain tensor and the second Piola-Kirchhoff stress tensor have the dyadic representations

$$\begin{aligned} \mathbf{E} &= \varepsilon_{IJ}\mathbf{e}^I\mathbf{e}^J = E_{I'J'}\mathbf{G}^{I'}\mathbf{G}^{J'} = E_{IJ}\mathbf{G}^I\mathbf{G}^J \\ \mathbf{s} &= S^{IJ}\mathbf{e}_I\mathbf{e}_J = s^{I'J'}\mathbf{G}_{I'}\mathbf{G}_{J'} = s^{IJ}\mathbf{G}_I\mathbf{G}_J \end{aligned} \tag{5.127}$$

in which the contravariant base vectors are defined in the usual manner. (Note that the S^{IJ} and s^{IJ} are Cartesian and general curvilinear components, respectively.) The transformation relations between the various strain and stress components, obtained by specializing Eqs. (2.49), (2.59), and (2.64), are[9]

$$\begin{aligned} \varepsilon_{IJ} &= A_I^{K'}A_J^{L'}E_{K'L'} = A_I^K A_J^L E_{KL} \\ s^{IJ} &= A_{K'}^I A_{L'}^J s^{K'L'} = A_K^I A_L^J S^{KL} \end{aligned} \tag{5.128}$$

where

$$\begin{aligned} A_I^{J'} &= \mathbf{e}_I\cdot\mathbf{G}^{J'} = \frac{\partial X^{J'}}{\partial Z^I} \\ A_I^J &= \mathbf{e}_I\cdot\mathbf{G}^J = \frac{\partial X^J}{\partial Z^I} \\ A_{I'}^J &= \mathbf{G}_{I'}\cdot\mathbf{G}^J = \frac{\partial X^J}{\partial X^{I'}}. \end{aligned} \tag{5.129}$$

With this groundwork, explicit stress-strain relations now are derived for isotropic, transversely isotropic, and orthotropic materials. The experimentally determined strain-energy density function is expressed most conveniently in terms of Cartesian strain components relative to the local material axes. Our goal, however, is to derive expressions for stresses in general curvilinear coordinates.

[9] For convenience and without confusion, all transformation tensor components here are represented by As, even for the contravariant stress components [see Eq. $(2.64)_2$].

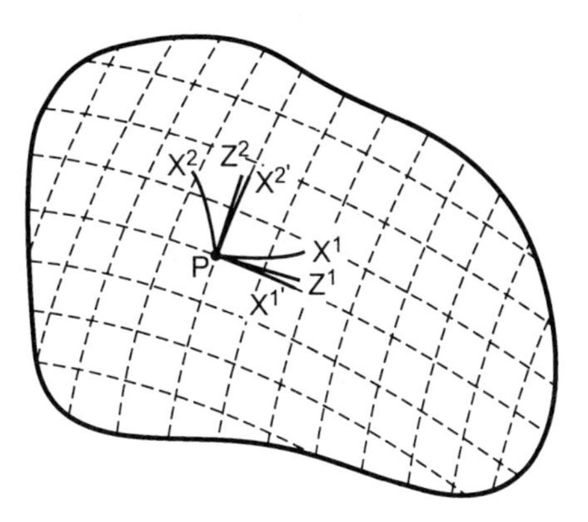

Fig. 5.4 Three coordinate systems defined at a point in an orthotropic material (two-dimensional case). The dashed lines represent principal material directions.

5.4.1 *Isotropic Material*

In deriving the stress-strain relations for an isotropic material, we first work in Cartesian coordinates. Then, the equations are converted into direct notation, valid for any coordinate system. And finally, the components with respect to general curvilinear coordinates are extracted.

For an isotropic material, Eq. (5.117) gives

$$W = W(I_1, I_2, I_3) \tag{5.130}$$

with Eqs. (3.142) providing the strain invariants[10]

$$\begin{aligned} I_1 &= c_{II} \\ I_2 &= \tfrac{1}{2}(I_1^2 - c_{IJ}c_{IJ}) \\ I_3 &= \det[c_{IJ}]. \end{aligned} \tag{5.131}$$

Here, the c_{IJ} are Cartesian components of the right Cauchy-Green deformation tensor $\mathbf{C}$; thus, Eq. $(3.30)_1$ yields

$$\varepsilon_{IJ} = \tfrac{1}{2}(c_{IJ} - \delta_{IJ}). \tag{5.132}$$

[10]Since we are working in Cartesian coordinates, superscript notation is suspended until Eq. (5.148).

The Cartesian stress-strain relations, given by Eq. $(5.94)_5$, are

$$\begin{aligned} S_{IJ} &= \frac{\partial W}{\partial \varepsilon_{IJ}} = 2\frac{\partial W}{\partial c_{IJ}} \\ &= 2\left(\frac{\partial W}{\partial I_1}\frac{\partial I_1}{\partial c_{IJ}} + \frac{\partial W}{\partial I_2}\frac{\partial I_2}{\partial c_{IJ}} + \frac{\partial W}{\partial I_3}\frac{\partial I_3}{\partial c_{IJ}}\right), \end{aligned} \tag{5.133}$$

for which the required derivatives of the strain invariants are computed as follows. The first two are relatively straightforward, with Eqs. $(5.131)_{1,2}$ giving

$$\frac{\partial I_1}{\partial c_{IJ}} = \frac{\partial c_{KK}}{\partial c_{IJ}} = \delta_{IK}\delta_{JK} = \delta_{IJ} \tag{5.134}$$

$$\begin{aligned} \frac{\partial I_2}{\partial c_{IJ}} &= \frac{1}{2}\frac{\partial}{\partial c_{IJ}}(I_1^2 - c_{KL}c_{KL}) \\ &= I_1\frac{\partial I_1}{\partial c_{IJ}} - \frac{1}{2}\left(\frac{\partial c_{KL}}{\partial c_{IJ}}c_{KL} + c_{KL}\frac{\partial c_{KL}}{\partial c_{IJ}}\right) \\ &= I_1\delta_{IJ} - c_{KL}\delta_{KI}\delta_{LJ} \\ &= I_1\delta_{IJ} - c_{IJ}. \end{aligned} \tag{5.135}$$

Computing $\partial I_3/\partial c_{IJ}$ begins with the Cayley-Hamilton theorem (2.78) in the form

$$\mathbf{C}^3 - I_1\mathbf{C}^2 + I_2\mathbf{C} - I_3\mathbf{I} = \mathbf{0}. \tag{5.136}$$

In terms of components, substituting $\mathbf{C} = c_{IJ}\mathbf{e}_I\mathbf{e}_J$ and $\mathbf{I} = \delta_{IJ}\mathbf{e}_I\mathbf{e}_J$ yields

$$c_{IL}c_{LK}c_{KJ} - I_1c_{IK}c_{KJ} + I_2c_{IJ} - I_3\delta_{IJ} = 0, \tag{5.137}$$

and setting $I = J$ produces the contracted form

$$c_{IL}c_{LK}c_{KI} - I_1c_{IK}c_{KI} + I_2c_{II} - 3I_3 = 0.$$

Finally, solving this equation for I_3, computing $\partial I_3/\partial c_{IJ}$, and noting the above relations for the invariants and their derivatives yields

$$\frac{\partial I_3}{\partial c_{IJ}} = I_2\delta_{IJ} - I_1c_{IJ} + c_{IK}c_{KJ}. \tag{5.138}$$

The details of these last manipulations are left to the reader (see Problem **5–8**).

In direct notation, Eqs. (5.134), (5.135), and (5.138) become

$$\begin{aligned}\frac{\partial I_1}{\partial \mathbf{C}} &= \mathbf{I} \\ \frac{\partial I_2}{\partial \mathbf{C}} &= I_1\mathbf{I} - \mathbf{C} \\ \frac{\partial I_3}{\partial \mathbf{C}} &= I_2\mathbf{I} - I_1\mathbf{C} + \mathbf{C}^2. \end{aligned} \tag{5.139}$$

The last derivative can be put in an alternate form by first writing it as

$$\frac{\partial I_3}{\partial \mathbf{C}} = \mathbf{C}^{-1} \cdot (I_2\mathbf{C} - I_1\mathbf{C}^2 + \mathbf{C}^3)$$

and then using Eq. (5.136) to get

$$\frac{\partial I_3}{\partial \mathbf{C}} = I_3\mathbf{C}^{-1}. \tag{5.140}$$

Now, substituting Eqs. (5.139) and (5.140) into (5.133) yields

$$\mathbf{s} = \frac{\partial W}{\partial \mathbf{E}} = 2[W_1\mathbf{I} + W_2(I_1\mathbf{I} - \mathbf{C}) + W_3 I_3 \mathbf{C}^{-1}] \tag{5.141}$$

where

$$\boxed{W_i \equiv \frac{\partial W}{\partial I_i}.} \tag{5.142}$$

With Eqs. (3.136) providing the strain invariants in terms of $\mathbf{C} = \mathbf{I} + 2\mathbf{E}$, double-dotting Eq. (5.141) with the appropriate dyads yields the stress-strain relations in any coordinate system of interest.

For the Cauchy stress tensor, Eqs. (4.30) and (5.141) give

$$\boldsymbol{\sigma} = 2J^{-1}\mathbf{F}\cdot[W_1\mathbf{I} + W_2(I_1\mathbf{I} - \mathbf{C}) + W_3 I_3 \mathbf{C}^{-1}]\cdot\mathbf{F}^T. \tag{5.143}$$

This expression can be simplified using the relations

$$\begin{aligned}\mathbf{F}^T\cdot\mathbf{F} &= \mathbf{C} = \mathbf{C}^T \\ \mathbf{F}\cdot\mathbf{F}^T &= \mathbf{B} = \mathbf{B}^T. \end{aligned} \tag{5.144}$$

In addition, direct substitution shows that $\mathbf{C}^{-1} = \mathbf{F}^{-1}\cdot\mathbf{F}^{-T}$ and that

$$\begin{aligned}\mathbf{F}\cdot\mathbf{C}\cdot\mathbf{F}^T &= \mathbf{B}^2 \\ \mathbf{F}\cdot\mathbf{C}^{-1}\cdot\mathbf{F}^T &= \mathbf{I}. \end{aligned} \tag{5.145}$$

Inserting these equations into (5.143) yields

$$\boxed{\boldsymbol{\sigma} = \alpha_0 \mathbf{I} + \alpha_1 \mathbf{B} + \alpha_2 \mathbf{B}^2} \tag{5.146}$$

where the response functions are (since $J = I_3^{1/2}$)

$$\boxed{\begin{aligned} \alpha_0 &= 2I_3^{1/2} W_3 \\ \alpha_1 &= 2I_3^{-1/2} (W_1 + I_1 W_2) \\ \alpha_2 &= -2I_3^{-1/2} W_2. \end{aligned}} \tag{5.147}$$

Lastly, we derive a component form of Eq. (5.146). Green and Zerna (1968) and Green and Adkins (1970) work in terms of the Cauchy stress components defined by

$$\boldsymbol{\sigma} = \sigma^{I^* J^*} \mathbf{g}_I \mathbf{g}_J \tag{5.148}$$

in which the $\mathbf{g}_I$ are base vectors convected from a set of arbitrary base vectors $\mathbf{G}_I$. To obtain explicit stress-strain relations for the $\sigma^{I^* J^*}$, we first substitute Eq. $(3.22)_1$ into $(5.144)_2$ to get

$$\begin{aligned} \mathbf{B} &= \mathbf{F} \cdot \mathbf{F}^T = (\mathbf{g}_I \mathbf{G}^I) \cdot (\mathbf{G}^J \mathbf{g}_J) \\ &= G^{IJ} \mathbf{g}_I \mathbf{g}_J \\ \mathbf{B}^2 &= \mathbf{B} \cdot \mathbf{B} = (G^{IJ} \mathbf{g}_I \mathbf{g}_J) \cdot (G^{KL} \mathbf{g}_K \mathbf{g}_L) \\ &= G^{IJ} G^{KL} g_{JK} \, \mathbf{g}_I \mathbf{g}_L \\ &= G^{IL} G^{KJ} g_{LK} \, \mathbf{g}_I \mathbf{g}_J. \end{aligned} \tag{5.149}$$

With these expressions, Eq. (5.146) gives

$$\begin{aligned} \sigma^{I^* J^*} &= \boldsymbol{\sigma} : \mathbf{g}^I \mathbf{g}^J = \mathbf{g}^I \cdot \boldsymbol{\sigma} \cdot \mathbf{g}^J \\ &= \mathbf{g}^I \cdot (\alpha_0 \mathbf{I} + \alpha_1 \mathbf{B} + \alpha_2 \mathbf{B}^2) \cdot \mathbf{g}^J \\ &= \alpha_0 g^{IJ} + \alpha_1 G^{IJ} + \alpha_2 G^{IL} G^{KJ} g_{LK}. \end{aligned}$$

Finally, substituting Eqs. (5.147) and rearranging yields

$$\boxed{\sigma^{I^* J^*} = \Phi G^{IJ} + \Psi Q^{IJ} - p g^{IJ}} \tag{5.150}$$

where

$$
\begin{aligned}
\Phi &= 2I_3^{-1/2}W_1 \\
\Psi &= 2I_3^{-1/2}W_2 \\
p &= -2I_3^{1/2}W_3 \\
Q^{IJ} &= I_1G^{IJ} - G^{IK}G^{JL}g_{KL}.
\end{aligned}
\tag{5.151}
$$

5.4.2 *Transversely Isotropic Material*

For a material that is transversely isotropic relative to the local Z^3-direction, Eq. (5.112) gives

$$
W = W(I_1, I_2, I_3, I_4, I_5) \tag{5.152}
$$

where Eqs. (5.131) provide the strain invariants I_1, I_2, and I_3. In addition, (5.113) gives

$$
\begin{aligned}
I_4 &= \varepsilon_{33} \\
I_5 &= \varepsilon_{31}^2 + \varepsilon_{23}^2,
\end{aligned}
\tag{5.153}
$$

with the ε_{IJ} being Cartesian strain components in the local material coordinate system. The corresponding Cartesian components of the second Piola-Kirchhoff stress tensor are

$$
S^{IJ} = \frac{1}{2}\left(\frac{\partial W}{\partial \varepsilon_{IJ}} + \frac{\partial W}{\partial \varepsilon_{JI}}\right). \tag{5.154}
$$

Differentiating Eq. (5.152) yields

$$
\begin{aligned}
\frac{\partial W}{\partial \varepsilon_{IJ}} &= \frac{\partial W}{\partial I_1}\frac{\partial I_1}{\partial \varepsilon_{IJ}} + \frac{\partial W}{\partial I_2}\frac{\partial I_2}{\partial \varepsilon_{IJ}} + \frac{\partial W}{\partial I_3}\frac{\partial I_3}{\partial \varepsilon_{IJ}} \\
&\quad + \frac{\partial W}{\partial I_4}\frac{\partial I_4}{\partial \varepsilon_{IJ}} + \frac{\partial W}{\partial I_5}\frac{\partial I_5}{\partial \varepsilon_{IJ}}.
\end{aligned}
\tag{5.155}
$$

The first three terms of this expression are equivalent to the right-hand-side of (5.133) and, therefore, lead to the stresses of Eq. (5.150). Thus, we only need to deal here with the last two terms, with Eq. (5.153) giving

$$\begin{aligned}
\frac{\partial I_4}{\partial \varepsilon_{IJ}} &= \frac{\partial \varepsilon_{33}}{\partial \varepsilon_{IJ}} = \delta_3^I \delta_3^J \\
\frac{\partial I_5}{\partial \varepsilon_{IJ}} &= 2\left(\varepsilon_{31}\frac{\partial \varepsilon_{31}}{\partial \varepsilon_{IJ}} + \varepsilon_{23}\frac{\partial \varepsilon_{23}}{\partial \varepsilon_{IJ}}\right) \\
&= 2(\varepsilon_{31}\delta_3^I\delta_1^J + \varepsilon_{23}\delta_2^I\delta_3^J).
\end{aligned} \tag{5.156}$$

Now, we are in position to compute the stress components relative to a general curvilinear coordinate system. First, substituting Eq. (5.154) into $(5.128)_2$ and noting (4.36) yields

$$\sigma^{I^*J^*} = J^{-1}s^{IJ} = \frac{1}{2J}\left(\frac{\partial W}{\partial \varepsilon_{KL}} + \frac{\partial W}{\partial \varepsilon_{LK}}\right) A_K^I A_L^J. \tag{5.157}$$

Next, using Eqs. (5.150), (5.155), and (5.156) transforms this relation into

$$\begin{aligned}
\sigma^{I^*J^*} &= \Phi G^{IJ} + \Psi Q^{IJ} - pg^{IJ} + (2J)^{-1}[W_4(\delta_3^K\delta_3^L + \delta_3^L\delta_3^K) \\
&\quad +2W_5(\varepsilon_{31}\delta_3^K\delta_1^L + \varepsilon_{23}\delta_2^K\delta_3^L + \varepsilon_{31}\delta_3^L\delta_1^K + \varepsilon_{23}\delta_2^L\delta_3^K)]A_K^I A_L^J \\
&= \Phi G^{IJ} + \Psi Q^{IJ} - pg^{IJ} + (2J)^{-1}[2W_4 A_3^I A_3^J \\
&\quad +2W_5(\varepsilon_{31}A_3^I A_1^J + \varepsilon_{23}A_2^I A_3^J + \varepsilon_{31}A_1^I A_3^J + \varepsilon_{23}A_3^I A_2^J)]
\end{aligned}$$

with W_4 and W_5 defined by Eq. (5.142). Thus, we can write

$$\boxed{\sigma^{I^*J^*} = \Phi G^{IJ} + \Psi Q^{IJ} - pg^{IJ} + \Theta M^{IJ} + \Lambda N^{IJ}} \tag{5.158}$$

where Φ, Ψ, p, and Q^{IJ} are defined by Eq. (5.151) and

$$\boxed{\begin{aligned}
\Theta &= I_3^{-1/2} W_4 \\
\Lambda &= I_3^{-1/2} W_5 \\
M^{IJ} &= A_3^I A_3^J \\
N^{IJ} &= (A_3^I A_\alpha^J + A_\alpha^I A_3^J)\varepsilon_{\alpha 3}
\end{aligned}} \tag{5.159}$$

with $\alpha = 1, 2$.

In direct form, Eq. (5.146) gives the first three terms of (5.158). To deal with the other terms, we note that Eq. $(5.129)_2$ gives

$$A_3^I = \mathbf{e}_3 \cdot \mathbf{G}^I. \tag{5.160}$$

With this relation, it is straightforward to show that (5.158) can be written

$$\boxed{\boldsymbol{\sigma} = \alpha_0 \mathbf{I} + \alpha_1 \mathbf{B} + \alpha_2 \mathbf{B}^2 + \Theta \mathbf{M} + \Lambda \mathbf{N}} \tag{5.161}$$

where

$$\begin{aligned} \mathbf{M} &= \mathbf{F} \cdot (\mathbf{e}_3 \mathbf{e}_3) \cdot \mathbf{F}^T \\ \mathbf{N} &= \mathbf{F} \cdot (\mathbf{e}_3 \mathbf{e}_\alpha + \mathbf{e}_\alpha \mathbf{e}_3) \cdot \mathbf{F}^T \, \varepsilon_{\alpha 3}. \end{aligned} \tag{5.162}$$

If we take $\boldsymbol{\sigma} = \sigma^{I^* J^*} \mathbf{g}_I \mathbf{g}_J$, $\mathbf{M} = M^{I^* J^*} \mathbf{g}_I \mathbf{g}_J$, and $\mathbf{N} = N^{I^* J^*} \mathbf{g}_I \mathbf{g}_J$, then Eqs. (5.159) give the components of $\mathbf{M}$ and $\mathbf{N}$ (see Problem **5–9**).

5.4.3 *Orthotropic Material*

The analysis for an orthotropic material is similar to that used for a transversely isotropic material. The strain-energy density function

$$W = W(\varepsilon_{11}, \varepsilon_{22}, \varepsilon_{33}, \varepsilon_{12}^2, \varepsilon_{23}^2, \varepsilon_{31}^2, I_3) \tag{5.163}$$

of Eq. (5.104) is substituted into (5.157). With the chain rule, the necessary derivatives are

$$\begin{aligned} \frac{\partial W}{\partial \varepsilon_{KL}} &= \frac{\partial W}{\partial \varepsilon_{11}} \frac{\partial \varepsilon_{11}}{\partial \varepsilon_{KL}} + \frac{\partial W}{\partial \varepsilon_{22}} \frac{\partial \varepsilon_{22}}{\partial \varepsilon_{KL}} + \frac{\partial W}{\partial \varepsilon_{33}} \frac{\partial \varepsilon_{33}}{\partial \varepsilon_{KL}} \\ &\quad + \frac{\partial W}{\partial \varepsilon_{12}^2} \frac{\partial \varepsilon_{12}^2}{\partial \varepsilon_{KL}} + \frac{\partial W}{\partial \varepsilon_{23}^2} \frac{\partial \varepsilon_{23}^2}{\partial \varepsilon_{KL}} + \frac{\partial W}{\partial \varepsilon_{31}^2} \frac{\partial \varepsilon_{31}^2}{\partial \varepsilon_{KL}} + \frac{\partial W}{\partial I_3} \frac{\partial I_3}{\partial \varepsilon_{KL}} \end{aligned} \tag{5.164}$$

where

$$\begin{aligned} \frac{\partial \varepsilon_{11}}{\partial \varepsilon_{KL}} &= \delta_1^K \delta_1^L \\ \frac{\partial \varepsilon_{12}^2}{\partial \varepsilon_{KL}} &= 2\varepsilon_{12} \frac{\partial \varepsilon_{12}}{\partial \varepsilon_{KL}} = 2\varepsilon_{12} \delta_1^K \delta_2^L \end{aligned} \tag{5.165}$$

and so on, with the I_3 term leading to the last term in Eq. (5.150). Substituting these and related expressions into (5.164) and the result into (5.157) yields

$$\sigma^{I^*J^*} = \frac{1}{2J}\left(2\frac{\partial W}{\partial \varepsilon_{11}}\delta_1^K\delta_1^L + \cdots + 2\varepsilon_{12}\frac{\partial W}{\partial \varepsilon_{12}^2}\delta_1^K\delta_2^L + \cdots\right)A_K^I A_L^J - pg^{IJ},$$

which simplifies to

$$\begin{aligned}\sigma^{I^*J^*} &= I_3^{-1/2}\left(\frac{\partial W}{\partial \varepsilon_{11}}A_1^I A_1^J + \frac{\partial W}{\partial \varepsilon_{22}}A_2^I A_2^J + \frac{\partial W}{\partial \varepsilon_{33}}A_3^I A_3^J\right.\\ &\left. + \varepsilon_{12}\frac{\partial W}{\partial \varepsilon_{12}^2}A_1^I A_2^J + \varepsilon_{23}\frac{\partial W}{\partial \varepsilon_{23}^2}A_2^I A_3^J + \varepsilon_{31}\frac{\partial W}{\partial \varepsilon_{31}^2}A_3^I A_1^J\right) - pg^{IJ}.\end{aligned} \tag{5.166}$$

5.4.4 *Curvilinear Anisotropy*

The strain-energy density function has been expressed thus far in terms of strain components referred to a local Cartesian set of material axes. Sometimes, however, it is useful to introduce orthogonal curvilinear coordinates $X^{I'}$ that are tangent to the principal material directions at each point in a body (Fig. 5.4). One example would be an orthotropic artery subjected to internal pressure, for which cylindrical polar coordinates are useful. In such problems, it is convenient to express W in terms of strains relative to the chosen curvilinear coordinates.

Within an infinitesimal neighborhood of a point in the undeformed body, the local Cartesian and curvilinear material axes coincide (Fig. 5.4). According to the principle of local action, therefore, it follows that we can obtain the appropriate form of W simply by replacing the Cartesian strains by the corresponding curvilinear strain components. However, since the Cartesian strains ε_{IJ} are dimensionless, they must be replaced by the dimensionless physical strain components $\hat{E}_{I'J'}$ relative to the $X^{I'}$ system. Thus, Eq. (5.99) becomes

$$W = W(\hat{E}_{1'1'}, \hat{E}_{2'2'}, \hat{E}_{3'3'}, \hat{E}_{1'2'}, \hat{E}_{2'3'}, \hat{E}_{3'1'}) \tag{5.167}$$

where Eq. (3.38) gives

$$\hat{E}_{I'J'} = \frac{E_{I'J'}}{\sqrt{G_{(I'I')}G_{(J'J')}}} \tag{5.168}$$

with $G_{(I'J')} = \mathbf{G}_{I'} \cdot \mathbf{G}_{J'}$ and the $E_{I'J'}$ defined by Eq. $(5.127)_1$.

The derivation of stress-strain relations for specific materials follows those discussed earlier. Care, however, must be taken to properly include the components of the metric tensor that enter by way of the definitions of the physical components of stress and strain. First, combining Eqs. $(5.94)_5$ and $(5.128)_2$ gives

$$\begin{aligned} s^{IJ} &= s^{K'L'} A^I_{K'} A^J_{L'} \\ &= \frac{\partial W}{\partial E_{K'L'}} A^I_{K'} A^J_{L'}, \end{aligned}$$

and substituting (5.168) yields

$$s^{IJ} = \frac{\partial W}{\partial \hat{E}_{K'L'}} \frac{A^I_{K'} A^J_{L'}}{\sqrt{G_{(K'K')} G_{(L'L')}}}.$$

This relation also can be written in the form

$$\boxed{\sigma^{I^*J^*} = J^{-1} s^{IJ} = \frac{1}{2J} \left(\frac{\partial W}{\partial \hat{E}_{K'L'}} + \frac{\partial W}{\partial \hat{E}_{L'K'}} \right) \hat{A}^I_{K'} \hat{A}^J_{L'}} \tag{5.169}$$

where Eq. $(5.129)_3$ gives

$$\hat{A}^I_{J'} \equiv \frac{A^I_{J'}}{\sqrt{G_{(J'J')}}} = \frac{1}{\sqrt{G_{(J'J')}}} \frac{\partial X^I}{\partial X^{J'}}. \tag{5.170}$$

Now, comparing Eqs. (5.157) and (5.169) reveals that the constitutive relations for curvilinear anisotropy can be obtained from those written in Cartesian material coordinates by making the following replacements:

$$\varepsilon_{IJ} \to \hat{E}_{I'J'} \quad \text{and} \quad A^I_J \to \hat{A}^I_{J'}. \tag{5.171}$$

Finally, we derive a general constitutive relation in terms of physical components for both stress and strain. Adapting Eq. (4.26) provides the physical Cauchy stress components

$$\hat{\sigma}^{I^*J^*} = \sqrt{\frac{G_{(JJ)}}{G^{(II)}}}\, \sigma^{I^*J^*}, \tag{5.172}$$

and substituting Eq. (5.169) yields

$$\begin{aligned}\hat{\sigma}^{I^*J^*} &= \frac{1}{2J}\left(\frac{\partial W}{\partial \hat{E}_{K'L'}} + \frac{\partial W}{\partial \hat{E}_{L'K'}}\right)\sqrt{\frac{G_{(JJ)}}{G^{(II)}}}\,\hat{A}^I_{K'}\hat{A}^J_{L'} \\ &= \frac{1}{2J}\left(\frac{\partial W}{\partial \hat{E}_{K'L'}} + \frac{\partial W}{\partial \hat{E}_{L'K'}}\right)\sqrt{\frac{G_{(JJ)}}{G^{(II)}}}\,\frac{A^I_{K'}A^J_{L'}}{\sqrt{G_{(K'K')}G_{(L'L')}}}\end{aligned} \tag{5.173}$$

where Eq. (5.170) has been used. This equation does not appear to be convenient to use, but significant simplification occurs if the principal material directions coincide with the reference curvilinear coordinates for the problem, i.e., $X^{I'} = X^I$. In this case, Eq. $(5.129)_3$ gives

$$A^I_{J'} = \frac{\partial X^I}{\partial X^{J'}} = \frac{\partial X^I}{\partial X^J} = \delta^I_J,$$

and Eq. (5.173) becomes

$$\begin{aligned}\hat{\sigma}^{I^*J^*} &= \frac{1}{2J}\left(\frac{\partial W}{\partial \hat{E}_{KL}} + \frac{\partial W}{\partial \hat{E}_{LK}}\right)\sqrt{\frac{G_{(JJ)}}{G^{(II)}}}\,\frac{\delta^I_K\delta^J_L}{\sqrt{G_{(KK)}G_{(LL)}}} \\ &= \frac{1}{2J}\left(\frac{\partial W}{\partial \hat{E}_{IJ}} + \frac{\partial W}{\partial \hat{E}_{JI}}\right)\frac{1}{\sqrt{G^{(II)}G_{(II)}}}.\end{aligned}$$

Since the material coordinates are assumed to be orthogonal, we have $G^{(II)}G_{(II)} = 1$ and the above relation becomes

$$\hat{\sigma}^{I^*J^*} = \frac{1}{2J}\left(\frac{\partial W}{\partial \hat{E}_{IJ}} + \frac{\partial W}{\partial \hat{E}_{JI}}\right). \tag{5.174}$$

Moreover, with Eqs. (5.168) and (5.172), it can be shown that

$$\sigma^{I^*J^*} = \frac{1}{2J}\left(\frac{\partial W}{\partial E_{IJ}} + \frac{\partial W}{\partial E_{JI}}\right). \tag{5.175}$$

Again, we emphasize that these last two equations are valid only if the curvilinear coordinates in the undeformed configuration coincide with principal material directions that are orthogonal. An example is axisymmetric deformation of an artery.

Example 5.7 Using the replacements (5.171), write the stress-strain relations in terms of general strain components for a curvilinearly transversely isotropic material.

Solution. First, we note that, as discussed in Section 2.7.2, the invariants I_1, I_2, and I_3 do not change when expressed in terms of physical components, i.e., $I'_1 = I_1$, $I'_2 = I_2$, and $I'_3 = I_3$, with prime denoting invariants expressed in terms of strains relative to the curvilinear material axes $X^{I'}$. Thus, Eq. (5.112) becomes

$$W = W(I_1, I_2, I_3, I'_4, I'_5) \tag{5.176}$$

where

$$\begin{aligned} I'_4 &= \hat{E}_{3'3'} \\ I'_5 &= \hat{E}^2_{3'1'} + \hat{E}^2_{2'3'}. \end{aligned} \tag{5.177}$$

Now, Eqs. (5.158), (5.159), and (5.171) yield

$$\sigma^{I^*J^*} = \Phi G^{IJ} + \Psi Q^{IJ} - pg^{IJ} + \Theta' M'^{IJ} + \Lambda' N'^{IJ} \tag{5.178}$$

where

$$\begin{aligned} \Theta' &= I_3^{-1/2} \frac{\partial W}{\partial I'_4} \\ \Lambda' &= I_3^{-1/2} \frac{\partial W}{\partial I'_5} \\ M'^{IJ} &= \hat{A}^I_{3'} \hat{A}^J_{3'} \\ N'^{IJ} &= (\hat{A}^I_{3'} \hat{A}^J_{\alpha'} + \hat{A}^I_{\alpha'} \hat{A}^J_{3'}) \hat{E}_{\alpha' 3'} \end{aligned} \tag{5.179}$$

for $\alpha = 1, 2$. ■

5.5 Incompressibility

As discussed earlier, soft tissues usually are assumed to be incompressible, with the deformation being constrained by the condition $J = 1$. Thus, the strain components are not all independent, and the constitutive relations must be modified. To determine the required modification, we first derive an alternate form for the incompressibility condition.

As a body deforms, a surface element da sweeps out a volume $\mathbf{v} \cdot \mathbf{n}\, da$ per unit time, where $\mathbf{v}$ is the velocity of the element and $\mathbf{n}$ is the unit normal to da.

Thus, the time rate of change of the volume v of the body is

$$\frac{dv}{dt} = \int_a \mathbf{v} \cdot \mathbf{n} \, da,$$

which becomes

$$\frac{dv}{dt} = \int_v \bar{\boldsymbol{\nabla}} \cdot \mathbf{v} \, dv \tag{5.180}$$

after application of the divergence theorem (2.191). For an infinitesimal element of volume δv, the **dilatation rate** is defined as

$$\Delta = \lim_{\delta v \to 0} \frac{1}{\delta v} \frac{d\,\delta v}{dt} = \lim_{\delta v \to 0} \frac{1}{\delta v} \int_{\delta v} \bar{\boldsymbol{\nabla}} \cdot \mathbf{v} \, dv$$

or

$$\Delta = \bar{\boldsymbol{\nabla}} \cdot \mathbf{v}. \tag{5.181}$$

In an incompressible body, Δ must vanish at each point for all time. Moreover, applying the definition of the trace (2.65) to Eq. $(3.215)_1$ yields

$$\begin{aligned} \mathrm{tr}\,\mathbf{D} &= \tfrac{1}{2}[(\bar{\boldsymbol{\nabla}} \cdot \mathbf{v})^T + \bar{\boldsymbol{\nabla}} \cdot \mathbf{v}] \\ &= \bar{\boldsymbol{\nabla}} \cdot \mathbf{v}. \end{aligned} \tag{5.182}$$

Thus, the incompressibility condition can be written in the form

$$\boxed{\Delta = \bar{\boldsymbol{\nabla}} \cdot \mathbf{v} = \mathrm{tr}\,\mathbf{D} = 0.} \tag{5.183}$$

Consider now the stress power. If we define a modified Cauchy stress tensor by

$$\boldsymbol{\sigma}^* = \boldsymbol{\sigma} - p\mathbf{I} \tag{5.184}$$

where p is an arbitrary function of position analogous to a hydrostatic pressure, then substitution into Eq. (5.22) yields the modified stress power

$$J\,\boldsymbol{\sigma}^* : \mathbf{D} = J\,(\boldsymbol{\sigma} - p\mathbf{I}) : \mathbf{D}. \tag{5.185}$$

Because $\mathbf{I} : \mathbf{D} = \mathrm{tr}\,\mathbf{D} = 0$ (see Table 2.4, page 32), this equation becomes

$$\boldsymbol{\sigma}^* : \mathbf{D} = \boldsymbol{\sigma} : \mathbf{D}. \tag{5.186}$$

In other words, a hydrostatic pressure does no net work on an incompressible material. The physical reason for this result is that the work done by a pressure load is due to the pressure acting through a change in volume. Since an incompressible material undergoes no volume change, the hydrostatic pressure imparts no net energy.

Consequently, for a given strain field in an incompressible material, the constitutive relations determine the state of stress only up to an arbitrary function, $-p\mathbf{I}$. For $J = 1$, therefore, Eq. $(5.86)_1$ must be modified to read

$$\boxed{\boldsymbol{\sigma} = \mathbf{F}\cdot\frac{\partial W}{\partial \mathbf{E}}\cdot\mathbf{F}^T - p\,\mathbf{I}} \tag{5.187}$$

which is the general constitutive relation for an incompressible hyperelastic material. Inserting this equation and $J = 1$ into Eqs. (4.31) then gives the Piola-Kirchhoff stress tensors

$$\boxed{\begin{aligned} \mathrm{t} &= \mathbf{F}^{-1}\cdot\boldsymbol{\sigma} = \frac{\partial W}{\partial \mathbf{E}}\cdot\mathbf{F}^T - p\,\mathbf{F}^{-1} \\ \mathrm{s} &= \mathbf{F}^{-1}\cdot\boldsymbol{\sigma}\cdot\mathbf{F}^{-T} = \frac{\partial W}{\partial \mathbf{E}} - p\,\mathbf{F}^{-1}\cdot\mathbf{F}^{-T}. \end{aligned}} \tag{5.188}$$

It is important to note that, although it may be useful to the think of the function p as being in some ways analogous to a hydrostatic pressure, this terminology often has been a source of confusion in the biomechanics literature. Several authors have associated p with a "tissue pressure" that can be measured with an appropriate transducer. In fact, however, p generally does not have a direct physical interpretation. Rather, it is a Lagrange multiplier that is needed to enforce the incompressibility constraint.

The component forms of the constitutive relations for a compressible material were derived in Section 5.2.6. To obtain the corresponding equations for an incompressible material, we need only examine the terms involving the Lagrange multiplier. The stress components of Eqs. (5.94) are defined by the dyadics

$$\begin{aligned} \boldsymbol{\sigma} &= \sigma^{ij}\mathbf{g}_i\mathbf{g}_j = \sigma^{I^*J^*}\mathbf{g}_I\mathbf{g}_J \\ \mathrm{t} &= t^{Ij}\mathbf{G}_I\mathbf{g}_j = t^I_{.j}\mathbf{G}_I\mathbf{g}^j \\ \mathrm{s} &= s^{IJ}\mathbf{G}_I\mathbf{G}_J. \end{aligned} \tag{5.189}$$

Thus, the terms involving p in Eqs. (5.187) and (5.188) require the relations

$$\begin{aligned}
\mathbf{g}^i \cdot \mathbf{I} \cdot \mathbf{g}^j &= \mathbf{g}^i \cdot \mathbf{g}^j = g^{ij} \\
\mathbf{g}^I \cdot \mathbf{I} \cdot \mathbf{g}^J &= \mathbf{g}^I \cdot \mathbf{g}^J = g^{IJ} \\
\mathbf{G}^I \cdot \mathbf{F}^{-1} \cdot \mathbf{g}^j &= \mathbf{G}^I \cdot (\mathbf{F}_k^K \mathbf{G}_K \mathbf{g}^k) \cdot \mathbf{g}^j \\
&= \mathbf{F}_k^K \delta_K^I g^{kj} = F^{Ij} \\
\mathbf{G}^I \cdot \mathbf{F}^{-1} \cdot \mathbf{g}_j &= \mathbf{G}^I \cdot (\mathbf{F}_k^K \mathbf{G}_K \mathbf{g}^k) \cdot \mathbf{g}_j \\
&= \mathbf{F}_k^K \delta_K^I \delta_j^k = F_j^I \\
\mathbf{G}^I \cdot (\mathbf{F}^{-1} \cdot \mathbf{F}^{-T}) \cdot \mathbf{G}^J &= \mathbf{G}^I \cdot (\mathbf{G}_K \mathbf{g}^K) \cdot (\mathbf{g}^L \mathbf{G}_L) \cdot \mathbf{G}^J \\
&= \delta_K^I g^{KL} \delta_L^J = g^{IJ}
\end{aligned}$$

in which Eq. $(3.22)_2$ has been used. Using these results in Eqs. (5.187) and (5.188) and noting Eqs. (5.94) yields (for $J = 1$)

$$\boxed{\begin{aligned}
\sigma^{ij} &= F_I^i F_J^j \frac{\partial W}{\partial E_{IJ}} - p g^{ij} \\
\sigma^{I^* J^*} &= s^{IJ} = \frac{\partial W}{\partial E_{IJ}} - p g^{IJ} \\
t^{Ij} &= F_J^j \frac{\partial W}{\partial E_{IJ}} - p F^{Ij} \\
t_{\cdot j}^I &= \frac{\partial W}{\partial F_I^j} - p F_j^I .
\end{aligned}} \tag{5.190}$$

We now make some observations. First, consistent with Eq. (4.36), $\sigma^{I^* J^*} = s^{IJ}$ for an incompressible material. Second, comparing the above expression for $\sigma^{I^* J^*}$ with Eqs. (5.150) and (5.151) reveals that the p term due to incompressibility replaces the $\partial W / \partial I_3$ term in the constitutive relations for a compressible material. Thus, with p defined as the Lagrange multiplier, Eqs. (5.150), (5.158), and (5.166) apply also to incompressible materials with isotropic, transversely isotropic, and orthotropic symmetry, respectively. Third, since $I_3 = 1$, incompressibility renders the $p = -2\, \partial W / \partial I_3$ term indeterminate from the constitutive behavior alone. In general, p is a function of position to be determined from the equilibrium equations and boundary conditions (see Chapter 6).

Finally, we consider the modified strain-energy density function

$$\boxed{W^* = W - p(J-1) = W - p(I_3^{1/2} - 1).} \tag{5.191}$$

Since $J = 1$, the added multiplier term does not alter the strain energy stored in the material. With this expression, we can use Eq. $(5.88)_2$ to compute a first Piola-Kirchhoff stress tensor

$$\mathbf{t}^* = \frac{\partial W^*}{\partial \mathbf{F}^T} = \frac{\partial W}{\partial \mathbf{F}^T} - p\,\frac{\partial J}{\partial \mathbf{F}^T}. \tag{5.192}$$

The last term of this equation can be transformed by noting that, because $J = \det \mathbf{F}$, the last formula in Table 2.5 (page 33) gives

$$\frac{\partial J}{\partial \mathbf{F}} = \frac{\partial \det \mathbf{F}}{\partial \mathbf{F}} = (\det \mathbf{F})\mathbf{F}^{-T} = J\mathbf{F}^{-T},$$

which, with $J = 1$, yields

$$\frac{\partial J}{\partial \mathbf{F}^T} = \mathbf{F}^{-1}. \tag{5.193}$$

Now, Eq. (5.192) becomes

$$\mathbf{t}^* = \frac{\partial W}{\partial \mathbf{F}^T} - p\mathbf{F}^{-1},$$

or, by Eq. $(5.88)_2$,

$$\mathbf{t}^* = \frac{\partial W}{\partial \mathbf{F}^T} - p\mathbf{F}^{-1} = \frac{\partial W}{\partial \mathbf{E}}\cdot\mathbf{F}^T - p\mathbf{F}^{-1}.$$

This result agrees with that for the first Piola-Kirchhoff stress tensor of Eq. $(5.188)_1$.

Thus, the constitutive relations for an incompressible material also can be obtained by replacing W by W^* in Eqs. (5.88) to get

$$\boxed{\begin{aligned} \boldsymbol{\sigma} &= \mathbf{F}\cdot\frac{\partial W^*}{\partial \mathbf{E}}\cdot\mathbf{F}^T \\ \mathbf{t} &= \frac{\partial W^*}{\partial \mathbf{E}}\cdot\mathbf{F}^T = \frac{\partial W^*}{\partial \mathbf{F}^T} \\ \mathbf{s} &= \frac{\partial W^*}{\partial \mathbf{E}} \end{aligned}} \tag{5.194}$$

with W^* defined by Eq. (5.191).

5.6 Linear Elastic Material

For a linear material, the strain-energy density function is a quadratic function of the strain components. As shown in Appendix A (Section A.4.3), characterizing the mechanical behavior of a general anisotropic linear material requires 21 independent elastic constants. In the following, the results of Section 5.3 are used to derive the appropriate stress-strain relations for isotropic, transversely isotropic, and orthotropic materials. For simplicity, only Cartesian coordinates are considered.

5.6.1 *Isotropic Material*

For an isotropic material, W depends only on the strain invariants I_1, I_2, and I_3 of Eqs. (3.142). However, because $I_3 = \det[\delta^I_J + 2E^I_J]$ is a cubic function of the strains, it can be deleted from the list of arguments for a linear material. For the subsequent analysis, it is convenient to replace I_1 and I_2 by the alternate strain invariants

$$\begin{aligned} J_1 &= \tfrac{1}{2}(I_1 - 3) \\ &= \varepsilon_{11} + \varepsilon_{22} + \varepsilon_{33} \\ J_2 &= \tfrac{1}{4}(I_2 - 2I_1 + 3) \\ &= \varepsilon_{11}\varepsilon_{22} + \varepsilon_{22}\varepsilon_{33} + \varepsilon_{33}\varepsilon_{11} - \varepsilon_{12}^2 - \varepsilon_{23}^2 - \varepsilon_{31}^2 \end{aligned} \qquad (5.195)$$

which are linear and quadratic functions of the strain components, respectively.

In terms of the alternate strain invariants above, the quadratic strain-energy density function has the form

$$W = C_1 J_1^2 + C_2 J_2 \qquad (5.196)$$

where C_1 and C_2 are material constants. Equation (5.98) now provides the constitutive relation

$$\sigma_{ij} = \frac{1}{2}\left(\frac{\partial W}{\partial \varepsilon_{ij}} + \frac{\partial W}{\partial \varepsilon_{ji}}\right), \qquad (5.197)$$

which gives

$$\begin{aligned}\sigma_{11} &= \frac{\partial W}{\partial \varepsilon_{11}} = 2C_1(\varepsilon_{11}+\varepsilon_{22}+\varepsilon_{33}) + C_2(\varepsilon_{22}+\varepsilon_{33}) \\ &\vdots \\ \sigma_{12} &= \frac{1}{2}\left(\frac{\partial W}{\partial \varepsilon_{12}} + \frac{\partial W}{\partial \varepsilon_{21}}\right) = -C_2\varepsilon_{12} \\ &\vdots\end{aligned}$$

On setting

$$\begin{aligned}C_1 &= \tfrac{1}{2}(\lambda + 2\mu) \\ C_2 &= -2\mu\end{aligned}$$

where λ and μ are the Lamé constants, the above stress-strain relations become

$$\sigma_{ij} = \lambda\Delta\delta_{ij} + 2\mu\varepsilon_{ij} \tag{5.198}$$

or

$$\boxed{\boldsymbol{\sigma} = \lambda\Delta\mathbf{I} + 2\mu\mathbf{E}} \tag{5.199}$$

where

$$\Delta = \varepsilon_{ii} = \operatorname{tr}\mathbf{E} \tag{5.200}$$

is the dilatation, and $\mathbf{E} = \varepsilon_{ij}\mathbf{e}^i\mathbf{e}^j$ is the strain tensor. The two material constants λ and μ completely characterize the mechanical behavior of a linear isotropic elastic material.

Note that Eq. (5.199) is valid for large strains *if the material behavior remains linear*. In general, however, materials behave nonlinearly when the deformation becomes large. Thus, Eq. (5.199) usually is restricted to small strains.

5.6.2 *Transversely Isotropic Material*

According to Eq. (5.112), the strain-energy density function for a transversely isotropic material depends on the quantities I_4 and I_5 defined by (5.113), in addition to the strain invariants I_1, I_2, and I_3 (or J_1, J_2, and I_3). Modifying Eq. (5.196) to include all quadratic combinations of the strain components gives

$$W = C_1J_1^2 + C_2J_2 + C_3J_1I_4 + C_4I_4^2 + C_5I_5. \tag{5.201}$$

Thus, a linear transversely isotropic material is characterized by the five elastic constants C_i, with Eq. (5.197) providing the stress-strain relations.

5.6.3 *Orthotropic Material*

Equation (5.104) gives the strain dependencies of W for an orthotropic material. Writing all quadratic combinations of these terms yields

$$\begin{aligned} W &= C_1\varepsilon_{11}^2 + C_2\varepsilon_{22}^2 + C_3\varepsilon_{33}^2 \\ &\quad + C_4\varepsilon_{11}\varepsilon_{22} + C_5\varepsilon_{22}\varepsilon_{33} + C_6\varepsilon_{33}\varepsilon_{11} \\ &\quad + C_7\varepsilon_{12}^2 + C_8\varepsilon_{23}^2 + C_9\varepsilon_{31}^2, \end{aligned} \tag{5.202}$$

and Eq. (5.197) again provides the stress-strain relations. The above equation indicates that defining the mechanical behavior of a linear orthotropic material requires nine independent elastic constants.

5.7 Determining W for Soft Tissues

Determining constitutive relations for soft biological tissues has been and remains a subject of intensive ongoing research. This section examines some of the issues involved in this process.

5.7.1 *Microstructural and Phenomenological Approaches*

In general, two approaches are used to model the mechanical behavior of soft tissues: microstructural and phenomenological (Humphrey, 2002). In the **microstructural approach**, the geometric and mechanical properties of individual tissue components, e.g., collagen and elastin, are measured directly. Then, with assumptions concerning the interactions between these components, a macroscopic constitutive relation for the composite tissue is derived from first principles. This approach has several advantages, including providing insight into the mechanisms of the tissue behavior and the ability to describe remodeling due to microstructural changes under changing loading conditions. On the other hand, attempts to adequately describe such complex materials have met with relatively limited success, as judged by the predicted response in comparison to experimental measurements.

In the **phenomenological approach**, macroscopic constitutive relations are determined directly by fitting computed stress-strain curves to measured tissue-level data. As discussed below, this usually involves postulating a functional form for W and then determining the material constants by minimizing the difference between computed and measured data. This approach is relatively simple mathematically, but the global nature of the method may mask the underlying microstructural mechanisms. Often, however, the microstructure is considered in postulating the functional form of the constitutive relation. For example, the tissue may be assumed to be transversely isotropic or orthotropic, depending on the observed arrangement of constituent fibers.

Due to its relative simplicity and consistency within a continuum framework, the remainder of this section focuses on the phenomenological approach.

5.7.2 *Experimental Considerations*

For a linear elastic material, depending on the specific material symmetry, the general form of W is known *a priori*, i.e., W must be quadratic in the strains. For a linear isotropic material, the two material constants can be determined from uniaxial testing alone. On the other hand, determining constants for linear anisotropic materials, e.g., bone, generally requires various sets of experiments, such as tensile tests on samples cut from different directions relative to the material axes. These experiments, however, are still relatively straightforward.

In contrast, the form of W is not known *a priori* for nonlinear elastic materials, and determining mechanical properties requires data from various loading protocols. Unless the structure is one-dimensional, e.g., an actin microfilament, uniaxial testing alone is not adequate. At the very least, biaxial testing is required (Humphrey, 2002).

In most studies of soft tissues, the functional form of W is postulated, guided by microstructural considerations and the principles discussed earlier in this chapter. (Some popular choices for W are given in Section 5.3.) Then, for the postulated W, the boundary value problem is solved for an experimental loading protocol, and fitting theoretical results to experimental data provides values for the unknown material coefficients.

A disadvantage of this approach is that it is difficult to know whether the chosen form for W is correct even if it appears to fit the data well. One reason for this is that the range of strain combinations that can be covered experimentally is limited, and some deformations may exist outside this realm for which the postulated W is not appropriate. In biomechanics applications, however, the choice for W may not need to be precise as long as it predicts approximately the correct behavior within the deformation range of interest. Another approach is to use experimental data to determine the functional form of W, as well as the material parameters. To date, this method has seen rather limited use (Humphrey et al., 1990a; 1990b), but it certainly warrants further study.

5.7.3 *Some Types of Experiments*

Due to the complexity of the problem, most material testing of soft tissues has involved relatively simple geometry and loading protocols. To enable precise control of loading conditions, these tests often are conducted on specimens that are cut to convenient geometries. Experiments on myocardium, for example, have used thin rectangular samples cut from the heart (Demer and Yin, 1983; Humphrey and Yin, 1987; Yin et al., 1987; Humphrey et al., 1990a; Humphrey et al., 1990b; Humphrey, 2002). To facilitate biaxial stretching in the fiber and cross-fiber directions, the edges of the specimen are cut approximately parallel to and normal to the local muscle fiber direction.

In addition to experimental convenience, there is another reason that simple geometries often are preferred. Determining material coefficients generally involves an optimization procedure to find the best fit of theoretical results to experimental data. Hence, numerous iterations may be required to find convergence, and the availability of an analytic solution can greatly speed the process.

One disadvantage of using dissected specimens for experimentation is that the act of cutting introduces unphysiological boundary conditions that may alter the tissue microstructure. Preconditioning by repeated loading and unloading, however, likely restores the structure in part, and parameters such as temperature and pH can be controlled. But some factors that may influence the behavior of the tissue *in vivo* are difficult to simulate.

For this reason, some effort has been given to determining material properties *in vivo*. For example, investigators have measured strains in beating hearts by analyzing the recorded motions of markers implanted in the wall (Waldman et al., 1985; Waldman et al., 1988; McCulloch et al., 1989; Villarreal et al., 1991; Hashima et al., 1993; Taber et al., 1994). Fitting these results to wall strains predicted by a computational model for the beating heart yields material coefficients for a postulated form for W (Guccione et al., 1991; Omens et al., 1993). Although promising, this approach is relatively difficult to apply at the present time. The experiments are difficult, and the need for substantial computing power is great. Future developments in tissue tagging by magnetic resonance imaging, however, should allow accurate noninvasive measurements of strain (Young and Axel, 1992; Azhari et al., 1993; Moore et al., 2000; Kuijer et al., 2002). This,

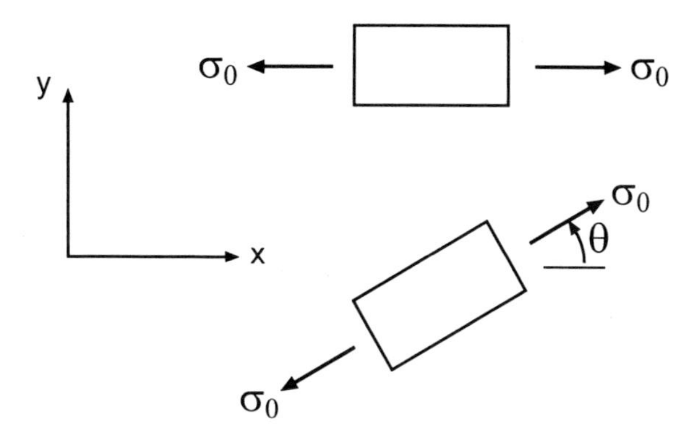

Fig. 5.5 Rotation of the stress tensor.

combined with even more powerful computers, likely will make *in vivo* measurements of material properties the wave of the future.

In the meantime, however, simple problems are the rule. Moreover, analytical solutions for benchmark problems always will be needed to validate computational codes. Hence, the next chapter presents solutions for a number of problems that are commonly used in experiments.

5.8 Problems

5–1 For a rigid-body rotation $\mathbf{Q}$, show that the Lagrangian and Eulerian strain tensors satisfy the transformation relations

$$\mathbf{E}^* = \mathbf{E}, \qquad \mathbf{e}^* = \mathbf{Q} \cdot \mathbf{e} \cdot \mathbf{Q}^T.$$

Explain the difference.

5–2 Consider a bar that is in a state of uniaxial stress $\boldsymbol{\sigma} = \sigma_0 \mathbf{e}_x \mathbf{e}_x$ relative to a fixed set of Cartesian coordinates. The bar then undergoes rotation through an angle θ in the xy-plane while maintaining the same uniaxial stress state (Fig. 5.5).

(a) Denote the principal directions of stress before and after rotation by $\mathbf{N}_i$ and $\mathbf{n}_i$, respectively. Using the formula $\boldsymbol{\Theta} = \mathbf{n}_i \mathbf{N}_i$, write the rotation tensor in terms of components relative to the fixed coordinate system.

(b) Compute the stress tensor $\boldsymbol{\sigma}^*$ in the rotated bar as it would be viewed from the fixed system.

5–3 Consider a stress tensor $\boldsymbol{\tau}$ defined by

$$\boldsymbol{\tau} = \tfrac{1}{2}(\mathbf{s} \cdot \mathbf{U} + \mathbf{U} \cdot \mathbf{s})$$

where $\mathbf{s}$ is the second Piola-Kirchhoff stress tensor, and $\mathbf{U}$ is the right stretch tensor.

(a) Show that $\boldsymbol{\tau}$ is symmetric.
(b) Show that the stress power can be written in the form $\boldsymbol{\tau} : \dot{\mathbf{U}}$.

5–4 An incompressible neo-Hookean material is characterized by the strain-energy density function $W = C(I_1 - 3)$. Ignoring shear terms, show that $C = E_Y/6$ in the limit of small strain, where E_Y is Young's modulus. *Hint:* For a linear isotropic material, the Lamé constants are related to Young's modulus and Poisson's ratio ν by the expressions

$$\begin{aligned} \lambda &= \frac{E_Y \nu}{(1-2\nu)(1+\nu)} \\ \mu &= \frac{E_Y}{2(1+\nu)}. \end{aligned}$$

In addition, $\nu = 0.5$ for an incompressible material.

5–5 Using the expressions

$$\mathbf{t} = \frac{\partial W}{\partial \mathbf{E}} \cdot \mathbf{F}^T \qquad \text{and} \qquad \mathbf{E} = \tfrac{1}{2}(\mathbf{F}^T \cdot \mathbf{F} - \mathbf{I}),$$

show that

$$\mathbf{t} = \frac{\partial W}{\partial \mathbf{F}^T}.$$

5–6 Show that $\varepsilon_{12}\varepsilon_{23}\varepsilon_{31}$ can be expressed in terms of $I_3 = \det(\delta_{IJ} + 2\varepsilon_{IJ})$ and the other arguments in Eq. (5.103).

5–7 Show that the expressions for I_5 given by Eqs. (5.113) and (5.114) are equivalent.

5–8 Derive Eq. (5.138).

5–9 Consider the tensors $\mathbf{M}$ and $\mathbf{N}$ defined by Eqs. (5.162). Show that the components $M^{I^*J^*}$ and $N^{I^*J^*}$ of these tensors are given by the right-hand sides of Eqs. $(5.159)_{3,4}$.

5–10 An incompressible material is characterized by the Mooney-Rivlin strain-energy density function

$$W = b_1(I_1 - 3) + b_2(I_2 - 3).$$

Write the constitutive relations for s^{11}, s^{12}, t^{1^*1}, t^{1^*2}, t^{2^*1}, and σ^{11} in terms of components of the Lagrangian strain tensor and the deformation gradient tensor.

5–11 A compressible material is characterized by the Blatz-Ko strain-energy density function

$$W = (\mu/2)(I_2/I_3 + 2I_3^{1/2} - 5).$$

Write W in terms of the principal stretch ratios Λ_i $(i = 1, 2, 3)$. Then, write the constitutive relations for the principal stresses s_2, t_2, and σ_2. *Hint:* For an isotropic material, the principal directions of stress and strain coincide.

5–12 For the material of Problem **5–11**, show that the constitutive relation for the Cauchy stress tensor can be written in the form

$$\boldsymbol{\sigma} = \mu \left(\mathbf{I} - I_3^{-1/2} \mathbf{B}^{-1} \right)$$

where $\mathbf{B}$ is the left Cauchy-Green deformation tensor.

5–13 Consider a strain-energy density function for a compressible material written in the form $W(\Lambda_1, \Lambda_2, \Lambda_3)$, where the Λ_i are principal stretch ratios. Show that the constitutive relations for the stress tensors can be written in the spectral forms

$$\mathsf{s} = \sum_{i=1}^{3} \frac{1}{\Lambda_i} \frac{\partial W}{\partial \Lambda_i} \mathbf{N}_i \mathbf{N}_i, \quad \mathsf{t} = \sum_{i=1}^{3} \frac{\partial W}{\partial \Lambda_i} \mathbf{N}_i \mathbf{n}_i, \quad \boldsymbol{\sigma} = J^{-1} \sum_{i=1}^{3} \Lambda_i \frac{\partial W}{\partial \Lambda_i} \mathbf{n}_i \mathbf{n}_i,$$

where the unit vectors $\mathbf{N}_i$ and $\mathbf{n}_i$ are principal directions in the undeformed and deformed body, respectively. *Hint:* See Eq. (3.135).

5–14 A body is composed of a compressible, transversely isotropic material with the strain-energy density function

$$W = \frac{a}{b} e^{b(I_1 - 3)} + c\,\varepsilon_{33}^2$$

in which ε_{33} is the strain relative to the fiber direction $\mathbf{e}_3$ in the undeformed body. In Cartesian coordinates, the body undergoes the deformation defined by

$$\begin{aligned} z_1 &= 0.2Z_1(1+2Z_1+Z_2) \\ z_2 &= 0.1Z_2(1+Z_2) \\ z_3 &= 0.8Z_3. \end{aligned}$$

For the point originally located at (1,2,1), compute the Cauchy stress components $\sigma^{1^*1^*}$ and $\sigma^{1^*2^*}$ using (a) Eq. (5.158) and (b) Eq. (5.161).

5–15 Suppose the undeformed body of Problem **5–14** is a rectangular block that is connected at one end to an isotropic circular cylinder such that the cylinder axis coincides with the fiber direction $\mathbf{e}_3$ in the block. For a particular set of loading and boundary conditions, it is convenient to work in polar cylindrical coordinates X^I, with $X^3 = Z^3$. If the block undergoes the deformation of Problem **5–14**, find $\sigma^{3^*3^*}$ at the point originally located at $(Z^1, Z^2, Z^3) = (1,2,1)$ using (a) Eq. (5.158) and (b) Eq. (5.161).

5–16 Consider a simple model for the left ventricle that consists of a thick-walled cylindrical tube composed of layers of incompressible, orthotropic myocardium (see Fig. 6.16b). At a particular point in the wall, the principal material directions of the undeformed myocardium are defined by the Cartesian unit vectors

$$\begin{aligned} \mathbf{e}_1 &= \mathbf{e}_R \\ \mathbf{e}_2 &= \mathbf{e}_\Theta \cos\beta + \mathbf{e}_Z \sin\beta \\ \mathbf{e}_3 &= -\mathbf{e}_\Theta \sin\beta + \mathbf{e}_Z \cos\beta \end{aligned}$$

where (R, Θ, Z) are cylindrical polar coordinates, and β is the muscle-fiber angle relative to the circumferential direction. In terms of strain components relative to the local material directions, the strain-energy density function is

$$\begin{aligned} W &= C(e^Q - 1) \\ Q &= a_1\varepsilon_{11}^2 + a_2\varepsilon_{22}^2 + a_3\varepsilon_{33}^2 + 2a_4\varepsilon_{11}\varepsilon_{22} + 2a_5\varepsilon_{22}\varepsilon_{33} + 2a_6\varepsilon_{33}\varepsilon_{11} \\ &\quad + 2a_7\varepsilon_{12}^2 + 2a_8\varepsilon_{23}^2 + 2a_9\varepsilon_{31}^2. \end{aligned}$$

The second Piola-Kirchhoff stress tensor can be written in the form $\mathbf{S} = S^{IJ}\mathbf{G}_I\mathbf{G}_J$, where the $\mathbf{G}_I$ represent base vectors of the polar coordinate system. Write the constitutive relations for the stress components S^{22} and S^{13} in terms of the ε_{IJ}.

Chapter 6

Biomechanics Applications

Finite element and other computational methods now make it possible to solve virtually any problem in biomechanics, no matter the complexity of the geometry, deformation, and material properties. Nevertheless, "exact solutions" remain useful for several reasons. First, they provide insight into the fundamental nonlinear behavior of soft tissues. Second, they provide benchmarks for checking the accuracy of numerical solutions. Third, as discussed in the previous chapter, exact solutions can be used in combination with experiments to determine material properties. Such experiments often involve relatively simple geometries and deformations, making the use of exact solutions convenient.

This chapter considers a number of fundamental problems in soft tissue biomechanics. Besides integrating the material presented in previous chapters, we use these problems to illustrate the effects of compressibility and anisotropy. Although soft tissues usually are assumed to be incompressible, extracellular fluid flow can produce significant compressibility effects, and microstructural considerations suggest that many soft tissues can be treated as orthotropic or transversely isotropic materials. Realistic models must take these features into account. For simplicity, gravity and other body forces, as well as inertia effects, are ignored in all problems of this chapter.

6.1 Boundary Value Problems

The governing equations, along with appropriate boundary and initial conditions, define a boundary value problem in the theory of elasticity. As derived in the previous chapters, the basic equations of nonlinear elasticity consist of the following:

Kinematic Relations:

$$\begin{aligned}
\mathbf{F} &= (\nabla\mathbf{r})^T \\
\mathbf{E} &= \tfrac{1}{2}(\mathbf{F}^T \cdot \mathbf{F} - \mathbf{I}) \\
&= \tfrac{1}{2}\left[\nabla\mathbf{u} + (\nabla\mathbf{u})^T + (\nabla\mathbf{u}) \cdot (\nabla\mathbf{u})^T\right]
\end{aligned} \tag{6.1}$$

Stresses:

$$\boldsymbol{\sigma} = J^{-1}\mathbf{F} \cdot \mathbf{t} = J^{-1}\mathbf{F} \cdot \mathbf{s} \cdot \mathbf{F}^T \tag{6.2}$$

Equations of Motion:

$$\begin{aligned}
\bar{\nabla} \cdot \boldsymbol{\sigma} + \mathbf{f} &= \rho\mathbf{a} \\
\nabla \cdot \mathbf{t} + \mathbf{f}_0 &= \rho_0\mathbf{a} \\
\nabla \cdot (\mathbf{s} \cdot \mathbf{F}^T) + \mathbf{f}_0 &= \rho_0\mathbf{a}
\end{aligned} \tag{6.3}$$

Constitutive Relations:

$$\begin{aligned}
\boldsymbol{\sigma} &= \mathbf{F} \cdot \frac{\partial W}{\partial \mathbf{E}} \cdot \mathbf{F}^T - p\,\mathbf{I} \\
\mathbf{t} &= \frac{\partial W}{\partial \mathbf{E}} \cdot \mathbf{F}^T - p\mathbf{F}^{-1} \\
\mathbf{s} &= \frac{\partial W}{\partial \mathbf{E}} - p\mathbf{F}^{-1} \cdot \mathbf{F}^{-T}
\end{aligned} \tag{6.4}$$

Incompressibility:

$$J = \det \mathbf{F} = 1 \tag{6.5}$$

The boundary conditions consist of specified tractions and displacements over the surface of the body. In dynamic problems, initial conditions also must be stipulated. These consist of the specified displacement and velocity of all points in the body at $t = 0$.

The specific equations and variables used to formulate and solve a boundary value problem depend on the geometry, the types of loads, the solution method (e.g., numerical or analytical), and personal preference. Moreover, elasticity problems can be formulated either by specifying the applied loads and computing the

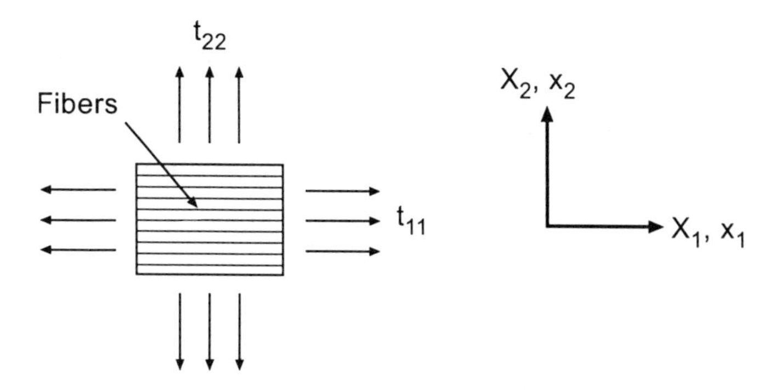

Fig. 6.1 Uniform extension of a transversely isotropic block.

resulting deformation or by specifying the deformation and computing the loads needed to produce it. The latter method, called the *inverse method*, often is more convenient in nonlinear problems. Sometimes, however, it is not reasonable to specify the entire deformation *a priori*. In such cases, a *semi-inverse method* may be useful in which only part of the deformation is specified, with the rest to be determined as part of the solution procedure. In linear elasticity, any method gives the same solution, as the solution is guaranteed to be unique by the uniqueness theorem. In nonlinear elasticity, however, multiple solutions are possible, and different approaches may yield different solutions, each of which may be valid. For example, investigating bifurcations in stability problems may benefit from combining various solution methods. In general, problems in this chapter are solved using either inverse or semi-inverse methods.

6.2 Extension and Compression of Soft Tissue

For simplicity, most experimental studies of the constitutive behavior of soft tissues involve uniaxial or biaxial tests. Example 3.2 (page 79) examined the kinematics of a rectangular block of tissue undergoing uniform extension. Here, we compute the tractions that must be applied to the block to maintain this deformation. For homogeneous stress and strain, these tractions must be uniform over the faces of the block (Fig. 6.1), and the equilibrium equations (6.3) (with $\mathbf{a} = \mathbf{0}$) are satisfied identically if there are no body forces.

We assume that the tissue is transversely isotropic, with fibers aligned in the

X^1-direction, which also is taken as a loading direction (Fig. 6.1). Skeletal muscle, heart muscle, ligaments, and tendons often are taken as transversely isotropic.

Hence, due to symmetry, the coordinate axes correspond to the principal directions of stress and strain. For convenience and since we are working entirely in principal Cartesian coordinates, we do not distinguish here between the X^I and x^i, between subscripts and superscripts, nor between lower and upper case indices.

6.2.1 *Governing Equations*

Kinematics. Relative to the Cartesian coordinate systems X_I and x_i (Fig. 6.1), Eqs. (3.45) and (3.48) give the deformation gradients and Lagrangian strain components

$$\begin{aligned} \left[F_I^i\right] &= [F_{iI}] = \text{diag}\,[\lambda_1, \lambda_2, \lambda_3] \\ [E_{IJ}] &= \text{diag}\,\left[\tfrac{1}{2}(\lambda_1^2 - 1), \tfrac{1}{2}(\lambda_2^2 - 1), \tfrac{1}{2}(\lambda_3^2 - 1)\right] \end{aligned} \tag{6.6}$$

where the λ_i are stretch ratios. Here, we assume that the λ_i are specified variables.

Constitutive Relations. Since the x_i are principal axes, there is no particular disadvantage in working with the first Piola-Kirchhoff stress tensor, which generally is not symmetric if shear stresses enter the analysis. In fact, $\mathbf{t}$ often is the stress tensor of choice in problems not *directly* involving shear. (Of course, shear stresses may occur relative to axes not aligned with the principal directions.) In this case, the constitutive relation $(6.4)_2$ yields

$$t_{II} = \frac{\partial W}{\partial E_{II}} F_{II} - p(F_{II})^{-1} \qquad (I \text{ not summed}),$$

with $t_{IJ} = 0$ for $I \neq J$. The term involving W can be simplified by noting, for example, that

$$\frac{\partial W}{\partial \lambda_1} = \frac{\partial W}{\partial E_{11}} \frac{\partial E_{11}}{\partial \lambda_1} = \frac{\partial W}{\partial E_{11}} \lambda_1 = \frac{\partial W}{\partial E_{11}} F_{11}$$

by (6.6). Thus, we can write

$$t_{II} = \frac{\partial W}{\partial \lambda_I} - \frac{p}{\lambda_I} \qquad (I \text{ not summed}). \tag{6.7}$$

This form is used often in nonlinear elasticity problems that are formulated in principal coordinates.

For a compressible material, Eq. (6.7) remains valid if we set $p = 0$. In this case, if $W(\lambda_I)$ is known and all of the λ_I are specified, then the force applied to each face of the block is given by multiplying the t_{II} by the corresponding *undeformed* area of the face, thereby completing the solution. For an incompressible block, however, all of the λ_I cannot be specified independently, since they are constrained by the incompressibility condition

$$\det \mathbf{F} = \lambda_1 \lambda_2 \lambda_3 = 1. \tag{6.8}$$

For example, if λ_1 and λ_2 are specified, then this relation gives λ_3 independently of the forces applied in the x_3-direction. Thus, a well-posed problem requires specifying one of the surface tractions, say t_{33}, rather than λ_3. (For the membrane to remain planar, t_{33} must be applied to both faces normal to x_3, in opposite directions.) Then, Eq. (6.7) gives

$$p = \lambda_3 \left(\frac{\partial W}{\partial \lambda_3} - t_{33} \right), \tag{6.9}$$

which is uniformly valid for a homogeneous deformation. The other stress components now can be determined using Eqs. (6.7) and (6.8).

6.2.2 *Biaxial Stretching of a Membrane*

Next, consider the special case of biaxial stretching of a thin homogeneous membrane in the x_1x_2-plane (Fig. 6.2). Biaxial loading has been used to determine the material properties of skin, pericardium, and myocardium, which experience multiaxial loading *in vivo*. We assume that the in-plane stretch ratios λ_1 and λ_2 are known, and the surfaces at $X_3 = \pm H/2$ are traction free (H = undeformed membrane thickness). In the following, we examine separately the problems for membranes composed of incompressible and compressible materials. These two cases require quite different solution methods.

Incompressible Membrane. If the membrane is composed of an incompressible material, the Lagrange multiplier p can be determined directly from the boundary conditions. Since the membrane is not loaded in the x_3-direction, we assume a condition of plane stress ($t_{33} = 0$ everywhere), and Eq. (6.9) gives

$$p = \lambda_3 \frac{\partial W}{\partial \lambda_3}. \tag{6.10}$$

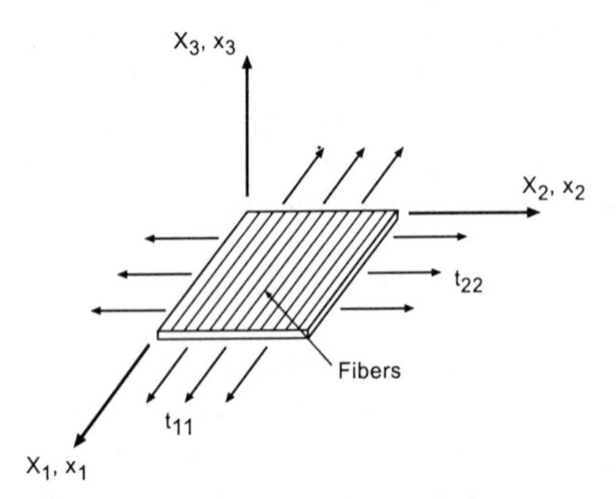

Fig. 6.2 Biaxial stretching of a transversely isotropic membrane.

Substitution into Eq. (6.7) then yields

$$t_{11} = \frac{\partial W}{\partial \lambda_1} - \frac{\lambda_3}{\lambda_1}\frac{\partial W}{\partial \lambda_3}$$

$$t_{22} = \frac{\partial W}{\partial \lambda_2} - \frac{\lambda_3}{\lambda_2}\frac{\partial W}{\partial \lambda_3}. \tag{6.11}$$

If the membrane is transversely isotropic relative to the x_1-axis (Fig. 6.2), then W is a function of the five strain invariants

$$\begin{aligned}
I_1 &= 3 + 2(E_{11} + E_{22} + E_{33}) = \lambda_1^2 + \lambda_2^2 + \lambda_3^2 \\
I_2 &= 3 + 4(E_{11} + E_{22} + E_{33} + E_{11}E_{22} + E_{22}E_{33} + E_{33}E_{11}) \\
&= \lambda_1^2\lambda_2^2 + \lambda_2^2\lambda_3^2 + \lambda_3^2\lambda_1^2 \\
I_3 &= (1 + 2E_{11})(1 + 2E_{22})(1 + 2E_{33}) = \lambda_1^2\lambda_2^2\lambda_3^2 = 1 \\
I_4 &= E_{11} = \tfrac{1}{2}(\lambda_1^2 - 1) \\
I_5 &= E_{12}^2 + E_{13}^2 = 0
\end{aligned} \tag{6.12}$$

as provided by Eqs. (3.142), (5.113), and (6.6). Since $I_3 = 1$ (incompressibility) and $I_5 = 0$, Eq. (5.112) reduces to

$$W = W(I_1, I_2, I_4). \tag{6.13}$$

These equations give the derivatives (with $W_i \equiv \partial W / \partial I_i$)

$$\begin{aligned}
\frac{\partial W}{\partial \lambda_1} &= W_1 \frac{\partial I_1}{\partial \lambda_1} + W_2 \frac{\partial I_2}{\partial \lambda_1} + W_4 \frac{\partial I_4}{\partial \lambda_1} \\
&= W_1(2\lambda_1) + W_2(2\lambda_1\lambda_2^2 + 2\lambda_1\lambda_3^2) + W_4(\lambda_1) \\
&= \lambda_1[2W_1 + 2W_2(\lambda_2^2 + \lambda_3^2) + W_4] \\
\frac{\partial W}{\partial \lambda_2} &= \lambda_2[2W_1 + 2W_2(\lambda_1^2 + \lambda_3^2)] \\
\frac{\partial W}{\partial \lambda_3} &= \lambda_3[2W_1 + 2W_2(\lambda_1^2 + \lambda_2^2)].
\end{aligned}$$

Inserting these expressions and $\lambda_3 = 1/\lambda_1\lambda_2$ into Eqs. (6.11) yields

$$\begin{aligned}
t_{11} &= 2\lambda_1 \left(1 - \frac{1}{\lambda_1^4 \lambda_2^2}\right)(W_1 + \lambda_2^2 W_2) + \lambda_1 W_4 \\
t_{22} &= 2\lambda_2 \left(1 - \frac{1}{\lambda_1^2 \lambda_2^4}\right)(W_1 + \lambda_1^2 W_2).
\end{aligned} \tag{6.14}$$

The Cauchy stress components, given by Eq. (6.2), are $\sigma_{II} = \lambda_I t_{II}$ (I not summed).

For illustration, suppose the membrane is composed of a material with

$$W = C_1(I_1 - 3) + C_4 I_4^2 \tag{6.15}$$

where the ratio $C_4/C_1 \equiv \beta$ correlates with the fiber stiffness. Substitution into Eqs. (6.14) then gives the stresses as functions of λ_1 and λ_2. For equibiaxial stretch ($\lambda_1 = \lambda_2$), t_{11} is plotted in Fig. 6.3 as a function of the stretch ratio for various values of β. Since the fibers do not contribute to the stress in the x_2-direction, the curve for $\beta = 0$ also is the result for t_{22} for all β. For relatively small strain ($\lambda_1 = \lambda_2 < 1.05$), all of the solutions nearly coincide. For large strain, however, the curves diverge significantly, with the curve shape changing from primarily concave downward to concave upward as β increases. The curves for $\beta = 0$ and 0.5 are typical for incompressible rubber, while the curves for larger β are similar to those generated by tests on most soft biological tissues (Fung, 1993).

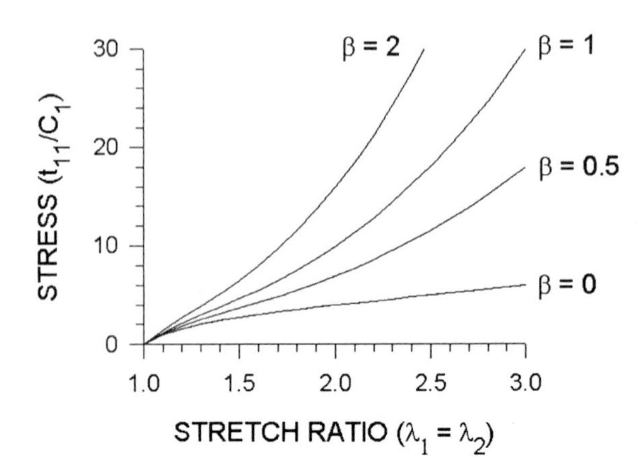

Fig. 6.3 Stress-stretch curves for equibiaxial stretching of a transversely isotropic, incompressible membrane for various values of the fiber modulus ($\beta = C_4/C_1$). The fibers are oriented in the x_1-direction.

Results also are shown for an isotropic membrane ($\beta = 0$, Fig. 6.4). Here the membrane is stretched in the x_1-direction, while the x_2-direction is held fixed at $\lambda_2 = 1, 1.5$, and 2. Note that t_{22} remains relatively constant as λ_1 increases while λ_2 is held at the value 2.

Compressible Membrane. If the membrane is composed of a compressible material, then the deformation is not constrained by Eq. (6.8), and we set the Lagrange multiplier $p = 0$. In this case, the plane stress condition $t_{33} = 0$ provides an equation to be solved for λ_3 (rather than p). However, since t_{33} depends on W, the value of λ_3 depends on the specific properties of the membrane material.

Consider, for example, a membrane composed of *isotropic* tissue that is characterized by the Blatz-Ko strain-energy density function

$$\begin{aligned} W &= C_2\left[I_2/I_3 - 3 + \tfrac{1-2\nu}{\nu}\left(I_3^{\nu/(1-2\nu)} - 1\right)\right] \\ &= C_2\left\{\lambda_1^{-2} + \lambda_2^{-2} + \lambda_3^{-2} - 3 + \tfrac{1-2\nu}{\nu}\left[(\lambda_1\lambda_2\lambda_3)^{2\nu/(1-2\nu)} - 1\right]\right\} \end{aligned} \tag{6.16}$$

as provided by Eqs. (6.12) and (5.124) with $\alpha = 0$ and $\mu = 2C_2$. With $p = 0$,

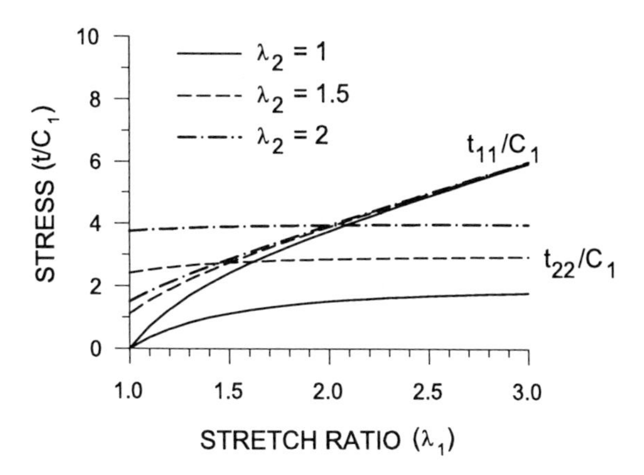

Fig. 6.4 Stress-stretch curves for biaxial stretching of an isotropic incompressible membrane. The stretch ratio λ_2 is held fixed at the three values indicated.

Eq. (6.7) yields

$$
\begin{aligned}
t_{11} &= 2C_2 \left[\lambda_2\lambda_3(\lambda_1\lambda_2\lambda_3)^{-(1-4\nu)/(1-2\nu)} - \lambda_1^{-3}\right] \\
t_{22} &= 2C_2 \left[\lambda_3\lambda_1(\lambda_1\lambda_2\lambda_3)^{-(1-4\nu)/(1-2\nu)} - \lambda_2^{-3}\right] \\
t_{33} &= 2C_2 \left[\lambda_1\lambda_2(\lambda_1\lambda_2\lambda_3)^{-(1-4\nu)/(1-2\nu)} - \lambda_3^{-3}\right],
\end{aligned}
\tag{6.17}
$$

and Eq. (6.2) gives the Cauchy stress components $\sigma_{11} = t_{11}/\lambda_2\lambda_3$, $\sigma_{22} = t_{22}/\lambda_3\lambda_1$, and $\sigma_{33} = t_{33}/\lambda_1\lambda_2$. For $t_{33} = 0$, Eq. $(6.17)_3$ gives

$$\lambda_3 = (\lambda_1\lambda_2)^{-\nu/(1-\nu)} \tag{6.18}$$

and then the other relations in (6.17) give

$$
\begin{aligned}
t_{11} &= 2C_2 \left[\left(\lambda_1^{-1+3\nu}\lambda_2^{2\nu}\right)^{1/(1-\nu)} - \lambda_1^{-3}\right] \\
t_{22} &= 2C_2 \left[\left(\lambda_1^{2\nu}\lambda_2^{-1+3\nu}\right)^{1/(1-\nu)} - \lambda_2^{-3}\right].
\end{aligned}
\tag{6.19}
$$

For small strain, the material parameter ν can be identified with Poisson's ratio. For the special case $\nu = 1/4$,

$$t_{11} = 2C_2 \left[\left(\frac{\lambda_2^2}{\lambda_1} \right)^{\frac{1}{3}} - \frac{1}{\lambda_1^3} \right]$$

$$t_{22} = 2C_2 \left[\left(\frac{\lambda_1^2}{\lambda_2} \right)^{\frac{1}{3}} - \frac{1}{\lambda_2^3} \right]. \qquad (6.20)$$

On the other hand, setting $\nu = 0.5$ and $I_3 = 1$ yields the incompressible case. In this instance, Eq. (6.18) gives $\lambda_3 = (\lambda_1 \lambda_2)^{-1}$ and Eq. (6.16) reduces to $W = C_2(I_2 - 3)$. Substitution into (6.14) with $W_4 = 0$ then reproduces Eqs. (6.19) if we set $\nu = 0.5$.

It is important to note that for some forms of W, the equation $t_{33} = 0$ may have multiple roots. Thus, it is theoretically possible that, for the same λ_1 and λ_2, the membrane can have different thicknesses, although it is unlikely that all solutions would be stable. The possibility of multiple solutions always must be kept in mind when dealing with nonlinear problems.

Equibiaxial loading curves computed from Eqs. (6.19) for various values of ν show that the effects of compressibility can be quite large (Fig. 6.5). The membrane stiffness increases markedly with ν, especially as ν approaches 0.5, corresponding to an incompressible membrane.

6.2.3 *Uniaxial Extension of a Bar*

Papillary muscles, actin microfilaments, ligaments, and articular cartilage are examples of biological structures that are subjected primarily to uniaxial loading conditions. If a tissue or cytoskeletal component is stretched by tractions applied only in the x_1 direction, then $t_{22} = t_{33} = 0$. In addition, symmetry demands that $\lambda_2 = \lambda_3$. For an incompressible material, Eq. (6.8) gives $\lambda_2 = \lambda_3 = \lambda_1^{-1/2}$, and substitution into $(6.14)_2$ reveals that the condition $t_{22} = 0$ is satisfied identically. (The condition $t_{33} = 0$ already has been used to determine p.) Hence, Eq. $(6.14)_1$ gives

$$t_{11} = 2\lambda_1 \left(1 - \frac{1}{\lambda_1^3} \right) \left(W_1 + \frac{1}{\lambda_1} W_2 \right) + \lambda_1 W_4. \qquad (6.21)$$

For a compressible (Blatz-Ko) material, setting $t_{22} = 0$ in Eq. $(6.19)_2$ gives

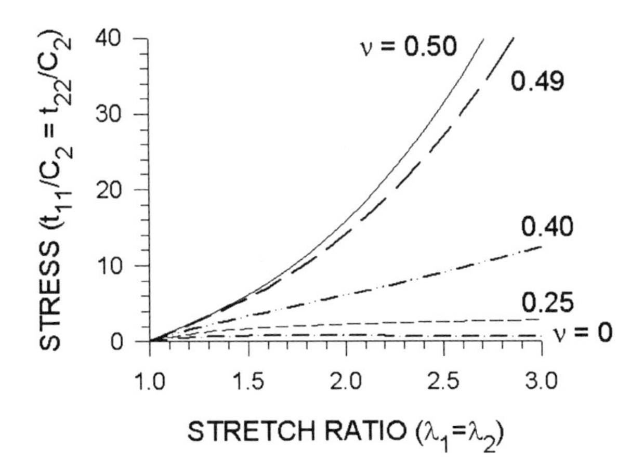

Fig. 6.5 Stress-stretch curves for equibiaxial stretching of an isotropic compressible membrane for various values of the material constant ν.

$\lambda_2 = \lambda_1^{-\nu}$. Then, $(6.19)_1$ yields

$$t_{11} = 2C_2 \left(\lambda_1^{-1+2\nu} - \lambda_1^{-3} \right). \tag{6.22}$$

For $\nu = 1/4$, the Cauchy stress is

$$\sigma_{11} = \frac{t_{11}}{\lambda_2 \lambda_3} = \lambda_1^{1/2} t_{11} = 2C_2 \left(1 - \lambda_1^{-5/2} \right). \tag{6.23}$$

The loading behavior for various values of ν are illustrated in Fig. 6.6. In contrast to the biaxial stress-stretch curves (Fig. 6.5), the curves for $\nu = 0.5$ and $\nu = 0.49$ are practically indistinguishable. (The curve for $\nu = 0.49$ is not shown.) Note also the strong asymmetry in the curves for tensile versus compressive loading.

6.3 Simple Shear of Soft Tissue

Shear deformation has received relatively little attention in the biomechanics literature. Problems exist, however, in which shear is an important issue. For example, transverse shear of the heart wall is thought to play a role in proper functioning of the left ventricle. In this section, we examine the forces required to sustain sim-

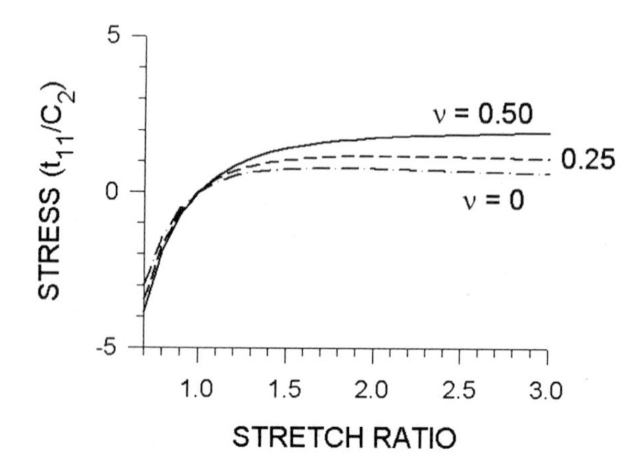

Fig. 6.6 Stress-stretch curves for uniaxial loading of an isotropic compressible bar for various values of the material constant ν.

ple shear of a homogeneous block of tissue (see Example 3.3 on page 81). As in the stretching problem, surface tractions are assumed to be distributed uniformly, giving a homogeneous deformation. Hence, the equilibrium equations (6.3) are satisfied identically (with $\mathbf{f} = \mathbf{a} = \mathbf{0}$). Although these equations were disposed of similarly in the stretching problem, this is where the similarity ends. The problem of simple shear illustrates a number of important points regarding the use of nonorthogonal coordinate systems in solving elasticity problems.

6.3.1 *Governing Equations*

Kinematics. For convenience, some of the results of Example 3.3 are listed again here. In terms of the undeformed coordinates X^I, the Cartesian coordinates of a point in the deformed block are

$$x^1 = X^1 + kX^2, \qquad x^2 = X^2, \qquad x^3 = X^3 \tag{6.24}$$

where the parameter k defines the magnitude of the shear (Fig. 6.7). This mapping produces the convected base vectors

$$\mathbf{g}_{1^*} = \mathbf{e}_1, \qquad \mathbf{g}_{2^*} = k\mathbf{e}_1 + \mathbf{e}_2, \qquad \mathbf{g}_{3^*} = \mathbf{e}_3 \tag{6.25}$$

as given by (3.53), and Eq. (2.17) provides the contravariant base vectors

$$\mathbf{g}^{1^*} = \mathbf{e}_1 - k\mathbf{e}_2, \qquad \mathbf{g}^{2^*} = \mathbf{e}_2, \qquad \mathbf{g}^{3^*} = \mathbf{e}_3. \tag{6.26}$$

The directions of these base vectors are shown in Fig. 6.7. In addition, with Eq. $(3.55)_1$, the Cartesian components of the Lagrangian strain tensor are

$$[E_{IJ}] = \tfrac{1}{2}[C_{IJ} - \delta_{IJ}] = \begin{bmatrix} 0 & k/2 & 0 \\ k/2 & k^2/2 & 0 \\ 0 & 0 & 0 \end{bmatrix}. \tag{6.27}$$

Note that Eq. (3.51) yields $J = \det \mathbf{F} = 1$ for any value of k. Thus, the deformation described by (6.24) is isochoric, regardless of the composition of the block.

Later, we will need components of the metric tensors. Since $\mathbf{G}_I = \mathbf{G}^I = \mathbf{e}_I$,

$$[G_{IJ}] = [G^{IJ}] = [\mathbf{e}_I \cdot \mathbf{e}_J] = \begin{bmatrix} 1 & 0 & 0 \\ 0 & 1 & 0 \\ 0 & 0 & 1 \end{bmatrix}, \tag{6.28}$$

and Eqs. (6.25) and (6.26) give

$$[g_{IJ}] = [\mathbf{g}_I \cdot \mathbf{g}_J] = \begin{bmatrix} 1 & k & 0 \\ k & 1+k^2 & 0 \\ 0 & 0 & 1 \end{bmatrix}$$

$$[g^{IJ}] = [\mathbf{g}^I \cdot \mathbf{g}^J] = \begin{bmatrix} 1+k^2 & -k & 0 \\ -k & 1 & 0 \\ 0 & 0 & 1 \end{bmatrix}. \tag{6.29}$$

Constitutive Relations. We want to compute the stresses in the block and the surface tractions required to sustain the specified deformation. The geometry suggests that the stress components $\sigma^{I^*J^*}$ are appropriate for this problem, as they act in the directions defined by the faces of the deformed block (Fig. 6.7). This can be seen from the dyadic representation $\boldsymbol{\sigma} = \sigma^{I^*J^*}\mathbf{g}_I\mathbf{g}_J$, with $\mathbf{g}_J$ giving the direction of action. If we stipulate that no loads are applied on the $\pm X^3$ faces of

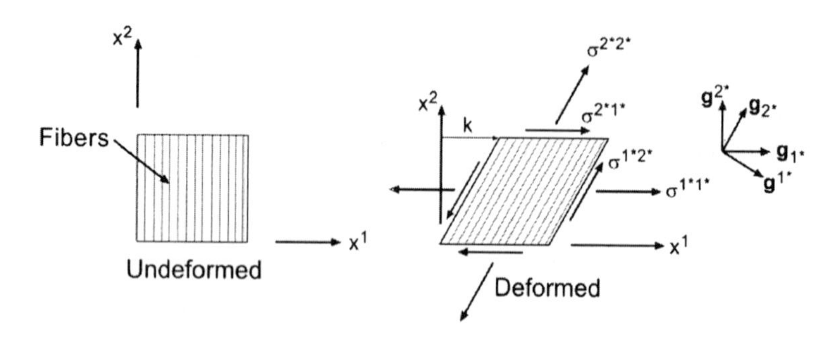

Fig. 6.7 Simple shear of a transversely isotropic block.

the block, then we have the plane stress conditions $\sigma^{3^*1^*} = \sigma^{3^*2^*} = \sigma^{3^*3^*} = 0$, which must hold throughout the block because the stress field is homogeneous. It is important to note that not all materials can satisfy plane stress and the plane strain condition $E_{33} = 0$ simultaneously. For example, if the anisotropy is such that the given shear in the X^1X^2-plane generates stresses on planes normal to X^3, then traction must be exerted on the $\pm X^3$ faces.

Here, we consider a block composed of a transversely isotropic material with fibers oriented originally in the X^2-direction (Fig. 6.7). For such a material, Eq. (5.112) gives

$$W = W(I_1, I_2, I_3, I_4, I_5). \tag{6.30}$$

Because $E^I_J = G^{IK}E_{KJ} = \delta^{IK}E_{KJ} = E_{IJ}$, Eqs. (3.142), (5.113), and (6.27) yield the strain invariants

$$\begin{aligned}
I_1 &= 3 + 2(E_{11} + E_{22} + E_{33}) \\
&= 3 + k^2 \\
I_2 &= 3 + 4(E_{11} + E_{22} + E_{33} + E_{11}E_{22} + E_{22}E_{33} + E_{33}E_{11} \\
&\qquad - E_{12}^2 - E_{23}^2 - E_{31}^2) \\
&= 3 + k^2 \\
I_3 &= \det(\delta_{IJ} + 2E_{IJ}) = 1 \\
I_4 &= E_{22} = k^2/2 \\
I_5 &= E_{12}^2 + E_{23}^2 = k^2/4.
\end{aligned} \tag{6.31}$$

Equations (4.36) and $(5.190)_2$ provide the constitutive equations

$$J\sigma^{I^*J^*} = s^{IJ} = \frac{\partial W}{\partial E_{IJ}} - pg^{IJ}. \tag{6.32}$$

If the material is incompressible, then the Lagrange multiplier p can be determined using the condition $\sigma^{3^*3^*} = 0$, i.e.,

$$p = \frac{1}{g^{3^*3^*}} \frac{\partial W}{\partial E_{33}} = \frac{\partial W}{\partial E_{33}}. \tag{6.33}$$

Then, given the strains of (6.27), all stress components can be computed from Eq. (6.32). On the other hand, if the block consists of a compressible material, then $p = 0$ and Eq. (6.32) stipulates that W must be of a form compatible with the constraint $\sigma^{3^*3^*} = \partial W/\partial E_{33} = 0$ when $E_{33} = 0$.

With W expressed in terms of the strain invariants, Eq. (5.158) gives the stress-strain relation

$$\sigma^{I^*J^*} = \Phi G^{IJ} + \Psi Q^{IJ} - pg^{IJ} + \Theta M^{IJ} + \Lambda N^{IJ} \tag{6.34}$$

for a transversely isotropic material. The Q^{IJ} in this equation are computed from (5.151). Moreover, because the principal material directions coincide with the X^I-axes, $\varepsilon_{IJ} = E_{IJ}$ and $A_I^J = \partial X^J/\partial Z^I = \partial X^J/\partial X^I = \delta_I^J$, and Eqs. (5.159) give (for fibers in the X^2-direction)

$$\begin{aligned} M^{IJ} &= \delta_2^I \delta_2^J \\ N^{IJ} &= (\delta_2^I \delta_1^J + \delta_1^I \delta_2^J) E_{12}. \end{aligned}$$

In matrix form, the results are

$$[Q^{IJ}] = \begin{bmatrix} 2+k^2 & -k & 0 \\ -k & 2 & 0 \\ 0 & 0 & 2+k^2 \end{bmatrix}$$

$$[M^{IJ}] = \begin{bmatrix} 0 & 0 & 0 \\ 0 & 1 & 0 \\ 0 & 0 & 0 \end{bmatrix}, \qquad [N^{IJ}] = \begin{bmatrix} 0 & 1 & 0 \\ 1 & 0 & 0 \\ 0 & 0 & 0 \end{bmatrix} E_{12}. \tag{6.35}$$

With these relations, Eq. (6.34) yields the Cauchy stress components

$$\begin{aligned}
\sigma^{1^*1^*} &= \Phi + (2+k^2)\Psi - (1+k^2)p \\
\sigma^{2^*2^*} &= \Phi + 2\Psi - p + \Theta \\
\sigma^{3^*3^*} &= \Phi + (2+k^2)\Psi - p \\
\sigma^{1^*2^*} &= k(-\Psi + p + \Lambda/2) \\
\sigma^{1^*3^*} &= \sigma^{2^*3^*} = 0
\end{aligned} \tag{6.36}$$

where Eqs. (5.151) and (5.159) give

$$\Phi = 2\,\frac{\partial W}{\partial I_1}, \qquad \Psi = 2\,\frac{\partial W}{\partial I_2}, \qquad p = -2\,\frac{\partial W}{\partial I_3}$$

$$\Theta = \frac{\partial W}{\partial I_4}, \qquad \Lambda = \frac{\partial W}{\partial I_5} \tag{6.37}$$

since $I_3 = 1$ in this problem.

Surface Tractions. With the components $\sigma^{I^*J^*}$ of the Cauchy stress tensor given above, we can compute the true traction vector on the surfaces of the block using

$$\mathbf{T}^{(\mathbf{n}_i)} = \mathbf{n}_i \cdot \boldsymbol{\sigma} \tag{6.38}$$

where $\mathbf{n}_i$ is the unit vector normal to the deformed x_i-surface. The geometry (Fig. 6.7), along with Eqs. (6.26) and (6.29), gives

$$\mathbf{n}_1 = \frac{\mathbf{g}^{1^*}}{\sqrt{g^{1^*1^*}}} = \frac{\mathbf{e}_1 - k\mathbf{e}_2}{\sqrt{1+k^2}}, \qquad \mathbf{n}_2 = \frac{\mathbf{g}^{2^*}}{\sqrt{g^{2^*2^*}}} = \mathbf{e}_2. \tag{6.39}$$

Hence, we have

$$\begin{aligned}
\mathbf{T}^{(\mathbf{n}_1)} &= \frac{\mathbf{g}^{1^*}}{\sqrt{g^{1^*1^*}}} \cdot \left(\sigma^{I^*J^*}\mathbf{g}_I\mathbf{g}_J\right) = \frac{\sigma^{I^*J^*}}{\sqrt{g^{1^*1^*}}}\delta_I^1\mathbf{g}_J = \frac{\sigma^{1^*J^*}}{\sqrt{g^{1^*1^*}}}\mathbf{g}_J \\
&= \frac{1}{\sqrt{g^{1^*1^*}}}\left(\sigma^{1^*1^*}\mathbf{g}_{1^*} + \sigma^{1^*2^*}\mathbf{g}_{2^*}\right) \\
\mathbf{T}^{(\mathbf{n}_2)} &= \frac{\mathbf{g}^{2^*}}{\sqrt{g^{2^*2^*}}} \cdot \left(\sigma^{I^*J^*}\mathbf{g}_I\mathbf{g}_J\right) = \frac{\sigma^{I^*J^*}}{\sqrt{g^{2^*2^*}}}\delta_I^2\mathbf{g}_J = \frac{\sigma^{2^*J^*}}{\sqrt{g^{2^*2^*}}}\mathbf{g}_J \\
&= \frac{1}{\sqrt{g^{2^*2^*}}}\left(\sigma^{2^*1^*}\mathbf{g}_{1^*} + \sigma^{2^*2^*}\mathbf{g}_{2^*}\right).
\end{aligned} \tag{6.40}$$

Because the $\mathbf{T}^{(\mathbf{n}_i)}$ are vectors, the stress components normal and tangential to the surfaces of the block can be determined by dotting with the corresponding unit vectors (see the geometry in Fig. 6.7). On the x_1-face, these components are

$$\begin{aligned}
\mathbf{T}^{(\mathbf{n}_1)}_{\text{norm}} &= \frac{\mathbf{g}^{1^*}}{\sqrt{g^{1^*1^*}}} \cdot \mathbf{T}^{(\mathbf{n}_1)} = \frac{\sigma^{1^*1^*}}{g^{1^*1^*}} \\
&= \frac{\sigma^{1^*1^*}}{1+k^2} \\
\mathbf{T}^{(\mathbf{n}_1)}_{\text{tang}} &= \frac{\mathbf{g}_{2^*}}{\sqrt{g_{2^*2^*}}} \cdot \mathbf{T}^{(\mathbf{n}_1)} = \frac{1}{\sqrt{g^{1^*1^*} g_{2^*2^*}}} \left(\sigma^{1^*1^*} g_{2^*1^*} + \sigma^{1^*2^*} g_{2^*2^*} \right) \\
&= \frac{k}{1+k^2} \sigma^{1^*1^*} + \sigma^{1^*2^*},
\end{aligned} \tag{6.41}$$

while on the x_2-face,

$$\begin{aligned}
\mathbf{T}^{(\mathbf{n}_2)}_{\text{norm}} &= \frac{\mathbf{g}^{2^*}}{\sqrt{g^{2^*2^*}}} \cdot \mathbf{T}^{(\mathbf{n}_2)} = \frac{\sigma^{2^*2^*}}{\sqrt{g^{2^*2^*}}} \\
&= \sigma^{2^*2^*} \\
\mathbf{T}^{(\mathbf{n}_2)}_{\text{tang}} &= \frac{\mathbf{g}_{1^*}}{\sqrt{g_{1^*1^*}}} \cdot \mathbf{T}^{(\mathbf{n}_2)} = \frac{1}{\sqrt{g_{1^*1^*}\, g^{2^*2^*}}} \left(\sigma^{2^*1^*} g_{1^*1^*} + \sigma^{2^*2^*} g_{1^*2^*} \right) \\
&= \sigma^{2^*1^*} + k\sigma^{2^*2^*}.
\end{aligned} \tag{6.42}$$

The applied forces needed to produce the specified deformation can be computed by multiplying these tractions by the appropriate deformed surface areas. (Note that, for our solution to be "exact," these forces must be distributed uniformly over the surfaces of the block.) Suppose the areas of the undeformed and deformed faces of the block are dA_i and da_i, respectively. Then, a glance at the geometry (Fig. 6.7) reveals that dA_2 does not change during deformation, but dA_1 increases by a factor of $\sqrt{1+k^2}$, i.e., by the change in length of the originally vertical side. Alternatively, we can use Eq. (3.128), which gives (for $J = 1$)

$$d\mathbf{a}_i = d\mathbf{A}_i \cdot \mathbf{F}^{-1} \tag{6.43}$$

where

$$\begin{aligned}
d\mathbf{A}_1 &= \mathbf{e}_1\, dA_1, \qquad & d\mathbf{A}_2 &= \mathbf{e}_2\, dA_2 \\
d\mathbf{a}_1 &= \mathbf{n}_1\, da_1, \qquad & d\mathbf{a}_2 &= \mathbf{n}_2\, da_2.
\end{aligned} \tag{6.44}$$

With Eqs. (3.22) and (6.26) and $\mathbf{G}_I = \mathbf{e}_I$, the inverse of the deformation gradient tensor for this problem can be written in the form

$$\mathbf{F}^{-1} = \mathbf{G}_I\mathbf{g}^I = \mathbf{e}_1(\mathbf{e}_1 - k\mathbf{e}_2) + \mathbf{e}_2\mathbf{e}_2 + \mathbf{e}_3\mathbf{e}_3. \tag{6.45}$$

Now, inserting Eqs. (6.39), (6.44), and (6.45) into (6.43) gives

$$da_1 = \sqrt{1+k^2}\, dA_1, \qquad da_2 = dA_2, \tag{6.46}$$

in agreement with our previous result.

Next, we examine results for incompressible and compressible blocks composed of tissues with specific properties.

6.3.2 *Solution*

Incompressible Tissue. For an incompressible material, the incompressibility condition $I_3 = 1$ renders $\partial W/\partial I_3$ indeterminate, and p in Eq. (6.37) becomes a Lagrange multiplier. Setting $\sigma^{3^*3^*} = 0$ in (6.36) yields

$$p = \Phi + (2+k^2)\Psi, \tag{6.47}$$

and so Eqs. (6.36) provide the nonzero components of $\boldsymbol{\sigma}$ in the form

$$\begin{aligned}
\sigma^{1^*1^*} &= -k^2[\Phi + (2+k^2)\Psi] \\
\sigma^{2^*2^*} &= -k^2\Psi + \Theta \\
\sigma^{1^*2^*} &= k[\Phi + (1+k^2)\Psi + \Lambda/2].
\end{aligned} \tag{6.48}$$

Substituting these relations into Eqs. (6.41) and (6.42) yields the surface tractions

$$\begin{aligned}
T^{(\mathbf{n}_1)}_{\text{norm}} &= -\frac{k^2}{1+k^2}[\Phi + (2+k^2)\Psi] \\
T^{(\mathbf{n}_1)}_{\text{tang}} &= \frac{k}{1+k^2}(\Phi + \Psi) + \frac{k\Lambda}{2} \\
T^{(\mathbf{n}_2)}_{\text{norm}} &= -k^2\Psi + \Theta \\
T^{(\mathbf{n}_2)}_{\text{tang}} &= k(\Phi + \Psi + \Lambda/2 + \Theta).
\end{aligned} \tag{6.49}$$

To illustrate the behavior of this solution, we consider a block composed of soft tissue with the strain-energy density function of Eq. (6.15), in which the strain invariants are given by Eq. (6.31). In this case, $\Phi = 2C_1$, $\Theta = C_4 k^2$, and $\Psi = \Lambda = 0$. For this material, Eqs. (6.48) give the stresses

$$\sigma^{1^*1^*} = -2C_1 k^2, \qquad \sigma^{2^*2^*} = C_4 k^2, \qquad \sigma^{1^*2^*} = 2C_1 k, \tag{6.50}$$

and Eqs. (6.49) yield the surface tractions

$$\mathbf{T}^{(\mathbf{n}_1)}_{\text{norm}} = -\frac{2C_1 k^2}{1+k^2}, \qquad \mathbf{T}^{(\mathbf{n}_1)}_{\text{tang}} = \frac{2C_1 k}{1+k^2}$$

$$\mathbf{T}^{(\mathbf{n}_2)}_{\text{norm}} = C_4 k^2, \qquad \mathbf{T}^{(\mathbf{n}_2)}_{\text{tang}} = 2C_1 k + C_4 k^3. \tag{6.51}$$

Note that, unlike in the linear theory, normal forces must be exerted on the block to maintain simple shear. For the material considered here, the normal stress is compressive on the x_1-face and tensile on the x_2-face. However, if the material is isotropic ($C_4 = 0$), the normal force on the x_2-face is zero. In fact, Eq. $(6.49)_3$ shows that $\mathbf{T}^{(\mathbf{n}_2)}_{\text{norm}}$ can be compressive for certain transversely isotropic materials. For $k << 1$, these results reduce to those of the linear theory, with the tangential tractions becoming equal to the shear stress $\sigma^{1^*2^*}$, and the normal stresses [O(k^2)] can be neglected compared to the shear stresses [O(k)].

Compressible Tissue. If the block is composed of a compressible material, Eqs. $(6.36)_3$ and (6.37) give the plane-stress condition

$$\sigma^{3^*3^*} = 2\left[\frac{\partial W}{\partial I_1} + (2+k^2)\frac{\partial W}{\partial I_2} + \frac{\partial W}{\partial I_3}\right] = 0. \tag{6.52}$$

This relation constrains the composition of the block to materials that can sustain both the plane strain ($E_{33} = 0$) and the plane stress ($\sigma^{3^*3^*} = 0$) conditions simultaneously for shear in the X^1X^2-plane. One material that does satisfy these conditions consists of fibers initially aligned in the X^2-direction that are embedded in a matrix of the Blatz-Ko type. Here, we consider a material characterized by

$$W = C_1\left[I_1 - 3 + \tfrac{1-2\nu}{\nu}\left(I_3^{-\nu/(1-2\nu)} - 1\right)\right] + C_4 I_4^2, \tag{6.53}$$

which is an extended form of Eq. (5.124) with $\alpha = 1$ and reduces to Eq. (6.15) when $I_3 = 1$. Direct substitution shows that Eq. (6.52) is satisfied identically, and Eqs. (6.36) yield stress components identical to those of Eq. (6.50). This is the expected result, because the specified deformation is isochoric, and hence material compressibility should not affect the solution.

6.4 Extension and Torsion of a Papillary Muscle

Papillary muscles play a vital role in the pumping efficiency of the heart. These cylindrically shaped muscles are attached at one end to the wall of the left or right ventricle and at the other end to the mitral or tricuspid valve, respectively. During ventricular systole, the increasing pressure in the ventricles tends to push the valves upward into the atria, which would allow backflow of blood during ejection. The papillary muscles prevent this inversion by contracting and holding the valves closed. Due to their relatively simple geometry and muscle fibers aligned along their axes, papillary muscles have been popular in studies of the passive and active mechanical properties of heart muscle (Pinto and Fung, 1973; Humphrey et al., 1992; Fung, 1993; Criscione et al., 1999). For determining normal and shear properties, combined extension and torsion is a useful loading protocol.

As a model for a papillary muscle, consider a solid cylinder of undeformed radius b_0 and length ℓ_0 (Fig. 6.8). The unloaded cylinder is composed of material that is transversely isotropic relative to the axial direction Z. The ends are subjected to tractions that supply a net twisting moment M and normal force N, while the curved surface is traction free. Given the angle of twist per unit undeformed length ψ and the (uniform) axial stretch ratio λ, we want to compute M, N, and the stress distribution in the cylinder. Note that, unlike in the previous problems in this chapter, torsional deformation of a cylinder is not homogeneous.[1]

6.4.1 *Governing Equations*

Kinematics. This problem is a special case of that considered in Example 3.4 (page 84), where we analyzed the deformation of a hollow cylinder undergoing simultaneous extension, inflation, and torsion. As in that example, we use the cylindrical polar coordinates $(X^1, X^2, X^3) = (R, \Theta, Z)$ and $(x^1, x^2, x^3) = (r, \theta, z)$. If each cross section remains circular and planar, the deformation is described by

$$r = r(R), \qquad \theta = \Theta + \psi Z, \qquad z = \lambda Z \tag{6.54}$$

[1]This book considers only passive behavior of muscle tissue. It also is possible, however, to use a pseudoelastic analysis for contracting muscle if the constitutive equations take into account changes in muscle stiffness and the zero-stress configuration during activation (Taber, 1991a).

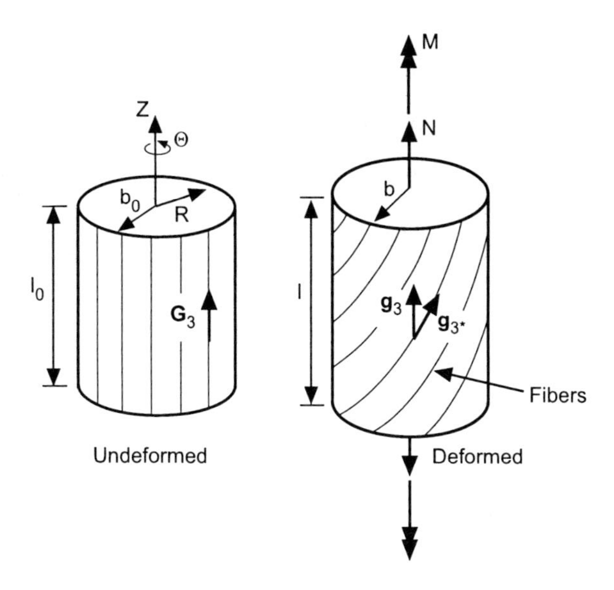

Fig. 6.8 Extension and torsion of a cylindrical model for a papillary muscle due to an axial force $\mathbf{N}$ and twisting moment $\mathbf{M}$.

as given by Eq. (3.56). The covariant base vectors, provided by Eqs. (3.59), are

$$[\mathbf{G}_I] = \begin{bmatrix} \mathbf{e}_R \\ R\mathbf{e}_\Theta \\ \mathbf{e}_Z \end{bmatrix}, \quad [\mathbf{g}_i] = \begin{bmatrix} \mathbf{e}_r \\ r\mathbf{e}_\theta \\ \mathbf{e}_z \end{bmatrix}, \quad [\mathbf{g}_I] = \begin{bmatrix} \dot{r}\mathbf{e}_r \\ r\mathbf{e}_\theta \\ \psi r\,\mathbf{e}_\theta + \lambda\mathbf{e}_z \end{bmatrix}, \tag{6.55}$$

and Eqs. (2.17) give the contravariant base vectors

$$[\mathbf{G}^I] = \begin{bmatrix} \mathbf{e}_R \\ \dfrac{1}{R}\mathbf{e}_\Theta \\ \mathbf{e}_Z \end{bmatrix}, \quad [\mathbf{g}^i] = \begin{bmatrix} \mathbf{e}_r \\ \dfrac{1}{r}\mathbf{e}_\theta \\ \mathbf{e}_z \end{bmatrix}, \quad [\mathbf{g}^I] = \begin{bmatrix} \dfrac{1}{\dot{r}}\mathbf{e}_r \\ \dfrac{1}{r}\mathbf{e}_\theta - \dfrac{\psi}{\lambda}\mathbf{e}_z \\ \dfrac{1}{\lambda}\mathbf{e}_z \end{bmatrix}. \tag{6.56}$$

To prevent confusion with primed quantities used later in this section, we use dot to denote differentiation with respect to R. With these equations, the various components of the metric tensor are

$$[G_{IJ}] = [\mathbf{G}_I \cdot \mathbf{G}_J] = \begin{bmatrix} 1 & 0 & 0 \\ 0 & R^2 & 0 \\ 0 & 0 & 1 \end{bmatrix}, \qquad [g_{ij}] = [\mathbf{g}_i \cdot \mathbf{g}_j] = \begin{bmatrix} 1 & 0 & 0 \\ 0 & r^2 & 0 \\ 0 & 0 & 1 \end{bmatrix},$$

$$[g_{IJ}] = [\mathbf{g}_I \cdot \mathbf{g}_J] = \begin{bmatrix} \dot{r}^2 & 0 & 0 \\ 0 & r^2 & \psi r^2 \\ 0 & \psi r^2 & \lambda^2 + \psi^2 r^2 \end{bmatrix} \tag{6.57}$$

$$[G^{IJ}] = [\mathbf{G}^I \cdot \mathbf{G}^J] = \begin{bmatrix} 1 & 0 & 0 \\ 0 & \dfrac{1}{R^2} & 0 \\ 0 & 0 & 1 \end{bmatrix}, \qquad [g^{ij}] = [\mathbf{g}^i \cdot \mathbf{g}^j] = \begin{bmatrix} 1 & 0 & 0 \\ 0 & \dfrac{1}{r^2} & 0 \\ 0 & 0 & 1 \end{bmatrix},$$

$$[g^{IJ}] = [\mathbf{g}^I \cdot \mathbf{g}^J] = \begin{bmatrix} \dfrac{1}{\dot{r}^2} & 0 & 0 \\ 0 & \dfrac{1}{r^2} + \dfrac{\psi^2}{\lambda^2} & -\dfrac{\psi}{\lambda^2} \\ 0 & -\dfrac{\psi}{\lambda^2} & \dfrac{1}{\lambda^2} \end{bmatrix}. \tag{6.58}$$

The deformation gradient tensor can be written in several forms. With the above relations, Eq. $(3.22)_1$ gives

$$\begin{aligned} \mathbf{F} &= \mathbf{g}_I \mathbf{G}^I \\ &= \mathbf{g}_{1*} \mathbf{G}^1 + \mathbf{g}_{2*} \mathbf{G}^2 + \mathbf{g}_{3*} \mathbf{G}^3 \\ &= \dot{r} \mathbf{e}_r \mathbf{e}_R + \frac{r}{R} \mathbf{e}_\theta \mathbf{e}_\Theta + \lambda \mathbf{e}_z \mathbf{e}_Z + \psi r \mathbf{e}_\theta \mathbf{e}_Z \\ &= \dot{r} \mathbf{g}_1 \mathbf{G}^1 + \mathbf{g}_2 \mathbf{G}^2 + \lambda \mathbf{g}_3 \mathbf{G}^3 + \psi \mathbf{g}_2 \mathbf{G}^3. \end{aligned} \tag{6.59}$$

Note that, with the F^i_I given by Eq. (3.63), the last expression is consistent with the form $\mathbf{F} = F^i_I \, \mathbf{g}_i \mathbf{G}^I$. (Recall again the distinction between the base vectors $\mathbf{g}_{1*}$ and $\mathbf{g}_1$, etc.) For relatively simple problems such as this, many authors find it convenient to work entirely with physical components, as given by the third line of (6.59). Here, however, we stay with tensor components. Of course, physical components always can be computed as desired.

Finally, Eq. (3.68) gives the Lagrangian strain components

$$[E_{IJ}] = \frac{1}{2}\begin{bmatrix} \dot{r}^2 - 1 & 0 & 0 \\ 0 & r^2 - R^2 & \psi r^2 \\ 0 & \psi r^2 & \lambda^2 + \psi^2 r^2 - 1 \end{bmatrix}, \tag{6.60}$$

and (3.146) and (5.177) yield the strain invariants

$$\begin{aligned} I_1 &= \dot{r}^2 + \frac{r^2}{R^2} + \lambda^2 + \psi^2 r^2 \\ I_2 &= \dot{r}^2 \left(\frac{r^2}{R^2} + \lambda^2 + \psi^2 r^2 \right) + \frac{\lambda^2 r^2}{R^2} \\ I_3 &= \frac{\lambda^2 \dot{r}^2 r^2}{R^2} \\ I_4' &= \hat{E}_{33} = \tfrac{1}{2}(\lambda^2 + \psi^2 r^2 - 1) \\ I_5' &= \hat{E}_{13}^2 + \hat{E}_{23}^2 = \frac{\psi^2 r^4}{4R^2}. \end{aligned} \tag{6.61}$$

The modified invariants I_4' and I_5' are needed for a curvilinear transversely isotropic material (see Example 5.7 on page 232), with the physical strain components $\hat{E}_{13}$ and $\hat{E}_{23}$ provided by Eq. (3.69).

Constitutive Relations. The stress analysis can be simplified by taking advantage of the inherent symmetry in this problem. For example, from the symmetry of the geometry and loading, we can show that the conditions $\sigma^{12} = \sigma^{13} = 0$ or $\sigma^{1^*2^*} = \sigma^{1^*3^*} = 0$ must be satisfied. Further simplification can be achieved by working with the stress components σ^{ij}, rather than $\sigma^{I^*J^*}$ (or s^{IJ}). The reason for this will become apparent when we write down the equilibrium equations.[2] In addition, the twist causes $\mathbf{t}$ to be asymmetric, making t^{IJ} an inconvenient choice.

To determine the σ^{ij}, however, we first must effectively compute the $\sigma^{I^*J^*}$, and then Eqs. (4.35) and (4.36) give the σ^{ij}. Because the principal material axes $X^{I'}$ coincide with the X^I-axes, there is no need to transform from the local material coordinates to the global polar coordinates. Thus, the constitutive equation

[2]Green and Zerna (1968) circumvented complications in using $\sigma^{I^*J^*}$ by taking the undeformed base vector $\mathbf{G}_3 = \mathbf{G}_Z$ to be at an angle relative to the vertical axis so that $\mathbf{g}_3 = \mathbf{g}_Z$ is vertical in the *deformed* cylinder.

is

$$\sigma^{I^* J^*} = \frac{\partial W}{\partial E_{IJ}} - p g^{IJ}, \tag{6.62}$$

as given by Eq. $(5.190)_2$, where $p = 0$ if the material is compressible. Alternatively, in terms of invariants, Eq. (5.178) gives

$$\sigma^{I^* J^*} = \Phi G^{IJ} + \Psi Q^{IJ} - p g^{IJ} + \Theta' M'^{IJ} + \Lambda' N'^{IJ} \tag{6.63}$$

where

$$\Phi = 2\, I_3^{-1/2} \frac{\partial W}{\partial I_1}, \qquad \Psi = 2\, I_3^{-1/2} \frac{\partial W}{\partial I_2}, \qquad p = -2\, I_3^{1/2} \frac{\partial W}{\partial I_3}$$

$$\Theta' = I_3^{-1/2} \frac{\partial W}{\partial I_4'}, \qquad \Lambda' = I_3^{-1/2} \frac{\partial W}{\partial I_5'}. \tag{6.64}$$

Here, the M'^{IJ} are functions of the $\hat{A}^I_{J'}$ defined by Eq. (5.170), i.e.,

$$\hat{A}^I_{J'} = \frac{1}{\sqrt{G_{(J'J')}}} \frac{\partial X^I}{\partial X^{J'}} = \frac{1}{\sqrt{G_{(J'J')}}} \delta^I_J = \begin{bmatrix} 1 & 0 & 0 \\ 0 & R^{-1} & 0 \\ 0 & 0 & 1 \end{bmatrix}. \tag{6.65}$$

Hence, Eqs. (5.151) and (5.179) yield

$$[Q^{IJ}] = \begin{bmatrix} \dfrac{r^2}{R^2} + \lambda^2 + \psi^2 r^2 & 0 & 0 \\ 0 & \dfrac{1}{R^2}\left(\dot{r}^2 + \lambda^2 + \psi^2 r^2\right) & -\dfrac{r^2}{R^2}\psi \\ 0 & -\dfrac{r^2}{R^2}\psi & \dot{r}^2 + \dfrac{r^2}{R^2} \end{bmatrix}$$

$$[M'^{IJ}] = \begin{bmatrix} 0 & 0 & 0 \\ 0 & 0 & 0 \\ 0 & 0 & 1 \end{bmatrix}, \qquad [N'^{IJ}] = \begin{bmatrix} 0 & 0 & 0 \\ 0 & 0 & \dfrac{\psi r^2}{2R^2} \\ 0 & \dfrac{\psi r^2}{2R^2} & 0 \end{bmatrix} \tag{6.66}$$

in which Eq. (3.69) provided the $\hat{E}_{IJ}$ required by N'^{IJ}. Inserting these relations into (6.63) gives

$$\begin{aligned}
\sigma^{1^*1^*} &= \Phi + \left(\frac{r^2}{R^2} + \lambda^2 + \psi^2 r^2\right)\Psi - \frac{1}{\dot{r}^2}\,p \\
\sigma^{2^*2^*} &= \frac{1}{R^2}\Phi + \frac{1}{R^2}\left(\dot{r}^2 + \lambda^2 + \psi^2 r^2\right)\Psi - \left(\frac{1}{r^2} + \frac{\psi^2}{\lambda^2}\right)p \\
\sigma^{3^*3^*} &= \Phi + \left(\dot{r}^2 + \frac{r^2}{R^2}\right)\Psi - \frac{1}{\lambda^2}\,p + \Theta' \\
\sigma^{2^*3^*} &= \psi\left(-\frac{r^2}{R^2}\Psi + \frac{1}{\lambda^2}\,p + \frac{r^2}{2R^2}\Lambda'\right) \\
\sigma^{1^*2^*} &= \sigma^{1^*3^*} = 0
\end{aligned} \tag{6.67}$$

which are consistent with the symmetry requirement $\sigma^{1^*2^*} = \sigma^{1^*3^*} = 0$.

The Cauchy stress components relative to the $\mathbf{g}_i$ basis now can be computed using Eqs. (4.35) and (4.36), which give

$$\sigma^{ij} = F^i_I F^j_J \,\sigma^{I^*J^*}.$$

With the F^i_I given by the last of (6.59), this equation yields

$$[\sigma^{ij}] = \begin{bmatrix} \dot{r}^2\sigma^{1^*1^*} & 0 & 0 \\ 0 & \sigma^{2^*2^*} + 2\psi\sigma^{2^*3^*} + \psi^2\sigma^{3^*3^*} & \lambda(\sigma^{2^*3^*} + \psi\sigma^{3^*3^*}) \\ 0 & \lambda(\sigma^{2^*3^*} + \psi\sigma^{3^*3^*}) & \lambda^2\sigma^{3^*3^*} \end{bmatrix}. \tag{6.68}$$

Now, substituting Eq. (6.62) produces

$$\begin{aligned}
\sigma^{11} &= \dot{r}^2\frac{\partial W}{\partial E_{11}} - p \\
\sigma^{22} &= \frac{\partial W}{\partial E_{22}} + 2\psi\frac{\partial W}{\partial E_{23}} + \psi^2\frac{\partial W}{\partial E_{33}} - \frac{p}{r^2} \\
\sigma^{33} &= \lambda^2\frac{\partial W}{\partial E_{33}} - p \\
\sigma^{23} &= \lambda\left(\frac{\partial W}{\partial E_{23}} + \psi\frac{\partial W}{\partial E_{33}}\right) \\
\sigma^{12} &= \sigma^{13} = 0
\end{aligned} \tag{6.69}$$

in which $(6.58)_3$ has been used, while Eqs. (6.63) give

$$\begin{aligned}
\sigma^{11} &= \dot{r}^2\left[\Phi + \left(\frac{r^2}{R^2} + \lambda^2 + \psi^2 r^2\right)\Psi\right] - p \\
\sigma^{22} &= \left(\frac{1}{R^2} + \psi^2\right)\Phi + \left(\frac{\dot{r}^2}{R^2} + \frac{\lambda^2}{R^2} + \psi^2\dot{r}^2\right)\Psi - \frac{p}{r^2} + \frac{\psi^2 r^2}{R^2}\Lambda' + \psi^2\Theta' \\
\sigma^{33} &= \lambda^2\left[\Phi + \left(\dot{r}^2 + \frac{r^2}{R^2}\right)\Psi + \Theta'\right] - p \\
\sigma^{23} &= \lambda\psi\left(\Phi + \dot{r}^2\Psi + \frac{r^2}{2R^2}\Lambda' + \Theta'\right) \\
\sigma^{12} &= \sigma^{13} = 0.
\end{aligned} \tag{6.70}$$

Either set of constitutive relations [(6.69) or (6.70)] can be used during the solution process. For illustration, we consider both.

Equilibrium. Since the stress field in this problem is not uniform, the equations of equilibrium are not all satisfied trivially as they were in the previous problems. Several choices are available for the form of these equations; see Eqs. (4.80) and (4.92). All of these forms are equivalent, but some may be more convenient than others. For example, since $\mathbf{t}$ is not symmetric for this problem (due to the shear), we would like to take advantage of the symmetry in the stress tensor $\boldsymbol{\sigma}$ or $\mathbf{s}$. In many problems, the form $\nabla \cdot (\mathbf{s} \cdot \mathbf{F}^T) = 0$ has advantages, because derivatives are taken with respect to the undeformed coordinates, which are known *a priori*. In the current problem, however, the relatively simple geometry makes the choice $\bar{\nabla} \cdot \boldsymbol{\sigma} = \mathbf{0}$ more convenient. In this case, Eqs. (4.96) apply.

Due to symmetry, the Cauchy stress components are expected to be functions of r alone. Hence, for $\sigma^{12} = \sigma^{13} = 0$, the last two of Eqs. (4.96) are satisfied identically, and the first equation becomes

$$r\frac{d\sigma^{11}}{dr} + \sigma^{11} - r^2\sigma^{22} = 0, \tag{6.71}$$

which corresponds to equilibrium in the radial direction.

Boundary Conditions. We seek a solution for which tractions are applied only at the ends of the cylinder. On the deformed curved surface, the unit normal is

$$\mathbf{n}_1 = \frac{\mathbf{g}^1}{\sqrt{g^{11}}} = \mathbf{g}^1,$$

and so the true traction vector on this surface is

$$\begin{aligned} \mathbf{T}^{(\mathbf{n}_1)} &= \mathbf{n}_1 \cdot \boldsymbol{\sigma} \\ &= \mathbf{g}^1 \cdot (\sigma^{ij} \mathbf{g}_i \mathbf{g}_j) = \sigma^{ij} \delta_i^1 \mathbf{g}_j \\ &= \sigma^{1j} \mathbf{g}_j. \end{aligned}$$

Thus, for a stress-free surface at the deformed radius b, the appropriate boundary conditions are

$$r = b: \quad \sigma^{11} = \sigma^{12} = \sigma^{13} = 0. \tag{6.72}$$

As discussed previously, σ^{12} and σ^{13} already are assumed to be zero everywhere, and the condition on σ^{11} provides the boundary condition required by the differential equation (6.71).

Similarly, on the ends of the cylinder, the traction is

$$\mathbf{T}^{(\mathbf{n}_3)} = \mathbf{n}_3 \cdot \boldsymbol{\sigma} = \mathbf{g}^3 \cdot \boldsymbol{\sigma} = \sigma^{3j} \mathbf{g}_j,$$

and the resultant force and moment applied to the ends are[3]

$$\begin{aligned} \mathbf{N} &= \int_0^{2\pi} \int_0^b \mathbf{T}^{(\mathbf{n}_3)}\, r\, dr\, d\theta \\ \mathbf{M} &= \int_0^{2\pi} \int_0^b \mathbf{r} \times \mathbf{T}^{(\mathbf{n}_3)}\, r\, dr\, d\theta, \end{aligned} \tag{6.73}$$

where $\mathbf{r} = r\mathbf{e}_r$ is the radius vector from the x^3-axis. With the $\mathbf{g}_i$ given by (6.55), these expressions become

$$\begin{aligned} \mathbf{N} &= \int_0^{2\pi} \int_0^b \left(\sigma^{31}\, \mathbf{e}_r + \sigma^{32}\, r\mathbf{e}_\theta + \sigma^{33}\, \mathbf{e}_z\right) r\, dr\, d\theta \\ \mathbf{M} &= \int_0^{2\pi} \int_0^b (r\mathbf{e}_\mathbf{r}) \times \left(\sigma^{31}\, \mathbf{e}_r + \sigma^{32}\, r\mathbf{e}_\theta + \sigma^{33}\, \mathbf{e}_z\right) r\, dr\, d\theta. \end{aligned}$$

These equations simplify if we note that the stresses are functions of r and that $\mathbf{e}_r = \mathbf{e}_x \cos\theta + \mathbf{e}_y \sin\theta$ and $\mathbf{e}_\theta = -\mathbf{e}_x \sin\theta + \mathbf{e}_y \cos\theta$. Then, since $\int_0^{2\pi} \sin\theta\, d\theta = \int_0^{2\pi} \cos\theta\, d\theta = 0$, we have

$$\begin{aligned} \mathbf{N} &= 2\pi \mathbf{e}_z \int_0^b \sigma^{33}\, r\, dr \equiv N\mathbf{e}_z \\ \mathbf{M} &= 2\pi \mathbf{e}_z \int_0^b \sigma^{32}\, r^3\, dr \equiv M\mathbf{e}_z, \end{aligned} \tag{6.74}$$

[3]Here, $\mathbf{N}$ is a force rather than a unit normal vector.

which provide a resultant axial force N and twisting moment M applied at the ends of the cylinder.

6.4.2 *Solution*

In the following, we develop explicit solutions for cylindrical specimens composed of incompressible and compressible transversely isotropic muscle.

Incompressible Muscle. If the papillary muscle is assumed to be incompressible, the deformation must satisfy the condition

$$J = I_3^{1/2} = \lambda \frac{r}{R}\frac{dr}{dR} = 1 \tag{6.75}$$

as provided by Eq. $(6.61)_3$. Integrating this equation yields

$$\lambda \frac{r^2}{2} = \frac{R^2}{2} + C,$$

and the condition $r(0) = 0$ gives $C = 0$. Thus,

$$r(R) = \frac{R}{\sqrt{\lambda}} \tag{6.76}$$

which provides the deformed radius $b = r(b_0) = b_0/\sqrt{\lambda}$. With λ and ψ specified, this relation completes the solution for the deformation field, and Eq. (6.60) gives the Lagrangian strain components

$$[E_{IJ}] = \frac{1}{2}\begin{bmatrix} \dfrac{1}{\lambda} - 1 & 0 & 0 \\ 0 & \left(\dfrac{1}{\lambda} - 1\right) R^2 & \dfrac{\psi R^2}{\lambda} \\ 0 & \dfrac{\psi R^2}{\lambda} & \lambda^2 + \dfrac{\psi^2 R^2}{\lambda} - 1 \end{bmatrix}. \tag{6.77}$$

In addition, substituting Eq. (6.76), with $\dot{r} = 1/\sqrt{\lambda}$, into Eqs. (6.70) gives the nonzero stress components

$$\begin{aligned} \sigma^{11} &= \frac{1}{\lambda}\left[\Phi + \left(\frac{1}{\lambda} + \lambda^2 + \psi^2 r^2\right)\Psi\right] - p \\ r^2\sigma^{22} &= \left(\frac{1}{\lambda} + \psi^2 r^2\right)\Phi + \frac{1}{\lambda}\left(\frac{1}{\lambda} + \lambda^2 + \psi^2 r^2\right)\Psi \\ &\quad + \psi^2 r^2\left(\frac{1}{\lambda}\Lambda' + \Theta'\right) - p \end{aligned}$$

$$\sigma^{33} = \lambda^2 \left(\Phi + \frac{2}{\lambda}\Psi + \Theta' \right) - p$$

$$\sigma^{23} = \lambda\psi \left(\Phi + \frac{1}{\lambda}\Psi + \frac{1}{2\lambda}\Lambda' + \Theta' \right). \tag{6.78}$$

To complete the stress analysis, we need to find p. Because $\sigma^{11} - r^2\sigma^{22}$ is independent of p, inserting Eqs. $(6.69)_{1,2}$ or $(6.78)_{1,2}$ into (6.71) and integrating yields the alternative forms

$$p(r) = \frac{1}{\lambda}\frac{\partial W}{\partial E_{11}} + \int_r^b \left[r^2 \left(\frac{\partial W}{\partial E_{22}} + 2\psi \frac{\partial W}{\partial E_{23}} + \psi^2 \frac{\partial W}{\partial E_{33}} \right) - \frac{1}{\lambda}\frac{\partial W}{\partial E_{11}} \right] \frac{dr}{r} \tag{6.79}$$

or

$$p(r) = \frac{1}{\lambda}\left[\Phi + \left(\frac{1}{\lambda} + \lambda^2 + \psi^2 r^2 \right) \Psi \right] + \psi^2 \int_r^b \left(\Phi + \frac{1}{\lambda}\Lambda' + \Theta' \right) r\, dr. \tag{6.80}$$

The limits of integration in these equations were chosen to satisfy the boundary condition $\sigma^{11}(b) = 0$, as can be shown by direct substitution into Eq. $(6.69)_1$ or $(6.78)_1$. Given W, these relations can be integrated either in closed form (rarely) or numerically for the Lagrange multiplier at any point in the cylinder. The first form (6.79) is valid for any W that provides zero shear stress on the curved surface. The second form (6.80) is valid for the specific transversely isotropic cylinder studied here, i.e., with fibers oriented originally in the X^3-direction. The solution for an isotropic cylinder can be obtained by setting $\Lambda' = \Theta' = 0$. Note also that the integrals can be converted easily to integrations over R using Eq. (6.76).

With p known, the stress components are given by Eqs. (6.67), (6.69), or (6.78). In the last case, the nonzero σ^{ij} become

$$\hat{\sigma}^{11} = \sigma^{11} = -\psi^2 \int_r^b \left(\Phi + \frac{1}{\lambda}\Lambda' + \Theta' \right) r\, dr$$

$$\hat{\sigma}^{22} = r^2\sigma^{22} = \sigma^{11} + \psi^2 r^2 \left(\Phi + \frac{1}{\lambda}\Lambda' + \Theta' \right)$$

$$\hat{\sigma}^{33} = \sigma^{33} = \sigma^{11} + \left(\lambda^2 - \frac{1}{\lambda} \right)\Phi + \left(\lambda - \frac{1}{\lambda^2} - \frac{\psi^2 r^2}{\lambda} \right)\Psi + \lambda^2\Theta'$$

$$\hat{\sigma}^{23} = r\sigma^{23} = \psi r \left(\lambda\Phi + \Psi + \frac{1}{2}\Lambda' + \lambda\Theta' \right) \tag{6.81}$$

in which hat indicates physical stress components. Finally, the resultant force and moment at the ends of the cylinder can be computed using Eqs. (6.74).

For the special case of muscle modeled as a modified Mooney-Rivlin material with

$$W = C_1(I_1 - 3) + C_2(I_2 - 3) + C_4 I_4'^2, \tag{6.82}$$

Eqs. (6.64) and $(6.81)_1$ yield (with $I_3 = 1$)

$$\sigma^{11} = -\psi^2(b^2 - r^2)[C_1 + C_4(\lambda^2 - 1)/2 + C_4\psi^2(b^2 + r^2)/4], \tag{6.83}$$

and the other stress components can be computed easily from the rest of Eqs. (6.81). Substitution into (6.74) then yields the axial force and twisting moment

$$N = 2\pi b_0^2 \left\{ C_1 \left(\lambda - \frac{1}{\lambda^2} - \frac{\psi^2 b_0^2}{4\lambda^2} \right) + C_2 \left(1 - \frac{1}{\lambda^3} - \frac{\psi^2 b_0^2}{2\lambda^3} \right) \right.$$
$$\left. + \frac{C_4}{2} \left[\lambda^3 - \lambda + \left(1 + \frac{1}{\lambda^2} \right) \frac{\psi^2 b_0^2}{4} - \frac{\psi^4 b_0^4}{6\lambda^3} \right] \right\}$$

$$M = \frac{\pi b_0^4 \psi}{\lambda} \left[C_1 + \frac{C_2}{\lambda} + \frac{C_4}{4} \left(\lambda^2 - 1 + \frac{2\psi^2 b_0^2}{3\lambda} \right) \right]. \tag{6.84}$$

Note that, for an isotropic cylinder ($C_4 = 0$), M is proportional to ψ, as in the linear solution. Such is not the case, however, if the material is transversely isotropic relative to the axial direction ($C_4 \neq 0$).

Stress distributions for the special case of a neo-Hookean material ($C_2 = C_4 = 0$) reveal clear differences between the linear and nonlinear solutions (see Fig. 6.9a,b). According to the linear theory, for example, the only nonzero stress component is the shear stress $\hat{\sigma}^{23}$. In contrast, for large angles of twist, the nonlinear theory predicts comparable magnitudes for all three normal stress components (Fig. 6.9a). The results in Fig. 6.9a are for a cylinder held at its initial length. As shown in Fig. 6.9b, axial stretch increases the magnitude of the torsional shear stress $\hat{\sigma}^{23}$. Moreover, the distribution of axial stress $\hat{\sigma}^{33}$, which is independent of r for stretch without twist, becomes more and more nonuniform as the torsion increases (Fig. 6.9a). (Note also that $\hat{\sigma}^{33} = \hat{\sigma}^{11}$ when $\lambda = 1$.) Finally, the presence of axial fibers increases the magnitudes of the peak stresses (Fig. 6.9c).

If the cylinder is held fixed in length with $\lambda = 1$, the axial stress $\hat{\sigma}^{33}$ is compressive at all r (Fig. 6.9a). This stress distribution produces a net compressive force N that increases in magnitude with the twist (Fig. 6.10).[4] Hence, if N is not supplied, the cylinder elongates when twisted. This phenomenon is known

[4]For comparison with the following solution for a compressible cylinder, the results in Fig. 6.9a,b are for an isotropic material with $C_1 = C_4 = 0$. The case $C_2 = C_4 = 0$ is similar.

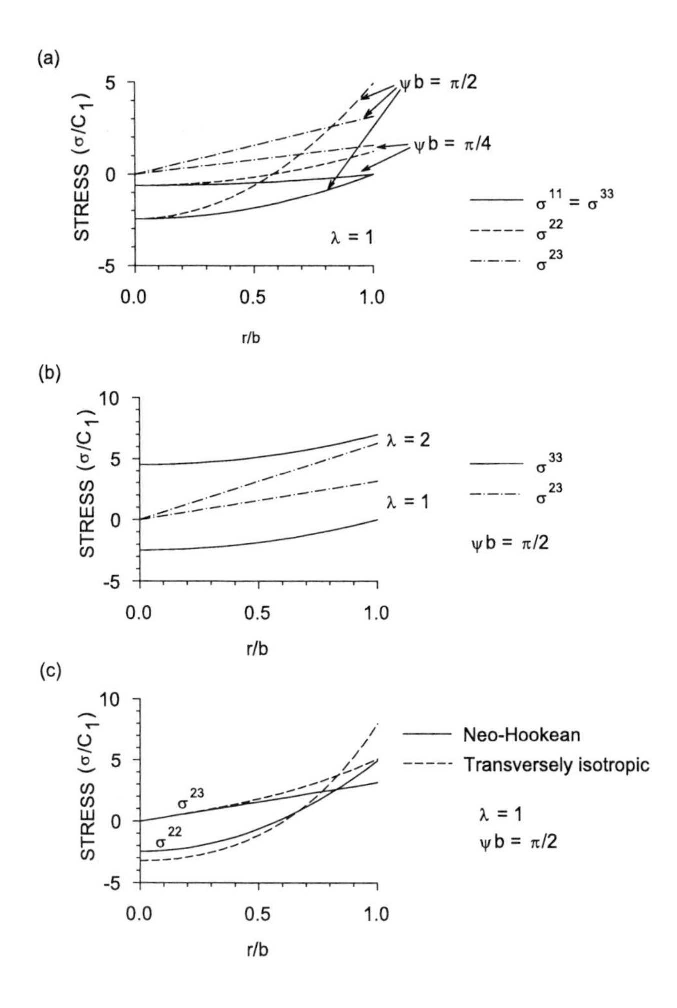

Fig. 6.9 Distributions of Cauchy stress (physical components) in a solid circular cylinder (papillary muscle) undergoing extension and torsion. (a,b) isotropic cylinder with $C_2 = C_4 = 0$; (c) transversely isotropic cylinder with $C_2 = 0$.

as the *Poynting effect*. For $\lambda = 2$, the effect of twist on N is not nearly as large (Fig. 6.10).

Compressible Muscle. If the muscle is assumed to be compressible, then the problem contains one fewer unknown (p). But since the deformation is no longer

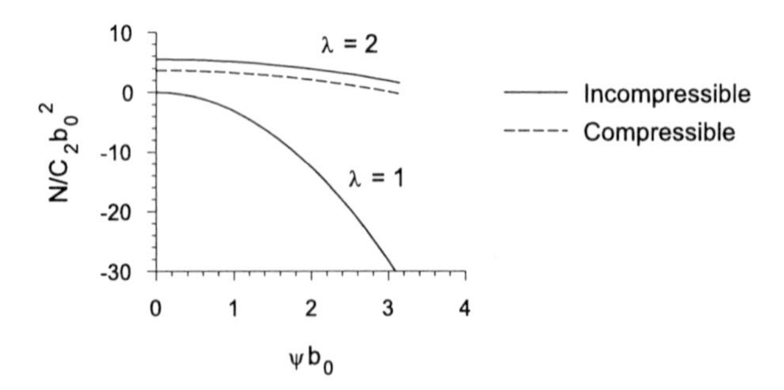

Fig. 6.10 Axial force generated in a solid isotropic circular cylinder undergoing extension and torsion ($C_1 = C_4 = 0$). The incompressible and compressible solutions are identical for $\lambda = 1$.

constrained by the incompressibility condition, the function $r(R)$ is not found as easily as in the incompressible case. For a given W, Eqs. (6.69) and (6.70) still apply, but p must be set to zero in the former equations and is computed directly from $(6.64)_3$ in the latter equations. If λ and ψ are specified, then substituting the constitutive relations into the equilibrium equation (6.71) provides a differential equation to be solved for $r(R)$. In general, the resulting equation must be solved numerically, e.g., by finite differences, which may be significantly more involved than the solution by quadratures in the incompressible case.

A relatively simple solution, however, can be found for passive muscle modeled by a specialized Blatz-Ko material (Beatty, 1987; Carroll and Horgan, 1990). In particular, we consider an isotropic material with

$$W = C_2 \left(\frac{I_2}{I_3} + 2I_3^{1/2} - 5 \right) \tag{6.85}$$

as given by Eq. (5.125) with $\mu = 2C_2$. Note that this relation is the compressible counterpart of Eq. (6.82) if $C_1 = C_4 = 0$. For this material, Eqs. (6.61), (6.64), and (6.70) yield the nonzero physical stress components

$$\begin{aligned}
\hat{\sigma}^{11} &= \sigma^{11} = 2C_2 \left(1 - \frac{R}{\lambda r \dot{r}^3} \right) \\
\hat{\sigma}^{22} &= r^2 \sigma^{22} = 2C_2 \left(1 - \frac{R^3}{\lambda r^3 \dot{r}} \right) \\
\hat{\sigma}^{33} &= \sigma^{33} = 2C_2 \left(1 - \frac{R(1 + \psi^2 R^2)}{\lambda^3 r \dot{r}} \right) \\
\hat{\sigma}^{23} &= r\sigma^{23} = 2C_2 \frac{\psi R^3}{\lambda^2 r^2 \dot{r}}
\end{aligned} \tag{6.86}$$

in which p has been set to zero.

Substituting these relations into Eq. (6.71) now gives the nonlinear differential equation

$$3R\frac{dR}{dr}\frac{d^2R}{dr^2} + \left(\frac{dR}{dr}\right)^3 - \left(\frac{R}{r}\right)^3 = 0 \tag{6.87}$$

to be solved for the *undeformed* coordinate $R(r)$. This equation, written in Eulerian form for convenience, admits a solution of the form

$$R = cr \tag{6.88}$$

where c is a constant. Inserting this relation into $(6.86)_1$ and applying the boundary condition $\sigma^{11}(b) = 0$ yields

$$c = \lambda^{1/4}, \tag{6.89}$$

and the stress components become

$$\begin{aligned}
\sigma^{33} &= 2C_2\left[1 - \lambda^{-5/2}(1 + \lambda^{1/2}\psi^2 r^2)\right] \\
r\sigma^{23} &= 2C_2\frac{\psi r}{\lambda} \\
\sigma^{11} &= \sigma^{22} = \sigma^{12} = \sigma^{13} = 0.
\end{aligned} \tag{6.90}$$

Finally, with these expressions, the resultant force and twisting moment computed from Eq. (6.74) are

$$\begin{aligned}
N &= \frac{2\pi}{\lambda^{1/2}}\int_0^{b_0} \sigma^{33}\, R\,dR = 2\pi C_2 b_0^2\left(\frac{1}{\lambda^{1/2}} - \frac{1}{\lambda^3} - \frac{\psi^2 b_0^2}{2\lambda^3}\right) \\
M &= \frac{2\pi}{\lambda}\int_0^{b_0} \sigma^{23} R^3\, dR = \frac{\pi C_2 b_0^4}{\lambda^2}\,\psi
\end{aligned} \tag{6.91}$$

in which Eq. (6.88) has provided the transformation from r to R.

Results are compared to those for the corresponding incompressible cylinder with $C_1 = C_4 = 0$, i.e., $W = C_2(I_2 - 3)$. For $\lambda = 1$, the deformation is isochoric ($r = R$ and $I_3 = 1$) even for the compressible cylinder, and so compressibility has no effect. Otherwise, compressibility decreases the magnitudes of the stresses and the applied axial force (Figs. 6.10 and 6.11).

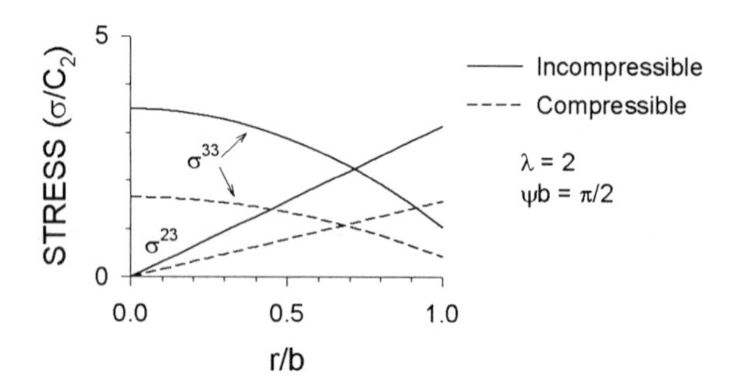

Fig. 6.11 Effects of material compressibility on Cauchy stresses (physical components) in a solid isotropic circular cylinder (papillary muscle) undergoing extension and torsion ($C_1 = C_4 = 0$).

6.5 Extension, Inflation, and Torsion of an Artery with Residual Stress

Arteries are the blood vessels that carry blood away from the heart. The wall of an artery consists of three layers. The inner layer (intima) consists of a single layer of epithelial cells, the middle layer (media) contains elastin and smooth muscle, and the outer layer (adventitia) is composed mainly of connective tissue.

Reflecting differences in function, the size and composition of arteries change with distance from the heart. Arteries near the heart, including the aorta, are relatively large and serve primarily as conduits of blood. The walls of these "elastic arteries" are dominated by elastin. Downstream, arteries gradually become smaller in diameter, and smooth muscle becomes more prominent. The small-caliber arterioles ("muscular arteries") regulate the flow of blood to the tissues via smooth-muscle contraction or relaxation. This response changes the diameter of the vessel and thereby alters resistance to flow.

Hence, arteries are not merely passive tubes. Besides actively changing their diameter, arteries grow (change size) and remodel (change material properties) to adapt to changes in flow and pressure. Developing theoretical models for these processes, while accounting for heterogeneity and viscoelasticity, is a challenging and ongoing endeavor (Rachev, 1997; Taber, 1998; Rachev et al., 1998; Humphrey, 2002). In this section, however, we ignore these complications and consider arteries as thick-walled passive, homogeneous, pseudoelastic tubes. We do include anisotropy, however.

Another feature that we include is **residual stress**, i.e., the stress that remains

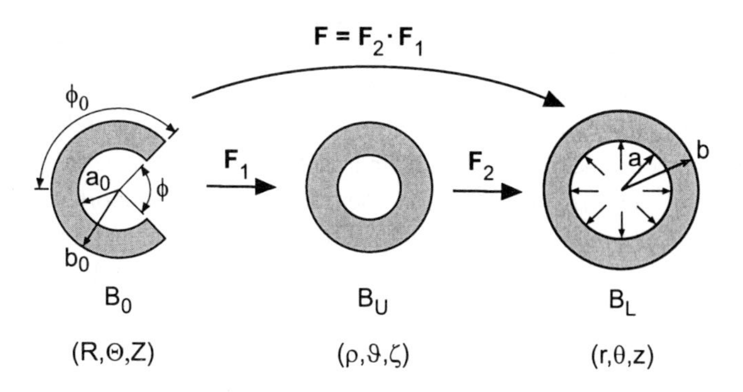

Fig. 6.12 Artery (or left ventricle) cross section in zero-stress (B_0), unloaded (B_U), and loaded (B_L) configurations.

in a body when all external loads are removed. Residual stress in artery walls can be produced by nonuniform growth, swelling, turnover of wall constituents, or other means. A popular method used to characterize the magnitude of residual stress in a soft tissue is to cut the tissue and measure the resulting deformation due to the release of stress. For example, when an unloaded section of an artery is cut transmurally, the section typically springs open as circumferential residual stress is relieved (Fig. 6.12). Measuring the opening angle ϕ provides an approximate way to characterize the amount of residual *strain* that was in the section before the cut. Residual stress then can be computed if material properties are known.

Most investigators have assumed that a single radial cut is enough to render an artery nearly free of stress. Recent studies have suggested, however, that a combination of radial and circumferential cuts may be needed (Vossoughi et al., 1993; Greenwald et al., 1997; Taber and Humphrey, 2001). For simplicity, we assume here that the zero-stress state is produced by one radial cut.

6.5.1 *Artery Model*

Consider the following experiment. A straight section of an artery is dissected and mounted in a testing machine that subjects the vessel to simultaneous extension, inflation, and torsion. The inner and outer radii, axial stretch, and angle of twist are recorded, in addition to the applied internal pressure P, axial force N, and twisting moment M. After testing under various loading protocols, a thin slice of the artery is dissected and cut once transmurally, and the opening angle is measured, along with the inner and outer radii of the opened section. One way to determine material properties would be to vary the material coefficients in our model until the theoretical results match as closely as possible the values of all of the measured variables under various loading conditions.

The artery is modeled as a straight thick-walled circular cylinder composed of a homogeneous, incompressible, cylindrically orthotropic material (Fig. 6.12). For simplicity, our analysis ignores all end effects near the supports and near the cut edges in B_0. In the unloaded configuration B_U, the principal material directions coincide with the radial, circumferential, and longitudinal directions of the vessel. The zero-stress configuration B_0, obtained by cutting B_U transmurally, is assumed to approximate the shape of a circular sector with an inner radius a_0, outer radius b_0, and opening angle ϕ. The inner and outer radii of the loaded configuration B_L are a and b, respectively. Due to material and geometric symmetry, the principal material directions remain radial, circumferential, and longitudinal in B_0 and B_L.

This problem bears many similarities to the papillary muscle problem. Hence, many aspects of the analysis are similar. However, to illustrate again that multiple approaches usually are possible in nonlinear elasticity, we vary some key aspects. For example, the formulation is based on a general strain-energy density function, rather than assuming a specific material symmetry at the outset. In addition, the entire analysis uses physical components of stress and strain, rather than first introducing tensor components.

6.5.2 *Governing Equations*

Kinematics. The deformation for this problem, excluding residual stress, was discussed in Example 3.4 (page 84). For further illustration, however, we present herein a complete analysis from scratch. It is convenient to divide the total deformation into two parts. First, the deformation $\mathbf{F}_1$ carries a point from the coordinates (R, Θ, Z) in B_0 to (ρ, ϑ, ζ) in B_U, and then $\mathbf{F}_2$ takes it to (r, θ, z) in B_L (Fig. 6.12). For axisymmetric deformation, the position vectors to the point in B_0, B_U, and B_L, respectively, are

$$\begin{aligned}
\mathbf{R} &= R\,\mathbf{e}_R + Z\,\mathbf{e}_Z \\
\boldsymbol{\rho} &= \rho\,\mathbf{e}_\rho + \zeta\,\mathbf{e}_\zeta \\
\mathbf{r} &= r\,\mathbf{e}_r + z\,\mathbf{e}_z.
\end{aligned} \tag{6.92}$$

In terms of a set of Cartesian base vectors, the unit vectors for each coordinate system are

$$\begin{aligned}
\mathbf{e}_R &= \mathbf{e}_x \cos\Theta + \mathbf{e}_y \sin\Theta \\
\mathbf{e}_\Theta &= -\mathbf{e}_x \sin\Theta + \mathbf{e}_y \cos\Theta \\
\mathbf{e}_Z &= \mathbf{e}_z \\
\\
\mathbf{e}_\rho &= \mathbf{e}_x \cos\vartheta + \mathbf{e}_y \sin\vartheta \\
\mathbf{e}_\vartheta &= -\mathbf{e}_x \sin\vartheta + \mathbf{e}_y \cos\vartheta \\
\mathbf{e}_\zeta &= \mathbf{e}_z \\
\\
\mathbf{e}_r &= \mathbf{e}_x \cos\theta + \mathbf{e}_y \sin\theta \\
\mathbf{e}_\theta &= -\mathbf{e}_x \sin\theta + \mathbf{e}_y \cos\theta \\
\mathbf{e}_z &= \mathbf{e}_z.
\end{aligned} \tag{6.93}$$

The deformation $\mathbf{F}_1$ describes bending of a circular bar, and the mapping from B_0 to B_U is assumed to have the form

$$\rho = \rho(R), \qquad \vartheta = \pi\Theta/\phi_0, \qquad \zeta = \Lambda Z \tag{6.94}$$

where ϕ_0 is the half-sector angle defined in Fig. 6.12, and Λ is the axial stretch ratio of B_U relative to B_0. Note that $\vartheta = 0$ when $\Theta = 0$, $\vartheta = \pi$ when $\Theta = \phi_0$, and the opening angle is given by $\phi = 2(\pi - \phi_0)$. The extension, inflation, and torsion from B_U to B_L is defined by the relations

$$r = r(\rho), \qquad \theta = \vartheta + \psi\zeta, \qquad z = \lambda\zeta \tag{6.95}$$

where ψ is the angle of twist per unit axial length of B_U, and λ is the axial stretch ratio of the loaded tube B_L relative to the unloaded tube B_U.

Equations (6.92) and (6.93) give the natural base vectors for the coordinate systems in B_0 and B_U as

$$\begin{aligned}
B_0:\quad & \mathbf{G}_R = \mathbf{R}_{,R} = \mathbf{e}_R & B_U:\quad & \mathbf{G}_\rho = \boldsymbol{\rho}_{,\rho} = \mathbf{e}_\rho \\
& \mathbf{G}_\Theta = \mathbf{R}_{,\Theta} = R\,\mathbf{e}_\Theta & & \mathbf{G}_\vartheta = \boldsymbol{\rho}_{,\vartheta} = \rho\,\mathbf{e}_\vartheta \\
& \mathbf{G}_Z = \mathbf{R}_{,Z} = \mathbf{e}_Z & & \mathbf{G}_\zeta = \boldsymbol{\rho}_{,\zeta} = \mathbf{e}_\zeta \\
\\
& \mathbf{G}^R = \mathbf{e}_R & & \mathbf{G}^\rho = \mathbf{e}_\rho \\
& \mathbf{G}^\Theta = R^{-1}\mathbf{e}_\Theta & & \mathbf{G}^\vartheta = \rho^{-1}\mathbf{e}_\vartheta \\
& \mathbf{G}^Z = \mathbf{e}_Z & & \mathbf{G}^\zeta = \mathbf{e}_\zeta
\end{aligned} \tag{6.96}$$

where Eq. (2.17) was used to compute the contravariant base vectors. In addition, with Eqs. (6.92)–(6.95), the convected base vectors are

$$
\begin{aligned}
B_0 \to B_U\colon \quad & \mathbf{g}_R = \boldsymbol{\rho}_{,R} = \rho' \mathbf{e}_\rho \\
& \mathbf{g}_\Theta = \boldsymbol{\rho}_{,\Theta} = \frac{\pi}{\phi_0} \rho\, \mathbf{e}_\vartheta \\
& \mathbf{g}_Z = \boldsymbol{\rho}_{,Z} = \Lambda \mathbf{e}_\zeta
\end{aligned} \tag{6.97}
$$

$$
\begin{aligned}
B_U \to B_L\colon \quad & \mathbf{g}_\rho = \mathbf{r}_{,\rho} = \dot{r} \mathbf{e}_r \\
& \mathbf{g}_\vartheta = \mathbf{r}_{,\vartheta} = r \mathbf{e}_\theta \\
& \mathbf{g}_\zeta = \mathbf{r}_{,\zeta} = \psi r \mathbf{e}_\theta + \lambda \mathbf{e}_z
\end{aligned}
$$

where prime and dot denote differentiation with respect to R and ρ, respectively.

The deformation gradient tensors now can be computed by way of Eq. $(3.22)_1$, which yields

$$
\begin{aligned}
\mathbf{F}_1 &= \mathbf{g}_R \mathbf{G}^R + \mathbf{g}_\Theta \mathbf{G}^\Theta + \mathbf{g}_Z \mathbf{G}^Z \\
&= \rho' \mathbf{e}_\rho \mathbf{e}_R + \frac{\pi \rho}{\phi_0 R} \mathbf{e}_\vartheta \mathbf{e}_\Theta + \Lambda \mathbf{e}_\zeta \mathbf{e}_Z
\end{aligned} \tag{6.98}
$$

$$
\begin{aligned}
\mathbf{F}_2 &= \mathbf{g}_\rho \mathbf{G}^\rho + \mathbf{g}_\vartheta \mathbf{G}^\vartheta + \mathbf{g}_\zeta \mathbf{G}^\zeta \\
&= \dot{r} \mathbf{e}_r \mathbf{e}_\rho + \frac{r}{\rho} \mathbf{e}_\theta \mathbf{e}_\vartheta + \lambda \mathbf{e}_z \mathbf{e}_\zeta + \psi r\, \mathbf{e}_\theta \mathbf{e}_\zeta .
\end{aligned} \tag{6.99}
$$

The total deformation from B_0 to B_L is described by (Fig. 6.12)

$$
\begin{aligned}
\mathbf{F} &= \mathbf{F}_2 \cdot \mathbf{F}_1 \\
&= \lambda_R \mathbf{e}_r \mathbf{e}_R + \lambda_\Theta \mathbf{e}_\theta \mathbf{e}_\Theta + \lambda_Z \mathbf{e}_z \mathbf{e}_Z + \gamma \mathbf{e}_\theta \mathbf{e}_Z
\end{aligned} \tag{6.100}
$$

where

$$
\begin{aligned}
\lambda_R &= r' \\
\lambda_\Theta &= \frac{\pi r}{\phi_0 R} \\
\lambda_Z &= \lambda \Lambda \\
\gamma &= \psi r \Lambda .
\end{aligned} \tag{6.101}
$$

Here, we have used the relation $\dot{r}\rho' = (\partial r/\partial \rho)(\partial \rho/\partial R) = \partial r/\partial R = r'$. Note that the λs in these terms represent stretch ratios only if $\gamma = 0$. Also, if there is no residual stress in B_U, then B_0 and B_U are identical. In this case, $\phi_0 = \pi$ and $\Lambda = 1$, and Eq. (6.100) reduces to Eq. (3.61) of Example 3.4, as it should.

Finally, with (6.100) and $\mathbf{I} = \mathbf{G}_R\mathbf{G}^R + \mathbf{G}_\Theta\mathbf{G}^\Theta + \mathbf{G}_Z\mathbf{G}^Z = \mathbf{e}_R\mathbf{e}_R + \mathbf{e}_\Theta\mathbf{e}_\Theta + \mathbf{e}_Z\mathbf{e}_Z$, the Lagrangian strain tensor of B_L relative to B_0 is

$$\begin{aligned}
\mathbf{E} &= \tfrac{1}{2}(\mathbf{F}^T \cdot \mathbf{F} - \mathbf{I}) \\
&= \hat{E}_{RR}\mathbf{e}_R\mathbf{e}_R + \hat{E}_{\Theta\Theta}\mathbf{e}_\Theta\mathbf{e}_\Theta + \hat{E}_{ZZ}\mathbf{e}_Z\mathbf{e}_Z \\
&\qquad + \hat{E}_{\Theta Z}\mathbf{e}_\Theta\mathbf{e}_Z + \hat{E}_{Z\Theta}\mathbf{e}_Z\mathbf{e}_\Theta
\end{aligned} \tag{6.102}$$

where

$$\begin{aligned}
\hat{E}_{RR} &= \tfrac{1}{2}\left(\lambda_R^2 - 1\right) \\
\hat{E}_{\Theta\Theta} &= \tfrac{1}{2}\left(\lambda_\Theta^2 - 1\right) \\
\hat{E}_{ZZ} &= \tfrac{1}{2}\left(\lambda_Z^2 + \gamma^2 - 1\right) \\
\hat{E}_{\Theta Z} &= \hat{E}_{Z\Theta} = \tfrac{1}{2}\gamma\lambda_\Theta.
\end{aligned} \tag{6.103}$$

As indicated by Eq. (3.41), the $\hat{E}_{IJ}$ are physical components of $\mathbf{E}$.

Constitutive Relations. With the artery assumed to be incompressible and pseudoelastic, Eq. (5.187) gives the constitutive equation for the Cauchy stress tensor as

$$\boldsymbol{\sigma} = \mathbf{F} \cdot \frac{\partial W}{\partial \mathbf{E}} \cdot \mathbf{F}^T - p\,\mathbf{I} \tag{6.104}$$

for a general W. As in the papillary muscle problem, we take advantage of simplifications by writing the equilibrium equations in terms of stress components relative to the basis associated with the $(r, \theta, z) = (x^1, x^2, x^3)$ coordinates in the deformed body. Since these coordinates are orthogonal, Eq. (4.99) gives

$$\boldsymbol{\sigma} = \hat{\sigma}^{ij}\mathbf{e}_i\mathbf{e}_j \tag{6.105}$$

in which the $\hat{\sigma}^{ij}$ are physical components of $\boldsymbol{\sigma}$. Substituting Eqs. (6.100), (6.102), and (6.105), along with $\mathbf{I} = \mathbf{e}_r\mathbf{e}_r + \mathbf{e}_\theta\mathbf{e}_\theta + \mathbf{e}_z\mathbf{e}_z$, into (6.104) yields

$$\begin{aligned}
\hat{\sigma}^{rr} &= \bar{\sigma}^{rr} - p \\
\hat{\sigma}^{\theta\theta} &= \bar{\sigma}^{\theta\theta} - p \\
\hat{\sigma}^{zz} &= \bar{\sigma}^{zz} - p \\
\\
\hat{\sigma}^{\theta z} &= \lambda_Z \left(\lambda_\Theta \frac{\partial W}{\partial \hat{E}_{\Theta Z}} + \gamma \frac{\partial W}{\partial \hat{E}_{ZZ}} \right) \\
\hat{\sigma}^{r\theta} &= \lambda_R \left(\lambda_\Theta \frac{\partial W}{\partial \hat{E}_{R\Theta}} + \gamma \frac{\partial W}{\partial \hat{E}_{ZR}} \right) \\
\hat{\sigma}^{zr} &= \lambda_R \lambda_Z \frac{\partial W}{\partial \hat{E}_{ZR}}
\end{aligned} \tag{6.106}$$

where

$$\begin{aligned}
\bar{\sigma}^{rr} &= \lambda_R^2 \frac{\partial W}{\partial \hat{E}_{RR}} \\
\bar{\sigma}^{\theta\theta} &= \lambda_\Theta^2 \frac{\partial W}{\partial \hat{E}_{\Theta\Theta}} + \gamma \lambda_\Theta \left(\frac{\partial W}{\partial \hat{E}_{\Theta Z}} + \frac{\partial W}{\partial \hat{E}_{Z\Theta}} \right) + \gamma^2 \frac{\partial W}{\partial \hat{E}_{ZZ}} \\
\bar{\sigma}^{zz} &= \lambda_Z^2 \frac{\partial W}{\partial \hat{E}_{ZZ}}.
\end{aligned} \tag{6.107}$$

Note the coupling between normal stress and shear strain components for a general anisotropic material.

Equilibrium. With end effects ignored, the symmetry in this problem demands that stress must be independent of θ and z. In the absence of inertial and body forces, Eqs. (4.103) reduce to

$$\begin{aligned}
\frac{\partial \hat{\sigma}^{rr}}{\partial r} + \frac{\hat{\sigma}^{rr} - \hat{\sigma}^{\theta\theta}}{r} &= 0 \\
\frac{\partial}{\partial r}(r^2 \hat{\sigma}^{r\theta}) &= 0 \\
\frac{\partial}{\partial r}(r \hat{\sigma}^{rz}) &= 0.
\end{aligned} \tag{6.108}$$

Boundary Conditions. The inner wall of the artery is subjected to an internal pressure P, the outer wall is free of loads, and the testing machine exerts a resultant force N and twisting moment M at the ends. With the exception of the pressure, the boundary conditions match those of the papillary muscle problem.

At the deformed inner surface, the unit normal is $\mathbf{n} = -\mathbf{e}_r = -\mathbf{e}_1 = -\mathbf{e}^1$, and the true traction vector is

$$\begin{aligned} \mathbf{T}^{(\mathbf{n})} &= \mathbf{n} \cdot \boldsymbol{\sigma} \\ &= -\mathbf{e}^1 \cdot (\hat{\sigma}^{ij} \mathbf{e}_i \mathbf{e}_j) \\ &= -\hat{\sigma}^{ij} \delta_i^1 \mathbf{e}_j \\ &= -\hat{\sigma}^{1j} \mathbf{e}_j. \end{aligned}$$

This traction must match the applied traction $P\,\mathbf{e}_r = P\,\mathbf{e}_1$ at $r = a$, i.e., $-\hat{\sigma}^{1j}\mathbf{e}_j = -(\hat{\sigma}^{11}\mathbf{e}_1 + \hat{\sigma}^{12}\mathbf{e}_2 + \hat{\sigma}^{13}\mathbf{e}_3) = P\,\mathbf{e}_1$. Hence, adding this condition to those given by slightly modified versions of Eqs. (6.72) and (6.74) yields

$$\begin{aligned} r &= a: \quad \hat{\sigma}^{rr} = -P, \quad \hat{\sigma}^{r\theta} = \hat{\sigma}^{rz} = 0 \\ r &= b: \quad \hat{\sigma}^{rr} = \hat{\sigma}^{r\theta} = \hat{\sigma}^{rz} = 0 \end{aligned} \tag{6.109}$$

$$\begin{aligned} \pi a^2 P + N &= 2\pi \int_a^b \hat{\sigma}^{zz}\, r\, dr \\ M &= 2\pi \int_a^b \hat{\sigma}^{z\theta}\, r^2\, dr. \end{aligned} \tag{6.110}$$

The pressure term in Eq. $(6.110)_1$ comes from the force exerted by P on the supports at the ends of the artery (see Problem **6–11**). Note that the equations for N and M were derived from (6.74) using the relations between tensor and physical stress components of Eq. (4.100).

6.5.3 *Solution*

Because the stresses depend only on r, Eqs. $(6.108)_{2,3}$ can be integrated to obtain

$$\hat{\sigma}^{r\theta} = \frac{C_1}{r^2}, \qquad \hat{\sigma}^{rz} = \frac{C_2}{r} \tag{6.111}$$

where C_1 and C_2 are constants. Applying the boundary conditions (6.109) gives $C_1 = C_2 = 0$, and thus $\hat{\sigma}^{r\theta} = \hat{\sigma}^{rz} = 0$ everywhere. In the papillary muscle problem, we used symmetry arguments to establish this result *a priori*, and so the circumferential and axial equilibrium equations $(6.108)_{2,3}$ were satisfied identically. Moreover, Eqs. (6.106) show that, in order for these shear stresses to be zero, the artery must be composed of tissue with a strain-energy density function that satisfies the conditions

$$\frac{\partial W}{\partial \hat{E}_{R\Theta}} = \frac{\partial W}{\partial \hat{E}_{ZR}} = 0. \tag{6.112}$$

Otherwise, the stipulated deformation cannot be realized without applying shear stresses to the curved surfaces of the tube.

Because the artery wall is assumed to be incompressible, the matrix form of Eq. (6.99) gives the constraint

$$J = \det \mathbf{F} = \lambda_R \lambda_\Theta \lambda_Z = 1.$$

Substituting (6.101) yields

$$\left(\frac{\pi}{\phi_0}\lambda\Lambda\right)\frac{r}{R}\frac{dr}{dR} = 1$$

which can be integrated to obtain

$$r^2 = \frac{\phi_0}{\pi\lambda\Lambda}(R^2 - a_0^2) + a^2 \tag{6.113}$$

in which the integration constant was found from the condition $r(a_0) = a$. If the opening angle ϕ, the axial stretch ratio λ, and the deformed inner radius a are known, then the deformed outer radius can be computed using the relation

$$b = r(b_0) = \left[\frac{\phi_0}{\pi\lambda\Lambda}(b_0^2 - a_0^2) + a^2\right]^{\frac{1}{2}}. \tag{6.114}$$

With $r(R)$ now known, Eqs. (6.101) and (6.103) provide the deformation gradient and strain components. Then, with $W(\hat{E}_{IJ})$ specified, Eqs. (6.106) give the stress components in terms of the yet unknown Lagrange multiplier p. To determine p, we substitute $(6.106)_{1,2}$ into the radial equilibrium equation $(6.108)_1$ and integrate to get

$$p(r) = \bar{\sigma}^{rr} + \int_r^b \left(\bar{\sigma}^{\theta\theta} - \bar{\sigma}^{rr}\right)\frac{dr}{r}. \tag{6.115}$$

The upper limit on this integral was chosen to satisfy the boundary condition $\hat{\sigma}^{rr}(b) = 0$, as can be verified by substitution into $(6.106)_1$. The internal pressure required to maintain the deformation now can be found by substituting the resulting expression for $\hat{\sigma}^{rr}$ into the final boundary condition on the curved surfaces, $\hat{\sigma}^{rr}(a) = -P$, to obtain

$$P = \int_a^b \left(\bar{\sigma}^{\theta\theta} - \bar{\sigma}^{rr}\right) \frac{dr}{r}. \tag{6.116}$$

Finally, the applied end loads M and N can be computed from Eqs. (6.110).

For a general form of W, the above integrals must be evaluated numerically. Computing N, therefore, involves evaluating the above integral for p within the integral in Eq. $(6.110)_1$. Although this procedure is relatively straightforward, it may be more convenient to write the equation for N in a form that does not involve p. This can be done by first writing Eq. $(6.110)_1$ as

$$\pi a^2 P + N = 2\pi \left[\int_a^b \left(\hat{\sigma}^{zz} - \hat{\sigma}^{rr}\right) r\, dr + \int_a^b \hat{\sigma}^{rr}\, r\, dr\right].$$

Integrating the last integral by parts and using (6.106) and $(6.108)_1$ yields

$$\begin{aligned}\int_a^b \hat{\sigma}^{rr}\, r\, dr &= \left[\frac{r^2}{2}\hat{\sigma}^{rr}\right]_a^b - \int_a^b \frac{d\hat{\sigma}^{rr}}{dr}\frac{r^2}{2} dr \\ &= \frac{b^2}{2}\hat{\sigma}^{rr}(b) - \frac{a^2}{2}\hat{\sigma}^{rr}(a) + \int_a^b \left(\frac{\bar{\sigma}^{rr} - \bar{\sigma}^{\theta\theta}}{r}\right)\frac{r^2}{2}\, dr.\end{aligned}$$

Now, applying the boundary conditions $\hat{\sigma}^{rr}(a) = -P$ and $\hat{\sigma}^{rr}(b) = 0$ and substituting the result into the above relation for N yields

$$\begin{aligned}\pi a^2 P + N &= \pi a^2 P + \pi \int_a^b \left(2\hat{\sigma}^{zz} - \hat{\sigma}^{rr} - \hat{\sigma}^{\theta\theta}\right) r\, dr \\ &= \pi a^2 P + \pi \int_a^b \left(2\bar{\sigma}^{zz} - \bar{\sigma}^{rr} - \bar{\sigma}^{\theta\theta}\right) r\, dr. \end{aligned} \tag{6.117}$$

This expression does not contain p.

6.5.4 *Geometric and Material Properties*

The results presented here are based on the parameters used by Chuong and Fung (1986) for the rabbit aorta. They assumed that the artery wall is cylindrically orthotropic, with the strain-energy density function having the form of Eq. (5.105), i.e.,

$$\begin{aligned} W &= C(e^Q - 1) \\ Q &= a_1 \hat{E}_{RR}^2 + a_2 \hat{E}_{\Theta\Theta}^2 + a_3 \hat{E}_{ZZ}^2 \\ &\quad +2\left(a_4 \hat{E}_{RR}\hat{E}_{\Theta\Theta} + a_5 \hat{E}_{\Theta\Theta}\hat{E}_{ZZ} + a_6 \hat{E}_{ZZ}\hat{E}_{RR}\right) \\ &\quad +a_8\left(\hat{E}_{\Theta Z}^2 + \hat{E}_{Z\Theta}^2\right). \end{aligned} \tag{6.118}$$

Note that the $\hat{E}_{R\Theta}$ and $\hat{E}_{ZR}$ terms have been dropped in order to satisfy the requirement (6.112). Using experimental data, the authors determined the following values for the material constants:

$$\begin{array}{lll} C = 11.2\,\text{kPa} & & \\ a_1 = 0.0499 & a_4 = 0.0042 & a_8 = 0.100 \\ a_2 = 1.0672 & a_5 = 0.0903 & \\ a_3 = 0.4775 & a_6 = 0.0585. & \end{array} \tag{6.119}$$

Because Chuong and Fung (1986) did not examine torsion, they actually set $a_8 = 0$. The value $a_8 = 0.1$ is used here only for illustration. In addition, measurements of vessel geometry before and after a transmural cut yielded the following approximate parameter values:

$$\begin{aligned} a_0 &= 3.9\,\text{mm} \\ b_0 &= 4.5\,\text{mm} \\ a &= 1.4\,\text{mm (unloaded)} \\ b &= 2.0\,\text{mm (unloaded)} \\ \phi &= 220^\circ \quad (\phi_0 = 70^\circ) \\ \Lambda &= 1. \end{aligned} \tag{6.120}$$

For the unloaded intact artery with $\lambda = 1$, these values satisfy Eq. (6.114) approximately.

6.5.5 *Residual Stress in Unloaded Artery*

Deformation from the zero-stress configuration B_0 to the unloaded intact configuration B_U (Fig. 6.12) produces residual stress. To compute this stress, we specialize our solution by setting $N = M = P = 0$. In addition, we take $\mathbf{F}_2 = \mathbf{I}$,

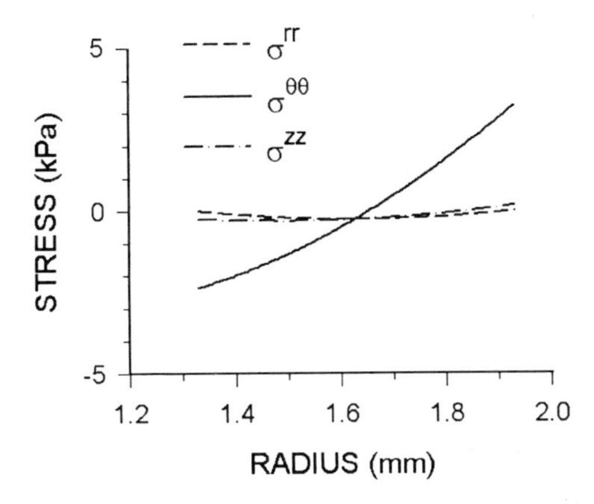

Fig. 6.13 Residual stress distributions in unloaded artery (physical components).

which implies that B_L is the same as B_U with $\rho = r$, $\vartheta = \theta$ ($\psi = 0$), and $\zeta = z$ ($\lambda = 1$). Hence, if experimental measurements provide a_0 and ϕ_0, Eqs. (6.94) and (6.113) give the deformation field in terms of the pair of unknowns a and Λ. These variables can be determined by solving the integral equations $(6.110)_1$ and (6.116) simultaneously with $N = P = 0$. Numerically, this essentially amounts to solving two nonlinear algebraic equations.

For illustration purposes, it is simpler to compute an approximate solution by assuming that $\Lambda = 1$ and solving Eq. (6.116) for a. This procedure yields $a = 1.33$ mm, which is relatively close to the measured value of 1.4 mm (Chuong and Fung, 1986), considering all of the assumptions made.

Transmural distributions of residual stress in B_U (Fig. 6.13) illustrate a number of points. First, the circumferential stress $\hat{\sigma}^{\theta\theta}$ is the largest stress component, being compressive in the inner region and tensile in the outer region of the wall. This distribution resembles the expected stress distribution for bending of a curved beam. Second, $\hat{\sigma}^{rr}$ vanishes at the inner and outer radii, consistent with the boundary conditions for zero pressure. Third, $\hat{\sigma}^{zz}$ is quite small compared to $\hat{\sigma}^{\theta\theta}$, justifying the plane strain approximation $\Lambda = 1$ in place of the average plane stress condition $N = 0$.

Finally, it is important to note that, because the artery is unloaded, residual stresses must be self-equilibrating. In other words, the (integrated) resultant forces due to $\hat{\sigma}^{\theta\theta}$ and $\hat{\sigma}^{zz}$ must be zero. The results in Fig. 6.13 are consistent with this requirement.

6.5.6 *Loaded Artery*

The deformation of the loaded artery is determined completely by the deformed inner radius a, the twist ψ, and the axial stretch λ. Figure 6.14 illustrates the nonlinear coupling between these deformation modes.

Pressure-radius curves for $\psi = 0$ show that the apparent artery stiffness increases with λ (Fig. 6.14a). This behavior is due to the exponential form for W, which characterizes a material that stiffens with increasing strain. Adding twist ($\psi = 0.5$ rad/mm) essentially shifts the pressure-radius curve upward, indicating that the pressure required to maintain a given radius increases, but the stiffness (slope) changes relatively little. This curve also shows that the vessel radius at zero pressure decreases with twist.

For $\psi = 0$ and λ held at a fixed value, the axial force N remains relatively constant when circumferential stretch is small, but, depending on the value of λ, N increases or decreases as the radius becomes much larger than the unloaded radius of 1.4 mm (Fig. 6.14b). This complex behavior is caused by the coupling between axial and circumferential strain. For example, pressure induces tension in both the axial and circumferential directions, but circumferential stretch alone is accompanied by axial shortening. In addition, for $\psi > 0$ and λ fixed, the twisting moment M increases with radius (Fig. 6.14c). The torque-radius curve for $a_8 = 0$ shows that the artery can sustain a twisting moment even if all shear strains are dropped from W [see Eq. (6.118)]. The shear stress $\hat{\sigma}^{\theta z}$ that provides M is furnished by the coupling term involving $\partial W/\partial \hat{E}_{ZZ}$ in Eq. $(6.106)_4$.

The global response curves of Fig. 6.14 are relatively unaffected by the presence of residual stress. In contrast, even though residual stresses are relatively small (Fig. 6.13), local stress distributions undergo dramatic changes in the loaded vessel (Fig. 6.15). In the absence of residual stress ($\phi = 0$), strong concentrations of circumferential and axial stress occur in the inner layers of the wall (Fig. 6.15a), although transmural differences in strain are relatively modest (Fig. 6.15b). This behavior is due to the strong material nonlinearity between stress and strain, i.e., relatively small changes in strain can produce large changes in stress. The bending that accompanies the deformation from B_0 to B_U (Fig. 6.12) lowers the strain at the inner wall and raises it at the outer wall (Fig. 6.15b, $\phi = 220°$), causing corresponding but relatively large changes in stress in the loaded artery (Fig. 6.15a). According to these results, the presence of residual stress and strain in arteries

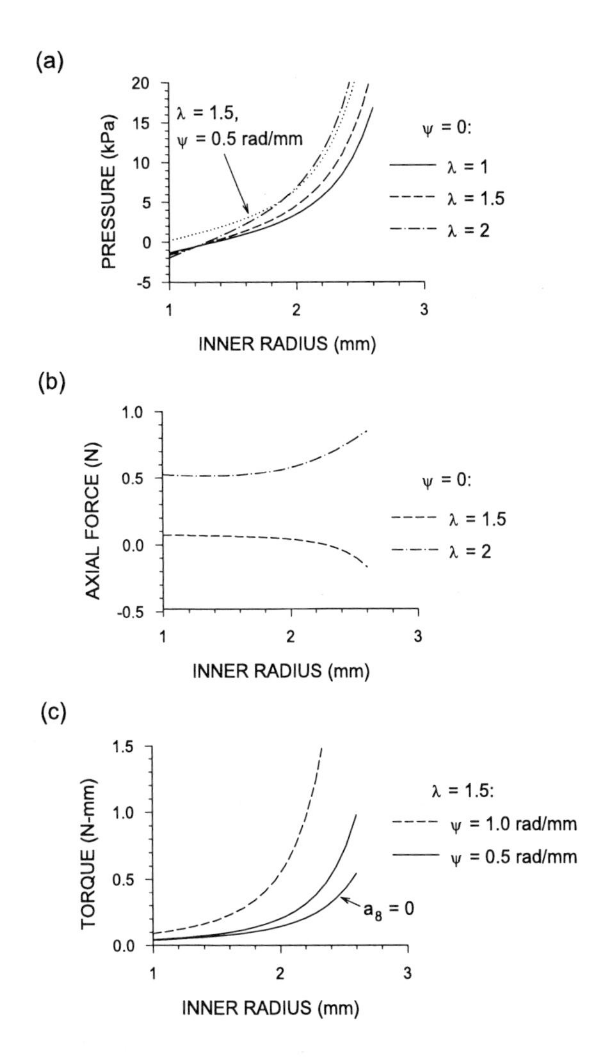

Fig. 6.14 Pressure, axial force, and torque versus deformed inner radius in artery.

tends to homogenize transmural stress distributions in the loaded vessel, thereby increasing the efficiency of the artery as a load-bearing structure.

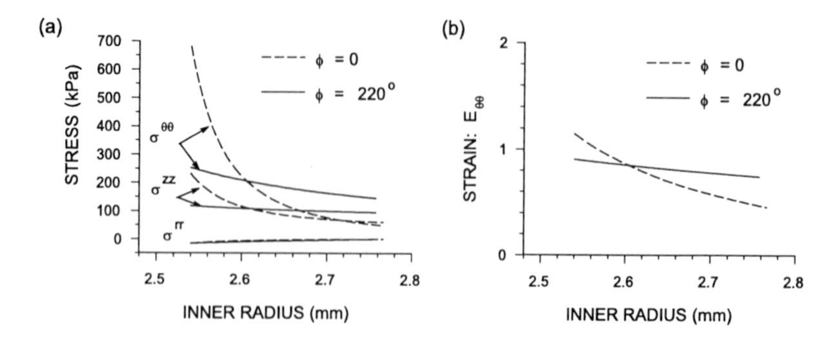

Fig. 6.15 Stress and strain distributions (physical components) in loaded artery with ($\phi = 220°$) and without ($\phi = 0$) residual stress.

6.6 Passive Filling of the Left Ventricle

The left ventricle (LV) is the heart chamber that pumps blood to the tissues and organs of the body. Due to the relatively high hydraulic resistance of the systemic vascular system, the LV experiences higher pressures than the other three chambers and, therefore, does the most work. Being composed of cardiac muscle (myocardium), the LV responds like other muscles to increased workload, i.e., it grows. The wall of the LV, therefore, is thicker than the walls of the atria and right ventricle, and it becomes even thicker during hypertension (high blood pressure). And because more blood is needed to supply this greater mass, the LV is more prone to myocardial infarction than the other heart chambers. (Infarctions usually are caused by blockage of a coronary artery feeding the heart muscle.) It is believed that wall stress plays a role in adaptation of the heart to altered loading, as well as in pathological conditions including infarction. For more than a century, therefore, researchers have proposed models to compute wall stress in the heart (Woods, 1892; Mirsky, 1974; Yin, 1981; McCulloch et al., 1992).

6.6.1 *Model for the Left Ventricle*

This section considers a model for inflation of the *passive* LV. Passive filling of the LV has received a great deal of attention, as diastolic dysfunction can lead to severe pathological consequences. However, it must be remembered that the highest wall stresses occur during systolic contraction, a topic that is beyond the scope of this book.

To a first approximation, the human LV can be treated as a thick-walled el-

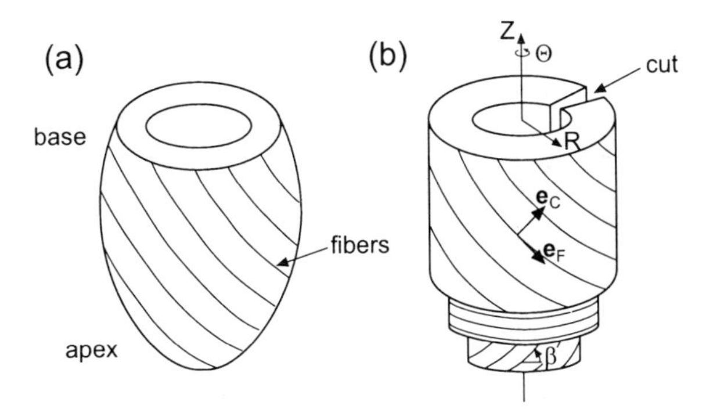

Fig. 6.16 Models for left ventricle. (a) Ellipsoid of revolution. (b) Thick-walled circular cylinder in approximately zero-stress configuration showing fiber orientation at epicardium, midwall, and endocardium.

lipsoidal shell of revolution (Fig. 6.16a). Near the base of the LV, moreover, a circular cylinder suffices as an even simpler representation. Here, we consider a cylindrical tube (Fig. 6.16b) that is fixed at its upper end (base) and free at its lower end (apex). Both ends are assumed to be closed to maintain an internal blood pressure P, but end effects and all other external loads are ignored. In addition, like arteries, the LV contains residual stress, as illustrated by ventricular cross sections that open when cut transmurally (Omens and Fung, 1990). Hence, this LV model is similar to that for the artery, and we can build on the analysis in the previous section. Like the artery, the LV is assumed to have the zero-stress, unloaded, and loaded configurations depicted in Fig. 6.12.

Although the geometry has been simplified considerably, it is important to account for the anisotropic muscle fiber architecture. The wall of the LV is composed of highly ordered layers of muscle connected by a network of collagen fibers (LeGrice et al., 1995; 1997). The muscle fibers in these layers are organized into helices that wind around the lumen (Fig. 6.16). Relative to the circumferential direction, the fiber helix angle β varies continuously across the wall from about 60° at the inner radius (a_0, endocardium) to nearly circumferential at midwall to -60° at the outer radius (b_0, epicardium) (see Fig. 6.16b). A rough approximation for the transmural distribution in the zero-stress configuration is given by the relation

$$\beta(R) = \beta_0 \left[\frac{2R - a_0 - b_0}{a_0 - b_0} \right]^n \tag{6.121}$$

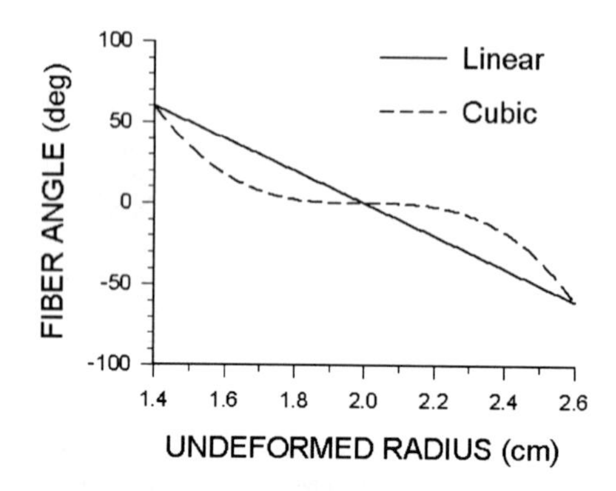

Fig. 6.17 Transmural fiber angle distributions for model of left ventricle.

where β_0 is the endocardial value of β, and n is an odd integer. Here, we consider the cases $n = 1$ and $n = 3$ (Fig. 6.17), with the latter value being more consistent with the classical measurements of Streeter (1979) for the dog LV.

In summary, our model for the LV is a thick-walled cylindrical tube with closed ends and an internal pressure P (Fig. 6.16b). The wall is assumed to be composed of muscle that is transversely isotropic relative to the local fiber direction, which varies across the wall according to Eq. (6.121).

6.6.2 *Analysis*

When subjected to internal pressure, the passive LV stretches in the longitudinal and circumferential directions and, due to the helical fiber geometry, twists about its axis of symmetry. Because this type of deformation was studied for an artery in the previous section, we assume that the deformation of the LV model also is described by Eqs. (6.94) and (6.95). Therefore, with the following modifications, the analysis of the artery also applies to the LV.

First, because the apex is not constrained, we set the twisting moment and axial force to zero, i.e., $M = N = 0$. With these specified loads, Eqs. $(6.110)_2$ and (6.117) provide two equations to be solved simultaneously for ψ and λ. Moreover, if we specify a, as in the artery problem, then Eq. (6.116) provides P. Otherwise, we could stipulate the value of P and add (6.116) as a third simultaneous equation for the unknown a. In practice, solving three nonlinear equations is not much more difficult than solving two. Hence, we choose to specify the value of P, along with $M = N = 0$, and solve for a, λ, and ψ.

Second, we need to modify the constitutive relations (6.106) to account for the fiber geometry. For this purpose, it is convenient to express W in terms of strain components relative to local Cartesian axes (R, F, C) in the zero-stress configuration, where F and C represent the fiber and cross-fiber directions, respectively. The unit vectors $\mathbf{e}_F$ and $\mathbf{e}_C$ are obtained by rotating $\mathbf{e}_\Theta$ and $\mathbf{e}_Z$ about $\mathbf{e}_R$ through the helix angle β, giving (Fig. 6.16b)

$$\begin{aligned} \mathbf{e}_F &= \mathbf{e}_\Theta \cos\beta + \mathbf{e}_Z \sin\beta \\ \mathbf{e}_C &= -\mathbf{e}_\Theta \sin\beta + \mathbf{e}_Z \cos\beta. \end{aligned} \tag{6.122}$$

Relative to this local material basis, the Lagrangian strain tensor is

$$\mathbf{E} = \hat{E}_{RR}\mathbf{e}_R\mathbf{e}_R + \hat{E}_{FF}\mathbf{e}_F\mathbf{e}_F + \hat{E}_{CC}\mathbf{e}_C\mathbf{e}_C + \hat{E}_{FC}\mathbf{e}_F\mathbf{e}_C + \hat{E}_{CF}\mathbf{e}_C\mathbf{e}_F, \tag{6.123}$$

and the strain-energy density function has the form

$$W = W(\hat{E}_{RR}, \hat{E}_{FF}, \hat{E}_{CC}, \hat{E}_{FC}, \hat{E}_{CF}). \tag{6.124}$$

Note that, as in the artery problem, the stipulated geometry demands that $\hat{E}_{R\Theta} = \hat{E}_{RZ} = 0$, which implies $\hat{E}_{RF} = \hat{E}_{RC} = 0$ [see Eq. (6.102)]. Similarly, the artery analysis shows that the Cauchy stress tensor for this problem can be written

$$\boldsymbol{\sigma} = \hat{\sigma}^{rr}\mathbf{e}_r\mathbf{e}_r + \hat{\sigma}^{\theta\theta}\mathbf{e}_\theta\mathbf{e}_\theta + \hat{\sigma}^{zz}\mathbf{e}_z\mathbf{e}_z + \hat{\sigma}^{\theta z}\mathbf{e}_\theta\mathbf{e}_z + \hat{\sigma}^{z\theta}\mathbf{e}_z\mathbf{e}_\theta \tag{6.125}$$

relative to the deformed coordinates (r, θ, z).

Now, substituting Eqs. (6.100) and (6.123)–(6.125), along with the relation $\mathbf{I} = \mathbf{e}_r\mathbf{e}_r + \mathbf{e}_\theta\mathbf{e}_\theta + \mathbf{e}_z\mathbf{e}_z$, into (6.104) provides the appropriate constitutive relations. The manipulations are straightforward but lengthy because of the coordinate transformations involved. To make the computations more manageable, we first note that Eqs. (2.153) and (6.123) give

$$\frac{\partial W}{\partial \mathbf{E}} = W_{RR}\mathbf{e}_R\mathbf{e}_R + W_{FF}\mathbf{e}_F\mathbf{e}_F + W_{CC}\mathbf{e}_C\mathbf{e}_C + W_{FC}\mathbf{e}_F\mathbf{e}_C + W_{CF}\mathbf{e}_C\mathbf{e}_F \tag{6.126}$$

where

$$W_{IJ} \equiv \frac{\partial W}{\partial \hat{E}_{IJ}}. \tag{6.127}$$

Next, Eqs. (6.122) are used to write the above dyads in terms of $\mathbf{e}_\Theta$ and $\mathbf{e}_Z$, i.e.,

$$\begin{aligned}
\mathbf{e}_F\mathbf{e}_F &= \mathbf{e}_\Theta\mathbf{e}_\Theta \cos^2\beta + (\mathbf{e}_\Theta\mathbf{e}_Z + \mathbf{e}_Z\mathbf{e}_\Theta)\sin\beta\cos\beta + \mathbf{e}_Z\mathbf{e}_Z\sin^2\beta \\
\mathbf{e}_C\mathbf{e}_C &= \mathbf{e}_\Theta\mathbf{e}_\Theta \sin^2\beta - (\mathbf{e}_\Theta\mathbf{e}_Z + \mathbf{e}_Z\mathbf{e}_\Theta)\sin\beta\cos\beta + \mathbf{e}_Z\mathbf{e}_Z\cos^2\beta \\
\mathbf{e}_F\mathbf{e}_C &= \sin\beta\cos\beta(\mathbf{e}_Z\mathbf{e}_Z - \mathbf{e}_\Theta\mathbf{e}_\Theta) + \mathbf{e}_\Theta\mathbf{e}_Z\cos^2\beta - \mathbf{e}_Z\mathbf{e}_\Theta\sin^2\beta \\
\mathbf{e}_C\mathbf{e}_F &= \sin\beta\cos\beta(\mathbf{e}_Z\mathbf{e}_Z - \mathbf{e}_\Theta\mathbf{e}_\Theta) - \mathbf{e}_\Theta\mathbf{e}_Z\sin^2\beta + \mathbf{e}_Z\mathbf{e}_\Theta\cos^2\beta.
\end{aligned} \tag{6.128}$$

Inserting Eqs. (6.100), (6.126), and (6.128) into (6.104) reveals that all dot products now involve the orthogonal system $(\mathbf{e}_R, \mathbf{e}_\Theta, \mathbf{e}_Z)$, hence eliminating these base vectors and leaving only $(\mathbf{e}_r, \mathbf{e}_\theta, \mathbf{e}_z)$, which is the basis used for $\boldsymbol{\sigma}$ in (6.125). This observation allows us to express Eq. (6.104) in matrix form without confusion.

The matrix forms of Eqs. (6.100) and (6.125) are

$$[\mathbf{F}] = \begin{bmatrix} \lambda_R & 0 & 0 \\ 0 & \lambda_\Theta & \gamma \\ 0 & 0 & \lambda_Z \end{bmatrix}, \qquad [\boldsymbol{\sigma}] = \begin{bmatrix} \hat{\sigma}^{rr} & 0 & 0 \\ 0 & \hat{\sigma}^{\theta\theta} & \hat{\sigma}^{\theta z} \\ 0 & \hat{\sigma}^{z\theta} & \hat{\sigma}^{zz} \end{bmatrix},$$

and putting (6.128) into (6.126) yields

$$\left[\frac{\partial W}{\partial \mathbf{E}}\right] = \begin{bmatrix} W_{RR} & 0 & 0 \\ 0 & \left\{\begin{matrix} c^2W_{FF} + s^2W_{CC} \\ -sc(W_{FC} + W_{CF}) \end{matrix}\right\} & \left\{\begin{matrix} sc(W_{FF} - W_{CC}) \\ +c^2W_{FC} - s^2W_{CF}) \end{matrix}\right\} \\ 0 & \left\{\begin{matrix} sc(W_{FF} - W_{CC}) \\ +c^2W_{FC} - s^2W_{CF}) \end{matrix}\right\} & \left\{\begin{matrix} s^2W_{FF} + c^2W_{CC} \\ +sc(W_{FC} + W_{CF}) \end{matrix}\right\} \end{bmatrix} \tag{6.129}$$

where

$$c \equiv \cos\beta, \qquad s \equiv \sin\beta. \tag{6.130}$$

Finally, in terms of these matrices, Eq. (6.104) becomes

$$[\boldsymbol{\sigma}] = [\mathbf{F}]\left[\frac{\partial W}{\partial \mathbf{E}}\right][\mathbf{F}]^T - p\,[\mathbf{I}]\,. \tag{6.131}$$

For $\beta = 0$, we have $\mathbf{e}_F = \mathbf{e}_\Theta$ and $\mathbf{e}_C = \mathbf{e}_Z$, and these constitutive relations reduce to Eqs. (6.106). (Recall that $\hat{\sigma}^{r\theta} = \hat{\sigma}^{zr} = 0$.)

This procedure effectively provides the Cauchy stress components as functions of r to be used in Eqs. $(6.110)_2$, (6.116), and (6.117). Given $M = N = 0$ and the value of P, these equations can be solved numerically for a, λ, and ψ.

To complete the solution, we take $\Lambda = 1$ as in the artery problem. Then, Eq. (6.113) gives $r(R)$, and (6.101) and (6.103) give the strains relative to the (R, Θ, Z) system. Next, the strains in the rotated (R, F, C) system are computed by standard coordinate transformation of tensor components, Eq. (6.115) provides p, and (6.131) yields the stress components relative to (r, θ, z).

It often is of interest to determine the load carried by the muscle fibers along their lengths. The fiber stress $\hat{\sigma}^{ff}$ can be computed from the relation

$$\hat{\sigma}^{ff} = \mathbf{e}_f \cdot \boldsymbol{\sigma} \cdot \mathbf{e}_f \tag{6.132}$$

where $\mathbf{e}_f$ is the unit vector in the deformed fiber direction. This vector is given by

$$\mathbf{e}_f = \frac{\mathbf{F} \cdot \mathbf{e}_F}{|\mathbf{F} \cdot \mathbf{e}_F|} \tag{6.133}$$

in which Eqs. (6.100) and (6.122) provide $\mathbf{F}$ and $\mathbf{e}_F$, respectively.

6.6.3 *Geometric and Material Properties*

Representative results are presented for the dog LV. For locally transversely isotropic myocardium, the strain-energy density function is assumed to have the form

$$\begin{aligned} W &= C(e^Q - 1) \\ Q &= 2a_1(\hat{E}_{RR} + \hat{E}_{FF} + \hat{E}_{CC}) + a_2\hat{E}_{FF}^2 \\ &\qquad + a_3(\hat{E}_{FC}^2 + \hat{E}_{CF}^2), \end{aligned} \tag{6.134}$$

which is consistent with Eq. (5.112). Considering the data of Guccione et al. (1991), we use the following representative material constants:

$$\begin{aligned} C &= 0.5\,\text{kPa} \\ a_1 &= 2, \qquad a_2 = 15, \qquad a_3 = 10. \end{aligned}$$

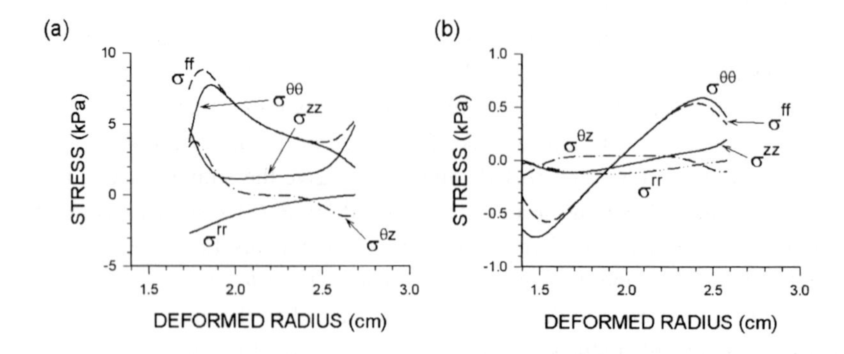

Fig. 6.18 Transmural distributions of physical wall stresses in model for left ventricle with cubic fiber angle distribution and opening angle $\phi = 50°$. (a) Stresses for $P = 2.67$ kPa. (b) Residual stresses ($P = 0$).

In addition, we use the following geometrical parameters:

$$\begin{aligned} a_0 &= 1.7 \text{ cm}, \qquad b_0 = 2.9 \text{ cm} \\ a &= 1.4 \text{ cm}, \qquad b = 2.6 \text{ cm} \qquad \text{(unloaded)} \\ \phi &= 50° \quad (\phi_0 = 155°), \qquad \Lambda = 1. \end{aligned}$$

6.6.4 *Results*

Stress distributions are shown for cubic distributions of fiber angles ($n = 3$ and $\beta_0 = 60°$, see Fig. 6.17) at an inflation pressure $P = 2.67$ kPa $= 20$ mmHg (Fig. 6.18a). Unlike the results for the artery (Fig. 6.15a), the stress components in the LV generally do not vary monotonically across the wall. Rather, the peak fiber stress occurs in the subendocardial layers of the wall. This behavior reflects the effects of the fiber architecture. For example, near midwall, where the fibers are oriented essentially in the circumferential direction (Fig. 6.17), the fiber and circumferential stresses are nearly identical. Near the endocardium and epicardium, however, the longitudinal stress contributes significantly to the fiber stress.

Exhibiting similar complexity, residual stresses are an order of magnitude smaller than the wall stresses in the loaded ventricle (Fig. 6.18b). As discussed in the artery problem, these stresses must be self-equilibrating. This requirement provides a valuable check on the accuracy of our solution.

The effects of fiber angle distribution and residual strain on fiber stress are

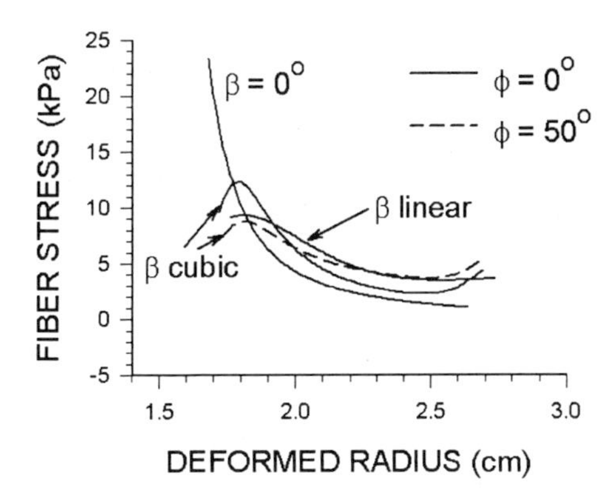

Fig. 6.19 Effects of fiber angle β and opening angle ϕ on transmural distributions of physical fiber stress in model for left ventricle ($P = 2.67$ kPa).

illustrated in Fig. 6.19, where various distributions of the fiber angle β are considered. Note the following:

1 For circumferential fibers ($\beta = 0°$) and no residual stress ($\phi = 0°$), a strong stress concentration in $\hat{\sigma}^{ff} = \hat{\sigma}^{\theta\theta}$ occurs near the endocardium, similar to the behavior in an artery without residual stress (see Fig. 6.15a).
2 Varying the fiber orientation across the wall decreases peak fiber stress by about half, regardless of the precise transmural distribution of β.
3 Because stresses are significantly reduced by fiber architecture alone and because the opening angle ($\phi = 50°$) is smaller, residual strain has a much smaller effect on wall stress in the LV than it does in arteries (see Fig. 6.15a).

The influence of residual strain and fiber angle distribution on global variables is shown in Fig. 6.20. Pressure-volume curves are affected only at relatively high pressures, and the axial stretch ratio λ changes only slightly. The twist ψ is affected the most, especially by residual strain. With increasing pressure, the LV twists in one direction ($\psi > 0$) and then reverses course. Interestingly, when residual strain is included ($\phi = 50°$), the LV is twisted even when $P = 0$. This behavior is caused by interaction between the fiber architecture and the redistribution of wall stresses required for self-equilibrium (see Fig. 6.18b).

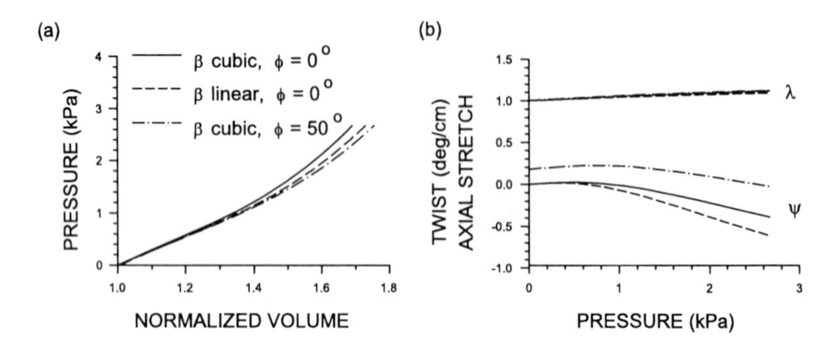

Fig. 6.20 Effects of fiber geometry and residual stress on global behavior of model for left ventricle. (a) Pressure-volume curves. (b) Axial stretch λ and twist ψ versus pressure.

6.7 Blastula with Internal Pressure

Embryos undergo dramatic changes in form during development. In general, the deformations that produce three-dimensional form are extremely complex and require computational methods for analysis. There are, however, a number of relatively simple problems of morphogenesis that are amenable to analytical solution. Here, we consider a problem that occurs in the early embryo.

A fertilized egg initially undergoes a series of divisions that produces a group of cells. Next, a fluid pocket forms to create a hollow ball of cells called a blastula. Then, the process of gastrulation rearranges the cells into the primary germ layers, which later form specialized tissues and organs (Gilbert, 2003). During gastrulation, cells move from the outer layers to the interior of the blastula, forming the primitive gut. These events are crucial for proper development of the embryo. Considering a wide range of experimental observations, Beloussov (1998) has suggested that the fluid in the blastula exerts an outward pressure on the cells, which then respond to the added stress by moving and deforming in ways that tend to restore stresses back toward their initial levels. He speculates that this response partially drives gastrulation. This section examines the stresses in a sea urchin blastula, which is essentially a thick-walled spherical shell filled with fluid (Fig. 6.21).

For simplicity, we ignore residual stress and assume that the cells in the blastula wall are isotropic and incompressible. Symmetry then demands that the shell maintains a spherical shape as it deforms under an internal pressure P. Let a_0 and

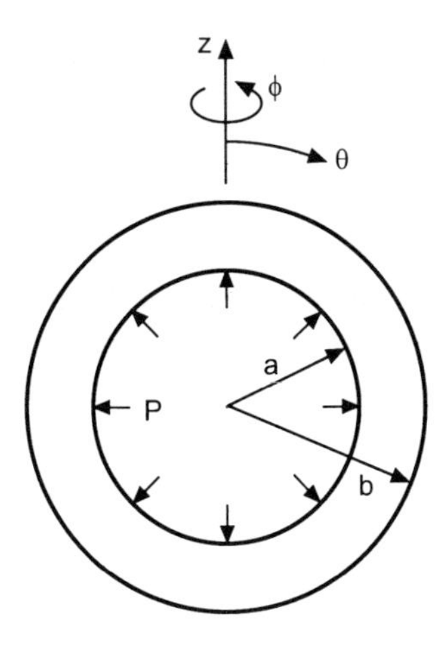

Fig. 6.21 Model for sea urchin blastula.

b_0 be the undeformed inner and outer radii, respectively, and a and b the corresponding deformed radii.

6.7.1 *Governing Equations*

Kinematics. Let (R, Θ, Φ) and (r, θ, ϕ) represent spherical polar coordinates in the reference (unloaded, zero-stress) configuration and the deformed configuration, respectively. In terms of Cartesian base vectors, the unit base vectors for these coordinate systems are (see Appendix B)

$$\begin{aligned}
\mathbf{e}_R &= \mathbf{e}_x \sin\Theta\cos\Phi + \mathbf{e}_y \sin\Theta\sin\Phi + \mathbf{e}_z \cos\Theta \\
\mathbf{e}_\Theta &= \mathbf{e}_x \cos\Theta\cos\Phi + \mathbf{e}_y \cos\Theta\sin\Phi - \mathbf{e}_z \sin\Theta \\
\mathbf{e}_\Phi &= -\mathbf{e}_x \sin\Phi + \mathbf{e}_y \cos\Phi
\end{aligned}$$

$$\begin{aligned}
\mathbf{e}_r &= \mathbf{e}_x \sin\theta\cos\phi + \mathbf{e}_y \sin\theta\sin\phi + \mathbf{e}_z \cos\theta \\
\mathbf{e}_\theta &= \mathbf{e}_x \cos\theta\cos\phi + \mathbf{e}_y \cos\theta\sin\phi - \mathbf{e}_z \sin\theta \\
\mathbf{e}_\phi &= -\mathbf{e}_x \sin\phi + \mathbf{e}_y \cos\phi.
\end{aligned} \qquad (6.135)$$

For spherically symmetric deformation, the position vectors before and after deformation are, respectively,

$$\mathbf{R} = R\,\mathbf{e}_R(\Theta, \Phi), \qquad \mathbf{r} = r\,\mathbf{e}_r(\theta, \phi), \tag{6.136}$$

with

$$r = r(R), \qquad \theta = \Theta, \qquad \phi = \Phi. \tag{6.137}$$

Hence, $\mathbf{e}_R = \mathbf{e}_r$, $\mathbf{e}_\Theta = \mathbf{e}_\theta$, and $\mathbf{e}_\Phi = \mathbf{e}_\phi$ for this problem. The corresponding undeformed natural base vectors are

$$\begin{aligned}
\mathbf{G}_R &= \mathbf{R}_{,R} = \mathbf{e}_R & \mathbf{G}^R &= \mathbf{e}_R \\
\mathbf{G}_\Theta &= \mathbf{R}_{,\Theta} = R\,\mathbf{e}_{R,\Theta} = R\,\mathbf{e}_\Theta & \mathbf{G}^\Theta &= R^{-1}\mathbf{e}_\Theta \\
\mathbf{G}_\Phi &= \mathbf{R}_{,\Phi} = R\mathbf{e}_{R,\Phi} = R\sin\Theta\,\mathbf{e}_\Phi & \mathbf{G}^\Phi &= (R\sin\Theta)^{-1}\,\mathbf{e}_\Phi
\end{aligned} \tag{6.138}$$

and the convected base vectors are

$$\begin{aligned}
\mathbf{g}_R &= \mathbf{r}_{,R} = r'\mathbf{e}_r & \mathbf{g}^R &= (r')^{-1}\mathbf{e}_r \\
\mathbf{g}_\Theta &= \mathbf{r}_{,\Theta} = r\,\mathbf{e}_\theta & \mathbf{g}^\Theta &= r^{-1}\mathbf{e}_\theta \\
\mathbf{g}_\Phi &= \mathbf{r}_{,\Phi} = r\,\sin\theta\,\mathbf{e}_\phi & \mathbf{g}^\Phi &= (r\sin\theta)^{-1}\,\mathbf{e}_\phi
\end{aligned} \tag{6.139}$$

where prime denotes differentiation with respect to R. (Note that the base vectors are mutually orthogonal.) With these relations, Eq. $(3.22)_1$ gives the deformation gradient tensor

$$\begin{aligned}
\mathbf{F} &= \mathbf{g}_I\mathbf{G}^I = \mathbf{g}_R\mathbf{G}^R + \mathbf{g}_\Theta\mathbf{G}^\Theta + \mathbf{g}_\Phi\mathbf{G}^\Phi \\
&= \lambda_R\mathbf{e}_r\mathbf{e}_R + \lambda_\Theta\mathbf{e}_\theta\mathbf{e}_\Theta + \lambda_\Phi\mathbf{e}_\phi\mathbf{e}_\Phi
\end{aligned} \tag{6.140}$$

where

$$\lambda_R = r', \qquad \lambda_\Theta = \lambda_\Phi = \frac{r}{R}. \tag{6.141}$$

Because the expression for $\mathbf{F}$ does not contain shear terms, these components represent stretch ratios.

Constitutive Relations. Using Eqs. (5.88), we write the constitutive equation (6.104) in the form

$$\boldsymbol{\sigma} = \mathbf{F}\cdot\frac{\partial W}{\partial \mathbf{F}^T} - p\,\mathbf{I} \tag{6.142}$$

where $W = W(\mathbf{F}) = W(\lambda_R, \lambda_\Theta, \lambda_\Phi)$. Substituting (6.140) and the dyadic representations $\mathbf{I} = \mathbf{e}_r\mathbf{e}_r + \mathbf{e}_\theta\mathbf{e}_\theta + \mathbf{e}_\phi\mathbf{e}_\phi$ and

$$\boldsymbol{\sigma} = \hat{\sigma}^{rr}\mathbf{e}_r\mathbf{e}_r + \hat{\sigma}^{\theta\theta}\mathbf{e}_\theta\mathbf{e}_\theta + \hat{\sigma}^{\phi\phi}\mathbf{e}_\phi\mathbf{e}_\phi \tag{6.143}$$

into (6.142) yields

$$\begin{aligned}
\hat{\sigma}^{rr} &= \lambda_R \frac{\partial W}{\partial \lambda_R} - p \\
\hat{\sigma}^{\theta\theta} &= \lambda_\Theta \frac{\partial W}{\partial \lambda_\Theta} - p \\
\hat{\sigma}^{\phi\phi} &= \lambda_\Phi \frac{\partial W}{\partial \lambda_\Phi} - p.
\end{aligned} \tag{6.144}$$

Equilibrium. Symmetry indicates that the stress field depends only on R. In fact, we also can set $\hat{\sigma}^{\theta\theta} = \hat{\sigma}^{\phi\phi}$ and $\lambda_\Theta = \lambda_\Phi$, but we keep these quantities distinct for now. In spherical coordinates, therefore, the equation of radial equilibrium reduces to

$$\frac{\partial \hat{\sigma}^{rr}}{\partial r} + \frac{1}{r}\left(2\hat{\sigma}^{rr} - \hat{\sigma}^{\theta\theta} - \hat{\sigma}^{\phi\phi}\right) = 0 \tag{6.145}$$

as given in Appendix B (section B.2), with equilibrium in the other coordinate directions satisfied identically.

Boundary Conditions. For an internal pressure P, the boundary conditions are similar to those of Eqs. (6.109) for an artery. In particular, we have

$$\begin{aligned}
r &= a: \quad \hat{\sigma}^{rr} = -P \\
r &= b: \quad \hat{\sigma}^{rr} = 0.
\end{aligned} \tag{6.146}$$

6.7.2 *Solution*

With Eqs. (6.141), enforcing incompressibility yields

$$J = \det \mathbf{F} = \lambda_R \lambda_\Theta \lambda_\Phi = \frac{r^2}{R^2}\frac{dr}{dR} = 1. \tag{6.147}$$

Integrating this relation and using the condition $r(a_0) = a$ gives

$$r^3 = R^3 - a_0^3 + a^3. \tag{6.148}$$

Hence, if the deformed inner radius a is specified, then the outer radius $b = r(b_0)$ is

$$b = \left(b_0^3 - a_0^3 + a^3\right)^{1/3}. \tag{6.149}$$

The Lagrange multiplier is determined by inserting (6.144) into (6.145) and integrating. This procedure yields

$$p(r) = \lambda_R \frac{\partial W}{\partial \lambda_R} + \int_r^b \left(\lambda_\Theta \frac{\partial W}{\partial \lambda_\Theta} + \lambda_\Phi \frac{\partial W}{\partial \lambda_\Phi} - 2\lambda_R \frac{\partial W}{\partial \lambda_R}\right) \frac{dr}{r}, \tag{6.150}$$

which satisfies the boundary condition $\hat{\sigma}^{rr}(b) = 0$. Substituting this expression into Eq. $(6.144)_1$ and using the boundary condition $\hat{\sigma}^{rr}(a) = -P$ gives

$$P = \int_a^b \left(\lambda_\Theta \frac{\partial W}{\partial \lambda_\Theta} + \lambda_\Phi \frac{\partial W}{\partial \lambda_\Phi} - 2\lambda_R \frac{\partial W}{\partial \lambda_R}\right) \frac{dr}{r}. \tag{6.151}$$

If a is specified, then Eqs. (6.141) and (6.148) provide the stretch ratios, and the above equations give p and P. Then, the stress components can be computed from (6.144).

If the shell wall is "thin," then the blastula can be treated as a pressurized **membrane**. In this case, $a_0/h_0 >> 1$ with $h_0 = b_0 - a_0$ being the undeformed wall thickness, and the above integrals can be simplified by assuming that stress and strain vary relatively little across the wall. As the deformed wall thickness $h = b - a$ approaches zero, Eqs. (6.150) and (6.151) become

$$\begin{aligned} p &\cong \lambda_R \frac{\partial W}{\partial \lambda_R} \\ P &\cong \frac{h}{a}\left(\lambda_\Theta \frac{\partial W}{\partial \lambda_\Theta} + \lambda_\Phi \frac{\partial W}{\partial \lambda_\Phi} - 2\lambda_R \frac{\partial W}{\partial \lambda_R}\right). \end{aligned} \tag{6.152}$$

Hence, with Eqs. (6.144), the pressure can be written

$$P = \frac{h}{a}\left(\hat{\sigma}^{\theta\theta} + \hat{\sigma}^{\phi\phi}\right),$$

and using symmetry gives

$$\hat{\sigma}^{\theta\theta} = \hat{\sigma}^{\phi\phi} = \frac{Pa}{2h}. \tag{6.153}$$

This equation, which also can be derived from equilibrium considerations alone, is known as the **Law of Laplace** for a pressurized spherical membrane.

Note further that Eqs. $(6.144)_1$ and $(6.152)_1$ give $\hat{\sigma}^{rr} \cong 0$ for a membrane. Actually, the radial stress is not zero; rather it is small compared to the other stress components. We can see this by considering the boundary conditions (6.146), which indicate that the value of $\hat{\sigma}^{rr}$ varies from $-P$ at $r = a$ to 0 at $r = b$, i.e., $\hat{\sigma}^{rr} = O(P)$. Hence, for a thin membrane with $a/h >> 1$, Eq. (6.153) shows that $\hat{\sigma}^{\theta\theta} = \hat{\sigma}^{\phi\phi} >> \hat{\sigma}^{rr}$. Typically, a shell is considered thin when the *undeformed* radius-to-thickness ratio a_0/h_0 is greater than about 20. For large deformation, however, an incompressible shell becomes thinner as it inflates, i.e., a/h grows larger. Thus, even if the undeformed shell is relatively thick, the membrane approximation may become valid as its radius increases.

6.7.3 *Geometric and Material Properties*

We assume that the strain-energy density function for the blastula wall has the form

$$W = \frac{C}{\alpha}\left[e^{\alpha(I_1-3)} - 1\right] \tag{6.154}$$

where C and α are material constants and

$$I_1 = \lambda_R^2 + \lambda_\Theta^2 + \lambda_\Phi^2$$

is a strain invariant. In the limit of small strain ($I_1 \to 3$), it can be shown that $C = E_Y/6$ (see Problem **5–4**), where E_Y is the small-strain elastic modulus. From small-strain measurements for the sea urchin blastula (Davidson et al., 1999), we take the following parameter values:

$$\begin{aligned} C &= 0.2 \text{ kPa} \\ a_0 &= 50\ \mu\text{m} \\ b_0 &= 60\ \mu\text{m}. \end{aligned}$$

Because the nonlinear properties apparently are unknown, we examine various values for α. In addition, it is instructive to study the effects of the radius-to-thickness ratio a_0/h_0.

6.7.4 *Results*

Pressure-radius curves for three values of α are shown in Fig. 6.22a. As α increases, the blastula stiffens considerably at large radii. For $\alpha \to 0$, Eq. (6.154) reduces to the form of a neo-Hookean material, i.e., $W \to C(I_1 - 3)$. For this case, the pressure reaches a peak near $a/a_0 = 1.3$, indicating a limit-point instability. If pressure is prescribed for $\alpha = 0$, then the radius would become infinite,

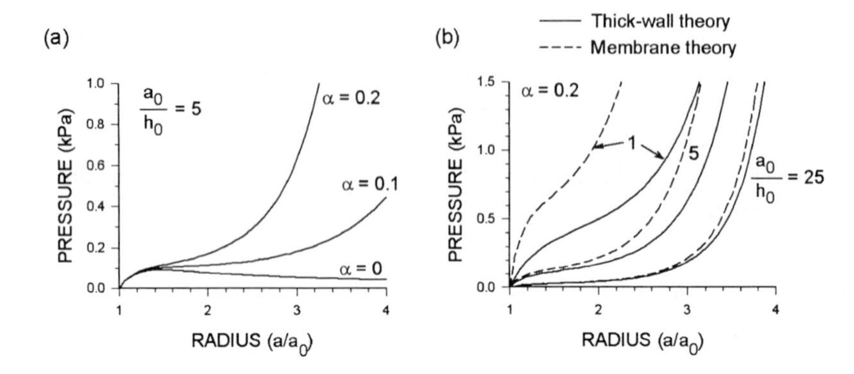

Fig. 6.22 Pressure-radius curves for blastula model. (a) Effects of material constant α. (b) Effects of ratio of undeformed inner radius to wall thickness. Results for a membrane approximation also are shown.

i.e., the blastula would burst, when the pressure is greater than about 0.1 kPa. In contrast, the curves for larger values of α remain stable, which should benefit the embryo. This type of analysis also has applications in studies of aneurysm rupture (Humphrey, 2002).

For $\alpha = 0.2$, Fig. 6.22b illustrates the effects of the relative wall thickness. Results are shown for the "exact" thick-walled theory (6.151) and for thin-walled membrane theory (6.152). As expected, the shell stiffens as the shell becomes thicker (a_0/h_0 decreases). Also as expected, the accuracy of membrane theory increases dramatically for thinner shells.

As in the artery without residual stress (Fig. 6.15a), circumferential stress decreases from the inner to the outer radius (Fig. 6.23). The results shown correspond to a deformation $a = 4a_0$.[5]

As the material becomes more nonlinear (increasing α), a strong stress concentration develops in the inner layers of the wall. In early work in cardiac mechanics, Mirsky (1973) modeled the left ventricle as a thick-walled isotropic spherical shell and found an extremely large stress concentration near the endocardium. As Fung (1991; 1997) later pointed out, however, these results do not make much sense physiologically, because the inner layers of the ventricular wall

[5]This exceptionally large deformation is used here only for illustration. It is not likely that blastulae actually deform this much due to internal pressure. During gastrulation, however, regional strains may exceed 100%.

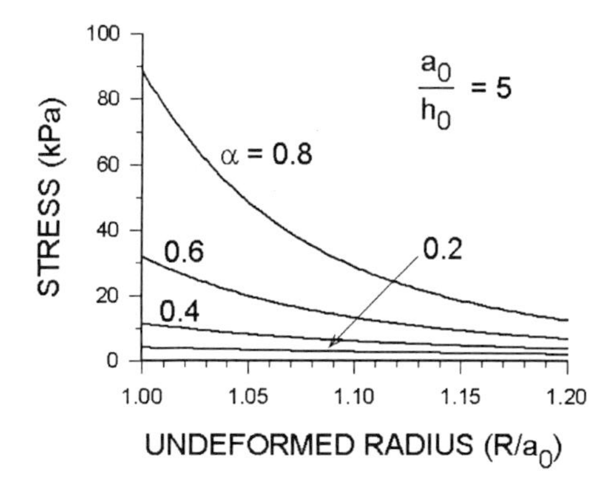

Fig. 6.23 Computed distribution of circumferential stress across blastula wall for $a = 0.4a_0$. A stress concentration develops as the material constant α increases.

would carry much more of the load than the outer layers. Such a design would make the heart an inefficient load-bearing structure. This observation led in part to Fung's emphasis on the importance of residual stress (Fung, 1991), which can reduce stress concentrations dramatically (Fig. 6.15a).

6.8 Bending of the Embryonic Heart

The heart is the first functioning organ in the embryo. During the early stages of development, a pair of membranes fold and merge near the midline of the embryo to create a single thick-walled cardiac tube composed of an outer layer of myocardium, a middle layer of extracellular matrix called cardiac jelly, and an inner layer of endocardium. Soon thereafter, the first contractions occur in the myocardium and the morphogenetic process of **cardiac looping** begins, as the heart tube bends and twists into a curved tube to lay out the basic plan of the future four-chambered pump (Taber, 2001).

Studies suggest that the forces that drive looping arise from sources that are both intrinsic and extrinsic to the heart. During the later stages of looping, for example, the body of the embryo likely plays a role in bending the heart tube by exerting loads on its ends. This section examines a relatively simple model for bending of the embryonic heart by end loads.

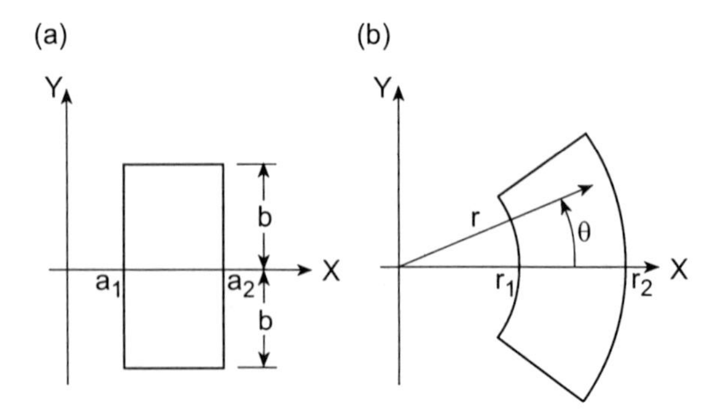

Fig. 6.24 Beam model for bending of the embryonic heart.

Bending a thick-walled tube is similar to bending a beam. As a first approximation, therefore, the heart is modeled as a rectangular "beam" that bends into a circular sector. Furthermore, ignoring the layered structure of the heart, we treat the beam as if it were composed entirely of homogeneous myocardium.

It is convenient here to introduce two coordinate systems (Fig. 6.24). The undeformed geometry is described in terms of Cartesian coordinates (X, Y, Z), while the deformed geometry is referred to cylindrical polar coordinates (r, θ, z). Initially, the beam is bounded by the planes $X = a_1, a_2$; $Y = \pm b$; and $Z = \pm c$ (Fig. 6.24a). After bending, the beam is assumed to be symmetric about the z-axis, with the planes $X = a_1, a_2$ becoming the circular cylindrical surfaces $r = r_1, r_2$ (Fig. 6.24b). In this simple "flattened" representation of the heart tube, the X- and Y-directions correspond to the circumferential and longitudinal directions, respectively, in the undeformed heart.

6.8.1 *Governing Equations*

Kinematics. The undeformed and deformed position vectors to a point in the beam are

$$\begin{aligned} \mathbf{R} &= X\mathbf{e}_X + Y\mathbf{e}_Y + Z\mathbf{e}_Z \\ \mathbf{r} &= r\mathbf{e}_r + z\mathbf{e}_z \end{aligned} \tag{6.155}$$

where the unit base vectors are related by

$$\begin{aligned}\mathbf{e}_r &= \mathbf{e}_X \cos\theta + \mathbf{e}_Y \sin\theta \\ \mathbf{e}_\theta &= -\mathbf{e}_X \sin\theta + \mathbf{e}_Y \cos\theta \\ \mathbf{e}_z &= \mathbf{e}_Z. \end{aligned} \tag{6.156}$$

For simplicity, this analysis does not allow for anticlastic curvature, i.e., the heterogeneous deformation that occurs in the z-direction during bending. Rather, tractions are applied on the sides of the beam to maintain a uniform stretch in the z-direction. Hence, deformation into a circular beam is defined by the transformation

$$r = r(X), \qquad \theta = kY, \qquad z = \lambda Z, \tag{6.157}$$

in which k and λ are constants. These relations give the deformation gradient tensor

$$\begin{aligned}\mathbf{F} &= (\nabla \mathbf{r})^T \\ &= \left[\left(\mathbf{e}_X \frac{\partial}{\partial X} + \mathbf{e}_Y \frac{\partial}{\partial Y} + \mathbf{e}_Z \frac{\partial}{\partial Z}\right) [r(\mathbf{e}_X \cos\theta + \mathbf{e}_Y \sin\theta) + z\mathbf{e}_Z]\right]^T \\ &= \lambda_X \mathbf{e}_r \mathbf{e}_X + \lambda_Y \mathbf{e}_\theta \mathbf{e}_Y + \lambda_Z \mathbf{e}_z \mathbf{e}_Z \end{aligned} \tag{6.158}$$

where

$$\lambda_X = \frac{\partial r}{\partial X}, \qquad \lambda_Y = kr, \qquad \lambda_Z = \lambda \tag{6.159}$$

are stretch ratios.

Constitutive Relations. We assume that the heart is composed of incompressible material with principal material directions that align with the (X, Y, Z) axes in the undeformed configuration. For pure bending, therefore, there should be no shear stress relative to the (r, θ, z) coordinates, and so the Cauchy stress tensor has the form

$$\boldsymbol{\sigma} = \hat{\sigma}^{rr} \mathbf{e}_r \mathbf{e}_r + \hat{\sigma}^{\theta\theta} \mathbf{e}_\theta \mathbf{e}_\theta + \hat{\sigma}^{zz} \mathbf{e}_z \mathbf{e}_z. \tag{6.160}$$

Expressions similar to those of Eqs. (6.144) provide the constitutive relations

$$\begin{aligned}\hat{\sigma}^{rr} &= \lambda_X \frac{\partial W}{\partial \lambda_X} - p \\ \hat{\sigma}^{\theta\theta} &= \lambda_Y \frac{\partial W}{\partial \lambda_Y} - p \\ \hat{\sigma}^{zz} &= \lambda_Z \frac{\partial W}{\partial \lambda_Z} - p.\end{aligned} \tag{6.161}$$

Equilibrium. Relative to the deformed polar coordinates, the only nontrivial equilibrium equation, given by Eq. $(6.108)_1$, is

$$\frac{\partial \hat{\sigma}^{rr}}{\partial r} + \frac{\hat{\sigma}^{rr} - \hat{\sigma}^{\theta\theta}}{r} = 0. \tag{6.162}$$

Boundary Conditions. In the present model, the embryo applies loads only to the ends of the heart tube (in addition to the tractions needed at $z = \pm\lambda c$). Hence, the curved surfaces are assumed to be traction free, with the appropriate boundary conditions given by

$$\hat{\sigma}^{rr}(r_1) = 0, \qquad \hat{\sigma}^{rr}(r_2) = 0. \tag{6.163}$$

Because of symmetry, we need only consider boundary conditions at one end of the beam. At $\theta = kb$ $(Y = b)$, Eq. (6.160) gives the true traction vector

$$\mathbf{T}^{(\mathbf{e}_\theta)} = \mathbf{e}_\theta \cdot \boldsymbol{\sigma} = \hat{\sigma}^{\theta\theta}\mathbf{e}_\theta. \tag{6.164}$$

Hence, the resultant force $\mathbf{N}$ and moment $\mathbf{M}$ (about the origin) applied to the ends are given by the integrals

$$\begin{aligned}\mathbf{N} &= \int_{-\lambda c}^{\lambda c} \int_{r_1}^{r_2} \mathbf{T}^{(\mathbf{e}_\theta)} \, dr \, dz \\ \mathbf{M} &= \int_{-\lambda c}^{\lambda c} \int_{r_1}^{r_2} \mathbf{r} \times \mathbf{T}^{(\mathbf{e}_\theta)} \, dr \, dz\end{aligned} \tag{6.165}$$

where $\mathbf{r}$ is the position vector relative to the origin. Note that we also could compute a force resultant due to $\hat{\sigma}^{zz}$ on the z-faces of the beam, but this is not of interest here. Substituting Eqs. (6.155) and (6.164) and integrating over z yields

$$\begin{aligned} \mathbf{N} &= 2\lambda c\,\mathbf{e}_\theta \int_{r_1}^{r_2} \hat{\sigma}^{\theta\theta}\,dr \equiv N\mathbf{e}_\theta \\ \mathbf{M} &= 2\lambda c\,\mathbf{e}_z \int_{r_1}^{r_2} r\,\hat{\sigma}^{\theta\theta}\,dr \equiv M\mathbf{e}_z \end{aligned} \tag{6.166}$$

with N and M being the normal end force and bending moment, respectively. Note, however, that global equilibrium for the entire deformed beam demands that $N = 0$. This observation provides a check on the accuracy of any numerical solution.

6.8.2 *Solution*

With Eqs. (6.159), the incompressibility condition is

$$\det \mathbf{F} = \lambda_X \lambda_Y \lambda_Z = k\lambda r \frac{\partial r}{\partial X} = 1,$$

which, upon integration, yields

$$r(X) = [(2/k\lambda)(X - A)]^{1/2} \tag{6.167}$$

where A is a constant. Applying the conditions $r(a_1) = r_1$ and $r(a_2) = r_2$ (see Fig. 6.24) gives

$$\begin{aligned} k\lambda &= \frac{2\,(a_2 - a_1)}{r_2^2 - r_1^2} \\ A &= \frac{a_1 r_2^2 - a_2 r_1^2}{r_2^2 - r_1^2}. \end{aligned} \tag{6.168}$$

Thus, Eqs. (6.159), (6.167), and (6.168) give the stretch ratios in terms of the variables r_1, r_2, and λ.

To solve the governing equations, we first substitute Eqs. (6.161) into (6.162) and integrate to get

$$p = \lambda_X \frac{\partial W}{\partial \lambda_X} + \int_r^{r_2} \left(\lambda_Y \frac{\partial W}{\partial \lambda_Y} - \lambda_X \frac{\partial W}{\partial \lambda_X} \right) \frac{dr}{r} \tag{6.169}$$

where the constant of integration has been chosen to satisfy the boundary condition $\hat{\sigma}^{rr}(r_2) = 0$. With this expression for p, Eq. $(6.161)_1$ and the boundary condition $\hat{\sigma}^{rr}(r_1) = 0$ yield

$$\int_{r_1}^{r_2} \left(\lambda_Y \frac{\partial W}{\partial \lambda_Y} - \lambda_X \frac{\partial W}{\partial \lambda_X} \right) \frac{dr}{r} = 0, \tag{6.170a}$$

which is to be solved for r_2 if r_1 and λ are specified. With the stretch ratios now known in terms of these variables, Eqs. (6.161) and (6.169) give the stress components, and Eqs. (6.166) provide N (which should be zero) and M.

6.8.3 *Geometric and Material Properties*

Little is currently known about the material properties of the embryonic heart during looping. It *is* known, however, that the heart tube contains circumferentially oriented actin filaments (Itasaki et al., 1989; Shiraishi et al., 1992), suggesting that the myocardium is transversely isotropic. Therefore, we assume that the strain-energy density function has the form given by Eq. (6.15), i.e.,

$$W = C_1(I_1 - 3) + C_4 I_4^2 \tag{6.171}$$

where Eqs. (6.12) give

$$\begin{aligned} I_1 &= \lambda_X^2 + \lambda_Y^2 + \lambda_Z^2 \\ I_4 &= E_{11} = \tfrac{1}{2}(\lambda_F^2 - 1), \end{aligned} \tag{6.172}$$

in which λ_F is the stretch ratio in the fiber direction. For illustration, we consider three cases: (1) isotropic ($C_4 = 0$); (2) transversely isotropic relative to the circumferential direction ($\lambda_F = \lambda_X$); and (3) transversely isotropic relative to the longitudinal direction ($\lambda_F = \lambda_Y$). The material constants, based primarily on estimates from pressure-volume curves for somewhat older embryonic chick hearts (Taber et al., 1992; Lin and Taber, 1994), are taken as $C_1 = 20$ Pa and $C_4 = 40$ Pa. In addition, we take $\lambda = 1$ and $a_2 - a_1 = 2c = 0.2$ mm. (The actual values of a_1 and a_2 do not matter here, and the circular cross section is represented by a square.)

6.8.4 *Results*

The bending moment M is plotted as a function of the average beam curvature

$$\kappa = \frac{2}{r_1 + r_2}$$

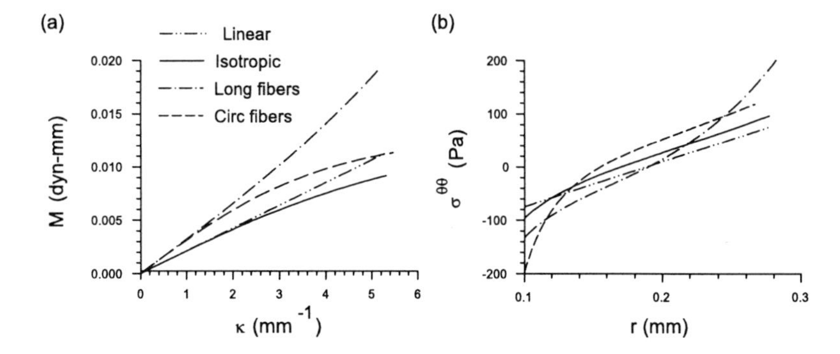

Fig. 6.25 Beam model for looping of the embryonic heart. (a) Bending moment versus curvature κ. The linear solution for an isotropic beam is shown for comparison. (b) Bending stress distributions.

in Fig. 6.25a. Also shown is the bending moment given by the linear theory for cylindrical bending of an isotropic *plate*.[6] For a plate, the total bending moment is (Szilard, 1974)

$$(M)_{\text{lin}} = \frac{E_Y I \kappa}{1 - \nu^2}.$$

In the present problem with $C_4 = 0$, Young's modulus is given by $E_Y = 6C_1$ (see Problem **5–4**), Poisson's ratio is $\nu = 0.5$ (incompressible material), and the cross-sectional area moment of inertia is $I = (a_2 - a_1)(2c)^3/12$. The linear theory also gives the bending stress (Szilard, 1974)

$$(\hat{\sigma}^{\theta\theta})_{\text{lin}} = \frac{E_Y \kappa (r - r_0)}{1 - \nu^2}.$$

For the isotropic beam, the linear and nonlinear solutions for M agree until $\kappa \gtrsim 1.5\ \text{mm}^{-1}$ (Fig. 6.25a). For larger bending, however, the magnitude of M drops below the linear value.

As expected, including fibers increases M, with longitudinal fibers having the greater effect. For this reason, Nakamura et al. (1980) speculated that the circumferential fiber alignment in the heart tube facilitates looping by offering less resistance to bending than fibers of other orientations.

[6] In cylindrical bending of a plate, the length of the plate in the z-direction constrains the deformation so that $\lambda_z \cong 1$, consistent with our beam analysis. Thus, a plate solution is more appropriate than a beam solution for comparison purposes.

Distributions of bending stress ($\hat{\sigma}^{\theta\theta}$) are shown for $\kappa \approx 5$ (Fig. 6.25b). The linear and nonlinear solutions for the isotropic beam are similar, but note the shift in the location of the neutral axis, as defined by the point where $\hat{\sigma}^{\theta\theta} = 0$. In contrast, the transversely isotropic beams contain relatively strong stress concentrations — at the inner curvature for circumferential fibers and at the outer curvature for longitudinal fibers.

The possible implications of these results for cardiac looping are not clear. Stress concentrations certainly can affect growth and remodeling (Taber, 1995), but the relative simplicity of this model must be kept in mind when interpreting these results. Nevertheless, this problem illustrates how models can suggest avenues for further experimental investigation.

6.9 Problems

6–1 Consider the governing equations for *linear* elastic deformation of an isotropic solid:

$$\begin{aligned} \boldsymbol{\nabla} \cdot \boldsymbol{\sigma} + \mathbf{f} &= \rho \ddot{\mathbf{u}} \\ \mathbf{E} &= \tfrac{1}{2}\left[\boldsymbol{\nabla}\mathbf{u} + (\boldsymbol{\nabla}\mathbf{u})^T\right] \\ \boldsymbol{\sigma} &= \lambda\Delta\mathbf{I} + 2\mu\mathbf{E}. \end{aligned}$$

Here, $\mathbf{u}$ is the displacement vector, $\Delta = \operatorname{tr}\mathbf{E}$, and dot denotes time differentiation. Derive Navier's displacement equation of motion

$$(\lambda + \mu)\boldsymbol{\nabla}(\boldsymbol{\nabla} \cdot \mathbf{u}) + \mu\boldsymbol{\nabla}^2\mathbf{u} + \mathbf{f} = \rho\ddot{\mathbf{u}}.$$

6–2 Consider uniaxial extension and equibiaxial extension of rectangular pieces of the following isotropic, incompressible materials:

(a) Rubber: $W = C[(I_1 - 3) + \gamma(I_2 - 3)]$
(b) Sclera: $W = (C/\gamma)\, e^{\gamma(I_1 - 3)}$

For both types of loading, plot t_{11}/C and σ_{11}/C vs. the stretch ratio λ for $0.6 \leq \lambda \leq 1.6$. Use $\gamma = 0$ and 0.2 for rubber and $\gamma = 1$ and 2 for sclera (the white of the eye).

6–3 A rectangular bar has the undeformed dimensions $20 \times 5 \times 4$ mm. It is composed of an isotropic compressible material with the strain-energy density function

$$W = C(I_1 + I_3^{-1/2} - 4)$$

where $C = 20$ kPa is a material constant. An axial load F stretches the long dimension of the bar to the length 28 mm.

(a) Determine the stretch ratios λ_1, λ_2, and λ_3.
(b) Determine the Cauchy stress σ_{11} in the loaded bar.
(c) Determine F.

6–4 Consider biaxial stretch of an incompressible, rectangular elastic membrane in the Cartesian X_1X_2-plane with

$$W = \sum_{n=1}^{N} a_n(\lambda_1^{b_n} + \lambda_2^{b_n} + \lambda_3^{b_n} - 3).$$

Compute the Cauchy stress components σ_{11} and σ_{22} in terms of the stretch ratios λ_1 and λ_2.

6–5 A rectangular block of isotropic, incompressible tissue has undeformed dimensions a_0, b_0, and c_0 in the x_1, x_2, and x_3 directions, respectively. The strain-energy density function per unit undeformed volume is

$$W = C_1(I_1 - 3)^2 + C_2(I_2 - 3)^3$$

where C_1 and C_2 are material constants. With the width of the specimen held fixed in the x_2-direction, a testing machine stretches the tissue to a new length a in the x_1-direction. In terms of known quantities, compute the following:

(a) The deformed thickness (in the x_3-direction).
(b) The principal Lagrange strains.
(c) The load P applied by the testing machine in the x_1 and x_2 directions.
(d) The true (Cauchy) principal stresses far from the edges of the specimen.

6–6 For the problem of simple shear (section 6.3), compute the stress components $\sigma^{I^*}_{\cdot J^*}$ and $\sigma^{\cdot I^*}_{J^*}$. Sketch these stresses on the deformed block.

6–7 Consider simple shear of a block composed of transversely isotropic Blatz-Ko type material defined by Eq. (6.53).

(a) Show that the plane stress condition $\sigma^{3^*3^*}$ is satisfied identically.
(b) Show that the convected Cauchy stress components are given by Eqs. (6.50).

6–8 In Cartesian coordinates, combined extension and simple shear of a block is described by the mapping

$$\begin{aligned} x_1 &= \lambda_1 X_1 + k\lambda_2 X_2 \\ x_2 &= \lambda_2 X_2 \\ x_3 &= \lambda_3 X_3 \end{aligned}$$

where the λ_i and k are constants. If the block is composed of compressible, isotropic material with a general form for W, determine the surface tractions required to maintain the block in equilibrium.

6–9 The deformation of a compressible unit cube is described by the relations

$$\begin{aligned} x_1 &= \lambda X_1 \\ x_2 &= aX_1^2 + bX_2 \\ x_3 &= X_3 \end{aligned}$$

where the X_i and x_i are undeformed and deformed Cartesian coordinates of a point, respectively. The constants a, b, and λ are all positive with $b > a$.

(a) Sketch the deformed shape of the block in the x_1x_2-plane.
(b) Determine the ratio of deformed to undeformed length of a line segment that is located at the center of the cube before deformation and is parallel to the x_1-axis *after* deformation.
(c) The block is composed of an elastic material with a strain-energy density function of the form

$$W = C_1(F_{11}^2 + F_{22}^2 + F_{33}^2 - 3) + C_2(F_{12}^2 + F_{21}^2)$$

where the F_{ij} are Cartesian components of the deformation gradient tensor. Compute the distributions of the first Piola-Kirchhoff stresses t_{12} and t_{21} in the deformed cube as functions of the X_i.

6–10 For extension and torsion of a papillary muscle, derive Eqs. (6.84) and (6.87).

6–11 Using a force balance on an artery in an inflation test system, derive Eq. $(6.110)_1$.

6–12 For extension, inflation, and torsion of an artery, derive Eqs. (6.106).

6–13 For the model of the left ventricle, show that Eqs. (6.129) and (6.131) reduce to Eqs. (6.106) for $\beta = 0$.

6–14 Consider inflation of a compressible spherical shell composed of Blatz-Ko material with W given by Eq. (6.85). Show that the deformed radial coordinate $r(R)$ is governed by the nonlinear ordinary differential equation (Chung et al., 1986)

$$3R^2r^3\frac{d^2r}{dR^2} - 2Rr^3\frac{dr}{dR} + 2R^4\left(\frac{dr}{dR}\right)^4 = 0.$$

6–15 In the undeformed state, a spherical membrane has a radius a_0 and wall thickness h_0. It is composed of an isotropic, incompressible, elastic material with a strain-energy density function W.

(a) When the membrane is subjected to an internal pressure P (per unit deformed area), show that

$$P = \frac{4h_0}{\lambda a_0}\left(1 - \frac{1}{\lambda^6}\right)\left(\frac{\partial W}{\partial I_1} + \lambda^2\frac{\partial W}{\partial I_2}\right)$$

where $\lambda = \lambda_\Theta = \lambda_\Phi$ by symmetry and I_1 and I_2 are strain invariants. *Hint:* Assume that $\sigma^{rr} \cong 0$ and use Laplace's law (6.153).

(b) For $W = C[I_1 - 3 + \alpha(I_2 - 3)]$ with $\alpha = 0, 0.1, 0.2$, plot Pa_0/Ch_0 and $\hat{\sigma}^{\theta\theta}/C$ as functions of λ for $1 \leq \lambda \leq 5$.

6–16 A thin-walled spherical membrane of undeformed radius a_0 and wall thickness h_0 is loaded by an internal pressure P. The membrane is composed of Blatz-Ko material with W given by Eq. (6.85). Using the approximations $\hat{\sigma}^{rr} = 0$ and Laplace's law (6.153), write P as a function of the stretch ratio $\lambda = \lambda_\Theta = \lambda_\Phi$. Plot Pa_0/C_2h_0 vs. λ for $1 \leq \lambda \leq 5$.

6–17 Unlike linear problems, nonlinear elasticity problems can possess multiple equilibrium solutions for a given loading. (Of course, some of these solutions may be unstable.) Consider an unloaded hemispherical shell of inner radius a_0 and outer radius b_0 that is composed of isotropic, incompressible material. Obviously, the undeformed state is an equilibrium configuration for no applied loads. Another is the inverted state, i.e., the shell is turned inside out. In spherical coordinates, consider the inverted configuration described by the mapping (Fig. 6.26)

$$\begin{aligned} r &= r(R) \\ \theta &= \pi - \Theta \\ \phi &= \Phi. \end{aligned}$$

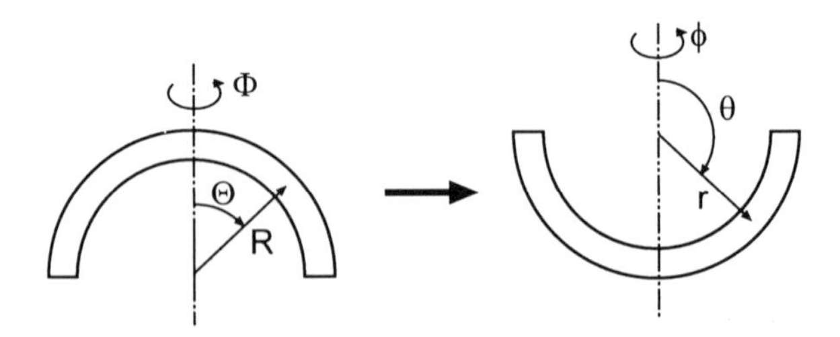

Fig. 6.26 Inversion of a hemispherical shell.

(a) Show that the normal components of **F** are given by $\lambda_R = dr/dR$, $\lambda_\Theta = -r/R$, and $\lambda_\Phi = r/R$.
(b) With W given by Eq. (6.154), set up and solve the boundary value problem numerically and plot the physical Cauchy stress components across the wall. Take $C = 2$ kPa, $\alpha = 0.2$, $a_0 = 6$ mm, and $b_0 = 10$ mm. Note that if the undeformed shell radii satisfy the relation $R_2 > R_1$, then $r_2 < r_1$ in the inverted shell.
(c) What tractions must be applied to the inverted shell to maintain a hemispherical shape? If the edge loads are removed, the shell remains inverted, but the shape changes locally near the edge. Describe the deformed shape in qualitative terms.

6–18 Consider cylindrical bending of a compressible block composed of Blatz-Ko material with W given by Eq. (6.85).

(a) With the stretch ratios given by Eqs. (6.159), show that the physical Cauchy stress components can be written in the form (Carroll and Horgan, 1990)

$$\begin{aligned}
\hat{\sigma}^{rr} &= 2C_2\left[1 - \frac{1}{\lambda k r}\left(\frac{dX}{dr}\right)^3\right] \\
\hat{\sigma}^{\theta\theta} &= 2C_2\left(1 - \frac{1}{\lambda k^3 r^3}\frac{dX}{dr}\right) \\
\hat{\sigma}^{zz} &= 2C_2\left(1 - \frac{1}{\lambda^3 k r}\frac{dX}{dr}\right).
\end{aligned}$$

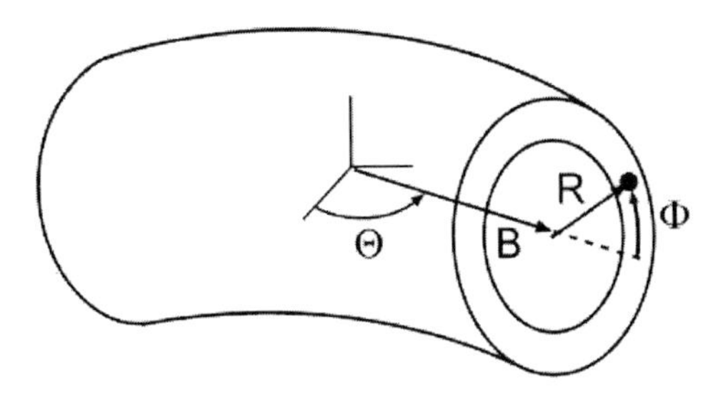

Fig. 6.27 Section of a toroidal tube in undeformed state.

(b) Show that equilibrium leads to the equation

$$\frac{dX}{dr}\frac{d^2X}{dr^2} = \frac{1}{3k^2r^3}.$$

(c) Integrate this equation to get

$$X(r) = \frac{1}{k\sqrt{3}}\left(\sqrt{\alpha r^2 - 1} - \sec^{-1} r\sqrt{\alpha} + \beta\right)$$

where α and β are constants of integration.

(d) Discuss how the boundary conditions can be used to determine the unknown parameters.

6–19 The aortic arch and the early embryonic heart are examples of pressurized curved tubes. Consider the related problem of a complete thick-walled, incompressible, isotropic toroidal shell with an internal pressure P (Fig. 6.27), e.g., an inner tube. In toroidal coordinates (see Appendix B), assume that the deformation geometry is described by the relations

$$\begin{aligned} r &= r(R, \Phi) \\ \theta &= \Theta \\ \phi &= \phi(R, \Phi). \end{aligned}$$

(a) Determine the toroidal tensor components of the Lagrangian strain tensor.

(b) Derive the equilibrium equations in terms of the convected Cauchy stress components $\sigma^{I^*J^*}$. (This is a lot of work.)

(c) Complete the formulation of this problem by listing appropriate constitutive relations and boundary conditions.

6–20 Write a computer program to solve the left ventricle problem discussed in section 6.6.

Appendix A

Linear Theory of Elasticity

The linear theory of elasticity deals with problems in which deformations, displacements, and rotations are "small." In such problems, the undeformed and deformed configurations are virtually identical, the governing equations assume relatively simple forms, and the principle of superposition applies. In biomechanics, the linear theory usually is accurate only in the analysis of bone and other hard tissue. However, this theory sometimes may be useful in studying qualitative trends in problems involving large deformation of soft tissue, if strains are no more than about 20%.

To fully appreciate the complexities of nonlinear elasticity theory, it is useful to be familiar with the linear theory. This appendix is intended as a review and an aid to those readers who have only a modest familiarity with linear elasticity theory. (*Some* familiarity is assumed, however.) Because some readers may find it helpful to read this material before delving into the more involved presentation in the main text, the following development is relatively self-contained. The order of presentation also is similar, and it may be instructive to compare the development of the nonlinear theory with that in this appendix.

An important restriction in this appendix is the almost exclusive use of rectangular Cartesian coordinates. This allows us to focus on fundamental concepts without getting bogged down in the added complexity of curvilinear coordinates.

A.1 Mathematical Preliminaries

Our presentation begins with a brief discussion of the mathematical background required in this appendix. First, we introduce a shorthand notation that is used throughout the book. According to the **summation convention**, when an index

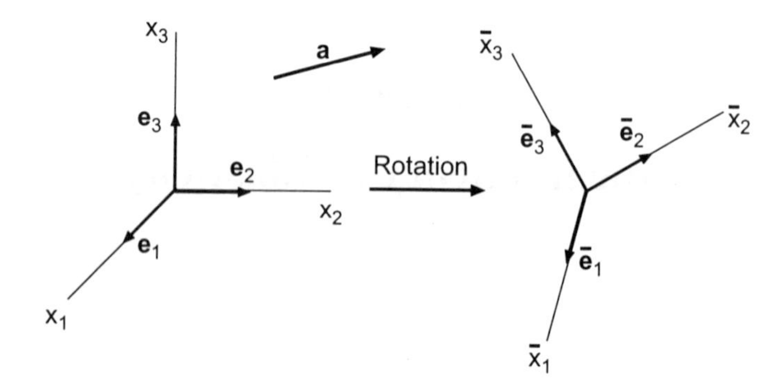

Fig. A.1 Cartesian coordinate systems and base vectors.

is repeated in a term of an expression, summation over 1, 2, 3 for that index is automatically implied (unless stated otherwise or the subscripts are placed in parentheses).[7] Thus, we write

$$\sum_{i=1}^{3} a_i b_i = a_i b_i$$

where the repeated index i is called a "dummy" index, since it can be replaced by any letter without altering the meaning, i.e., $a_i b_i = a_j b_j$. Summation is *not* implied, however, for subscripts surrounded by parentheses, e.g., $a_{(i)} b_{(i)}$. Furthermore, an index that is not repeated implies a set of equations. For example,

$$a_i b_{ij} = \sum_{i=1}^{3} a_i b_{ij}$$

gives a separate equation for each $j = 1, 2, 3$. Note also that the indices in an equation must "balance" in that all terms must contain the same number of non-dummy indices. Thus, $a_i + b_{ij} c_j = 0$ is a valid equation, but $a_{ik} + b_{ij} c_j = 0$ is not since the first term contains an extra k. Moreover, three or more of the same index in a single term is not allowed (unless some of the indices are placed in parentheses). The following development assumes an elementary knowledge of vector analysis.

[7]The summation convention introduced here is modified slightly from that used in Chapter 2. The difference is that here, because of the focus on Cartesian coordinates, repetition of subscripts in a term is allowed.

A.1.1 *Base Vectors*

Consider a set of Cartesian coordinate axes (x_1, x_2, x_3) fixed in Euclidean space, and let $(\mathbf{e}_1, \mathbf{e}_2, \mathbf{e}_3)$ represent unit vectors directed along these axes (Fig. A.1). Since any vector can be expressed in terms of these unit vectors, the $\mathbf{e}_i$ are called **base vectors**. Being mutually orthogonal, they satisfy the conditions

$$\boxed{\begin{aligned} \mathbf{e}_i \cdot \mathbf{e}_j &= \delta_{ij} \\ \mathbf{e}_i \times \mathbf{e}_j &= \epsilon_{ijk}\, \mathbf{e}_k \end{aligned}} \qquad \text{(A.1)}$$

where

$$\boxed{\delta_{ij} = \begin{cases} 1 \text{ for } i = j \\ 0 \text{ for } i \neq j \end{cases}} \qquad \text{(A.2)}$$

is the **Kronecker delta** and

$$\boxed{\epsilon_{ijk} = \begin{cases} +1 & \text{if } (i, j, k) \text{ is an even permutation} \\ -1 & \text{if } (i, j, k) \text{ is an odd permutation} \\ 0 & \text{if two or more indices are equal} \end{cases}} \qquad \text{(A.3)}$$

is the **permutation symbol**. For example, Eqs. $(\text{A.1})_2$ and (A.3) give $\mathbf{e}_2 \times \mathbf{e}_1 = \epsilon_{21k}\mathbf{e}_k = \epsilon_{211}\mathbf{e}_1 + \epsilon_{212}\mathbf{e}_2 + \epsilon_{213}\mathbf{e}_3 = -\mathbf{e}_3$.

The permutation symbol can be expressed directly in terms of the $\mathbf{e}_i$ by dotting both sides of Eq. $(\text{A.1})_2$ by $\mathbf{e}_l$ to get

$$\mathbf{e}_l \cdot (\mathbf{e}_i \times \mathbf{e}_j) = \mathbf{e}_l \cdot (\epsilon_{ijk}\, \mathbf{e}_k) = \epsilon_{ijk}\, \mathbf{e}_l \cdot \mathbf{e}_k = \epsilon_{ijk}\delta_{lk}$$

where Eqs. (A.1) have been used. For a given l, the only term on the right-hand-side that survives the summation over k is the term for $k = l$, i.e.,

$$\epsilon_{ijk}\delta_{lk} = \epsilon_{ij1}\delta_{l1} + \epsilon_{ij2}\delta_{l2} + \epsilon_{ij3}\delta_{l3} = \epsilon_{ijl}.$$

In other words, we can replace i by j or j by i in terms containing δ_{ij} and remove the Kronecker delta. This operation, which can be used generally to simplify expressions, is called **contraction**. Thus, after renaming the subscripts in the above expressions, we obtain

$$\boxed{\epsilon_{ijk} = \mathbf{e}_i \cdot (\mathbf{e}_j \times \mathbf{e}_k).} \tag{A.4}$$

The Kronecker delta and the permutation symbol are related through the **ϵ-δ identity**

$$\boxed{\epsilon_{ijk}\epsilon_{irs} = \delta_{jr}\delta_{ks} - \delta_{js}\delta_{kr}.} \tag{A.5}$$

A.1.2 *Vectors, Dyadics, and Tensors*

A vector $\mathbf{a}$ has a magnitude and direction. Relative to the coordinate system x_i, we can write

$$\mathbf{a} = a_i \mathbf{e}_i \tag{A.6}$$

where the a_i are the components of $\mathbf{a}$ in the $\mathbf{e}_i$ directions. It is important to note that if the coordinate system changes (see the $\bar{x}_i$ system in Fig. A.1), the components of $\mathbf{a}$ change, but the vector $\mathbf{a}$ does not. A vector, therefore, is geometrically invariant, and the form of a vector equation is independent of the coordinate system.

There are three types of vector product. In terms of components, the **dot product** of two vectors $\mathbf{a} = a_i \mathbf{e}_i$ and $\mathbf{b} = b_i \mathbf{e}_i$ is

$$\mathbf{a} \cdot \mathbf{b} = (a_i \mathbf{e}_i) \cdot (b_j \mathbf{e}_j) = a_i b_j (\mathbf{e}_i \cdot \mathbf{e}_j) = a_i b_j \delta_{ij}$$

or

$$\boxed{\mathbf{a} \cdot \mathbf{b} = a_i b_i.} \tag{A.7}$$

Note that the dummy indices in $\mathbf{b}$ have been changed from i to j, since more than two of any subscript in a term is meaningless. The **cross product** is

$$\mathbf{a} \times \mathbf{b} = (a_i \mathbf{e}_i) \times (b_j \mathbf{e}_j) = a_i b_j (\mathbf{e}_i \times \mathbf{e}_j)$$

or

$$\mathbf{a} \times \mathbf{b} = a_i b_j \epsilon_{ijk} \, \mathbf{e}_k \tag{A.8}$$

by Eq. $(A.1)_2$. Finally, the **tensor product** is defined by

$$\mathbf{ab} = (a_i \mathbf{e}_i)(b_j \mathbf{e}_j)$$

or

$$\mathbf{ab} = a_i b_j \mathbf{e}_i \mathbf{e}_j. \tag{A.9}$$

The tensor product of two vectors is called a **dyad**, and a linear combination of dyads, e.g., the right-hand-side of Eq. (A.9), is called a **dyadic**. A useful operation involving dyads is the **double-dot (scalar) product**, which is defined by

$$\mathbf{ab}{:}\mathbf{cd} = (\mathbf{a}\cdot\mathbf{c})(\mathbf{b}\cdot\mathbf{d}). \tag{A.10}$$

The left-hand-sides of Eqs. (A.7)–(A.9) are written in **direct notation**, valid for *any* coordinate system. The right-hand-sides are written in **indicial notation**, or component form, here being specific to Cartesian coordinates. Direct notation promotes physical understanding, but indicial notation often makes manipulations easier. Throughout this book, therefore, we emphasize the importance of being able to move back and forth between the two approaches.

At this point, we note several things. First, the vector dot, cross, and tensor products yield a scalar, a vector, and a dyad, respectively. Second, the dot product is commutative, but the cross and tensor products are not. Thus, $\mathbf{a}\cdot\mathbf{b} = \mathbf{b}\cdot\mathbf{a}$, while $\mathbf{a} \times \mathbf{b} = -\mathbf{b} \times \mathbf{a}$, and all we can say about the tensor product is that $\mathbf{ab} \neq \mathbf{ba}$ in general. Third, just as the a_i are the components of the vector $\mathbf{a}$ with respect to the basis $\{\mathbf{e}_i\}$, $a_i b_j$ are the components of the dyad $\mathbf{ab}$ with respect to the basis $\{\mathbf{e}_i \mathbf{e}_j\}$ [see Eq. (A.9)]. Since a dyad is composed of vectors, the components of $\mathbf{ab}$ change if the basis changes, but the dyad itself does not. Finally, *as long as care is taken to preserve the order of the vectors*, operations with dyadics are straightforward. For example, $(\mathbf{ab})\cdot\mathbf{c} = \mathbf{a}(\mathbf{b}\cdot\mathbf{c}) = (\mathbf{b}\cdot\mathbf{c})\mathbf{a}$ because $\mathbf{b}\cdot\mathbf{c}$ is a scalar.

Next, we introduce the concept of a **tensor**. A second-order tensor $\mathbf{T}$ is defined as a linear vector function that, when dotted with a vector, transforms the vector into another vector, i.e., $\mathbf{T}\cdot\mathbf{a} = \mathbf{b}$ transforms $\mathbf{a}$ into $\mathbf{b}$.[8] A dyad satisfies this requirement; since $(\mathbf{ab})\cdot\mathbf{c} = (\mathbf{b}\cdot\mathbf{c})\mathbf{a}$, the dyad $\mathbf{ab}$ transforms the vector $\mathbf{c}$ into the vector $(\mathbf{b}\cdot\mathbf{c})\mathbf{a}$. Thus, a dyad is a tensor. In fact, any second-order tensor can be expressed as a dyadic, i.e.,

$$\boxed{\mathbf{T} = T_{ij}\mathbf{e}_i\mathbf{e}_j} \tag{A.11}$$

where the T_{ij} are the components of $\mathbf{T}$ with respect to the basis $\{\mathbf{e}_i\mathbf{e}_j\}$. Higher order tensors can be written analogously. Like a vector or dyad, a tensor is geometrically invariant; when the basis changes, the components of $\mathbf{T}$ change, but $\mathbf{T}$ itself does not. (A vector is a tensor of the first order.)

The double-dot product of two tensors yields a scalar. Consider, for example, the scalar product of $\mathbf{A} = A_{ij}\mathbf{e}_i\mathbf{e}_j$ and $\mathbf{B} = B_{ij}\mathbf{e}_i\mathbf{e}_j$. By Eq. (A.10), we have

$$\begin{aligned} \mathbf{A}:\mathbf{B} &= (A_{ij}\mathbf{e}_i\mathbf{e}_j):(B_{kl}\mathbf{e}_k\mathbf{e}_l) \\ &= A_{ij}B_{kl}(\mathbf{e}_i\cdot\mathbf{e}_k)(\mathbf{e}_j\cdot\mathbf{e}_l) \\ &= A_{ij}B_{kl}\delta_{ik}\delta_{jl} \end{aligned}$$

and contraction yields

$$\boxed{\mathbf{A}:\mathbf{B} = A_{ij}B_{ij}.} \tag{A.12}$$

For computational purposes, it often is convenient to write a second-order tensor in the form of a matrix, i.e.,

$$\mathbf{T} = [T_{ij}] = \begin{bmatrix} T_{11} & T_{12} & T_{13} \\ T_{21} & T_{22} & T_{23} \\ T_{31} & T_{32} & T_{33} \end{bmatrix} \tag{A.13}$$

where the position (i, j) in the matrix corresponds to the base dyad $\mathbf{e}_i\mathbf{e}_j$. From this representation, it follows that the transpose of a tensor is obtained by switching the order of the base vectors to get

$$\mathbf{T}^T = T_{ij}\mathbf{e}_j\mathbf{e}_i = T_{ji}\mathbf{e}_i\mathbf{e}_j. \tag{A.14}$$

[8] In this section, lower-case bold letters denote first-order tensors (vectors), and upper-case bold letters denote second-order tensors (dyadics).

Moreover, we can show that

$$\mathbf{T}\cdot\mathbf{a} = \mathbf{a}\cdot\mathbf{T}^T \tag{A.15}$$

by the manipulations

$$\begin{aligned} (T_{ij}\mathbf{e}_i\mathbf{e}_j)\cdot(a_k\mathbf{e}_k) &= (a_k\mathbf{e}_k)\cdot(T_{ij}\mathbf{e}_j\mathbf{e}_i) \\ T_{ij}a_k\,\mathbf{e}_i\,\delta_{jk} &= a_kT_{ij}\delta_{kj}\mathbf{e}_i \\ T_{ij}a_j\,\mathbf{e}_i &= T_{ij}a_j\mathbf{e}_i. \end{aligned}$$

A tensor $\mathbf{T}$ is symmetric if $\mathbf{T} = \mathbf{T}^T$, i.e., $T_{ij} = T_{ji}$.

Finally, we note that the **identity tensor I** satisfies the relations $\mathbf{a}\cdot\mathbf{I} = \mathbf{I}\cdot\mathbf{a} = \mathbf{a}$ and $\mathbf{T}\cdot\mathbf{I} = \mathbf{I}\cdot\mathbf{T} = \mathbf{T}$. In dyadic form, the representation

$$\boxed{\mathbf{I} = \delta_{ij}\mathbf{e}_i\mathbf{e}_j = \mathbf{e}_i\mathbf{e}_i} \tag{A.16}$$

can be verified, for example, by the manipulations

$$\begin{aligned} \mathbf{a}\cdot\mathbf{I} &= (a_i\mathbf{e}_i)\cdot(\mathbf{e}_j\mathbf{e}_j) = a_i(\mathbf{e}_i\cdot\mathbf{e}_j)\mathbf{e}_j \\ &= a_i\delta_{ij}\mathbf{e}_j = a_i\mathbf{e}_i = \mathbf{a}. \end{aligned}$$

Eq. (A.16) shows that the Kronecker delta δ_{ij} represents the component of $\mathbf{I}$ with respect to the Cartesian dyad $\mathbf{e}_i\mathbf{e}_j$.

A.1.3 *Coordinate Transformation*

As mentioned earlier, the components of vectors and tensors change when the base vectors, i.e., the coordinates, change. For Cartesian coordinates, rotating the coordinate axes produces a new Cartesian coordinate system.

Consider translation and rotation of the coordinate system x_i with base vectors $\mathbf{e}_i$ into the system $\bar{x}_i$ with base vectors $\bar{\mathbf{e}}_i$ (Figs. A.1 and A.2). Any vector can be written in terms of components relative to either basis. Writing the vector $\bar{\mathbf{e}}_i$ in terms of the $\mathbf{e}_i$, for example, yields

$$\bar{\mathbf{e}}_i = A_{ij}\mathbf{e}_j \tag{A.17}$$

where the **transformation components** A_{ij} are the components of $\bar{\mathbf{e}}_i$ relative to the basis $\{\mathbf{e}_i\}$. Dotting this equation with $\mathbf{e}_k$ gives

$$\bar{\mathbf{e}}_i\cdot\mathbf{e}_k = A_{ij}\mathbf{e}_j\cdot\mathbf{e}_k = A_{ij}\delta_{jk} = A_{ik}$$

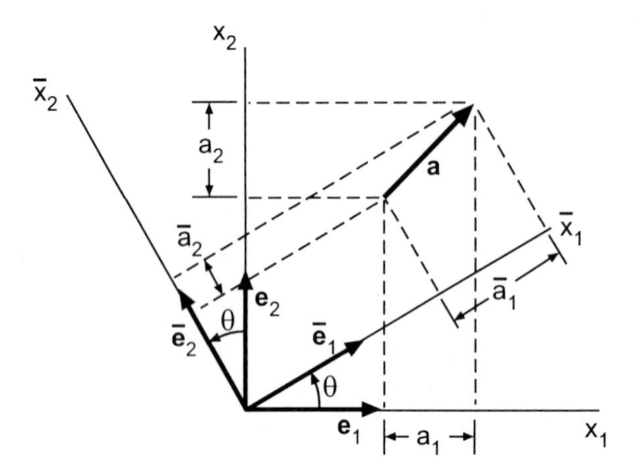

Fig. A.2 Components of a vector $\mathbf{a}$ relative to two Cartesian coordinate systems, x_i and $\bar{x}_i$, with base vectors $\mathbf{e}_i$ and $\bar{\mathbf{e}}_i$.

or

$$\boxed{A_{ij} = \bar{\mathbf{e}}_i \cdot \mathbf{e}_j,} \tag{A.18}$$

which indicates that the A_{ij} represent the direction cosines between the two sets of coordinate axes. For the geometry of Fig. A.2,

$$\begin{aligned} \bar{\mathbf{e}}_1 &= \mathbf{e}_1 \cos\theta + \mathbf{e}_2 \sin\theta \\ \bar{\mathbf{e}}_2 &= -\mathbf{e}_1 \sin\theta + \mathbf{e}_2 \cos\theta \\ \bar{\mathbf{e}}_3 &= \mathbf{e}_3, \end{aligned} \tag{A.19}$$

and so

$$[A_{ij}] = [\bar{\mathbf{e}}_i \cdot \mathbf{e}_j] = \begin{bmatrix} \cos\theta & \sin\theta & 0 \\ -\sin\theta & \cos\theta & 0 \\ 0 & 0 & 1 \end{bmatrix}. \tag{A.20}$$

With these relations, transforming vector and tensor components is straightforward. First, however, we note a convenient method for extracting the components of vectors and tensors. Dotting Eq. (A.6) with $\mathbf{e}_j$ gives

$$\mathbf{a} \cdot \mathbf{e}_j = (a_i \mathbf{e}_i) \cdot \mathbf{e}_j = a_i \delta_{ij} = a_j,$$

while double-dotting Eq. (A.11) with $\mathbf{e}_k\mathbf{e}_l$ yields

$$\begin{aligned}\mathbf{T}:(\mathbf{e}_k\mathbf{e}_l) &= (T_{ij}\mathbf{e}_i\mathbf{e}_j):(\mathbf{e}_k\mathbf{e}_l) \\ &= T_{ij}(\mathbf{e}_i\cdot\mathbf{e}_k)(\mathbf{e}_j\cdot\mathbf{e}_l) \\ &= T_{ij}\delta_{ik}\delta_{jl} = T_{kl}.\end{aligned}$$

where Eq. (A.10) has been used. Thus, we have

$$\boxed{\begin{aligned}a_i &= \mathbf{a}\cdot\mathbf{e}_i \\ T_{ij} &= \mathbf{T}:\mathbf{e}_i\mathbf{e}_j = \mathbf{e}_i\cdot\mathbf{T}\cdot\mathbf{e}_j.\end{aligned}} \tag{A.21}$$

In other words, dotting the vector $\mathbf{a}$ with $\mathbf{e}_i$ gives the component of $\mathbf{a}$ along $\mathbf{e}_i$, and double-dotting the dyadic $\mathbf{T}$ with $\mathbf{e}_i\mathbf{e}_j$ gives the component of $\mathbf{T}$ along $\mathbf{e}_i\mathbf{e}_j$.[9] The components with respect to any other basis can be extracted similarly.

To relate the components between two bases, we use the invariance properties and write (Figs. A.1 and A.2)

$$\begin{aligned}\mathbf{a} &= a_i\mathbf{e}_i = \bar{a}_i\bar{\mathbf{e}}_i \\ \mathbf{T} &= T_{ij}\mathbf{e}_i\mathbf{e}_j = \bar{T}_{ij}\bar{\mathbf{e}}_i\bar{\mathbf{e}}_j.\end{aligned} \tag{A.22}$$

Then, the components of $\mathbf{a}$ and $\mathbf{T}$ relative to the $\bar{x}_i$ system are

$$\begin{aligned}\bar{a}_i &= \mathbf{a}\cdot\bar{\mathbf{e}}_i = (a_j\mathbf{e}_j)\cdot\bar{\mathbf{e}}_i = a_j(\mathbf{e}_j\cdot\bar{\mathbf{e}}_i) \\ \bar{T}_{ij} &= \mathbf{T}:\bar{\mathbf{e}}_i\bar{\mathbf{e}}_j = (T_{kl}\mathbf{e}_k\mathbf{e}_l):(\bar{\mathbf{e}}_i\bar{\mathbf{e}}_j) = T_{kl}(\mathbf{e}_k\cdot\bar{\mathbf{e}}_i)(\mathbf{e}_l\cdot\bar{\mathbf{e}}_j),\end{aligned} \tag{A.23}$$

and using Eq. (A.18) gives

$$\boxed{\begin{aligned}\bar{a}_i &= A_{ij}a_j \\ \bar{T}_{ij} &= A_{ik}A_{jl}T_{kl}.\end{aligned}} \tag{A.24}$$

According to these relations, an A_{ij} is needed to transform each subscript to the new basis.

[9] In Cartesian coordinates, the components of a vector are the orthogonal projections of the vector on the coordinate axes (Fig. A.2). A physical interpretation for the components of a tensor cannot be visualized as easily.

A.1.4 *Vector and Tensor Calculus*

In Cartesian coordinates, the **gradient ("del") operator** $\boldsymbol{\nabla}$ is defined as

$$\boxed{\boldsymbol{\nabla} \equiv \mathbf{e}_i \frac{\partial}{\partial x_i}.} \tag{A.25}$$

Then, the **gradient** of a scalar function $\phi(x_i)$ is

$$\boldsymbol{\nabla}\phi = \left(\mathbf{e}_i \frac{\partial}{\partial x_i}\right) \phi$$

or

$$\boxed{\boldsymbol{\nabla}\phi = \mathbf{e}_i \phi_{,i}} \tag{A.26}$$

where $\phi_{,i} \equiv \partial\phi/\partial x_i$. Moreover, the gradient of a vector function $\mathbf{a}(x_i)$ is

$$\boldsymbol{\nabla}\mathbf{a} = \left(\mathbf{e}_i \frac{\partial}{\partial x_i}\right) (a_j \mathbf{e}_j) = \mathbf{e}_i (a_j \mathbf{e}_j)_{,i}$$

or

$$\boxed{\boldsymbol{\nabla}\mathbf{a} = a_{j,i}\, \mathbf{e}_i \mathbf{e}_j} \tag{A.27}$$

since the $\mathbf{e}_i$ are constant. Similarly, the **divergence** of $\mathbf{a}(x_i)$ is

$$\boldsymbol{\nabla}\cdot\mathbf{a} = \mathbf{e}_i \cdot (a_j \mathbf{e}_j)_{,i} = a_{j,i}\, \mathbf{e}_i \cdot \mathbf{e}_j = a_{j,i}\, \delta_{ij}$$

or

$$\boxed{\boldsymbol{\nabla}\cdot\mathbf{a} = a_{i,i}\,.} \tag{A.28}$$

In addition, the **curl** of $\mathbf{a}$ is

$$\boldsymbol{\nabla} \times \mathbf{a} = \mathbf{e}_i \times (a_j \mathbf{e}_j)_{,i} = a_{j,i}\, \mathbf{e}_i \times \mathbf{e}_j,$$

and using Eq. $(A.1)_2$ gives

$$\boxed{\boldsymbol{\nabla} \times \mathbf{a} = a_{j,i}\, \epsilon_{ijk} \mathbf{e}_k.} \tag{A.29}$$

Finally, we list some relations that are useful in transforming volume integrals into surface integrals and vice versa. In these equations, which are derived in

Section 2.8.4, $\mathbf{n}$ is a unit normal to the surface area A, which encloses the volume V.

Gradient Theorem:

$$\boxed{\int_V \nabla\phi \, dV = \int_A \mathbf{n}\phi \, dA \quad \text{or} \quad \int_V \phi_{,i} \, dV = \int_A n_i \phi \, dA} \tag{A.30}$$

Divergence Theorem:

$$\boxed{\int_V \nabla\cdot\mathbf{a} \, dV = \int_A \mathbf{n}\cdot\mathbf{a} \, dA. \quad \text{or} \quad \int_V a_{i,i} \, dV = \int_A n_i a_i \, dA} \tag{A.31}$$

Curl Theorem:

$$\boxed{\int_V \nabla\times\mathbf{a} \, dV = \int_A \mathbf{n}\times\mathbf{a} \, dA \quad \text{or} \quad \int_V a_{j,i} \, \epsilon_{ijk} \, dV = \int_A n_i a_j \epsilon_{ijk} \, dA} \tag{A.32}$$

Similar relations between surface and line integrals can be obtained by replacing V by A and A by the contour C (length element ds) in these equations.

A.2 Analysis of Deformation

When subjected to forces, a body transforms from one configuration to another. It is convenient to characterize this transformation relative to a **reference configuration**, which is the collective position of all particles of the body at any specified instant of time. A general mapping from one configuration to another consists of rigid-body motion and deformation. Rigid-body motion involves translation and rotation, while deformation involves extensional and shear strains.

A.2.1 *Strain*

Consider a body that deforms from the undeformed (reference) configuration B into the deformed configuration b (Fig. A.3a). An arbitrary point P in the body B is located by the position vector

$$\mathbf{R} = x_i \mathbf{e}_i \tag{A.33}$$

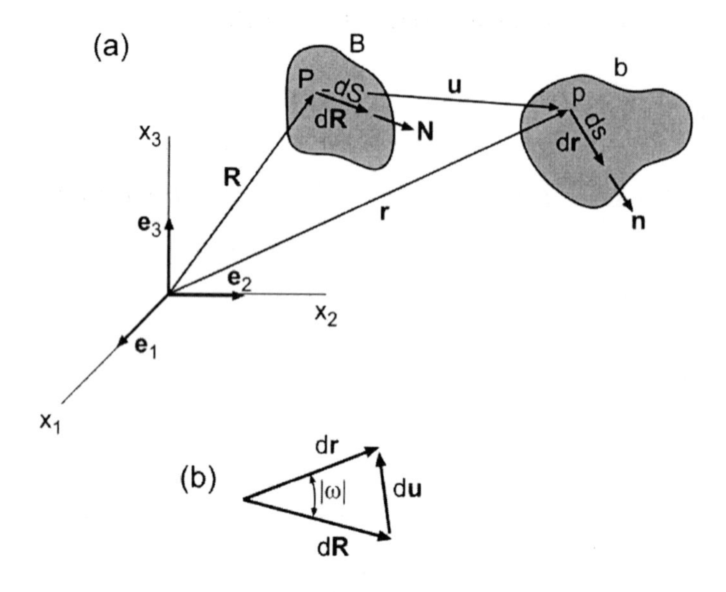

Fig. A.3 (a) Geometry of deformation. (b) Differential position and displacement vectors.

where the x_i are the Cartesian coordinates of P. In the body b, the position vector to p, the deformed image of P, is

$$\mathbf{r} = \mathbf{R} + \mathbf{u} \tag{A.34}$$

where $\mathbf{u}$ is the displacement vector from P to p. For a one-to-one mapping of B into b, we can write

$$\mathbf{r} = \mathbf{r}(\mathbf{R}, t), \qquad \mathbf{u} = \mathbf{u}(\mathbf{R}, t), \tag{A.35}$$

with t being time.

Displacement Gradient Tensor

The state of strain at a point in b is defined in terms of the deformation of the differential length element $d\mathbf{R}$ at P into the element $d\mathbf{r}$ at p (Fig. A.3a). Equation (A.34) relates these elements by

$$d\mathbf{r} = d\mathbf{R} + d\mathbf{u}. \tag{A.36}$$

To express $d\mathbf{u}$ in a different form, we write $\mathbf{u} = u_i \mathbf{e}_i$, and then Eqs. (A.27) and (A.33) and the chain rule yield

$$\begin{aligned} d\mathbf{R}\cdot(\nabla\mathbf{u}) &= (dx_i\,\mathbf{e}_i)\cdot(u_{k,j}\,\mathbf{e}_j\mathbf{e}_k) \\ &= dx_i\,u_{k,j}\,\delta_{ij}\mathbf{e}_k = dx_i\,u_{k,i}\,\mathbf{e}_k \\ &= dx_i\,\frac{\partial u_k}{\partial x_i}\,\mathbf{e}_k = du_k\,\mathbf{e}_k. \end{aligned}$$

Thus,

$$d\mathbf{u} = d\mathbf{R}\cdot(\nabla\mathbf{u}) = (\nabla\mathbf{u})^T\cdot d\mathbf{R} \tag{A.37}$$

where Eq. (A.15) has been used. The dyad $\nabla\mathbf{u}$ is called the **displacement gradient tensor**. Now, inserting this expression into Eq. (A.36) gives, since $\mathbf{I}^T = \mathbf{I}$,

$$\boxed{\begin{aligned} d\mathbf{r} &= d\mathbf{R}\cdot(\mathbf{I} + \nabla\mathbf{u}) \\ &= [\mathbf{I} + (\nabla\mathbf{u})^T]\cdot d\mathbf{R}. \end{aligned}} \tag{A.38}$$

This relation shows that $\nabla\mathbf{u}$ characterizes the deformation of $d\mathbf{R}$ into $d\mathbf{r}$.

Extensional Strain

To characterize the stretching of $d\mathbf{R}$ as it deforms into $d\mathbf{r}$, we must compare absolute lengths. Thus, we write

$$d\mathbf{R} = \mathbf{N}\,dS, \qquad d\mathbf{r} = \mathbf{n}\,ds \tag{A.39}$$

where dS and ds are the lengths of $d\mathbf{R}$ and $d\mathbf{r}$, which point in the directions of the unit vectors $\mathbf{N}$ and $\mathbf{n}$, respectively (Fig. A.3a). In large-deformation problems, it is convenient to define strain in terms of the difference in squared lengths. Thus, the Lagrangian *extensional strain* of $d\mathbf{R}$ is defined as

$$E_{(\mathrm{NN})} = \frac{1}{2}\left(\frac{ds^2 - dS^2}{dS^2}\right) = \frac{1}{2}\left(\frac{d\mathbf{r}\cdot d\mathbf{r} - d\mathbf{R}\cdot d\mathbf{R}}{dS^2}\right) \tag{A.40}$$

in which Eqs. (A.39) have been used, and substituting Eq. (A.38) yields

$$E_{(\mathrm{NN})} = \frac{1}{2}\frac{d\mathbf{R}}{dS}\cdot\left[(\mathbf{I} + \nabla\mathbf{u})\cdot(\mathbf{I} + (\nabla\mathbf{u})^T) - \mathbf{I}\right]\cdot\frac{d\mathbf{R}}{dS}.$$

Expanding the term in brackets and using Eq. $(\text{A.39})_1$ gives

$$E_{(\mathrm{NN})} = \mathbf{N}\cdot\mathbf{E}\cdot\mathbf{N} \tag{A.41}$$

where

$$\boxed{\mathbf{E} = \tfrac{1}{2}[\nabla\mathbf{u} + (\nabla\mathbf{u})^T + (\nabla\mathbf{u})\cdot(\nabla\mathbf{u})^T]} \tag{A.42}$$

is the **Lagrangian strain tensor**. Equations $(A.21)_2$ and (A.41) show that the extensional strain $E_{(\mathbf{NN})}$ is the component of the tensor $\mathbf{E}$ along the dyad $\mathbf{NN}$. In the following, we show that shear deformations also can be computed from $\mathbf{E}$.

Equation (A.42) is the direct form of the **strain-displacement relations**, valid for *arbitrarily large deformations in any coordinate system*. Writing these relations for a specific coordinate system, e.g., Cartesian or polar coordinates, requires simply expressing ∇ in the appropriate form. As shown later, the strain-displacement relation for the linear theory can be obtained by linearizing (A.42).

Shear Strain

So far, we have shown that the extensional strain for a line element oriented originally in the direction of $\mathbf{N}$ can be computed from $\mathbf{E}$. Since $\mathbf{N}$ is arbitrary, the extensional strain in any direction at a point in the body can be computed from this strain tensor. A complete description of the deformation, however, requires also the shear strains, which correspond to angle (shape) changes. Next, we show that $\mathbf{E}$ also contains this information.

Consider two orthogonal line elements at P described by (Fig. A.4)

$$d\mathbf{R}_1 = \mathbf{N}_1\,dS_1, \qquad d\mathbf{R}_2 = \mathbf{N}_2\,dS_2 \tag{A.43}$$

where dS_1 and dS_2 are the undeformed lengths of the elements oriented in the directions of the orthogonal unit vectors $\mathbf{N}_1$ and $\mathbf{N}_2$ in the undeformed body. According to Eq. (A.38), these elements deform into

$$\begin{aligned} d\mathbf{r}_1 &= \mathbf{n}_1\,ds_1 = d\mathbf{R}_1\cdot(\mathbf{I}+\nabla\mathbf{u}) = [\mathbf{I}+(\nabla\mathbf{u})^T]\cdot d\mathbf{R}_1 \\ d\mathbf{r}_2 &= \mathbf{n}_2\,ds_2 = d\mathbf{R}_2\cdot(\mathbf{I}+\nabla\mathbf{u}) = [\mathbf{I}+(\nabla\mathbf{u})^T]\cdot d\mathbf{R}_2 \end{aligned} \tag{A.44}$$

in the directions of the unit vectors $\mathbf{n}_1$ and $\mathbf{n}_2$, which are not orthogonal in general (Fig. A.4).

Now, if α is the angle between $d\mathbf{r}_1$ and $d\mathbf{r}_2$, then

$$\begin{aligned} \cos\alpha &= \mathbf{n}_1\cdot\mathbf{n}_2 = \frac{d\mathbf{r}_1\cdot d\mathbf{r}_2}{ds_1 ds_2} \\ &= \frac{d\mathbf{R}_1\cdot[(\mathbf{I}+\nabla\mathbf{u})\cdot(\mathbf{I}+(\nabla\mathbf{u})^T)]\cdot d\mathbf{R}_2}{ds_1 ds_2} \end{aligned}$$

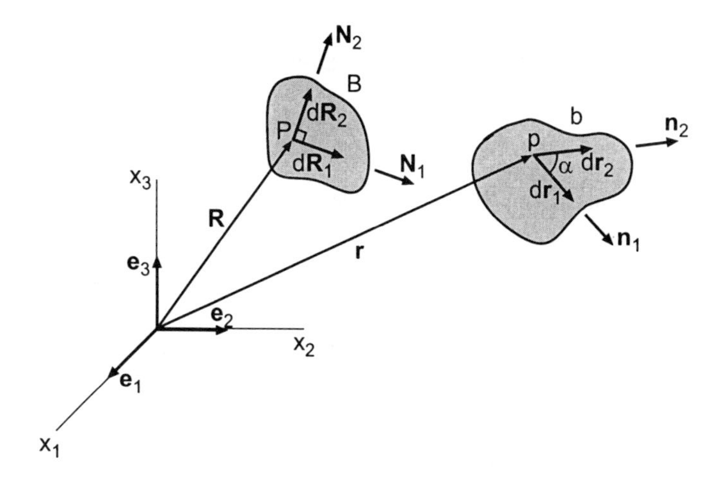

Fig. A.4 Geometry for two originally orthogonal line elements deforming into elements with an enclosed angle α.

and, since $d\mathbf{R}_1 \cdot d\mathbf{R}_2 = 0$, expanding the term in brackets leads to

$$\cos\alpha = \frac{2\mathbf{N}_1 \cdot \mathbf{E} \cdot \mathbf{N}_2}{\left(\dfrac{ds_1}{dS_1}\right)\left(\dfrac{ds_2}{dS_2}\right)}$$

where Eqs. (A.42) and (A.43) have been substituted. Finally, Eq. (A.40) gives $ds_1/dS_1 = (1 + 2E_{(\mathbf{N}_1\mathbf{N}_1)})^{1/2}$ and $ds_2/dS_2 = (1 + 2E_{(\mathbf{N}_2\mathbf{N}_2)})^{1/2}$, leading to the expression

$$\boxed{\cos\alpha = \frac{2\mathbf{N}_1 \cdot \mathbf{E} \cdot \mathbf{N}_2}{\left[1 + 2E_{(\mathbf{N}_1\mathbf{N}_1)}\right]^{1/2}\left[1 + 2E_{(\mathbf{N}_2\mathbf{N}_2)}\right]^{1/2}}} \tag{A.45}$$

where $E_{(\mathbf{N}_1\mathbf{N}_1)} = \mathbf{N}_1 \cdot \mathbf{E} \cdot \mathbf{N}_1$ and $E_{(\mathbf{N}_2\mathbf{N}_2)} = \mathbf{N}_2 \cdot \mathbf{E} \cdot \mathbf{N}_2$ by Eq. (A.41).

Thus, given $\mathbf{E}$, the angle α between the deformed images of $\mathbf{N}_1$ and $\mathbf{N}_2$ can be computed from Eq. (A.45), and $\pi/2 - \alpha$ gives a measure of the shear. The numerator of Eq. (A.45) is the component of $\mathbf{E}$ along $\mathbf{N}_1\mathbf{N}_2$, but the denominator shows that α also depends on the extensional strains along $\mathbf{N}_1$ and $\mathbf{N}_2$.

In summary, we have shown that the Lagrangian strain tensor $\mathbf{E}$ contains all of the information necessary to compute the extensional and shear strains for a differential element of any orientation. In general, the strains vary with location in the body, with the components of $\mathbf{E}$ (E_{ij}) at a given point defining the **state**

of strain at that point. Note that, while the E_{ij} themselves do not have direct physical interpretations, they are called Lagrangian normal strains for $i = j$ and shear strains for $i \neq j$.

Example A.1 The nonlinear strain-displacement relations for any coordinate system can be computed from Eq. (A.42). Derive the expressions for the Cartesian components of the Lagrangian strain tensor.

Solution. For $\mathbf{u} = u_i\mathbf{e}_i$, Eq. (A.27) gives

$$\boldsymbol{\nabla}\mathbf{u} = u_{j,i}\,\mathbf{e}_i\mathbf{e}_j, \qquad (\boldsymbol{\nabla}\mathbf{u})^T = u_{i,j}\,\mathbf{e}_i\mathbf{e}_j$$

and so

$$\begin{aligned}(\boldsymbol{\nabla}\mathbf{u})\cdot(\boldsymbol{\nabla}\mathbf{u})^T &= (u_{j,i}\,\mathbf{e}_i\mathbf{e}_j)\cdot(u_{k,l}\,\mathbf{e}_k\mathbf{e}_l)\\ &= u_{j,i}\,u_{k,l}\,(\mathbf{e}_j\cdot\mathbf{e}_k)\mathbf{e}_i\mathbf{e}_l\\ &= u_{j,i}\,u_{k,l}\,\delta_{jk}\mathbf{e}_i\mathbf{e}_l\\ &= u_{k,i}\,u_{k,l}\,\mathbf{e}_i\mathbf{e}_l\\ &= u_{k,i}\,u_{k,j}\,\mathbf{e}_i\mathbf{e}_j.\end{aligned}$$

Substituting these expressions into Eq. (A.42) yields

$$\mathbf{E} = \tfrac{1}{2}(u_{i,j} + u_{j,i} + u_{k,i}\,u_{k,j})\mathbf{e}_i\mathbf{e}_j. \tag{A.46}$$

Thus, with $\mathbf{E} = E_{ij}\mathbf{e}_i\mathbf{e}_j$,, the Cartesian strain components are given by

$$E_{ij} = \tfrac{1}{2}(u_{i,j} + u_{j,i} + u_{k,i}\,u_{k,j}). \tag{A.47}$$

Note that $E_{ij} = E_{ji}$, and so the strain tensor is symmetric, i.e., only six of the nine strain components are independent.

Given the E_{ij} at a point, the deformation can be computed for elements of any orientation. For example, the extensional strain for a fiber passing through a point that initially is parallel to the x_1-axis is obtained by setting $\mathbf{N} = \mathbf{e}_1$ in Eq. (A.41) to get $E_{(\mathbf{e}_1\mathbf{e}_1)} = \mathbf{e}_1\cdot\mathbf{E}\cdot\mathbf{e}_1 = E_{11}$, and Eq. (A.47) yields

$$E_{11} = \frac{\partial u_1}{\partial x_1} + \frac{1}{2}\left[\left(\frac{\partial u_1}{\partial x_1}\right)^2 + \left(\frac{\partial u_2}{\partial x_1}\right)^2 + \left(\frac{\partial u_3}{\partial x_1}\right)^2\right].$$

In addition, the deformed angle between fibers that are parallel to the x_1 and x_2 axes in the undeformed body is obtained by setting $\mathbf{N}_1 = \mathbf{e}_1$ and $\mathbf{N}_2 = \mathbf{e}_2$ in

Eq. (A.45). This procedure gives

$$\cos\alpha = \frac{2E_{12}}{(1+2E_{11})^{1/2}(1+2E_{22})^{1/2}}$$

where Eq. (A.47) provides

$$E_{12} = \frac{1}{2}\left(\frac{\partial u_1}{\partial x_2} + \frac{\partial u_2}{\partial x_1} + \frac{\partial u_1}{\partial x_1}\frac{\partial u_1}{\partial x_2} + \frac{\partial u_2}{\partial x_1}\frac{\partial u_2}{\partial x_2} + \frac{\partial u_3}{\partial x_1}\frac{\partial u_3}{\partial x_2}\right).$$

Similar relations can be obtained for the other strain components. ■

Example A.2 Consider the set of coordinate axes $\bar{x}_i$, which are obtained by rotating the Cartesian axes x_i by an angle θ about the x_3-axis (Fig. A.2, page 330). Write the strain components $\bar{E}_{ij}$ relative to the $\bar{x}_i$ system in terms of the components E_{ij} relative to the x_i system.

Solution. One way to do this is to use Eq. $(A.24)_2$ to write the tensor transformation

$$\bar{E}_{ij} = A_{ik}A_{jl}E_{kl} \tag{A.48}$$

with the A_{ij} provided by (A.20). For illustration, rather than simply sum over the repeated indices, we write Eq. (A.48) in matrix form. This requires some care, however, since matrix multiplication is not a commutative operation. To obtain the correct equation, we note that all components in a matrix equation must be referred to the same basis. Thus, we write

$$\begin{aligned}
\mathbf{E} &= E_{ij}\mathbf{e}_i\mathbf{e}_j = \bar{E}_{ij}\bar{\mathbf{e}}_i\bar{\mathbf{e}}_j \\
\bar{\mathbf{E}} &= \bar{E}_{ij}\mathbf{e}_i\mathbf{e}_j \\
\mathbf{A} &= A_{ij}\mathbf{e}_i\mathbf{e}_j
\end{aligned} \tag{A.49}$$

where $\mathbf{E}$ is the strain tensor expressed in terms of both sets of components, while $\bar{\mathbf{E}}$ is merely a tensor defined for computational convenience. Substituting these relations into the equation

$$\bar{\mathbf{E}} = \mathbf{A}\cdot\mathbf{E}\cdot\mathbf{A}^T \tag{A.50}$$

yields

$$\begin{aligned}
\bar{E}_{ij}\mathbf{e}_i\mathbf{e}_j &= (A_{ij}\mathbf{e}_i\mathbf{e}_j)\cdot(E_{kl}\mathbf{e}_k\mathbf{e}_l)\cdot(A_{mn}\mathbf{e}_n\mathbf{e}_m) \\
&= A_{ij}A_{mn}E_{kl}(\mathbf{e}_j\cdot\mathbf{e}_k)(\mathbf{e}_l\cdot\mathbf{e}_n)\mathbf{e}_i\mathbf{e}_m \\
&= A_{ij}A_{mn}E_{kl}\,\delta_{jk}\delta_{ln}\mathbf{e}_i\mathbf{e}_m \\
&= A_{ik}A_{ml}E_{kl}\mathbf{e}_i\mathbf{e}_m \\
&= A_{ik}A_{jl}E_{kl}\mathbf{e}_i\mathbf{e}_j .
\end{aligned}$$

Because this expression agrees with Eq. (A.48), Eq. (A.50) shows that the appropriate matrix transformation relation is

$$[\bar{E}_{ij}] = [A_{ik}][E_{kl}][A_{jl}]^T. \tag{A.51}$$

Substituting Eq. (A.20) now gives

$$\begin{bmatrix} \bar{E}_{11} & \bar{E}_{12} & \bar{E}_{13} \\ \bar{E}_{21} & \bar{E}_{22} & \bar{E}_{23} \\ \bar{E}_{31} & \bar{E}_{32} & \bar{E}_{33} \end{bmatrix} = \begin{bmatrix} \cos\theta & \sin\theta & 0 \\ -\sin\theta & \cos\theta & 0 \\ 0 & 0 & 1 \end{bmatrix} \begin{bmatrix} E_{11} & E_{12} & E_{13} \\ E_{21} & E_{22} & E_{23} \\ E_{31} & E_{32} & E_{33} \end{bmatrix} \cdot \begin{bmatrix} \cos\theta & -\sin\theta & 0 \\ \sin\theta & \cos\theta & 0 \\ 0 & 0 & 1 \end{bmatrix}, \tag{A.52}$$

and carrying out the matrix multiplication yields

$$\begin{aligned}
\bar{E}_{11} &= E_{11}\cos^2\theta + E_{22}\sin^2\theta + 2E_{12}\sin\theta\cos\theta \\
\bar{E}_{22} &= E_{11}\sin^2\theta + E_{22}\cos^2\theta - 2E_{12}\sin\theta\cos\theta \\
\bar{E}_{33} &= E_{33} \\
\bar{E}_{12} &= (E_{22} - E_{11})\sin\theta\cos\theta + E_{12}(\cos^2\theta - \sin^2\theta) \\
\bar{E}_{23} &= E_{23}\cos\theta - E_{31}\sin\theta \\
\bar{E}_{31} &= E_{23}\sin\theta + E_{31}\cos\theta.
\end{aligned} \tag{A.53}$$

■

Linear Strain Tensor

Equations (A.41), (A.42), and (A.45) are exact expressions, valid for arbitrarily large deformations. If the strains (displacement gradients) are small, then the nonlinear term of Eq. (A.42), i.e., the last term in brackets, can be neglected to obtain the **linear strain tensor**

$$\boxed{\mathbf{E}^* = \tfrac{1}{2}[\boldsymbol{\nabla}\mathbf{u} + (\boldsymbol{\nabla}\mathbf{u})^T].} \tag{A.54}$$

Moreover, if we define the linear shear strain as $E^*_{(\mathbf{N}_1\mathbf{N}_2)} = \frac{1}{2}(\pi/2 - \alpha) << 1$, then $\cos\alpha = \cos\left(\pi/2 - 2E^*_{(\mathbf{N}_1\mathbf{N}_2)}\right) = \sin 2E^*_{(\mathbf{N}_1\mathbf{N}_2)} \cong 2E^*_{(\mathbf{N}_1\mathbf{N}_2)}$. Thus, for $E^*_{(\mathbf{N}_1\mathbf{N}_1)} << 1$ and $E^*_{(\mathbf{N}_2\mathbf{N}_2)} << 1$, Eqs. (A.41) and (A.45) yield

$$\begin{aligned} E^*_{(\mathbf{NN})} &= \mathbf{N}\cdot\mathbf{E}^*\cdot\mathbf{N} \\ E^*_{(\mathbf{N}_1\mathbf{N}_2)} &= \mathbf{N}_1\cdot\mathbf{E}^*\cdot\mathbf{N}_2. \end{aligned} \tag{A.55}$$

Comparison with Eq. $(A.21)_2$ reveals that the linear extensional (normal) and shear strains are components of the linear strain tensor. (The shear no longer depends on the normal strains.) Thus, in general we can write

$$\boxed{\bar{E}^*_{ij} = \mathbf{E}^*{:}\bar{\mathbf{e}}_i\bar{\mathbf{e}}_j = \bar{\mathbf{e}}_i\cdot\mathbf{E}^*\cdot\bar{\mathbf{e}}_j} \tag{A.56}$$

where the $\bar{\mathbf{e}}_i$ are an orthogonal set of unit vectors of arbitrary orientation. If $i = j$, the $\bar{E}^*_{ij}$ are normal strain components; if $i \neq j$, the $\bar{E}^*_{ij}$ are shear strain components.

The scalar form of the linear strain-displacement relation in Cartesian coordinates is obtained by neglecting the nonlinear term $u_{k,i}\,u_{k,j}$ in Eq. (A.47) to get

$$\boxed{E^*_{ij} = \tfrac{1}{2}(u_{i,j} + u_{j,i}).} \tag{A.57}$$

Of course, the E^*_{ij} transform according to Eq. $(A.24)_2$.

As we have shown, the nonlinear strain components E_{ij} *characterize* the deformation at a point in a body; they have no direct physical meaning. The linear components E^*_{ij}, on the other hand, have the usual physical interpretations, i.e., the normal strains ($i = j$) represent relative extensions and the shear strains ($i \neq j$) represent angle changes. It must be remembered, however, that these interpretations are valid only when the deformation is "small."

Example A.3 Derive the linear strain-displacement relations in cylindrical polar coordinates.

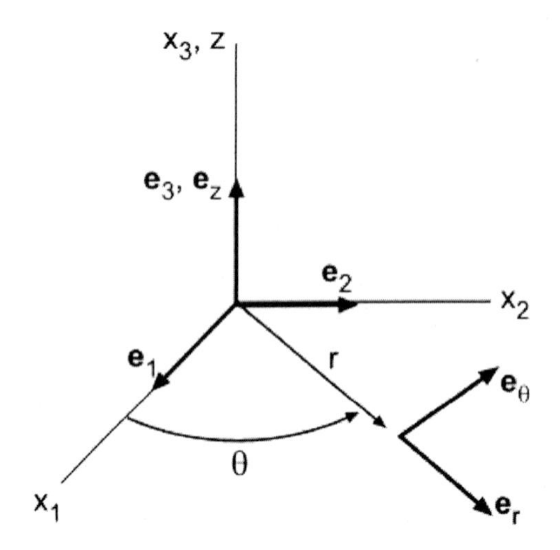

Fig. A.5 Cylindrical polar coordinate system.

Solution. The primary complication introduced by curvilinear coordinates is due to the fact that the base vectors may vary with position. In cylindrical polar coordinates, for example, a set of unit base vectors $(\mathbf{e}_r, \mathbf{e}_\theta, \mathbf{e}_z)$ can be defined in the (r, θ, z) directions, respectively. The vector $\mathbf{e}_z$ is constant, but $\mathbf{e}_r$ and $\mathbf{e}_\theta$ change direction (not magnitude) as θ changes (Fig. A.5). In terms of the Cartesian base vectors $\mathbf{e}_i$, we can write

$$\begin{aligned} \mathbf{e}_r(\theta) &= \mathbf{e}_1 \cos\theta + \mathbf{e}_2 \sin\theta \\ \mathbf{e}_\theta(\theta) &= -\mathbf{e}_1 \sin\theta + \mathbf{e}_2 \cos\theta \\ \mathbf{e}_z &= \mathbf{e}_3. \end{aligned} \tag{A.58}$$

Thus,

$$\begin{aligned} \frac{\partial \mathbf{e}_r}{\partial \theta} &= \mathbf{e}_{r,\theta} = -\mathbf{e}_1 \sin\theta + \mathbf{e}_2 \cos\theta = \mathbf{e}_\theta \\ \frac{\partial \mathbf{e}_\theta}{\partial \theta} &= \mathbf{e}_{\theta,\theta} = -\mathbf{e}_1 \cos\theta - \mathbf{e}_2 \sin\theta = -\mathbf{e}_r, \end{aligned} \tag{A.59}$$

and all derivatives of the unit vectors with respect to r and z vanish.

In cylindrical polar coordinates, the gradient operator has the form (see Section 2.8.3)

$$\nabla = \mathbf{e}_r \frac{\partial}{\partial r} + \mathbf{e}_\theta \frac{1}{r}\frac{\partial}{\partial \theta} + \mathbf{e}_z \frac{\partial}{\partial z}, \tag{A.60}$$

and the displacement vector is

$$\mathbf{u} = u_r \mathbf{e}_r + u_\theta \mathbf{e}_\theta + u_z \mathbf{e}_z. \tag{A.61}$$

Combining these relations yields the displacement gradient tensor

$$\begin{aligned}
\nabla \mathbf{u} \quad = \quad & \mathbf{e}_r(u_{r,r}\,\mathbf{e}_r + u_{\theta,r}\,\mathbf{e}_\theta + u_{z,r}\,\mathbf{e}_z) \\
& + \mathbf{e}_\theta \tfrac{1}{r}(u_{r,\theta}\,\mathbf{e}_r + u_r \mathbf{e}_{r,\theta} + u_{\theta,\theta}\,\mathbf{e}_\theta + u_\theta \mathbf{e}_{\theta,\theta} + u_{z,\theta}\,\mathbf{e}_z) \\
& + \mathbf{e}_z(u_{r,z}\,\mathbf{e}_r + u_{\theta,z}\,\mathbf{e}_\theta + u_{z,z}\,\mathbf{e}_z),
\end{aligned}$$

and substituting Eqs. (A.59) for the derivatives of the unit vectors gives the matrix representation

$$\nabla \mathbf{u} = \begin{bmatrix} u_{r,r} & u_{\theta,r} & u_{z,r} \\ \tfrac{1}{r}(u_{r,\theta} - u_\theta) & \tfrac{1}{r}(u_{\theta,\theta} + u_r) & \tfrac{1}{r}u_{z,\theta} \\ u_{r,z} & u_{\theta,z} & u_{z,z} \end{bmatrix} \tag{A.62}$$

with respect to the basis $\{\mathbf{e}_r, \mathbf{e}_\theta, \mathbf{e}_z\}$. With this expression, the linear strain tensor is

$$\mathbf{E}^* = \tfrac{1}{2}[\nabla\mathbf{u} + (\nabla\mathbf{u})^T] = \begin{bmatrix} E^*_{rr} & E^*_{r\theta} & E^*_{rz} \\ E^*_{\theta r} & E^*_{\theta\theta} & E^*_{\theta z} \\ E^*_{zr} & E^*_{z\theta} & E^*_{zz} \end{bmatrix}$$

where

$$\begin{aligned}
E^*_{rr} &= u_{r,r} \\
E^*_{\theta\theta} &= \tfrac{1}{r}(u_r + u_{\theta,\theta}) \\
E^*_{zz} &= u_{z,z} \\
E^*_{r\theta} = E^*_{\theta r} &= \tfrac{1}{2}\left[\tfrac{1}{r}(u_{r,\theta} - u_\theta) + u_{\theta,r}\right] \\
E^*_{\theta z} = E^*_{z\theta} &= \tfrac{1}{2}\left(\tfrac{1}{r}u_{z,\theta} + u_{\theta,z}\right) \\
E^*_{rz} = E^*_{zr} &= \tfrac{1}{2}(u_{r,z} + u_{z,r}).
\end{aligned} \tag{A.63}$$

■

A.2.2 *Rotation*

As Eq. (A.38) shows, the displacement gradient tensor describes the mapping of $d\mathbf{R}$ into $d\mathbf{r}$. Although translation does not affect this transformation, rotation changes $d\mathbf{R}$ and so is contained in $\nabla\mathbf{u}$. To separate the effects of deformation and rigid-body motion, we write

$$(\nabla\mathbf{u})^T = \mathbf{E}^* + \boldsymbol{\Theta}^* \tag{A.64}$$

where

$$\mathbf{E}^* = \tfrac{1}{2}[(\nabla\mathbf{u})^T + \nabla\mathbf{u}].$$

is the linear strain tensor of Eq. (A.54) and

$$\boxed{\boldsymbol{\Theta}^* = \tfrac{1}{2}[(\nabla\mathbf{u})^T - \nabla\mathbf{u}]} \tag{A.65}$$

is the **linear rotation tensor** (see below). Substituting Eq. (A.64) into (A.37) yields

$$d\mathbf{u} = (\mathbf{E}^* + \boldsymbol{\Theta}^*)\cdot d\mathbf{R}. \tag{A.66}$$

Consider now rigid-body motion. In this case, $\mathbf{E}^* = 0$ and Eq. (A.66) gives

$$d\mathbf{u} = \boldsymbol{\Theta}^*\cdot d\mathbf{R} = \boldsymbol{\omega} \times d\mathbf{R} \tag{A.67}$$

where $\boldsymbol{\omega}$ is the instantaneous **rotation vector**. The magnitude of $\boldsymbol{\omega}$ gives the rotation angle, while its direction gives the rotation axis. The right-hand-side of this expression follows from the geometry of Fig. A.3b (page 334), with $d\mathbf{R}$ rotating through an angle $|\boldsymbol{\omega}|$ as its tip moves a distance $|d\mathbf{u}| = |\boldsymbol{\omega}|\,|d\mathbf{R}|$. (Since $\boldsymbol{\omega}$ is orthogonal to $d\mathbf{R}$, $|\boldsymbol{\omega} \times d\mathbf{R}| = |\boldsymbol{\omega}|\,|d\mathbf{R}|\sin\frac{\pi}{2}$.) In terms of components, Eq. (A.67) can be written

$$(\Theta^*_{ij}\mathbf{e}_i\mathbf{e}_j)\cdot(dx_k\mathbf{e}_k) = (\omega_k\mathbf{e}_k) \times (dx_j\mathbf{e}_j)$$

or, with Eq. $(A.1)_2$,

$$\Theta^*_{ij}dx_k\delta_{jk}\,\mathbf{e}_i = \Theta^*_{ij}dx_j\,\mathbf{e}_i = \omega_k dx_j \epsilon_{kji}\,\mathbf{e}_i.$$

Thus,

$$\boxed{\Theta^*_{ij} = \epsilon_{kji}\,\omega_k} \tag{A.68}$$

which shows that $\Theta^*_{21} = \epsilon_{k12}\,\omega_k = \epsilon_{312}\,\omega_3 = \omega_3 = -\Theta^*_{12}$ represents the rotation about the x_3-axis, etc.

Another way to see the meaning of $\mathbf{\Theta}^*$ is to examine the component form of Eq. (A.65). Manipulations similar to those that led to Eq. (A.57) yield

$$\boxed{\Theta^*_{ij} = \tfrac{1}{2}(u_{i,j} - u_{j,i})} \tag{A.69}$$

which are the familiar rotation components from linear elasticity theory, with their usual geometric interpretations. In a deformable body, $\mathbf{\Theta}^*$ represents the average rotation of all fibers passing through a given point. Note that, whereas $\mathbf{E}^*$ is symmetric ($E^*_{ij} = E^*_{ji}$), $\mathbf{\Theta}^*$ is skew symmetric ($\Theta^*_{ij} = -\Theta^*_{ji}$).

In terms of the displacement vector, Eq. (A.65) gives $\mathbf{\Theta}^*$, and we can show that

$$\boxed{\boldsymbol{\omega} = \tfrac{1}{2}\boldsymbol{\nabla} \times \mathbf{u}.} \tag{A.70}$$

To see this, we use Eq. (A.29) to write this expression as

$$\omega_k \mathbf{e}_k = \tfrac{1}{2} u_{m,l}\, \epsilon_{lmk}\, \mathbf{e}_k,$$

and then multiplying both sides by ϵ_{ijk} yields

$$\omega_k \epsilon_{ijk} = \tfrac{1}{2} u_{m,l}\, \epsilon_{lmk}\epsilon_{ijk} = \tfrac{1}{2} u_{m,l}\, \epsilon_{klm}\epsilon_{kij}.$$

Next, using Eq. (A.5) gives

$$\omega_k \epsilon_{ijk} = \tfrac{1}{2} u_{m,l}\,(\delta_{li}\delta_{mj} - \delta_{lj}\delta_{mi}) = \tfrac{1}{2}(u_{j,i} - u_{i,j}).$$

Finally, reordering the subscripts in the permutation symbol and noting Eq. (A.69) gives

$$\omega_k \epsilon_{kji} = \tfrac{1}{2}(u_{i,j} - u_{j,i}) = \Theta^*_{ij}$$

which agrees with Eq. (A.68). Thus, the local rigid-body rotation can be described in terms of either the rotation tensor $\mathbf{\Theta}^*$, given by Eq. (A.65), or the rotation vector $\boldsymbol{\omega}$, given by Eq. (A.70).

A.2.3 *Principal Strains*

Consider the deformation part of Eq. (A.66), i.e.,

$$d\mathbf{u} = \mathbf{E}^* \cdot d\mathbf{R}. \tag{A.71}$$

We want to find the orientation of a differential element of the body for which the shear strains are zero. Since only extension occurs in this case, the vectors $d\mathbf{u}$ and $d\mathbf{R}$ align and we set $d\mathbf{u} = E^*\, d\mathbf{R}$, where E^* is the linear extensional strain. Combining this relation with Eq. (A.71) yields

$$E^*\, d\mathbf{R} = \mathbf{E}^* \cdot d\mathbf{R}$$

or, with $d\mathbf{R} = \mathbf{N}\, dS$,

$$\boxed{(\mathbf{E}^* - E^*\, \mathbf{I}) \cdot \mathbf{N} = \mathbf{0}} \tag{A.72}$$

which is in the form of an eigenvalue problem. The three eigenvalues E_i^* are the **principal strains**, and the corresponding **eigenvectors** $\mathbf{N}_i$ are the **principal directions of strain**. Since $\mathbf{E}^*$ is a symmetric tensor, the eigenvalues are real and the eigenvectors are mutually orthogonal.

Example A.4 Consider **simple shear** of a unit cube (Fig. A.6). During the deformation, each point of the cube moves parallel to the x_1-axis with the displacement vector given by

$$\mathbf{u} = kx_2\, \mathbf{e}_1 \tag{A.73}$$

where k is a constant. Compute the following:

(a) The Cartesian components of the linear strain tensor $\mathbf{E}^*$ and the linear rotation tensor $\boldsymbol{\Theta}^*$.
(b) The components of the nonlinear (Lagrangian) strain tensor $\mathbf{E}$.
(c) The components of $\mathbf{E}$ relative to axes $\bar{x}_i$ that are rotated through an angle $\theta = 45°$ about the x_3-axis (Fig. A.7).
(d) The principal strains and principal directions for $\mathbf{E}^*$.

Solution. (a) For this problem, the displacement gradient tensor is

$$\begin{aligned} \boldsymbol{\nabla}\mathbf{u} &= \left(\mathbf{e}_1 \frac{\partial}{\partial x_1} + \mathbf{e}_2 \frac{\partial}{\partial x_2} + \mathbf{e}_3 \frac{\partial}{\partial x_3} \right) (kx_2\, \mathbf{e}_1) \\ &= k\, \mathbf{e}_2 \mathbf{e}_1 = \begin{bmatrix} 0 & 0 & 0 \\ k & 0 & 0 \\ 0 & 0 & 0 \end{bmatrix} \end{aligned}$$

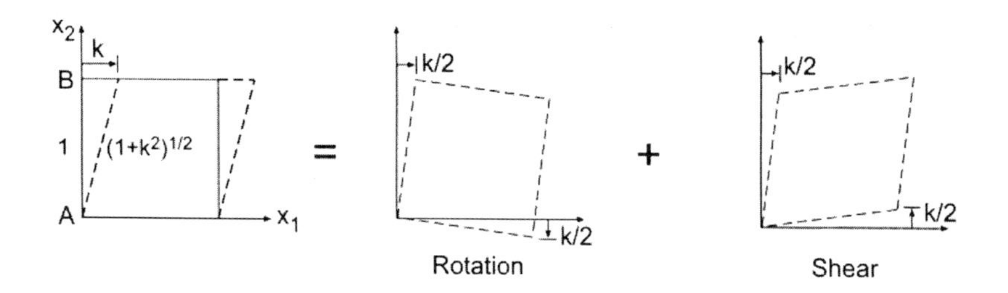

Fig. A.6 Simple shear of a cube.

and so

$$(\boldsymbol{\nabla}\mathbf{u})^T = k\,\mathbf{e}_1\mathbf{e}_2 = \begin{bmatrix} 0 & k & 0 \\ 0 & 0 & 0 \\ 0 & 0 & 0 \end{bmatrix}.$$

In terms of these quantities, Eqs. (A.54) and (A.65) provide the linear strain and rotation tensors

$$\mathbf{E}^* = \tfrac{1}{2}[(\boldsymbol{\nabla}\mathbf{u})^T + \boldsymbol{\nabla}\mathbf{u}] = \tfrac{1}{2}k\,(\mathbf{e}_1\mathbf{e}_2 + \mathbf{e}_2\mathbf{e}_1) = \begin{bmatrix} 0 & \frac{1}{2}k & 0 \\ \frac{1}{2}k & 0 & 0 \\ 0 & 0 & 0 \end{bmatrix}$$

$$\boldsymbol{\Theta}^* = \tfrac{1}{2}[(\boldsymbol{\nabla}\mathbf{u})^T - \boldsymbol{\nabla}\mathbf{u}] = \tfrac{1}{2}k\,(\mathbf{e}_1\mathbf{e}_2 - \mathbf{e}_2\mathbf{e}_1) = \begin{bmatrix} 0 & \frac{1}{2}k & 0 \\ -\frac{1}{2}k & 0 & 0 \\ 0 & 0 & 0 \end{bmatrix} \tag{A.74}$$

which are seen to be symmetric and skew symmetric tensors, respectively. Thus, for small deformation, the only nonzero strain components are $E^*_{12} = E^*_{21} = \frac{1}{2}k$, which is half the angle change, as expected (Fig. A.6). Because the strains and rotations are constant, the deformation of the cube is homogeneous.

As shown by Eqs. (A.36) and (A.66), the deformed configuration of the cube actually represents the superposition of a pure deformation and a rigid-body rotation. According to Eqs. (A.74), the deformation is a shear $k/2$, and the rotation has a magnitude of $k/2$ about the x_3-axis ($\Theta^*_{12} = -\Theta^*_{21} = k/2$). The rotation also can be computed from Eq. (A.70), i.e.,

$$\begin{aligned}\boldsymbol{\omega} &= \tfrac{1}{2}\boldsymbol{\nabla}\times\mathbf{u} = \frac{1}{2}\left(\mathbf{e}_1\frac{\partial}{\partial x_1} + \mathbf{e}_2\frac{\partial}{\partial x_2} + \mathbf{e}_3\frac{\partial}{\partial x_3}\right)\times(kx_2\,\mathbf{e}_1)\\ &= \tfrac{1}{2}k\,\mathbf{e}_2\times\mathbf{e}_1 = -\tfrac{1}{2}k\,\mathbf{e}_3 = \omega_3\mathbf{e}_3.\end{aligned}$$

Thus, $\omega_3 = -k/2$, consistent with a clockwise (negative) rotation about the x_3-axis. Moreover, Eq. (A.68) gives $\Theta^*_{12} = -\Theta^*_{21} = \epsilon_{321}\omega_3 = -\omega_3$, in agreement with this result. The deformational and rotational components for this problem are illustrated in Fig. A.6.

(b) The Lagrangian strain tensor requires the additional term

$$(\boldsymbol{\nabla}\mathbf{u})\cdot(\boldsymbol{\nabla}\mathbf{u})^T = (k\mathbf{e}_2\mathbf{e}_1)\cdot(k\mathbf{e}_1\mathbf{e}_2) = k^2\mathbf{e}_2\mathbf{e}_2,$$

and Eq. (A.42) yields

$$\begin{aligned}\mathbf{E} &= \tfrac{1}{2}\left[\boldsymbol{\nabla}\mathbf{u} + (\boldsymbol{\nabla}\mathbf{u})^T + (\boldsymbol{\nabla}\mathbf{u})\cdot(\boldsymbol{\nabla}\mathbf{u})^T\right]\\ &= \tfrac{1}{2}\left[k\,(\mathbf{e}_1\mathbf{e}_2 + \mathbf{e}_2\mathbf{e}_1) + k^2\mathbf{e}_2\mathbf{e}_2\right]\\ &= \begin{bmatrix} 0 & \tfrac{1}{2}k & 0 \\ \tfrac{1}{2}k & \tfrac{1}{2}k^2 & 0 \\ 0 & 0 & 0 \end{bmatrix}.\end{aligned} \tag{A.75}$$

The component E_{22} characterizes the stretch of a line element originally parallel to the x_2-axis. If the side AB of the undeformed cube has a length $dS = 1$, then its deformed length is $ds = \sqrt{1+k^2}$ (Fig. A.6). Equation (A.40) then gives $E_{22} = \frac{1}{2}(ds^2 - dS^2)/dS^2 = \frac{1}{2}k^2$, in agreement with the above result.

(c) For $\theta = 45°$, the unit vectors along the $\bar{x}_i$ axes in Fig. A.7 are

$$\begin{aligned}\bar{\mathbf{e}}_1 &= \tfrac{1}{\sqrt{2}}(\mathbf{e}_1 + \mathbf{e}_2) = \tfrac{1}{\sqrt{2}}[1,1,0]^T\\ \bar{\mathbf{e}}_2 &= \tfrac{1}{\sqrt{2}}(-\mathbf{e}_1 + \mathbf{e}_2) = \tfrac{1}{\sqrt{2}}[-1,1,0]^T\\ \bar{\mathbf{e}}_3 &= \mathbf{e}_3 = [0,0,1]^T\end{aligned} \tag{A.76}$$

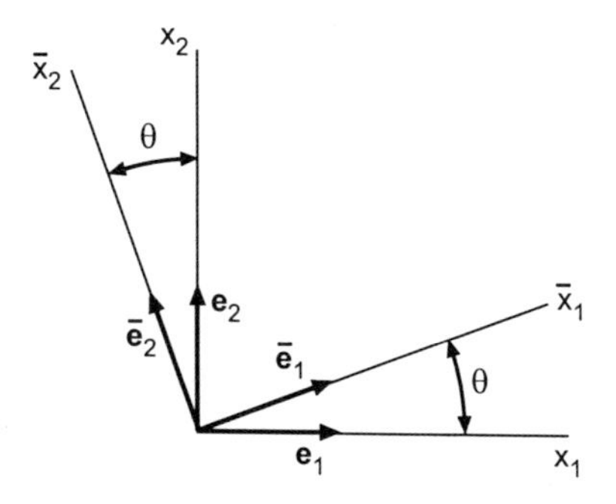

Fig. A.7 Rotation of coordinate axes in two dimensions.

in terms of the nonrotated unit vectors $\mathbf{e}_i$. Thus, Eqs. (A.21) and (A.75) give

$$\begin{aligned}
\bar{E}_{11} &= \bar{\mathbf{e}}_1 \cdot \mathbf{E} \cdot \bar{\mathbf{e}}_1 \\
&= \tfrac{1}{\sqrt{2}}[1,1,0] \cdot \tfrac{1}{2} \begin{bmatrix} 0 & k & 0 \\ k & k^2 & 0 \\ 0 & 0 & 0 \end{bmatrix} \cdot \tfrac{1}{\sqrt{2}} \begin{bmatrix} 1 \\ 1 \\ 0 \end{bmatrix} \\
&= \tfrac{1}{2}k + \tfrac{1}{4}k^2 \\
\bar{E}_{22} &= \bar{\mathbf{e}}_2 \cdot \mathbf{E} \cdot \bar{\mathbf{e}}_2 = -\tfrac{1}{2}k + \tfrac{1}{4}k^2 \\
\bar{E}_{12} &= \bar{\mathbf{e}}_1 \cdot \mathbf{E} \cdot \bar{\mathbf{e}}_2 = \tfrac{1}{4}k^2
\end{aligned} \tag{A.77}$$

and so on.

(d) To compute the principal strains and principal directions for the linear case, we use Eq. $(\text{A.74})_1$ and write Eq. (A.72) in matrix form

$$\begin{bmatrix} -E^* & \tfrac{1}{2}k & 0 \\ \tfrac{1}{2}k & -E^* & 0 \\ 0 & 0 & -E^* \end{bmatrix} \begin{bmatrix} N_1 \\ N_2 \\ N_3 \end{bmatrix} = \begin{bmatrix} 0 \\ 0 \\ 0 \end{bmatrix}. \tag{A.78}$$

Taking the determinant of the 3×3 matrix gives the characteristic equation

$$E^*(E^{*2} - \tfrac{1}{4}k^2) = 0,$$

which provides the principal strains

$$E_1^* = \tfrac{1}{2}k, \qquad E_2^* = -\tfrac{1}{2}k, \qquad E_3^* = 0. \tag{A.79}$$

The corresponding principal directions, obtained by substituting these equations into (A.78), are

$$\mathbf{N}_1 = \tfrac{1}{\sqrt{2}}[1,1,0]^T, \qquad \mathbf{N}_2 = \tfrac{1}{\sqrt{2}}[-1,1,0]^T, \qquad \mathbf{N}_3 = [0,0,1]^T. \tag{A.80}$$

Note that these directions are the same as those used in part (c) to transform the nonlinear strain components [see Eq. (A.76)], i.e., parallel to the diagonals of the undeformed cube. When linearized for $k << 1$, Eqs. (A.77) agree with the principal strains of Eq. (A.79), as they should, including the vanishing shear strain (relative to the magnitudes of E_{11} and E_{22}). But since the *nonlinear* shear strain is not zero, the principal directions for the linear and nonlinear problems are not the same. ■

A.2.4 *Compatibility Conditions*

If the strain field is specified, then the strain-displacement relations (A.57) provide six equations to solve for only three displacement components. Thus, the system is over-determined. In other words, the strains cannot be specified arbitrarily; they must satisfy some additional constraints called **compatibility conditions**.

The physical basis for this problem is the following. Suppose the undeformed body is cut into differential elements, and then the strain for each element is specified arbitrarily. It is quite possible that the deformed pieces would not fit together when the body is reassembled, i.e., the deformation may be incompatible. The compatibility conditions guarantee that, after deformation, the elements fit together without holes and with no further deformation.

Our derivation of the requisite equations makes use of the following identities, which we list without proof:

$$\begin{aligned}
\mathbf{I} \times (\boldsymbol{\nabla} \times \mathbf{a}) &= (\boldsymbol{\nabla}\mathbf{a})^T - \boldsymbol{\nabla}\mathbf{a} \\
\boldsymbol{\nabla} \times (\mathbf{I} \times \mathbf{a}) &= (\boldsymbol{\nabla}\mathbf{a})^T - \mathbf{I}\,\boldsymbol{\nabla}\cdot\mathbf{a} \\
\boldsymbol{\nabla}\cdot(\boldsymbol{\nabla} \times \mathbf{a}) &= 0 \\
\boldsymbol{\nabla} \times \boldsymbol{\nabla}\mathbf{a} &= \mathbf{0}.
\end{aligned} \tag{A.81}$$

We begin with Eq. (A.64) in the form

$$\begin{aligned}
\boldsymbol{\nabla}\mathbf{u} &= \mathbf{E}^{*T} + \boldsymbol{\Theta}^{*T} = \mathbf{E}^* - \boldsymbol{\Theta}^* \\
&= \mathbf{E}^* - \tfrac{1}{2}[(\boldsymbol{\nabla}\mathbf{u})^T - \boldsymbol{\nabla}\mathbf{u}]
\end{aligned}$$

where Eq. (A.65) and the symmetry properties of the strain and rotation tensors have been used. Then, applying Eq. $(A.81)_1$ yields

$$\begin{aligned} \boldsymbol{\nabla}\mathbf{u} &= \mathbf{E}^* - \tfrac{1}{2}\mathbf{I} \times (\boldsymbol{\nabla} \times \mathbf{u}) \\ &= \mathbf{E}^* - \mathbf{I} \times \boldsymbol{\omega} \end{aligned} \tag{A.82}$$

where $\boldsymbol{\omega}$ is given by Eq. (A.70). Taking the curl of this expression and using Eq. $(A.81)_2$ gives

$$\boldsymbol{\nabla} \times \boldsymbol{\nabla}\mathbf{u} = \boldsymbol{\nabla} \times \mathbf{E}^* - [(\boldsymbol{\nabla}\boldsymbol{\omega})^T - \mathbf{I}\,\boldsymbol{\nabla}\cdot\boldsymbol{\omega}].$$

Next, applying Eq. $(A.81)_4$ to the left-hand-side and $(A.81)_3$ to the right-hand-side [with (A.70) substituted] shows that the terms involving $\boldsymbol{\nabla}\mathbf{u}$ and $\boldsymbol{\nabla}\cdot\boldsymbol{\omega}$ vanish, and so

$$\boldsymbol{\nabla} \times \mathbf{E}^* = (\boldsymbol{\nabla}\boldsymbol{\omega})^T. \tag{A.83}$$

Finally, taking the transpose and curl of this equation and using Eq. $(A.81)_4$ yields

$$\boxed{\boldsymbol{\nabla} \times (\boldsymbol{\nabla} \times \mathbf{E}^*)^T = \mathbf{0}} \tag{A.84}$$

which is the compatibility condition in direct notation.

In component form,

$$\boldsymbol{\nabla} \times \mathbf{E}^* = \left(\mathbf{e}_k \frac{\partial}{\partial x_k}\right) \times (E^*_{ij}\mathbf{e}_i\mathbf{e}_j) = E^*_{ij},_k \epsilon_{kil}\, \mathbf{e}_l\mathbf{e}_j$$

by Eq. $(A.1)_2$, and so

$$\begin{aligned} \boldsymbol{\nabla} \times (\boldsymbol{\nabla} \times \mathbf{E}^*)^T &= \left(\mathbf{e}_m \frac{\partial}{\partial x_m}\right) \times (E^*_{ij},_k \epsilon_{kil}\, \mathbf{e}_j\mathbf{e}_l) \\ &= E^*_{ij},_{km}\, \epsilon_{kil}\epsilon_{mjn}\, \mathbf{e}_n\mathbf{e}_l = 0, \end{aligned}$$

which gives the nine scalar relations

$$\boxed{\epsilon_{kim}\epsilon_{ljn}E^*_{ij},_{kl} = 0.} \tag{A.85}$$

Due to the symmetry of the stress tensor, only six of these equations are independent. Moreover, using Eq. (A.3), we can show that the compatibility conditions can be written in the alternate form

$$\boxed{E^*_{ij},_{kl} + E^*_{kl},_{ij} - E^*_{lj},_{ki} - E^*_{ki},_{lj} = 0.} \tag{A.86}$$

Equation (A.84) is a necessary condition for strain compatibility. It also is a sufficient condition for simply connected bodies, i.e., bodies that have no holes. For a multiply connected body with n holes, the additional condition

$$\int_{C_i} d\mathbf{u} = 0, \qquad (i = 1, 2, \ldots, n) \tag{A.87}$$

must be satisfied to ensure single-valued displacements around the contours C_i of the holes.

A.3 Analysis of Stress

A.3.1 *Body and Contact Forces*

The forces that act on a body consist of two types: body forces and contact forces. **Body forces** act at a distance, e.g., gravitational and magnetic forces. We denote the body force per unit deformed volume by the vector $\mathbf{f}(\mathbf{r}, t)$. **Contact forces** act directly on surfaces, e.g., pressure and frictional forces. Internal forces in a body are contact forces that are generated by material adjacent to the imaginary surface of a volume element. If we let $d\mathbf{P}$ represent the force acting on a deformed surface-area element da with an outward-directed unit normal $\mathbf{n}$, then the **traction (stress) vector** $\mathbf{T}_{(\mathbf{n})}$ is defined as the force per unit deformed area, i.e.,

$$\mathbf{T}_{(\mathbf{n})}(\mathbf{r}, t) = \frac{d\mathbf{P}}{da}. \tag{A.88}$$

In the linear theory of elasticity, no distinction is made between the geometries of the undeformed and deformed bodies. Thus, although the forces actually act on the deformed body b, it is convenient to refer them to the undeformed body B. We, therefore, can write the body force as $\mathbf{f}(\mathbf{R}, t)$ per unit undeformed volume and the traction vector as

$$\mathbf{T}_{(\mathbf{N})}(\mathbf{R}, t) = \frac{d\mathbf{P}}{dA}. \tag{A.89}$$

where dA is the undeformed surface area of the element with unit normal $\mathbf{N} \cong \mathbf{n}$ (see Fig. A.8b). The nonlinear theory distinguishes between the undeformed and deformed geometries.

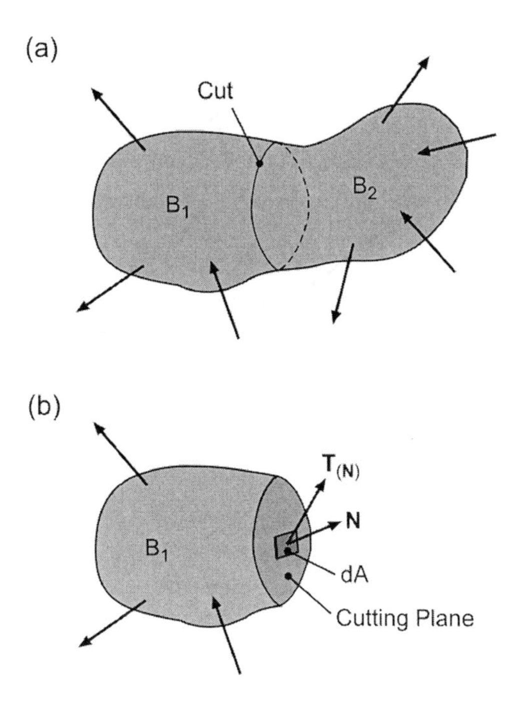

Fig. A.8 Forces acting on a body before and after cutting.

A.3.2 *Stress Tensor*

Consider a body B that is subjected to body and contact forces, and imagine that a **cutting plane** divides B into two sections, B_1 and B_2 (Fig. A.8a). Then, B_1 exerts forces on B_2 and vice versa. If we examine the force system on B_1 alone, then the forces exerted by B_2 are represented by tractions $\mathbf{T}_{(\mathbf{N})}$ acting on the cutting plane (Fig. A.8b). An infinite number of cutting planes can pass through any given point, each with a different traction vector. The set of tractions across all planes passing through a point in a body defines the **state of stress** at that point.

To relate the traction vectors at a point that act on planes of different orientations, we examine a differential element of B in the shape of a tetrahedron of volume dV (Fig. A.9). Three faces of the element are orthogonal to the coordinate axes (unit normals $-\mathbf{e}_i$ and areas dA_i), and the other face has an arbitrary orientation (unit normal $\mathbf{N}$ and area dA). A body force $\mathbf{f}$ acts on the volume of the element, and tractions $\mathbf{T}_{-i}$ and $\mathbf{T}_{(\mathbf{N})}$ act on the $-\mathbf{e}_i$ and $\mathbf{N}$ faces, respectively.

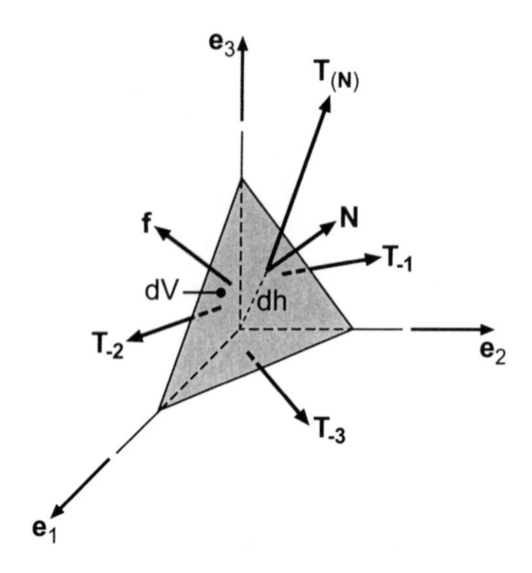

Fig. A.9 Forces acting on a differential tetrahedron.

According to Newton's second law of motion, the sum of the forces acting on a body is equal to the product of the mass of the body and the acceleration of its center of mass. For the differential element, Newton's law takes the form

$$\mathbf{T}_{(\mathbf{N})}\,dA + \mathbf{T}_{-1}\,dA_1 + \mathbf{T}_{-2}\,dA_2 + \mathbf{T}_{-3}\,dA_3 + \mathbf{f}\,dV = (\rho\,dV)\,\mathbf{a} \qquad \text{(A.90)}$$

where $\mathbf{a}$ is the acceleration and ρ is the mass density. The surface of the element adjacent to the $-\mathbf{e}_i$ surface has a unit normal $\mathbf{e}_i$ and traction $\mathbf{T}_i$, and by Newton's third law of action-reaction, $\mathbf{T}_{-i} = -\mathbf{T}_i$. Thus, Eq. (A.90) becomes

$$\mathbf{T}_{(\mathbf{N})}\,dA - \mathbf{T}_i\,dA_i + \mathbf{f}\,dV = \rho\mathbf{a}\,dV \qquad \text{(A.91)}$$

in which the summation convention applies. The surface areas and volume of the element are related by the geometrical relations

$$\begin{aligned} dA_i &= dA\cos(\mathbf{N},\mathbf{e}_i) = dA\,(\mathbf{N}\cdot\mathbf{e}_i) \\ dV &= \tfrac{1}{3}\,dh\,dA \end{aligned} \qquad \text{(A.92)}$$

where dh is the distance from the origin to the $\mathbf{N}$ face (Fig. A.9). Substituting Eqs. (A.92) into (A.91) gives

$$\mathbf{T}_{(\mathbf{N})}\,dA - \mathbf{T}_i\,dA\,(\mathbf{N}\cdot\mathbf{e}_i) + \mathbf{f}\,(\tfrac{1}{3}\,dh\,dA) = \rho\,\mathbf{a}(\tfrac{1}{3}\,dh\,dA).$$

Dividing through by dA and letting $dh \to 0$ yields

$$\mathbf{T}_{(\mathbf{N})} = (\mathbf{N}\cdot\mathbf{e}_i)\,\mathbf{T}_i, \tag{A.93}$$

i.e., as the volume of the element shrinks to zero, the body and inertia forces disappear. Equation (A.93) reveals that if the traction is known on three mutually orthogonal planes passing through a point, then it can be computed on any arbitrarily oriented plane. Putting $\mathbf{N} = N_j\mathbf{e}_j$ into this equation gives

$$\boxed{\mathbf{T}_{(\mathbf{N})} = N_i\mathbf{T}_i.} \tag{A.94}$$

Next, we write the traction vectors in the form

$$\mathbf{T}_i = \sigma_{ij}\mathbf{e}_j \tag{A.95}$$

where the σ_{ij} are the components of $\mathbf{T}_i$, i.e., the **stress components**, relative to the $\mathbf{e}_i$ system. Consistent with the standard convention, this expression shows that the first subscript on σ_{ij} indicates the orientation of the plane on which the stress component acts, and the second subscript indicates the direction of action. If $i = j$, σ_{ij} is a normal stress; if $i \neq j$, σ_{ij} is a shear stress.

Inserting Eq. (A.95) into (A.93) gives the **Cauchy stress formula**

$$\boxed{\mathbf{T}_{(\mathbf{N})} = \mathbf{N}\cdot\boldsymbol{\sigma}} \tag{A.96}$$

where

$$\boxed{\boldsymbol{\sigma} = \sigma_{ij}\mathbf{e}_i\mathbf{e}_j} \tag{A.97}$$

is the (linear) **stress dyadic** or **stress tensor**. The tensor character of $\boldsymbol{\sigma}$ is demonstrated by Eq. (A.96), which shows that $\boldsymbol{\sigma}$ transforms a vector into another vector. Thus, the components transform according to Eq. $(A.24)_2$, which is equivalent to the expression [see Eq. $(A.23)_2$]

$$\boxed{\bar{\sigma}_{ij} = \boldsymbol{\sigma}:\bar{\mathbf{e}}_i\bar{\mathbf{e}}_j = \bar{\mathbf{e}}_i\cdot\boldsymbol{\sigma}\cdot\bar{\mathbf{e}}_j.} \tag{A.98}$$

Here, the $\bar{\sigma}_{ij}$ are the components of $\boldsymbol{\sigma}$ with respect to any orthogonal basis $\{\bar{\mathbf{e}}_i\bar{\mathbf{e}}_j\}$. The nine components of the stress tensor *relative to any basis* completely define the state of stress at a point in a body.

Example A.5 Suppose the stress tensor at a point in a body is given by

$$\begin{aligned}
\boldsymbol{\sigma} &= 3\,\mathbf{e}_1\mathbf{e}_1 + 2\,\mathbf{e}_1\mathbf{e}_2 + 2\,\mathbf{e}_2\mathbf{e}_1 - 6\,\mathbf{e}_2\mathbf{e}_2 + 5\,\mathbf{e}_3\mathbf{e}_3 \\
&= \begin{bmatrix} 3 & 2 & 0 \\ 2 & -6 & 0 \\ 0 & 0 & 5 \end{bmatrix}
\end{aligned}$$

where the components are expressed in appropriate units relative to the Cartesian coordinates (x_1, x_2, x_3). Compute the traction vector and the normal and shear stresses across the plane passing through this point with unit normal

$$\mathbf{N} = \tfrac{1}{3}(\mathbf{e}_1 + 2\mathbf{e}_2 + 2\mathbf{e}_3).$$

Solution. Computing the traction vector is straightforward. Equation (A.96) gives

$$\begin{aligned}
\mathbf{T}_{(\mathbf{N})} &= \mathbf{N}\cdot\boldsymbol{\sigma} = \tfrac{1}{3}[1, 2, 2] \begin{bmatrix} 3 & 2 & 0 \\ 2 & -6 & 0 \\ 0 & 0 & 5 \end{bmatrix} \\
&= \tfrac{1}{3} \begin{bmatrix} 7 \\ -10 \\ 10 \end{bmatrix} = \tfrac{1}{3}(7\mathbf{e}_1 - 10\mathbf{e}_2 + 10\mathbf{e}_3).
\end{aligned}$$

To determine the normal and shear stresses, we define $\mathbf{S}$ to be the unit vector orthogonal to $\mathbf{N}$, i.e., in the plane, that points along the projection of $\mathbf{T}_{(\mathbf{N})}$ on the plane (Fig. A.10). Then, the traction vector can be written in the form

$$\mathbf{T}_{(\mathbf{N})} = \sigma_{(\mathbf{NN})}\mathbf{N} + \sigma_{(\mathbf{NS})}\mathbf{S}$$

where $\sigma_{(\mathbf{NN})}$ is the normal stress and $\sigma_{(\mathbf{NS})}$ the shear stress. Thus, the normal

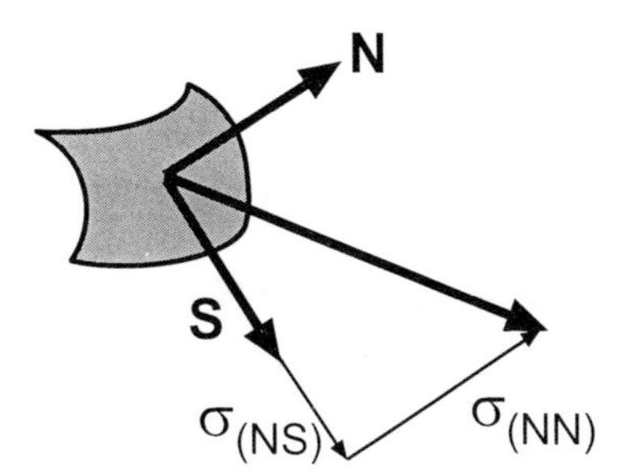

Fig. A.10 Normal and shear components of the traction vector.

stress is

$$\begin{aligned}\sigma_{(\mathbf{NN})} &= \mathbf{N}\cdot\mathbf{T}_{(\mathbf{N})} \\ &= \tfrac{1}{3}[1,2,2]\cdot\tfrac{1}{3}\begin{bmatrix} 7 \\ -10 \\ 10 \end{bmatrix} = \frac{7}{9}.\end{aligned}$$

Similarly, the shear stress is $\sigma_{(\mathbf{NS})} = \mathbf{S}\cdot\mathbf{T}_{(\mathbf{N})}$, but $\mathbf{S}$ must be defined carefully because an infinite number of vectors are orthogonal to $\mathbf{N}$. A straightforward way to compute $\sigma_{(\mathbf{NS})}$ is to note that

$$|\mathbf{T}_{(\mathbf{N})}|^2 = \mathbf{T}_{(\mathbf{N})}\cdot\mathbf{T}_{(\mathbf{N})} = \sigma^2_{(\mathbf{NN})} + \sigma^2_{(\mathbf{NS})},$$

which gives

$$\sigma^2_{(\mathbf{NS})} = \mathbf{T}_{(\mathbf{N})}\cdot\mathbf{T}_{(\mathbf{N})} - \sigma^2_{(\mathbf{NN})} = \frac{2192}{81}.$$

■

A.3.3 *Principal Stresses*

Since $\boldsymbol{\sigma}$ is a tensor, when the coordinate system $\bar{x}_i$ changes, the components of $\boldsymbol{\sigma}$ change, but $\boldsymbol{\sigma}$ itself does not. As in the analysis for principal strains (Section

A.2.3), we seek an orientation of the $\bar{x}_i$ for which the components of shear stress vanish.

If the traction vector $\mathbf{T}_{(\mathbf{N})}$ has no components of shear, then it aligns with the normal $\mathbf{N}$ to the surface, i.e.,

$$\mathbf{T}_{(\mathbf{N})} = \sigma \mathbf{N} \tag{A.99}$$

where σ is the normal stress component on the $\mathbf{N}$ surface. In the next subsection, we show that the stress tensor is symmetric, i.e., $\boldsymbol{\sigma} = \boldsymbol{\sigma}^T$. Thus, substituting Eq. (A.96) into (A.99) yields $\sigma\mathbf{N} = \mathbf{N}\cdot\boldsymbol{\sigma} = \boldsymbol{\sigma}^T\cdot\mathbf{N} = \boldsymbol{\sigma}\cdot\mathbf{N}$ or

$$\boxed{(\boldsymbol{\sigma} - \sigma\mathbf{I})\cdot\mathbf{N} = \mathbf{0}} \tag{A.100}$$

which is an eigenvalue problem similar to Eq. (A.72). Here, the three eigenvalues σ_i are the **principal stresses** and the corresponding eigenvectors $\mathbf{N}_i$ are the **principal directions of stress**. Since $\boldsymbol{\sigma}$ is symmetric, the eigenvalues are real and the eigenvectors are mutually orthogonal.

It is important to note that the principal axes of stress and strain do not coincide in general. In an anisotropic material, for example, normal stresses may produce shear strains [see Eq. (A.144) below]. In this case, the principal planes of stress would undergo shear deformation.

A.3.4 *Equations of Motion*

Thus far, we have used Newton's second law of motion to define the stress tensor at a *point*. Now, we apply Newton's law to a body on a global scale and then deduce the local equations of motion for an infinitesimal *element*. Each element of a body must satisfy these equations.

Consider first a particle of mass m that is in motion relative to a Newtonian frame of reference. The displacement of the mass at time t is $\mathbf{u}(\mathbf{R}, t)$ relative to its initial position $\mathbf{R}$ in the reference frame. Then, the velocity and acceleration of m are, respectively,

$$\begin{aligned} \mathbf{v}(\mathbf{R}, t) &= \frac{d\mathbf{u}}{dt} = \dot{\mathbf{u}} \\ \mathbf{a}(\mathbf{R}, t) &= \frac{d\mathbf{v}}{dt} = \dot{\mathbf{v}}, \end{aligned} \tag{A.101}$$

where dot denotes differentiation with respect to time. The law of conservation of

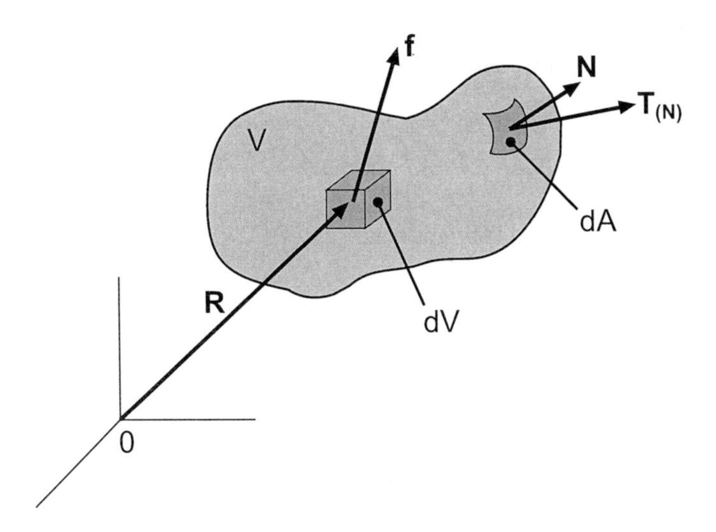

Fig. A.11 Body and surface forces acting on a body.

linear momentum for the particle is

$$\mathbf{F} = \frac{d}{dt}(m\mathbf{v}) \tag{A.102}$$

where $\mathbf{F}$ is the force acting on the particle and $m\mathbf{v}$ is the linear momentum. Because m is constant (ignoring relativistic effects), this relation becomes $\mathbf{F} = m\mathbf{a}$, i.e., Newton's second law.

Now, consider a body of mass density $\rho(\mathbf{R}, t)$ that is subjected to a traction $\mathbf{T}_{(\mathbf{N})}(\mathbf{R}, t)$ acting over the surface of area A and a body force $\mathbf{f}(\mathbf{R}, t)$ acting over the volume V (Fig. A.11). Here again, we neglect the difference between the geometries of the undeformed and deformed configurations. (The "configuration" is not affected by rigid-body motion.) Extending Eq. (A.102) to the entire body yields

$$\int_A \mathbf{T}_{(\mathbf{N})}\, dA + \int_V \mathbf{f}\, dV = \frac{d}{dt}\int_V \mathbf{v}\, \rho\, dV \tag{A.103}$$

where $\mathbf{v}(\mathbf{R}, t)$ is the velocity vector for the mass center of an element located originally at $\mathbf{R}$. The left-hand-side of this equation represents the total force acting on the body, and the right-hand-side is the time rate of change of the linear momentum for the body.

The surface integral in the first term of Eq. (A.103) can be transformed into a volume integral by first substituting Eq. (A.96) and then using the divergence theorem (A.31) to obtain

$$\int_A \mathbf{T}_{(\mathbf{N})}\, dA = \int_A \mathbf{N}\cdot\boldsymbol{\sigma}\, dA = \int_V \boldsymbol{\nabla}\cdot\boldsymbol{\sigma}\, dV.$$

In addition, differentiating the right-hand-side of (A.103) yields

$$\begin{aligned}
\frac{d}{dt}\int_V \mathbf{v}\,\rho\, dV &= \int_V \frac{d}{dt}(\mathbf{v}\,\rho\, dV) \\
&= \int_V \left[\frac{d\mathbf{v}}{dt}\,\rho\, dV + \mathbf{v}\frac{d}{dt}(\rho\, dV)\right] \\
&= \int_V \frac{d^2\mathbf{u}}{dt^2}\,\rho\, dV
\end{aligned}$$

where Eq. $(\text{A.101})_1$ has been used, and $\frac{d}{dt}(\rho\, dV) = 0$ since the mass $\rho\, dV$ of the element is constant. Thus, the global equation of motion (A.103) can be written

$$\int_V (\boldsymbol{\nabla}\cdot\boldsymbol{\sigma} + \mathbf{f} - \rho\ddot{\mathbf{u}})\, dV = \mathbf{0}. \qquad \text{(A.104)}$$

Because this relation must hold for an arbitrary volume in the body, it implies the local equation of motion

$$\boxed{\boldsymbol{\nabla}\cdot\boldsymbol{\sigma} + \mathbf{f} = \rho\ddot{\mathbf{u}}.} \qquad \text{(A.105)}$$

If a given problem is static or if inertia effects are small, i.e., the problem is *quasi-static*, then we can set $\ddot{\mathbf{u}} = \mathbf{0}$ and this relation becomes an **equilibrium equation**.

The component form of the equation of motion can be found by substituting $\boldsymbol{\sigma} = \sigma_{ij}\mathbf{e}_i\mathbf{e}_j$, $\mathbf{f} = f_i\mathbf{e}_i$, and $\mathbf{u} = u_i\mathbf{e}_i$ in Eq. (A.105). Then, the first term can be written

$$\begin{aligned}
\boldsymbol{\nabla}\cdot\boldsymbol{\sigma} &= \left(\mathbf{e}_k\frac{\partial}{\partial x_k}\right)\cdot(\sigma_{ij}\mathbf{e}_i\mathbf{e}_j) \\
&= \sigma_{ij},_k\, \delta_{ki}\mathbf{e}_j = \sigma_{ij},_i\, \mathbf{e}_j = \sigma_{ji},_j\, \mathbf{e}_i,
\end{aligned}$$

and Eq. (A.105) gives

$$\boxed{\sigma_{ji},_j + f_i = \rho\ddot{u}_i.} \qquad \text{(A.106)}$$

Equation (A.103) governs the linear motion of a body. The rotational motion must satisfy the law of conservation of angular momentum. For a particle, this principle has the form

$$\mathbf{R} \times \mathbf{F} = \frac{d}{dt}[m(\mathbf{R} \times \mathbf{v})] \tag{A.107}$$

where $\mathbf{R} \times \mathbf{F}$ is the applied moment and $m(\mathbf{R} \times \mathbf{v})$ the angular momentum about an arbitrarily chosen point. Extending this relation to a body gives

$$\int_A \mathbf{R} \times \mathbf{T}_{(\mathbf{N})}\, dA + \int_V \mathbf{R} \times \mathbf{f}\, dV = \frac{d}{dt}\int_V (\mathbf{R} \times \mathbf{v})\,\rho\, dV \tag{A.108}$$

where concentrated moments are neglected (nonpolar case), and $\mathbf{R}$ is the position vector to an arbitrary point in the body (Fig. A.11). The first term can be converted into a volume integral as follows:

$$\begin{aligned}
\int_A \mathbf{R} \times \mathbf{T}_{(\mathbf{N})}\, dA &= \int_A \mathbf{R} \times (\mathbf{N}\cdot\boldsymbol{\sigma})\, dA \\
&= \int_A (x_i\mathbf{e}_i) \times (N_j\mathbf{e}_j\cdot\sigma_{kl}\mathbf{e}_k\mathbf{e}_l)\, dA \\
&= \int_A x_i N_j \sigma_{kl}\, \mathbf{e}_i \times (\delta_{jk}\mathbf{e}_l)\, dA \\
&= \int_A x_i N_j \sigma_{jl} \epsilon_{ilm} \mathbf{e}_m\, dA \\
&= \int_V \epsilon_{ilm}(x_i\sigma_{jl})_{,j}\, \mathbf{e}_m\, dV.
\end{aligned}$$

In these manipulations, Eq. (A.96) was substituted in the first line, Eqs. (A.1) were used in the third and fourth lines, and Eq. (A.31) was used in the last line (with $a_j \to x_i\sigma_{jl}\epsilon_{ilm}\mathbf{e}_m$). The right-hand-side of (A.108) can be written

$$\begin{aligned}
\frac{d}{dt}\int_V (\mathbf{R} \times \mathbf{v})\,\rho\, dV &= \int_V (\dot{\mathbf{R}} \times \mathbf{v})\,\rho\, dV + \int_V (\mathbf{R} \times \dot{\mathbf{v}})\,\rho\, dV \\
&\quad + \int_V (\mathbf{R} \times \mathbf{v})\,\frac{d}{dt}(\rho\, dV) \\
&= \int_V \mathbf{R} \times \frac{d^2\mathbf{u}}{dt^2}\,\rho\, dV
\end{aligned}$$

in which the first integral on the right-hand-side of the first line vanishes since $\dot{\mathbf{R}} = \mathbf{v}$ and the last integral vanishes due to conservation of mass. With these expressions, $\mathbf{R} = x_i \mathbf{e}_i$, and Eq. (A.1)$_2$, the component form of Eq. (A.108) becomes

$$\int_V \epsilon_{ijk}[(x_i \sigma_{lj}),_l + x_i f_j - \rho x_i \ddot{u}_j]\, \mathbf{e}_k \, dV = \mathbf{0}$$

or

$$\int_V \epsilon_{ijk}[x_{i},_l \, \sigma_{lj} + x_i(\sigma_{lj},_l + f_j - \rho \ddot{u}_j)]\, \mathbf{e}_k \, dV = \mathbf{0}. \qquad \text{(A.109)}$$

The term in parentheses vanishes through the translational equation of motion (A.106), while $x_{i},_l = \partial x_i / \partial x_l = \delta_{il}$. Thus, for an arbitrary volume element, Eq. (A.109) implies

$$\epsilon_{ijk}\sigma_{ij}\, \mathbf{e}_k = \mathbf{0} \qquad \text{or} \qquad \epsilon_{jik}\sigma_{ji}\, \mathbf{e}_k = \mathbf{0}$$

where the second equation was obtained by exchanging i with j in the first equation. Now, since $\epsilon_{jik} = -\epsilon_{ijk}$, adding these equations gives

$$\epsilon_{ijk}(\sigma_{ij} - \sigma_{ji})\, \mathbf{e}_k = \mathbf{0}, \qquad \text{(A.110)}$$

and so $\sigma_{ij} = \sigma_{ji}$. Thus, the principle of angular momentum leads to the conclusion that the stress tensor is symmetric ($\boldsymbol{\sigma} = \boldsymbol{\sigma}^T$).

Example A.6 The stresses in the wedge of Fig. A.12 are given by

$$\boldsymbol{\sigma} = \begin{bmatrix} ax_1^2 & 2cx_1x_2 & 0 \\ 2cx_1x_2 & bx_2^2 & 0 \\ 0 & 0 & 0 \end{bmatrix} \qquad \text{(A.111)}$$

relative to the Cartesian coordinates (x_1, x_2, x_3). Compute the body forces and surface tractions required to hold the wedge in static equilibrium with the stipulated stress distribution.

Solution. First, consider local equilibrium. With $\ddot{u}_i = 0$, Eqs. (A.106) give the equilibrium equations

$$\begin{aligned} \frac{\partial \sigma_{11}}{\partial x_1} + \frac{\partial \sigma_{21}}{\partial x_2} + \frac{\partial \sigma_{31}}{\partial x_3} + f_1 &= 0 \\ \frac{\partial \sigma_{12}}{\partial x_1} + \frac{\partial \sigma_{22}}{\partial x_2} + \frac{\partial \sigma_{32}}{\partial x_3} + f_2 &= 0 \\ \frac{\partial \sigma_{13}}{\partial x_1} + \frac{\partial \sigma_{23}}{\partial x_2} + \frac{\partial \sigma_{33}}{\partial x_3} + f_3 &= 0, \end{aligned} \qquad \text{(A.112)}$$

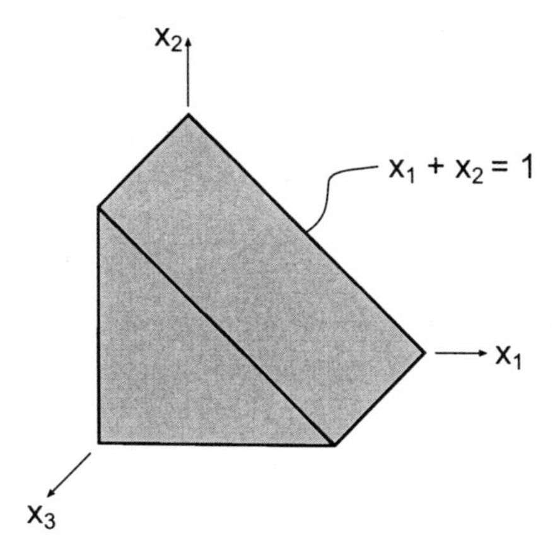

Fig. A.12 A solid wedge.

and substituting Eq. (A.111) yields

$$\begin{aligned} f_1 &= -2x_1(a+c) \\ f_2 &= -2x_2(b+c) \\ f_3 &= 0. \end{aligned} \tag{A.113}$$

These body force distributions are required to maintain the local equilibrium of each element of the body.

Global equilibrium of the body requires, in addition to the body forces, the application of surface tractions consistent with the stresses of Eq. (A.111). As $\sigma_{13} = \sigma_{23} = \sigma_{33} = 0$, no forces are required in the x_3-direction. On the vertical face, the unit (outward) normal is $\mathbf{N} = -\mathbf{e}_1$, and Eq. (A.96) gives

$$T_{(\mathbf{N})} = \mathbf{N}\cdot\boldsymbol{\sigma} = [-1, 0, 0] \begin{bmatrix} 0 & 0 & 0 \\ 0 & bx_2^2 & 0 \\ 0 & 0 & 0 \end{bmatrix} = \begin{bmatrix} 0 \\ 0 \\ 0 \end{bmatrix}$$

where we have set $x_1 = 0$. Similarly, setting $\mathbf{N} = -\mathbf{e}_2$ and $x_2 = 0$ gives

$$T_{(\mathbf{N})} = [0, -1, 0] \begin{bmatrix} ax_1^2 & 0 & 0 \\ 0 & 0 & 0 \\ 0 & 0 & 0 \end{bmatrix} = \begin{bmatrix} 0 \\ 0 \\ 0 \end{bmatrix}$$

on the horizontal face. These faces, therefore, are traction-free. On the inclined face, setting $\mathbf{N} = (\mathbf{e}_1 + \mathbf{e}_2)/\sqrt{2}$ and $x_2 = 1 - x_1$ yields

$$\begin{aligned} \mathbf{T}_{(\mathbf{N})} &= \frac{1}{\sqrt{2}}[1, 1, 0] \begin{bmatrix} ax_1^2 & 2cx_1(1-x_1) & 0 \\ 2cx_1(1-x_1) & b(1-x_1)^2 & 0 \\ 0 & 0 & 0 \end{bmatrix} \\ &= \frac{1}{\sqrt{2}} \begin{bmatrix} ax_1^2 + 2cx_1(1-x_1) \\ b(1-x_1)^2 + 2cx_1(1-x_1) \\ 0 \end{bmatrix}. \end{aligned} \tag{A.114}$$

The computed distributions of body forces and surface tractions must be applied to the body to maintain equilibrium if the internal stresses are to be those specified by Eq. (A.111). Overall equilibrium can be checked by summing the applied loads. For a wedge of unit thickness, doing this in the x_1-direction gives

$$\sum F_1 = \int_0^1 \int_0^{1-x_2} f_1 \, dx_1 dx_2 + \int_0^1 \sigma_{N1}(\sqrt{2}\, dx_1) \tag{A.115}$$

in which the body force f_1 is integrated over the volume of the wedge, and σ_{N1}, the x_1-component of $T_{(\mathbf{N})}$, is integrated over the area of the inclined surface. (Geometry gives the "arclength" $ds = \sqrt{2}\, dx_1$.) Substituting Eqs. (A.113) and (A.114) into (A.115) and carrying out the integrations confirms that $\sum F_1 = 0$. A similar calculation shows that $\sum F_2 = 0$. ■

Example A.7 Using dyadic analysis and Eq. (A.105), derive the scalar equations of motion in cylindrical polar coordinates.

Solution. In terms of the unit vectors $(\mathbf{e}_r, \mathbf{e}_\theta, \mathbf{e}_z)$ along the (r, θ, z) directions (Fig. A.5, page 342), the stress tensor, body force vector, and displacement vector can be written

$$
\begin{aligned}
\boldsymbol{\sigma} &= \sigma_{rr}\mathbf{e}_r\mathbf{e}_r + \sigma_{r\theta}\mathbf{e}_r\mathbf{e}_\theta + \sigma_{rz}\mathbf{e}_r\mathbf{e}_z \\
&\quad +\sigma_{\theta r}\mathbf{e}_\theta\mathbf{e}_r + \sigma_{\theta\theta}\mathbf{e}_\theta\mathbf{e}_\theta + \sigma_{\theta z}\mathbf{e}_\theta\mathbf{e}_z \\
&\quad +\sigma_{zr}\mathbf{e}_z\mathbf{e}_r + \sigma_{z\theta}\mathbf{e}_z\mathbf{e}_\theta + \sigma_{zz}\mathbf{e}_z\mathbf{e}_z \\
\mathbf{f} &= f_r\mathbf{e}_r + f_\theta\mathbf{e}_\theta + f_z\mathbf{e}_z \\
\mathbf{u} &= u_r\mathbf{e}_r + u_\theta\mathbf{e}_\theta + u_z\mathbf{e}_z.
\end{aligned} \tag{A.116}
$$

In addition, Eqs. (A.59) give the derivatives

$$
\begin{gathered}
\mathbf{e}_{r,\theta} = \mathbf{e}_\theta, \qquad \mathbf{e}_{\theta,\theta} = -\mathbf{e}_r \\
\mathbf{e}_{r,r} = \mathbf{e}_{r,z} = \mathbf{e}_{\theta,r} = \mathbf{e}_{\theta,z} = \mathbf{e}_{z,r} = \mathbf{e}_{z,\theta} = \mathbf{e}_{z,z} = 0.
\end{gathered} \tag{A.117}
$$

With $\boldsymbol{\nabla}$ given by Eq. (A.60), we have

$$
\begin{aligned}
\boldsymbol{\nabla}\cdot\boldsymbol{\sigma} &= \left(\mathbf{e}_r\frac{\partial}{\partial r} + \mathbf{e}_\theta\frac{1}{r}\frac{\partial}{\partial\theta} + \mathbf{e}_z\frac{\partial}{\partial z}\right)\cdot(\sigma_{rr}\mathbf{e}_r\mathbf{e}_r + \sigma_{r\theta}\mathbf{e}_r\mathbf{e}_\theta + \sigma_{rz}\mathbf{e}_r\mathbf{e}_z \\
&\quad +\sigma_{\theta r}\mathbf{e}_\theta\mathbf{e}_r + \sigma_{\theta\theta}\mathbf{e}_\theta\mathbf{e}_\theta + \sigma_{\theta z}\mathbf{e}_\theta\mathbf{e}_z + \sigma_{zr}\mathbf{e}_z\mathbf{e}_r + \sigma_{z\theta}\mathbf{e}_z\mathbf{e}_\theta + \sigma_{zz}\mathbf{e}_z\mathbf{e}_z) \\
&= \mathbf{e}_r\cdot\left(\frac{\partial\sigma_{rr}}{\partial r}\mathbf{e}_r\mathbf{e}_r + \frac{\partial\sigma_{r\theta}}{\partial r}\mathbf{e}_r\mathbf{e}_\theta + \frac{\partial\sigma_{rz}}{\partial r}\mathbf{e}_r\mathbf{e}_z + \cdots\right) \\
&\quad +\frac{\mathbf{e}_\theta}{r}\cdot(\sigma_{rr}\mathbf{e}_{r,\theta}\,\mathbf{e}_r + \sigma_{r\theta}\mathbf{e}_{r,\theta}\,\mathbf{e}_\theta + \sigma_{rz}\mathbf{e}_{r,\theta}\,\mathbf{e}_z \\
&\quad +\frac{\partial\sigma_{\theta r}}{\partial\theta}\mathbf{e}_\theta\mathbf{e}_r + \frac{\partial\sigma_{\theta\theta}}{\partial\theta}\mathbf{e}_\theta\mathbf{e}_\theta + \frac{\partial\sigma_{\theta z}}{\partial\theta}\mathbf{e}_\theta\mathbf{e}_z \\
&\quad +\sigma_{\theta r}\mathbf{e}_\theta\mathbf{e}_{r,\theta} + \sigma_{\theta\theta}\mathbf{e}_\theta\mathbf{e}_{\theta,\theta} + \cdots) \\
&\quad +\mathbf{e}_z\cdot\left(\frac{\partial\sigma_{zr}}{\partial z}\mathbf{e}_z\mathbf{e}_r + \frac{\partial\sigma_{z\theta}}{\partial z}\mathbf{e}_z\mathbf{e}_\theta + \frac{\partial\sigma_{zz}}{\partial z}\mathbf{e}_z\mathbf{e}_z + \cdots\right)
\end{aligned} \tag{A.118}
$$

where the dots represent terms that vanish after taking derivatives and the dot product. To complete the analysis, we simplify this expression using Eqs. (A.117) and then insert the result and Eqs. (A.116) into (A.105) to get a relation of the form

$$
F_r\mathbf{e}_r + F_\theta\mathbf{e}_\theta + F_z\mathbf{e}_z = 0.
$$

Setting $F_r = F_\theta = F_z = 0$ gives the equations of motion in the r, θ, and z

directions, respectively:

$$
\begin{aligned}
\frac{\partial \sigma_{rr}}{\partial r} + \frac{1}{r}\frac{\partial \sigma_{\theta r}}{\partial \theta} + \frac{\partial \sigma_{zr}}{\partial z} + \frac{\sigma_{rr} - \sigma_{\theta\theta}}{r} + f_r &= \rho \ddot{u}_r \\
\frac{\partial \sigma_{r\theta}}{\partial r} + \frac{1}{r}\frac{\partial \sigma_{\theta\theta}}{\partial \theta} + \frac{\partial \sigma_{z\theta}}{\partial z} + \frac{2\sigma_{r\theta}}{r} + f_\theta &= \rho \ddot{u}_\theta \\
\frac{\partial \sigma_{rz}}{\partial r} + \frac{1}{r}\frac{\partial \sigma_{\theta z}}{\partial \theta} + \frac{\partial \sigma_{zz}}{\partial z} + \frac{\sigma_{rz}}{r} + f_z &= \rho \ddot{u}_z. \qquad \text{(A.119)}
\end{aligned}
$$

The equations of motion in spherical coordinates, which can be derived through the same process, are given in Appendix B. ■

A.4 Constitutive Relations

All of the equations derived thus far are independent of material properties. Elastic, viscoelastic, and plastic bodies obey the same geometric relations and equations of motion. Stress and strain, however, are related through the constitutive relations, which are material-dependent. Although these equations must be determined experimentally, thermodynamic considerations restrict their form.

A.4.1 *Thermodynamics of Deformation*

The first law of thermodynamics is a statement of conservation of energy for a system. It can be written in the form

$$\dot{K} + \dot{U} = P + Q \qquad \text{(A.120)}$$

where K is the kinetic energy, U is the internal energy, P is the power input, and Q is the rate of heat input.

The kinetic energy for a body of mass density ρ and volume V is

$$K = \tfrac{1}{2}\int_V \rho\, \dot{\mathbf{u}}\cdot\dot{\mathbf{u}}\, dV \qquad \text{(A.121)}$$

where $\mathbf{u}(\mathbf{R}, t)$ is the displacement vector, and the internal energy is

$$U = \int_V \rho u\, dV \qquad \text{(A.122)}$$

in which u is the internal energy per unit mass. The power input is the rate of work done on the body. For applied surface tractions $\mathbf{T}_{(\mathbf{N})}$ and body forces $\mathbf{f}$,

$$P = \int_A \mathbf{T}_{(\mathbf{N})} \cdot \dot{\mathbf{u}} \, dA + \int_V \mathbf{f} \cdot \dot{\mathbf{u}} \, dV \tag{A.123}$$

where A is the surface area. Finally, the rate of heat added to the body is

$$Q = -\int_A \mathbf{q} \cdot \mathbf{N} \, dA + \int_V \rho r \, dV \tag{A.124}$$

where $\mathbf{q}$ is the outward-directed heat flux vector (per unit area), r is the rate of heat production (per unit mass) due to internal sources, and $\mathbf{N}$ is the (outward-directed) unit normal to the surface.

Next, we go through a series of manipulations in order to express the terms of Eq. (A.120) in alternate forms. First, Eq. (A.121) gives

$$\begin{aligned} \dot{K} &= \tfrac{1}{2} \int_V \left[\frac{d}{dt} (\rho \, dV) \dot{\mathbf{u}} \cdot \dot{\mathbf{u}} + \frac{d}{dt} (\dot{\mathbf{u}} \cdot \dot{\mathbf{u}}) \, \rho \, dV \right] \\ &= \tfrac{1}{2} \int_V (\dot{\mathbf{u}} \cdot \ddot{\mathbf{u}} + \ddot{\mathbf{u}} \cdot \dot{\mathbf{u}}) \, \rho \, dV = \int_V \dot{\mathbf{u}} \cdot \ddot{\mathbf{u}} \, \rho \, dV \\ &= \int_V \dot{\mathbf{u}} \cdot (\boldsymbol{\nabla} \cdot \boldsymbol{\sigma} + \mathbf{f}) \, dV \end{aligned} \tag{A.125}$$

in which the second line follows from the constancy of the mass $\rho \, dV$ of an element, and substituting the equation of motion (A.105) produces the third line. It often is convenient to develop tensor equations in component form and then convert the result back to direct notation, especially when the gradient operator is involved. With this procedure, the first term in the integrand of Eq. (A.125) becomes

$$\begin{aligned} \dot{\mathbf{u}} \cdot (\boldsymbol{\nabla} \cdot \boldsymbol{\sigma}) &= \dot{\mathbf{u}} \cdot \left(\mathbf{e}_k \frac{\partial}{\partial x_k} \cdot \sigma_{ij} \mathbf{e}_i \mathbf{e}_j \right) \\ &= \dot{\mathbf{u}} \cdot (\sigma_{ij},_k \, \delta_{ki} \mathbf{e}_j) = \dot{\mathbf{u}} \cdot (\sigma_{ij},_i \, \mathbf{e}_j) \\ &= (\dot{u}_k \mathbf{e}_k) \cdot (\sigma_{ij},_i \, \mathbf{e}_j) = \dot{u}_k \sigma_{ij},_i \, \delta_{kj} \\ &= \dot{u}_j \sigma_{ij},_i \\ &= (\dot{u}_j \sigma_{ij}),_i - \dot{u}_j,_i \, \sigma_{ij}. \end{aligned} \tag{A.126}$$

To convert this expression back to direct notation, we note the following:

$$\begin{aligned}
\boldsymbol{\nabla}\cdot(\boldsymbol{\sigma}\cdot\dot{\mathbf{u}}) &= \boldsymbol{\nabla}\cdot(\sigma_{ij}\mathbf{e}_i\mathbf{e}_j\cdot\dot{u}_k\mathbf{e}_k) = \boldsymbol{\nabla}\cdot(\sigma_{ij}\dot{u}_k\delta_{jk}\mathbf{e}_i) \\
&= \left(\mathbf{e}_k\frac{\partial}{\partial x_k}\right)\cdot(\sigma_{ij}\dot{u}_j\mathbf{e}_i) = (\sigma_{ij}\dot{u}_j)_{,k}\,\delta_{ki} = (\sigma_{ij}\dot{u}_j)_{,i} \\
\boldsymbol{\sigma}:\boldsymbol{\nabla}\dot{\mathbf{u}} &= (\sigma_{ij}\mathbf{e}_i\mathbf{e}_j):\left(\mathbf{e}_k\frac{\partial}{\partial x_k}\dot{u}_l\mathbf{e}_l\right) = \sigma_{ij}\dot{u}_{l,k}\,(\mathbf{e}_i\cdot\mathbf{e}_k)(\mathbf{e}_j\cdot\mathbf{e}_l) \\
&= \sigma_{ij}\dot{u}_{l,k}\,\delta_{ik}\delta_{jl} = \sigma_{ij}\dot{u}_{j,i}\,.
\end{aligned}$$

Thus, Eq. (A.126) can be written

$$\dot{\mathbf{u}}\cdot(\boldsymbol{\nabla}\cdot\boldsymbol{\sigma}) = \boldsymbol{\nabla}\cdot(\boldsymbol{\sigma}\cdot\dot{\mathbf{u}}) - \boldsymbol{\sigma}:\boldsymbol{\nabla}\dot{\mathbf{u}}. \tag{A.127}$$

Furthermore, Eq. (A.64) gives $\boldsymbol{\nabla}\mathbf{u} = \mathbf{E}^{*T} + \boldsymbol{\Theta}^{*T} = \mathbf{E}^* - \boldsymbol{\Theta}^*$ since $\mathbf{E}^{*T} = \mathbf{E}^*$ and $\boldsymbol{\Theta}^{*T} = -\boldsymbol{\Theta}^*$. Therefore, the last term of (A.127) takes the form

$$\boldsymbol{\sigma}:\boldsymbol{\nabla}\dot{\mathbf{u}} = \boldsymbol{\sigma}:(\dot{\mathbf{E}}^* - \dot{\boldsymbol{\Theta}}^*) = \boldsymbol{\sigma}:\dot{\mathbf{E}}^*, \tag{A.128}$$

which is a consequence of the fact that the double-dot product of a symmetric tensor ($\boldsymbol{\sigma}$) and a skew symmetric tensor ($\dot{\boldsymbol{\Theta}}^*$) is zero. Hence, substituting Eqs. (A.127) and (A.128) into (A.125) yields

$$\dot{K} = \int_V [\boldsymbol{\nabla}\cdot(\boldsymbol{\sigma}\cdot\dot{\mathbf{u}}) - \boldsymbol{\sigma}:\dot{\mathbf{E}}^* + \mathbf{f}\cdot\dot{\mathbf{u}}]\,dV. \tag{A.129}$$

The second term of Eq. (A.120), given by (A.122), is

$$\begin{aligned}
\dot{U} &= \int_V \left[\frac{d}{dt}(\rho\,dV)u + \dot{u}\,\rho\,dV\right] \\
&= \int_V \dot{u}\,\rho\,dV
\end{aligned} \tag{A.130}$$

where conservation of mass again has been used. Adding Eqs. (A.129) and (A.130) provides a volume integral for the left-hand-side of (A.120).

The surface integrals of Eqs. (A.123) and (A.124) now are converted into volume integrals. Substituting Eq. (A.96) into (A.123) and applying the divergence theorem (A.31) yields

$$\begin{aligned}
P &= \int_A \mathbf{N}\cdot\boldsymbol{\sigma}\cdot\dot{\mathbf{u}}\,dA + \int_V \mathbf{f}\cdot\dot{\mathbf{u}}\,dV \\
&= \int_V [\boldsymbol{\nabla}\cdot(\boldsymbol{\sigma}\cdot\dot{\mathbf{u}}) + \mathbf{f}\cdot\dot{\mathbf{u}}]\,dV.
\end{aligned} \tag{A.131}$$

Similarly, Eq. (A.124) becomes

$$Q = \int_V (\rho r - \boldsymbol{\nabla}\cdot\mathbf{q})\, dV. \tag{A.132}$$

Now, inserting Eqs. (A.129)–(A.132) into (A.120) and simplifying gives

$$\int_V (\rho\dot{u} - \boldsymbol{\sigma}:\dot{\mathbf{E}}^* - \rho r + \boldsymbol{\nabla}\cdot\mathbf{q})\, dV = 0.$$

For an arbitrary volume element, this relation implies

$$\boxed{\rho\dot{u} = \boldsymbol{\sigma}:\dot{\mathbf{E}}^* + \rho r - \boldsymbol{\nabla}\cdot\mathbf{q},} \tag{A.133}$$

which is the local form of the law of conservation of energy. According to this equation, the rate of increase in internal energy of a volume element ($\rho\dot{u}$) is equal to the sum of the rate of work done by the stresses on the element ($\boldsymbol{\sigma}:\dot{\mathbf{E}}^*$), the rate of internal heat production (ρr), and the rate of heat flow into the element ($-\boldsymbol{\nabla}\cdot\mathbf{q}$).

A.4.2 *Strain-Energy Density Function*

Equation (A.133) applies to general types of material. In the following, we focus on the deformation of elastic bodies. In addition to the first law, elastic deformations also must satisfy the second law of thermodynamics (see Section 5.1.2)

$$\rho T\dot{\eta} = \rho r - \boldsymbol{\nabla}\cdot\mathbf{q} \tag{A.134}$$

for reversible processes, where η is the specific entropy (per unit mass) and T is the absolute temperature. Combining this relation with Eq. (A.133) gives

$$\boxed{\boldsymbol{\sigma}:\dot{\mathbf{E}}^* = \rho(\dot{u} - T\dot{\eta}).} \tag{A.135}$$

We consider two special cases: (1) isentropic deformation ($\dot{\eta} = 0$) and (2) isothermal deformation ($\dot{T} = 0$). For an isentropic deformation, we define

$$W = \rho u \tag{A.136}$$

to be the **strain-energy density function** per unit volume. Then, for $\dot{\eta} = 0$ and constant density, Eq. (A.135) becomes

$$\boldsymbol{\sigma}:\dot{\mathbf{E}}^* = \dot{W}. \tag{A.137}$$

Moreover, if temperature effects are ignored, W can be taken as a function of deformation alone, i.e., $W = W(\mathbf{E}^*) = W(E^*_{ij})$.[10] In this case, Eq. (A.137) yields

$$\sigma_{ij}\dot{E}^*_{ij} = \frac{\partial W}{\partial E^*_{ij}}\dot{E}^*_{ij}$$

by virtue of the chain rule and Eq. (A.12). Thus, for arbitrary $\dot{E}^*_{ij}$, we have

$$\boxed{\sigma_{ij} = \frac{\partial W}{\partial E^*_{ij}},} \tag{A.138}$$

which are the **constitutive relations** for an elastic material. Once $W(E^*_{ij})$ is determined through appropriate experiments, this equation provides the stress-strain relations.

In the case of an isothermal deformation, we introduce the **free-energy function**

$$\boxed{\Psi = u - T\eta.} \tag{A.139}$$

For $\dot{T} = 0$, this relation gives $\dot{\Psi} = \dot{u} - T\dot{\eta}$, and so Eq. (A.135) becomes

$$\boldsymbol{\sigma}{:}\dot{\mathbf{E}}^* = \rho\dot{\Psi}. \tag{A.140}$$

Now, if an isothermal strain-energy density function is defined by

$$W = \rho\Psi, \tag{A.141}$$

then Eq. (A.140) takes the form of Eq. (A.137). Thus, for an elastic material with $W = W(E^*_{ij})$ under isothermal conditions, the constitutive equation again takes the form of Eq. (A.138).

In summary, when the deformation of an elastic body is isentropic or isothermal, the stress components can be computed as the derivatives of a strain-energy density function W with respect to the corresponding strain components. If the deformation is isentropic, W can be identified with the internal energy per unit volume. If the deformation is isothermal, W corresponds to the free energy per unit volume. The specific form of W must be determined experimentally for each particular material.

[10]In an *elastic material*, stress depends only on the current deformation.

A.4.3 *Linear Elastic Material*

For stress to depend linearly on strain, Eq. (A.138) shows that W must be a quadratic function of the strain components, i.e.,

$$\boxed{W = \tfrac{1}{2}\, C_{ijkl}\, E^*_{ij} E^*_{kl}} \tag{A.142}$$

where the C_{ijkl} are the components of a fourth-order tensor. The $3^4 = 81$ C_{ijkl} are **elastic constants** to be determined experimentally. Fortunately, their number can be reduced using the symmetry property of the strain tensor. For example, since Eq. (A.142) can be written $W = \frac{1}{2} C_{jikl} E^*_{ji} E^*_{kl} = \frac{1}{2} C_{jikl} E^*_{ij} E^*_{kl}$, then $C_{ijkl} = C_{jikl}$, and a similar argument shows that $C_{ijkl} = C_{ijlk}$, reducing the number of independent constants to 36. Furthermore, we can write $W = \frac{1}{2} C_{klij} E^*_{kl} E^*_{ij}$, and so $C_{ijkl} = C_{klij}$, i.e., the matrix of 36 constants is symmetric. This observation reduces the number of constants to 21. Thus, the mechanical properties of a general anisotropic, linear elastic material are characterized by 21 independent elastic constants with

$$C_{ijkl} = C_{jikl} = C_{ijlk} = C_{klij}. \tag{A.143}$$

When material symmetry is present, the number of independent elastic constants can be reduced still further (see Section 5.6). An orthotropic material, characterized by three mutually orthogonal planes of symmetry, has nine constants. A transversely isotropic material, which has an axis of symmetry, has five independent constants. And an isotropic material, with the same properties in all directions, has two constants. Note that, for certain anisotropic materials, extensional and shear effects are coupled.

To obtain the stress-strain relations, we substitute Eq. (A.142) into (A.138) and use (A.143) to obtain

$$\begin{aligned}
\sigma_{ij} &= \frac{\partial}{\partial E^*_{ij}} \left(\tfrac{1}{2}\, C_{klmn}\, E^*_{kl} E^*_{mn}\right) \\
&= \tfrac{1}{2}\, C_{klmn} \left(\frac{\partial E^*_{kl}}{\partial E^*_{ij}} E^*_{mn} + E^*_{kl} \frac{\partial E^*_{mn}}{\partial E^*_{ij}} \right) \\
&= \tfrac{1}{2}\, C_{klmn} \left(\delta_{ki}\delta_{lj} E^*_{mn} + E^*_{kl} \delta_{mi}\delta_{nj} \right) \\
&= \tfrac{1}{2} (C_{ijmn} E^*_{mn} + C_{klij} E^*_{kl}) \\
&= \tfrac{1}{2} (C_{ijkl} E^*_{kl} + C_{ijkl} E^*_{kl})
\end{aligned}$$

or

$$\boxed{\sigma_{ij} = C_{ijkl}E^*_{kl}} \tag{A.144}$$

which is called a **generalized Hooke's law**, valid for general anisotropy. In direct notation, this equation can be written in the form

$$\boxed{\boldsymbol{\sigma} = \mathbf{C}:\mathbf{E}^*} \tag{A.145}$$

where $\mathbf{C}$ is a fourth-order tensor. This can be seen by the manipulations

$$\begin{aligned}
\sigma_{ij}\mathbf{e}_i\mathbf{e}_j &= (C_{ijkl}\,\mathbf{e}_i\mathbf{e}_j\mathbf{e}_k\mathbf{e}_l):(E^*_{mn}\mathbf{e}_m\mathbf{e}_n) \\
&= C_{ijkl}E^*_{mn}\,\mathbf{e}_i\mathbf{e}_j\,(\mathbf{e}_k\cdot\mathbf{e}_m)(\mathbf{e}_l\cdot\mathbf{e}_n) \\
&= C_{ijkl}E^*_{mn}\,\mathbf{e}_i\mathbf{e}_j\,\delta_{km}\delta_{ln} \\
&= C_{ijkl}E^*_{kl}\mathbf{e}_i\mathbf{e}_j
\end{aligned}$$

which implies Eq. (A.144)

For an isotropic material, Eq. (A.145) can be written

$$\boxed{\boldsymbol{\sigma} = \lambda\Delta\mathbf{I} + 2\mu\mathbf{E}^*} \tag{A.146}$$

where λ and μ are the **Lamé constants**. In addition,

$$\Delta = \operatorname{tr}\mathbf{E}^* = E^*_{ii} \tag{A.147}$$

is the trace of the tensor $\mathbf{E}^*$, which represents the **dilatation**, i.e., the ratio of the change in volume to the undeformed volume of a material element. Of course, Eq. (A.146) also can be written in standard engineering form in terms of a Young's modulus, Poisson's ratio, and shear modulus. Again, only two of these material constants are independent.

A.5 Boundary Value Problems

The governing equations, along with appropriate boundary and initial conditions, define a boundary value problem in elasticity. As derived in this chapter, the basic equations of linear isotropic elasticity theory consist of the following:

Strain-Displacement Relations:

$$\mathbf{E}^* = \tfrac{1}{2}[\boldsymbol{\nabla}\mathbf{u} + (\boldsymbol{\nabla}\mathbf{u})^T] \qquad \text{or} \qquad E^*_{ij} = \tfrac{1}{2}(u_{i,j} + u_{j,i}) \tag{A.148}$$

Equations of Motion:

$$\nabla\cdot\boldsymbol{\sigma} + \mathbf{f} = \rho\ddot{\mathbf{u}} \qquad \text{or} \qquad \sigma_{ji},_j + f_i = \rho\ddot{u}_i \tag{A.149}$$

Constitutive Relations:

$$\boldsymbol{\sigma} = \lambda(\operatorname{tr}\mathbf{E}^*)\,\mathbf{I} + 2\mu\mathbf{E}^* \qquad \text{or} \qquad \sigma_{ij} = \lambda E^*_{kk}\delta_{ij} + 2\mu E^*_{ij} \tag{A.150}$$

Given the applied loads, the direct notation forms of these relations provide a system of three equations to be solved for $\mathbf{u}$, $\mathbf{E}^*$, and $\boldsymbol{\sigma}$. Due to the symmetry of the stress and strain tensors, the scalar forms represent a system of 15 equations for 15 unknowns: u_i (3), E^*_{ij} (6), and σ_{ij} (6). For static problems, $\ddot{\mathbf{u}} = 0$ and Eqs. (A.149) become equilibrium equations.

The boundary conditions consist of specified tractions $\hat{\mathbf{T}}_{(\mathbf{N})}$ and displacements $\hat{\mathbf{u}}$ over the surface of the body, where hat indicates a prescribed quantity. In dynamic problems, the initial conditions consist of the specified displacement $\hat{\mathbf{u}}^{(0)}$ and velocity $\hat{\mathbf{v}}^{(0)}$ of all points in the body at $t = 0$. With Eq. (A.96) and the unit surface normal $\mathbf{N} = N_i\mathbf{e}_i$, these conditions can be written in the following forms:

Boundary Conditions ($t \geq 0$)**:**

$$\mathbf{N}\cdot\boldsymbol{\sigma} = \hat{\mathbf{T}}_{(\mathbf{N})} \qquad \text{or} \qquad N_j\sigma_{ij} = \hat{T}_i \qquad \text{on } A_\sigma$$

$$\mathbf{u} = \hat{\mathbf{u}} \qquad \text{or} \qquad u_i = \hat{u}_i \qquad \text{on } A_u \tag{A.151}$$

Initial Conditions ($t = 0$)**:**

$$\mathbf{u} = \hat{\mathbf{u}}^{(0)} \qquad \text{or} \qquad u_i = \hat{u}_i^{(0)} \qquad \text{in } V$$

$$\dot{\mathbf{u}} = \hat{\mathbf{v}}^{(0)} \qquad \text{or} \qquad \dot{u}_i = \hat{v}_i^{(0)} \qquad \text{in } V \tag{A.152}$$

Here, A_σ and A_u represent the portions of the surface where tractions and displacements, respectively, are specified. In *linear* elasticity, the **uniqueness theorem** ensures that the solution to any particular boundary value problem is unique (Timoshenko and Goodier, 1969).

A.5.1 *Torsion of a Circular Cylinder*

Consider torsion of a solid circular cylinder of constant radius a and length L (Fig. A.13). The cylinder is subjected only to static end couples M about the longitudinal (z) axis. Our goal is to determine the stress and strain distributions.

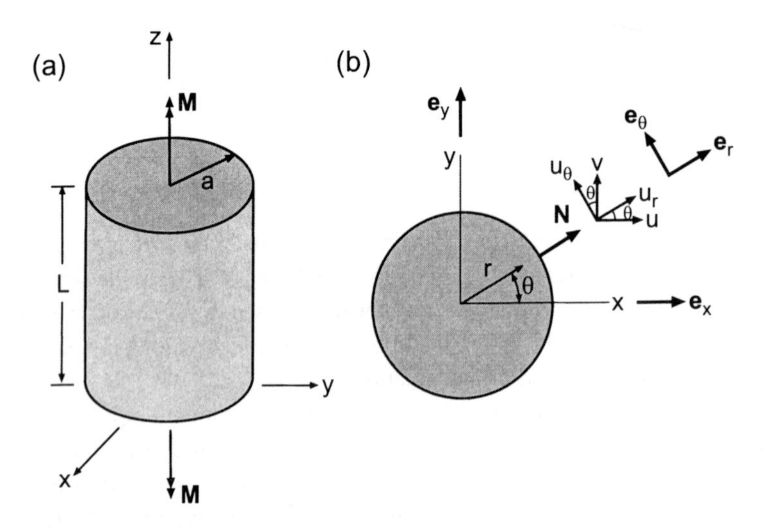

Fig. A.13 Torsion of a solid circular cylinder. (a) Three-dimensional view. (b) Cross section.

One technique used to solve elasticity problems is the **semi-inverse method**, which involves the following steps. First, some kinematic aspects of the solution are assumed. Second, from these assumptions, a solution is found that satisfies all of the governing equations and boundary conditions. Finally, *for the linear theory*, the uniqueness theorem guarantees that the obtained solution is the one and only solution to the given problem.

Following this procedure, we initially make some kinematic assumptions based on symmetry. First, since the cylinder and the loading are symmetric about the z-axis, we expect the solution to be independent of the circumferential coordinate θ. Second, we note that the cylinder and loading look the same when rotated $180°$ about the x or y-axis. Thus, plane cross sections must remain plane, i.e., there can be no out-of-plane warping, which would violate this symmetry. Third, we assume that each cross section rotates as a rigid body, with the rotation angle varying linearly along the length of the cylinder.

These assumptions imply that the displacement vector can be written

$$\begin{aligned} \mathbf{u} &= u_r\mathbf{e}_r + u_\theta\mathbf{e}_\theta + u_z\mathbf{e}_z \\ &= u\mathbf{e}_x + v\mathbf{e}_y + w\mathbf{e}_z \end{aligned} \qquad \text{(A.153)}$$

where the displacement components are

$$u_\theta = \beta z r, \qquad u_r = u_z = 0 \tag{A.154}$$

relative to the cylindrical polar coordinates (r, θ, z). The variable β can be identified as the angle of twist per unit length, which is constant for a uniform cylinder. The geometry linking the polar and Cartesian coordinates gives the relations (Fig. A.13b)

$$x = r\cos\theta, \qquad y = r\sin\theta$$

and

$$\begin{aligned}
\mathbf{e}_r &= \mathbf{e}_x \cos\theta + \mathbf{e}_y \sin\theta \\
\mathbf{e}_\theta &= -\mathbf{e}_x \sin\theta + \mathbf{e}_y \cos\theta.
\end{aligned} \tag{A.155}$$

Combining these expressions with Eqs. (A.153) and (A.154) and noting that $w = u_z$ yields

$$u = -\beta y z, \qquad v = \beta x z, \qquad w = 0. \tag{A.156}$$

With $(x_1, x_2, x_3) \equiv (x, y, z)$ and $(u_1, u_2, u_3) \equiv (u, v, w)$, substitution of Eqs. (A.156) into (A.148) provides the strain components

$$\begin{aligned}
E^*_{xz} &= E^*_{zx} = \frac{1}{2}\left(\frac{\partial u}{\partial z} + \frac{\partial w}{\partial x}\right) = -\tfrac{1}{2}\beta y \\
E^*_{yz} &= E^*_{zy} = \frac{1}{2}\left(\frac{\partial v}{\partial z} + \frac{\partial w}{\partial y}\right) = \tfrac{1}{2}\beta x \\
E^*_{xx} &= E^*_{yy} = E^*_{zz} = E^*_{xy} = E^*_{yx} = 0.
\end{aligned} \tag{A.157}$$

Inserting these relations into Eq. (A.150) then gives the stress components

$$\begin{aligned}
\sigma_{xz} &= \sigma_{zx} = -\mu\beta y \\
\sigma_{yz} &= \sigma_{zy} = \mu\beta x \\
\sigma_{xx} &= \sigma_{yy} = \sigma_{zz} = \sigma_{xy} = \sigma_{yx} = 0,
\end{aligned} \tag{A.158}$$

and so the stress dyadic is

$$\boldsymbol{\sigma} = \mu\beta(-y\,\mathbf{e}_x\,\mathbf{e}_z - y\,\mathbf{e}_z\mathbf{e}_x + x\,\mathbf{e}_y\mathbf{e}_z + x\mathbf{e}_z\mathbf{e}_y). \tag{A.159}$$

Direct substitution of Eqs. (A.158) reveals that these stresses identically satisfy the equilibrium equations given by Eqs. (A.149) with $f_i = \ddot{u}_i = 0$. Thus, if we can show that the boundary conditions are satisfied, then Eqs. (A.156)–(A.158) provide the exact solution to this problem.

In the torsion problem, the only loads acting on the cylinder are twisting moments applied at the ends. Thus, the lateral surface is stress free, and the net *force* on the ends must vanish. The corresponding boundary conditions are

$$\begin{aligned} r = a: &\qquad \mathbf{T}_{(\mathbf{N})} = \mathbf{0} \\ z = 0, L: &\qquad \mathbf{F} = \int_A \mathbf{T}_{(\mathbf{N})}\, dA = \mathbf{0} \\ &\qquad \mathbf{M} = M\,\mathbf{N} = \int_A \mathbf{r} \times \mathbf{T}_{(\mathbf{N})}\, dA \end{aligned} \tag{A.160}$$

where $\mathbf{F}$ is the resultant end force, $\mathbf{M}$ is the resultant end moment, $\mathbf{r}$ is the position vector in the cross section, $\mathbf{N}$ is the surface normal, and A is the cross-sectional area. We now examine each of these conditions.

On the lateral surface, $\mathbf{N} = \mathbf{e}_r$ and Eqs. (A.96), $(A.155)_1$, and (A.159) yield

$$\begin{aligned} \mathbf{T}_{(\mathbf{N})} &= \mathbf{N}\cdot\boldsymbol{\sigma} \\ &= (\mathbf{e}_x \cos\theta + \mathbf{e}_y \sin\theta)\cdot[\mu\beta(-y\,\mathbf{e}_x\mathbf{e}_z - y\,\mathbf{e}_z\mathbf{e}_x + x\,\mathbf{e}_y\mathbf{e}_z + x\,\mathbf{e}_z\mathbf{e}_y)] \\ &= \mu\beta(-y\cos\theta + x\sin\theta)\mathbf{e}_z \\ &= \mu\beta[-(r\sin\theta)\cos\theta + (r\cos\theta)\sin\theta]\mathbf{e}_z \\ &= \mathbf{0}. \end{aligned}$$

Thus, the first boundary condition of (A.160) is satisfied.

On the end $z = L$ ($z = 0$ is similar), $\mathbf{N} = \mathbf{e}_z$ and we get

$$\begin{aligned} \mathbf{T}_{(\mathbf{N})} = \mathbf{N}\cdot\boldsymbol{\sigma} &= \mathbf{e}_z\cdot[\mu\beta(-y\,\mathbf{e}_x\mathbf{e}_z - y\,\mathbf{e}_z\mathbf{e}_x + x\,\mathbf{e}_y\mathbf{e}_z + x\,\mathbf{e}_z\mathbf{e}_y)] \\ &= \mu\beta(-y\,\mathbf{e}_x + x\,\mathbf{e}_y). \end{aligned} \tag{A.161}$$

Since the cross section is symmetric about the x and y axes, this expression shows that the second boundary condition of (A.160), i.e., the vanishing end force, is satisfied. Substitution into the third condition of (A.160) gives

$$\begin{aligned} \mathbf{M} = M\,\mathbf{e}_z &= \mu\beta \int_A (x\,\mathbf{e}_x + y\,\mathbf{e}_y) \times (-y\,\mathbf{e}_x + x\,\mathbf{e}_y)\, dA \\ &= \mu\beta\mathbf{e}_z \int_A (x^2 + y^2)\, dA. \end{aligned}$$

This expression provides the torque-twist relation

$$M = \mu J \beta \tag{A.162}$$

where

$$J = \int_A (x^2 + y^2)\, dA = \int_A r^2\, dA \tag{A.163}$$

is the polar moment of inertia for the cross section. The quantity μJ is called the torsional rigidity, with $J = \pi a^4/2$ for a circular cross section. Given M, Eq. (A.162) provides β, and then Eqs. (A.156)–(A.158) give the displacements, strains, and stresses.

Thus, the postulated displacement field of Eq. (A.156) gives the exact solution to the torsion problem. Note, however, that this solution requires that the tractions on the ends of the cylinder be applied exactly according to the distribution of Eq. (A.161), which can be written (see geometry of Fig. A.13b)

$$\begin{aligned}
\mathbf{T}_{(\mathbf{N})} &= \mu\beta(-y\,\mathbf{e}_x + x\,\mathbf{e}_y) \\
&= \mu\beta[-(r\sin\theta)(\mathbf{e}_r\cos\theta - \mathbf{e}_\theta\sin\theta) + (r\cos\theta)(\mathbf{e}_r\sin\theta + \mathbf{e}_\theta\cos\theta)] \\
&= \mu\beta r\,\mathbf{e}_\theta.
\end{aligned} \tag{A.164}$$

The end traction, therefore, is a circumferential shear stress that increases linearly with r. Note also that we obtained this expression from the relation $\mathbf{T}_{(\mathbf{N})} = \mathbf{e}_z\cdot\boldsymbol{\sigma}$ [see Eq. (A.161)], and so the shear stress component is $\mathbf{T}_{(\mathbf{N})}\cdot\mathbf{e}_\theta = \mathbf{e}_z\cdot\boldsymbol{\sigma}\cdot\mathbf{e}_\theta = \sigma_{z\theta} = \mu\beta r$.

If the tractions do not satisfy Eq. (A.164) exactly but still furnish only a twisting moment M, then the present solution is not valid near the ends. By St. Venant's principle, however, the solution is a good approximation sufficiently far from the ends (Timoshenko and Goodier, 1969).

A.5.2 *Torsion of a Noncircular Cylinder*

If the cross section of a twisted cylinder is not circular, then our prior symmetry arguments are not valid. In this case, therefore, the problem is not axisymmetric, and warping of cross sections is possible. Here, we modify the solution of the previous example to allow a more general deformation.

Consider a solid cylinder of length L with a constant cross section that has a general contour C (Fig. A.14). Modifying Eqs. (A.156), we assume that the displacement components have the form

$$u = -\beta yz, \qquad v = \beta xz, \qquad w = \beta\,\psi(x, y) \tag{A.165}$$

where β again is a constant twist per unit length and ψ is a **warping function**.

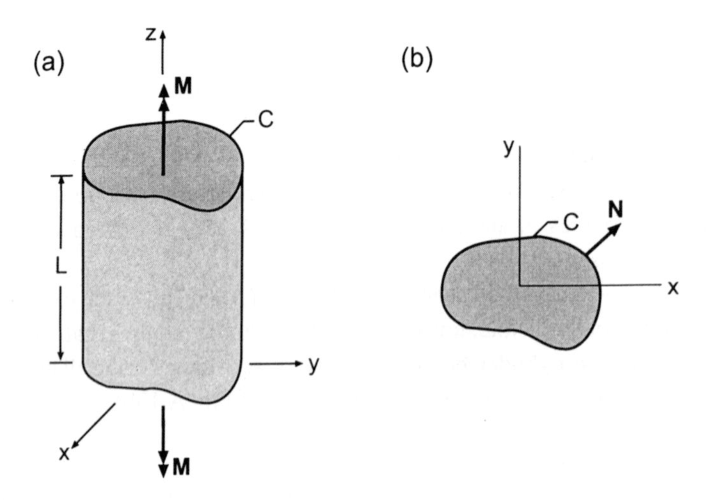

Fig. A.14 Torsion of a noncircular cylinder. (a) Three-dimensional view. (b) Cross section.

The strain components of Eqs. (A.157) now are replaced by

$$\begin{aligned}
E^*_{xz} &= E^*_{zx} = \frac{1}{2}\left(\frac{\partial u}{\partial z} + \frac{\partial w}{\partial x}\right) = \tfrac{1}{2}\beta\left(\frac{\partial \psi}{\partial x} - y\right) \\
E^*_{yz} &= E^*_{zy} = \frac{1}{2}\left(\frac{\partial v}{\partial z} + \frac{\partial w}{\partial y}\right) = \tfrac{1}{2}\beta\left(\frac{\partial \psi}{\partial y} + x\right) \\
E^*_{xx} &= E^*_{yy} = E^*_{zz} = E^*_{xy} = E^*_{yx} = 0,
\end{aligned} \qquad \text{(A.166)}$$

and the stress components of Eq. (A.158) become

$$\begin{aligned}
\sigma_{xz} &= \sigma_{zx} = \mu\beta\left(\frac{\partial \psi}{\partial x} - y\right) \\
\sigma_{yz} &= \sigma_{zy} = \mu\beta\left(\frac{\partial \psi}{\partial y} + x\right) \\
\sigma_{xx} &= \sigma_{yy} = \sigma_{zz} = \sigma_{xy} = \sigma_{yx} = 0.
\end{aligned} \qquad \text{(A.167)}$$

Note that, for $\psi = 0$, these relations reduce to those of Eqs. (A.157) and (A.158). The stresses of Eqs. (A.167) satisfy the equilibrium equations $\sigma_{ji},_j = 0$ identically for $i = 1, 2$. For $i = 3$, we have

$$\frac{\partial \sigma_{13}}{\partial x_1} + \frac{\partial \sigma_{23}}{\partial x_2} + \frac{\partial \sigma_{33}}{\partial x_3} = \frac{\partial \sigma_{xz}}{\partial x} + \frac{\partial \sigma_{yz}}{\partial y} + \frac{\partial \sigma_{zz}}{\partial z} = 0,$$

and substituting Eqs. (A.167) yields

$$\boxed{\frac{\partial^2 \psi}{\partial x^2} + \frac{\partial^2 \psi}{\partial y^2} = \boldsymbol{\nabla}^2 \psi = 0} \tag{A.168}$$

where $\boldsymbol{\nabla}^2 = \boldsymbol{\nabla}\cdot\boldsymbol{\nabla}$ is the Laplacian operator. Thus, $\psi(x, y)$ is a harmonic function.

Any solution of Eq. (A.168) satisfies the field equations of linear elasticity for the torsion problem, with geometric compatibility being ensured by Eqs. (A.165). For a particular geometry, however, ψ also must satisfy the boundary conditions, again given by Eqs. (A.160) but with the boundary $r = a$ replaced by the contour C.

Consider first the condition on the lateral surface. With Eqs. (A.167), the stress tensor can be written

$$\boldsymbol{\sigma} = \sigma_{xz}\,\mathbf{e}_x\mathbf{e}_z + \sigma_{zx}\,\mathbf{e}_z\mathbf{e}_x + \sigma_{yz}\,\mathbf{e}_y\mathbf{e}_z + \sigma_{zy}\,\mathbf{e}_z\mathbf{e}_y, \tag{A.169}$$

and for an arbitrary cross section, the unit normal to C is (Fig. A.14b)

$$\mathbf{N} = N_x\mathbf{e}_x + N_y\mathbf{e}_y \tag{A.170}$$

where the components N_x and N_y are known for a given geometry. Substituting these relations into Eq. (A.96) yields

$$\mathbf{T}_{(\mathbf{N})} = \mathbf{N}\cdot\boldsymbol{\sigma} = (N_x\sigma_{xz} + N_y\sigma_{yz})\mathbf{e}_z.$$

Thus, with Eqs. (A.167), the first boundary condition of Eq. (A.160) becomes

$$\boxed{\text{On } C: \qquad N_x\frac{\partial \psi}{\partial x} + N_y\frac{\partial \psi}{\partial y} = \mathbf{N}\cdot\boldsymbol{\nabla}\psi = yN_x - xN_y.} \tag{A.171}$$

For a given geometry, the right-hand-side of this equation is known, providing a Neumann type of boundary condition. Equations (A.168) and (A.171) provide a two-dimensional boundary value problem to solve for $\psi(x, y)$.

At the end $z = L$, Eqs. (A.96) and (A.169) yield

$$\begin{aligned}\mathbf{T}_{(\mathbf{N})} = \mathbf{N}\cdot\boldsymbol{\sigma} &= \mathbf{e}_z\cdot(\sigma_{xz}\,\mathbf{e}_x\mathbf{e}_z + \sigma_{zx}\,\mathbf{e}_z\mathbf{e}_x + \sigma_{yz}\,\mathbf{e}_y\mathbf{e}_z + \sigma_{zy}\,\mathbf{e}_z\mathbf{e}_y) \\ &= \sigma_{zx}\mathbf{e}_x + \sigma_{zy}\mathbf{e}_y.\end{aligned} \tag{A.172}$$

With this expression and Eqs. (A.167), the second boundary condition of (A.160) becomes

$$\int_A \mathbf{T}_{(\mathbf{N})}\, dA = \mu\beta \int_A \left[\left(\frac{\partial \psi}{\partial x} - y \right) \mathbf{e}_x + \left(\frac{\partial \psi}{\partial y} + x \right) \mathbf{e}_y \right] dA = \mathbf{0}. \quad \text{(A.173)}$$

Consider the first term in this equation. Because ψ satisfies Eq. (A.168) and since x and y are independent variables, this term can be written

$$\begin{aligned} \mu\beta\mathbf{e}_x \int_A \left(\frac{\partial \psi}{\partial x} - y \right) dA &= \mu\beta\mathbf{e}_x \int_A \left\{ \frac{\partial}{\partial x} \left[x \left(\frac{\partial \psi}{\partial x} - y \right) \right] \right. \\ &\quad \left. + \frac{\partial}{\partial y} \left[x \left(\frac{\partial \psi}{\partial y} + x \right) \right] \right\} dA \\ &= \mu\beta\mathbf{e}_x \int_A \boldsymbol{\nabla} \cdot \left[x \left(\frac{\partial \psi}{\partial x} - y \right) \mathbf{e}_x \right. \\ &\quad \left. + x \left(\frac{\partial \psi}{\partial y} + x \right) \mathbf{e}_y \right] dA \end{aligned}$$

where $\boldsymbol{\nabla} = \mathbf{e}_x \partial/\partial x + \mathbf{e}_y \partial/\partial y$. Applying the divergence theorem (A.31) now gives

$$\begin{aligned} \int_A \left(\frac{\partial \psi}{\partial x} - y \right) dA &= \int_C x\mathbf{N} \cdot \left[\left(\frac{\partial \psi}{\partial x} - y \right) \mathbf{e}_x + \left(\frac{\partial \psi}{\partial y} + x \right) \mathbf{e}_y \right] ds \\ &= \int_C x \left[\left(\frac{\partial \psi}{\partial x} - y \right) N_x + \left(\frac{\partial \psi}{\partial y} + x \right) N_y \right] ds \end{aligned}$$

since $N_x = \mathbf{N} \cdot \mathbf{e}_x$ and $N_y = \mathbf{N} \cdot \mathbf{e}_y$. Examining Eq. (A.171) reveals that this line integral vanishes. A similar manipulation shows that the second term of Eq. (A.173) also is zero, and so the second condition of (A.160) is satisfied.

The last boundary condition of (A.160) yields

$$\begin{aligned} \mathbf{M} = M\mathbf{e}_z &= \int_A \mathbf{r} \times \mathbf{T}_{(\mathbf{N})}\, dA \\ &= \int_A (x\mathbf{e}_x + y\mathbf{e}_y) \times (\sigma_{zx}\mathbf{e}_x + \sigma_{zy}\mathbf{e}_y)\, dA \\ &= \mathbf{e}_z \int_A (x\sigma_{zy} - y\sigma_{zx})\, dA \end{aligned}$$

where Eq. (A.172) has been used. Substituting Eqs. (A.167) gives

$$\boxed{M = \mu J^* \beta} \quad \text{(A.174)}$$

where

$$\boxed{J^* = J + \int_A \left(x\frac{\partial \psi}{\partial y} - y\frac{\partial \psi}{\partial x} \right) dA} \tag{A.175}$$

with J given by Eq. (A.163). The quantity μJ^* is the torsional rigidity of the shaft. For $\psi = 0$, Eq. (A.174) reduces to Eq. (A.162), as it should.

Once $\psi(x, y)$ is determined from Eqs. (A.168) and (A.171), Eqs. (A.174) and (A.175) provide β for a given M. Then, Eqs. (A.165)–(A.167) give the displacements, strains, and stresses for the torsion problem. Other forms for the solution to this problem can be found in standard texts on linear elasticity (Sokolnikoff, 1956; Timoshenko and Goodier, 1969).

Appendix B
Special Coordinate Systems

This appendix lists basic relations for some special orthogonal curvilinear coordinate systems. Coordinates and base vectors are expressed in terms of the Cartesian system (x, y, z). In addition, physical stress and deformation components are used.

B.1 Cylindrical Polar Coordinates

The geometry in shown in Fig. B.1.

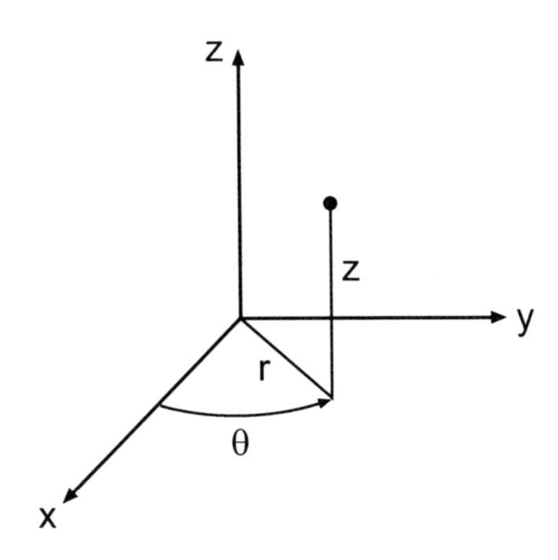

Fig. B.1 Cylindrical polar coordinate system.

Geometry:

$$(x^1, x^2, x^3) = (r, \theta, z)$$

$$\begin{aligned} x &= r\cos\theta, \qquad r = (x^2 + y^2)^{1/2} \\ y &= r\sin\theta, \qquad \theta = \tan^{-1}(y/x) \end{aligned}$$

$$\begin{aligned} \mathbf{g}_1 &= \mathbf{e}_x \cos\theta + \mathbf{e}_y \sin\theta = \mathbf{e}_r \\ \mathbf{g}_2 &= r(-\mathbf{e}_x \sin\theta + \mathbf{e}_y \cos\theta) = r\mathbf{e}_\theta \\ \mathbf{g}_3 &= \mathbf{e}_z \end{aligned}$$

$$\begin{aligned} g_{11} &= 1, \qquad g_{22} = r^2, \qquad g_{33} = 1 \\ g^{11} &= 1, \qquad g^{22} = r^{-2}, \qquad g^{33} = 1 \end{aligned}$$

$$\begin{aligned} \Gamma^1_{22} &= -r \\ \Gamma^2_{12} &= \Gamma^2_{21} = r^{-1} \\ \text{Other } \Gamma^k_{ij} &= 0 \end{aligned}$$

Differential operators:

$$\boldsymbol{\nabla} = \mathbf{e}_r \frac{\partial}{\partial r} + \mathbf{e}_\theta \frac{1}{r}\frac{\partial}{\partial \theta} + \mathbf{e}_z \frac{\partial}{\partial z}$$

$$\nabla^2 = \frac{\partial^2}{\partial r^2} + \frac{1}{r}\frac{\partial}{\partial r} + \frac{1}{r^2}\frac{\partial^2}{\partial \theta^2} + \frac{\partial^2}{\partial z^2}$$

Equations of motion:

$$\begin{aligned} \frac{\partial \hat{\sigma}^{rr}}{\partial r} + \frac{1}{r}\frac{\partial \hat{\sigma}^{\theta r}}{\partial \theta} + \frac{\partial \hat{\sigma}^{zr}}{\partial z} + \frac{1}{r}(\hat{\sigma}^{rr} - \hat{\sigma}^{\theta\theta}) + \hat{f}^r &= \rho \hat{a}^r \\ \frac{\partial \hat{\sigma}^{r\theta}}{\partial r} + \frac{1}{r}\frac{\partial \hat{\sigma}^{\theta\theta}}{\partial \theta} + \frac{\partial \hat{\sigma}^{z\theta}}{\partial z} + \frac{2\hat{\sigma}^{r\theta}}{r} + \hat{f}^\theta &= \rho \hat{a}^\theta \\ \frac{\partial \hat{\sigma}^{rz}}{\partial r} + \frac{1}{r}\frac{\partial \hat{\sigma}^{\theta z}}{\partial \theta} + \frac{\partial \hat{\sigma}^{zz}}{\partial z} + \frac{\hat{\sigma}^{rz}}{r} + \hat{f}^z &= \rho \hat{a}^z \end{aligned}$$

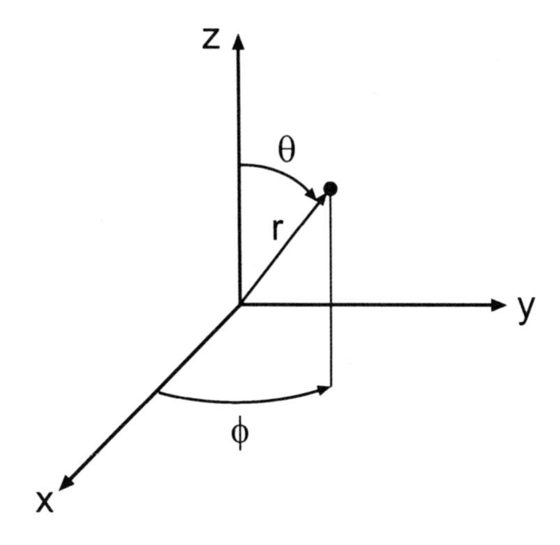

Fig. B.2 Spherical polar coordinate system.

Deformation gradient:

$$\mathbf{F}_{(\mathbf{e}_i\mathbf{e}_j)} = \begin{bmatrix} \dfrac{\partial r}{\partial R} & \dfrac{1}{R}\dfrac{\partial r}{\partial \Theta} & \dfrac{\partial r}{\partial Z} \\ r\dfrac{\partial \theta}{\partial R} & \dfrac{r}{R}\dfrac{\partial \theta}{\partial \Theta} & r\dfrac{\partial \theta}{\partial Z} \\ \dfrac{\partial z}{\partial R} & \dfrac{1}{R}\dfrac{\partial z}{\partial \Theta} & \dfrac{\partial z}{\partial Z} \end{bmatrix}$$

B.2 Spherical Polar Coordinates

The geometry in shown in Fig. B.2.

Geometry:

$$(x^1, x^2, x^3) = (r, \theta, \phi)$$

$$\begin{aligned} x &= r\sin\theta\cos\phi \\ y &= r\sin\theta\sin\phi \\ z &= r\cos\theta \end{aligned}$$

$$\begin{aligned} \mathbf{g}_1 &= \mathbf{e}_x\sin\theta\cos\phi + \mathbf{e}_y\sin\theta\sin\phi + \mathbf{e}_z\cos\theta = \mathbf{e}_r \\ \mathbf{g}_2 &= r(\mathbf{e}_x\cos\theta\cos\phi + \mathbf{e}_y\cos\theta\sin\phi - \mathbf{e}_z\sin\theta) = r\,\mathbf{e}_\theta \\ \mathbf{g}_3 &= r\sin\theta(-\mathbf{e}_x\sin\phi + \mathbf{e}_y\cos\phi) = r\sin\theta\,\mathbf{e}_\phi \end{aligned}$$

$$\begin{aligned} g_{11} &= 1, \qquad g_{22} = r^2, \qquad g_{33} = (r\sin\theta)^2 \\ g^{11} &= 1, \qquad g^{22} = r^{-2}, \qquad g^{33} = (r\sin\theta)^{-2} \end{aligned}$$

$$\begin{array}{ll} \Gamma^1_{22} = -r, & \Gamma^1_{33} = -r\sin^2\theta \\ \Gamma^2_{12} = \Gamma^2_{21} = r^{-1} & \Gamma^2_{33} = -\sin\theta\cos\theta \\ \Gamma^3_{13} = \Gamma^3_{31} = r^{-1} & \Gamma^3_{23} = \Gamma^3_{32} = \cot\theta \\ \text{Other } \Gamma^k_{ij} = 0 & \end{array}$$

Differential operators:

$$\boldsymbol{\nabla} = \mathbf{e}_r\frac{\partial}{\partial r} + \mathbf{e}_\theta\frac{1}{r}\frac{\partial}{\partial\theta} + \mathbf{e}_\phi\frac{1}{r\sin\theta}\frac{\partial}{\partial\phi}$$

$$\nabla^2 = \frac{1}{r^2}\frac{\partial}{\partial r}\left(r^2\frac{\partial}{\partial r}\right) + \frac{1}{r^2\sin\theta}\frac{\partial}{\partial\theta}\left(\sin\theta\frac{\partial}{\partial\theta}\right) + \frac{1}{r^2\sin^2\theta}\frac{\partial^2}{\partial\phi^2}$$

Equations of motion:

$$
\begin{aligned}
\frac{\partial \hat{\sigma}^{rr}}{\partial r} + \frac{1}{r}\frac{\partial \hat{\sigma}^{\theta r}}{\partial \theta} + \frac{1}{r\sin\theta}\frac{\partial \hat{\sigma}^{\phi r}}{\partial \phi} & \\
+\frac{1}{r}\left(2\hat{\sigma}^{rr} - \hat{\sigma}^{\theta\theta} - \hat{\sigma}^{\phi\phi} + \hat{\sigma}^{\theta r}\cot\phi\right) + \hat{f}^{r} &= \rho\hat{a}^{r} \\
\frac{\partial \hat{\sigma}^{r\theta}}{\partial r} + \frac{1}{r}\frac{\partial \hat{\sigma}^{\theta\theta}}{\partial \theta} + \frac{1}{r\sin\theta}\frac{\partial \hat{\sigma}^{\phi\theta}}{\partial \phi} & \\
+\frac{1}{r}\left[2\hat{\sigma}^{r\theta} + \hat{\sigma}^{\theta r} + (\hat{\sigma}^{\theta\theta} - \hat{\sigma}^{\phi\phi})\cot\theta\right] + \hat{f}^{\theta} &= \rho\hat{a}^{\theta} \\
\frac{\partial \hat{\sigma}^{r\phi}}{\partial r} + \frac{1}{r}\frac{\partial \hat{\sigma}^{\theta\phi}}{\partial \theta} + \frac{1}{r\sin\theta}\frac{\partial \hat{\sigma}^{\phi\phi}}{\partial \phi} & \\
+\frac{1}{r}(2\hat{\sigma}^{r\phi} + \hat{\sigma}^{\phi r} + 2\hat{\sigma}^{\theta\phi}\cot\theta + \hat{f}^{\phi} &= \rho\hat{a}^{\phi}
\end{aligned}
$$

Deformation gradient:

$$
\mathbf{F}_{(\mathbf{e}_i\mathbf{e}_j)} = \begin{bmatrix}
\dfrac{\partial r}{\partial R} & \dfrac{1}{R}\dfrac{\partial r}{\partial \Theta} & \dfrac{1}{R\sin\Theta}\dfrac{\partial r}{\partial \Phi} \\
r\dfrac{\partial \theta}{\partial R} & \dfrac{r}{R}\dfrac{\partial \theta}{\partial \Theta} & \dfrac{r}{R\sin\Theta}\dfrac{\partial \theta}{\partial \Phi} \\
r\sin\theta\dfrac{\partial \phi}{\partial R} & \dfrac{r\sin\theta}{R}\dfrac{\partial \phi}{\partial \Theta} & \dfrac{r\sin\theta}{R\sin\Theta}\dfrac{\partial \phi}{\partial \Theta}
\end{bmatrix}
$$

B.3 Toroidal Coordinates

Because of the complexity of some general expressions, only basic geometric relations are listed below (see Fig. B.3).

$$
(x^1, x^2, x^3) = (r, \theta, \phi)
$$

$$
\begin{aligned}
x &= (b + r\cos\phi)\cos\theta \\
y &= (b + r\cos\phi)\sin\theta \\
z &= r\sin\phi
\end{aligned}
$$

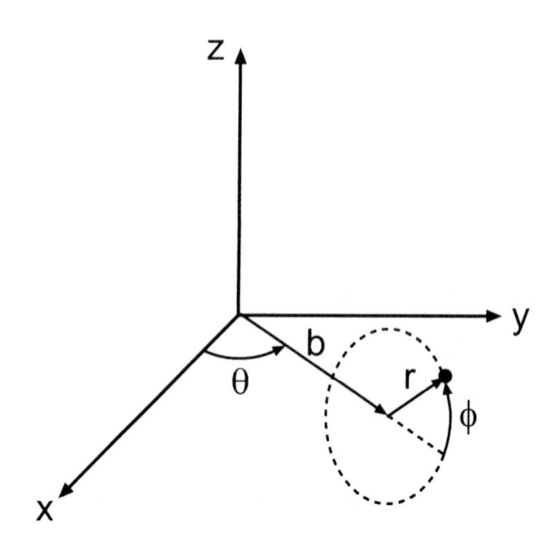

Fig. B.3 Toroidal coordinate system.

$$\begin{aligned}
\mathbf{g}_1 &= \cos\phi(\mathbf{e}_x\cos\theta + \mathbf{e}_y\sin\theta) + \mathbf{e}_z\sin\phi \\
\mathbf{g}_2 &= (b + r\cos\phi)(-\mathbf{e}_x\sin\theta + \mathbf{e}_y\cos\theta) \\
\mathbf{g}_3 &= -r\sin\phi(\mathbf{e}_x\cos\theta + \mathbf{e}_y\sin\theta) + \mathbf{e}_z r\cos\phi
\end{aligned}$$

$$\begin{aligned}
g_{11} &= 1, \qquad g_{22} = (b + r\cos\phi)^2, \qquad g_{33} = r^2 \\
g^{11} &= 1, \qquad g^{22} = (b + r\cos\phi)^{-2}, \qquad g^{33} = r^{-2}
\end{aligned}$$

$\Gamma^1_{22} = -(b + r\cos\phi)\cos\phi,$ $\quad \Gamma^1_{33} = -r$

$\Gamma^2_{12} = \Gamma^2_{21} = (b + r\cos\phi)^{-1}\cos\phi$ $\quad \Gamma^2_{23} = \Gamma^2_{32} = -(b + r\cos\phi)^{-1} r\sin\phi$

$\Gamma^3_{13} = \Gamma^3_{31} = r^{-1}$ $\quad \Gamma^3_{22} = (b + r\cos\phi) r^{-1}\sin\phi$

Other $\Gamma^k_{ij} = 0$

Bibliography

Armstrong CG, Lai WM, and Mow VC (1984). An analysis of the unconfined compression of articular cartilage, *J. Biomech. Eng.* **106**, 165-173.

Arts T, Meerbaum S, Reneman RS, and Corday E (1984). Torsion of the left ventricle during the ejection phase in the intact dog, *Cardiovasc. Res.* **18**, 183-193.

Arts T, Reneman RS, and Veenstra PC (1979). A model of the mechanics of the left ventricle, *Ann. Biomed. Eng.* **7**, 299-318.

Atkin RJ and Fox N (1980). *An Introduction to the Theory of Elasticity,* Longman, London.

Atluri SN (1984). Alternate stress and conjugate strain measures, and mixed variational formulations involving rigid rotations, for computational analyses of finitely deformed solids, with application to plates and shells — I. Theory, *Comput. Struct.* **18**, 93-116.

Azhari H, Weiss JL, Rogers WJ, Siu CO, Zerhouni EA, and Shapiro EP (1993). Noninvasive quantification of principal strains in normal canine hearts using tagged MRI images in 3-D, *Am. J. Physiol.* **264**, H205-H216

Bathe KJ (1996). *Finite Element Procedures,* Prentice-Hall, Englewood Cliffs, NJ.

Beatty MF (1987). Topics in finite elasticity: hyperelasticity of rubber, elastomers, and biological tissues — with examples, *Appl. Mech. Rev.* **40**, 1699-1734.

Beloussov LV (1998). *The Dynamic Architecture of a Developing Organism: An Interdisciplinary Approach to the Development of Organisms,* Kluwer, Dordrecht, the Netherlands.

Belytschko T, Liu WK, and Moran B (2000). *Nonlinear Finite Elements for Continua and Structures,* Wiley, Chichester, England.

Blatz PD and Ko WL (1962). Application of finite elasticity to the deformation of rubbery materials, *Trans. Soc. Rheology* **6**, 223-251.

Bowen RM (1989). *Introduction to Continuum Mechanics for Engineers,* Plenum Press, New York.

Carroll MM and Horgan CO (1990). Finite strain solutions for a compressible elastic solid, *Q. Appl. Math.* **48**, 767-780.

Chadwick P (1976). *Continuum Mechanics: Concise Theory and Problems,* Wiley, New York.

Chadwick RS (1982). Mechanics of the left ventricle, *Biophys. J.* **39**, 279-288.

Chandrasekharaiah DS and Debnath L (1994). *Continuum Mechanics,* Academic Press, Boston.

Chung DT, Horgan CO, and Abeyaratne R (1986). The finite deformation of internally pressurized hollow cylinders and spheres for a class of compressible elastic materials, *Int. J. Solids Struct.* **22**, 1557-1570.

Chung TJ (1988). *Continuum Mechanics,* Prentice-Hall, Englewood Cliffs, NJ.

Chuong CJ and Fung YC (1986). On residual stresses in arteries, *J. Biomech. Eng.* **108**, 189-192.

Criscione JC, Lorenzen-Schmidt I, Humphrey JD, and Hunter WC (1999). Mechanical contribution of endocardium during finite extension and torsion experiments on papillary muscles, *Ann. Biomed. Eng* **27**, 123-130.

Davidson LA, Oster GF, Keller RE, and Koehl MA (1999). Measurements of mechanical properties of the blastula wall reveal which hypothesized mechanisms of primary invagination are physically plausible in the sea urchin strongylocentrotus purpuratus, *Dev. Biol.* **209**, 221-238.

Demer LL and Yin FCP (1983). Passive biaxial mechanical properties of isolated canine myocardium, *J. Physiol.* **339**, 615-630.

Drew TB (1961). *Handbook of Vector and Polyadic Analysis,* Reinhold Pub. Corp, New York.

Eringen AC (1962). *Nonlinear Theory of Continuous Media,* McGraw-Hill, New York.

Eringen AC (1980). *Mechanics of Continua,* 2nd Ed., R. E. Krieger Pub. Co, Huntington, NY.

Flugge W (1972). *Tensor Analysis and Continuum Mechanics,* Springer, New York.

Fung YC (1965). *Foundations of Solid Mechanics,* Prentice-Hall, Englewood Cliffs, NJ.

Fung YC (1990). *Biomechanics: Motion, Flow, Stress, and Growth,* Springer, New York.

Fung YC (1991). What are the residual stresses doing in our blood vessels?, *Ann. Biomed. Eng.* **19**, 237-249.

Fung YC (1993). *Biomechanics: Mechanical Properties of Living Tissues,* 2nd Ed., Springer, New York.

Fung YC (1997). *Biodynamics: Circulation,* 2nd Ed., Springer, New York.

Fung YC, Fronek K, and Patitucci P (1979). Pseudoelasticity of arteries and the choice of its mathematical expression, *Am. J. Physiol.* **237**, H620-H631

Gilbert SF (2003). *Developmental Biology,* 7th Ed., Sinauer Associates, Sunderland, MA.

Green AE and Adkins JE (1970). *Large Elastic Deformations,* 2nd Ed., Oxford University Press, London.

Green AE and Zerna W (1968). *Theoretical Elasticity,* 2nd Ed., Oxford University Press, London.

Greenwald SE, Moore JEJ, Rachev A, Kane TPC, and Meister JJ (1997). Experimental investigation of the distribution of residual strains in the artery wall, *J. Biomech. Eng.* **119**, 438-444.

Guccione JM, McCulloch AD, and Waldman LK (1991). Passive material properties of intact ventricular myocardium determined from a cylindrical model, *J. Biomech. Eng.* **113**, 42-55.

Guccione JM, Waldman LK, and McCulloch AD (1993). Mechanics of active contraction in cardiac muscle : part II — Cylindrical models of the systolic left ventricle, *J. Biomech. Eng.* **115**, 82-90.

Hansen DE, Daughters GT, Alderman EL, Ingels NB, and Miller DC (1988). Torsional deformation of the left ventricular midwall in human hearts with intramyocardial markers: regional heterogeneity and sensitivity to the inotropic effects of abrupt rate changes, *Circ. Res.* **62**, 941-952.

Hashima AR, Young AA, McCulloch AD, and Waldman LK (1993). Nonhomogeneous analysis of epicardial strain distributions during acute myocardial ischemia in the dog, *J. Biomech.* **26**, 19-35.

Holmes MH (1986). Finite deformation of soft tissue: analysis of a mixture model in uni-axial compression, *J. Biomech.* **108**, 372-381.

Holzapfel GA (2000). *Nonlinear Solid Mechanics: A Continuum Approach for Engineering,* Wiley, New York.

Humphrey JD (2002). *Cardiovascular Solid Mechanics: Cells, Tissues, and Organs,* Springer, New York.

Humphrey JD, Barazotto RL, Jr., and Hunter WC (1992). Finite extension and torsion of papillary muscles: a theoretical framework, *J. Biomech.* **25**, 541-547.

Humphrey JD, Strumpf RK, and Yin FCP (1990a). Determination of a constitutive relation for passive myocardium: I. A new functional form, *J. Biomech. Eng.* **112**, 333-339.

Humphrey JD, Strumpf RK, and Yin FCP (1990b). Determination of a constitutive relation for passive myocardium: II. Parameter estimation, *J. Biomech. Eng.* **112**, 340-346.

Humphrey JD and Yin FCP (1987). On constitutive relations and finite deformations of passive cardiac tissue: I. A pseudostrain-energy function, *J. Biomech. Eng.* **109**, 298-304.

Huyghe JM, Arts T, van Campen DH, and Reneman RS (1992). Porous medium finite element model of the beating left ventricle, *Am. J. Physiol.* **262**, H1256-H1267

Ingels NB, Hansen DE, Daughters GT, Stinson EB, Alderman EL, and Miller DC (1989). Relation between longitudinal, circumferential, and oblique shortening and torsional deformation in the left ventricle of the transplanted human heart, *Circ. Res.* **64**, 915-927.

Itasaki N, Nakamura H, and Yasuda M (1989). Changes in the arrangement of actin bundles during heart looping in the chick embryo, *Anat. Embryol.* **180**, 413-420.

Kuijer JP, Marcus JT, Gotte MJ, van Rossum AC, and Heethaar RM (2002). Three-dimensional myocardial strains at end-systole and during diastole in the left ventricle of normal humans, *J. Cardiovasc. Magn Reson.* **4**, 341-351.

LeGrice IJ, Hunter PJ, and Smaill BH (1997). Laminar structure of the heart: a mathematical model, *Am. J. Physiol* **272**, H2466-H2476

LeGrice IJ, Smaill BH, Chai LZ, Edgar SG, Gavin JB, and Hunter PJ (1995). Laminar structure of the heart: ventricular myocyte arrangement and connective tissue architecture in the dog, *Am. J. Physiol* **269**, H571-H582

Leigh DC (1968). *Nonlinear Continuum Mechanics: An Introduction to the Continuum Physics and Mathematical Theory of the Nonlinear Mechanical Behavior of Materials,* McGraw-Hill, New York.

Lin IE and Taber LA (1994). Mechanical effects of looping in the embryonic chick heart, *J. Biomech.* **27**, 311-321.

Lurie AI (1990). *Nonlinear Theory of Elasticity,* North-Holland, Amsterdam.

Malvern LE (1969). *Introduction to the Mechanics of a Continuous Medium,* Prentice-Hall, Englewood Cliffs, NJ.

McCulloch A, Waldman L, and Rogers J (1992). Large-scale finite element analysis of the beating heart, *Crit. Rev. Biomed. Eng.* **20**, 427-449.

McCulloch AD, Smaill BH, and Hunter PJ (1989). Regional left ventricular epicardial deformation in the passive dog heart, *Circ. Res.* **64**, 721-733.

Mirsky I (1973). Ventricular and arterial wall stresses based on large deformation analyses, *Biophys. J.* **13**, 1141-1159.

Mirsky I (1974). Review of various theories for the evaluation of left ventricular wall stresses, in *Cardiac Mechanics: Physiological, Clinical, and Mathematical Considerations*, I Mirsky, DN Ghista, and H Sandler, Eds, Wiley, New York, 381-409.

Mooney M (1940). A theory of large elastic deformation, *J. Appl. Phys.* **11**, 582-592.

Moore CC, Lugo-Olivieri CH, McVeigh ER, and Zerhouni EA (2000). Three-dimensional systolic strain patterns in the normal human left ventricle: characterization with tagged mr imaging, *Radiology* **214**, 453-466.

Mow VC, Kwan MK, Lai WM, and Holmes MH (1986). A finite deformation theory for nonlinearly permeable soft hydrated biological tissues, in *Frontiers in Biomechanics*, GW Schmid-Schonbein, SLY Woo, and BW Zweifach, Eds, Springer-Verlag, New York, 153-179.

Nakamura A, Kulikowski RR, Lacktis JW, and Manasek FJ (1980). Heart looping: a regulated response to deforming forces, in *Etiology and Morphogenesis of Congenital Heart Disease*, R van Praagh and A Takao, Eds, Futura Publishing, Mount Kisco, NY, 81-98.

Narasimhan MNL (1993). *Principles of Continuum Mechanics,* Wiley, New York.

Ogden RW (1997). *Non-Linear Elastic Deformations,* Dover, New York.

Omens JH and Fung YC (1990). Residual strain in rat left ventricle, *Circ. Res.* **66**, 37-45.

Omens JH, MacKenna DA, and McCulloch AD (1993). Measurement of strain and analysis of stress in resting rat left ventricular myocardium, *J. Biomech.* **26**, 665-676.

Pinto JG and Fung YC (1973). Mechanical properties of the heart muscle in the passive state, *J. Biomech.* **6**, 597-616.

Rachev A (1997). Theoretical study of the effect of stress-dependent remodeling on arterial geometry under hypertensive conditions, *J. Biomech.* **30**, 819-827.

Rachev A, Stergiopulos N, and Meister JJ (1998). A model for geometric and mechanical adaptation of arteries to sustained hypertension, *J. Biomech. Eng.* **120**, 9-17.

Rivlin RS (1947). Torsion of a rubber cylinder, *J. Appl. Phys.* **18**, 444-449.

Rivlin RS (1956). Large elastic deformations, in *Rheology: Theory and Applications*, FR Eirich, Ed, Academic Press, New York, 351-385.

Rivlin RS and Saunders DW (1951). Large elastic deformations of isotropic materials VII. Experiments on the deformation of rubber, *Phil. Trans. Roy. Soc. London* **A243**, 251-288.

Rivlin RS, Barenblatt GI, and Joseph DD (1997). *Collected Papers of R.S. Rivlin,* Springer,

New York.

Sacks MS (1999). A method for planar biaxial mechanical testing that includes in-plane shear, *J. Biomech. Eng* **121**, 551-555.

Shiraishi I, Takamatsu T, Minamikawa T, and Fujita S (1992). 3-D observation of actin filaments during cardiac myofibrinogenesis in chick embryo using a confocal laser scanning microscope, *Anat. Embryol.* **185**, 401-408.

Simmonds JG (1994). *A Brief on Tensor Analysis,* 2nd Ed., Springer-Verlag, New York.

Simon BR and Gaballa MA (1988). Finite strain poroelastic finite element models for large arterial cross sections, in *Computational Methods in Bioengineering*, RL Spilker and BR Simon, Eds, ASME, New York, 325-333.

Sokolnikoff IS (1956). *Mathematical Theory of Elasticity,* 2nd Ed., McGraw-Hill, New York.

Spencer AJM (1980). *Continuum Mechanics,* Wiley, New York.

Spencer AJM (1984). Constitutive theory for strongly anisotropic solids, in *Continuum Theory of the Mechanics of Fibre-Reinforced Composites*, AJM Spencer, Ed, Springer-Verlag, New York, 1-32.

Streeter DD (1979). Gross morphology and fiber geometry of the heart, in *Handbook of Physiology, Section 2: The Cardiovascular System, Volume I: The Heart*, RM Berne, N Sperelakis, and SR Geiger, Eds, American Physiological Society, Bethesda, MD, 61-112.

Szilard R (1974). *Theory and Analysis of Plates,* Prentice-Hall, Englewood Cliffs, NJ.

Taber LA (1991a). On a nonlinear theory for muscle shells: I. Theoretical development, *J. Biomech. Eng.* **113**, 56-62.

Taber LA (1991b). On a nonlinear theory for muscle shells: II. Application to the active left ventricle, *J. Biomech. Eng.* **113**, 63-71.

Taber LA (1995). Biomechanics of growth, remodeling, and morphogenesis, *Appl. Mech. Rev.* **48**, 487-545.

Taber LA (1998). A model for aortic growth based on fluid shear and fiber stresses, *J. Biomech. Eng.* **120**, 348-354.

Taber LA (2001). Biomechanics of cardiovascular development, *Ann. Rev. Biomed. Eng.* **3**, 1-25.

Taber LA and Humphrey JD (2001). Stress-modulated growth, residual stress, and vascular heterogeneity, *J. Biomech. Eng.* **123**, 528-535.

Taber LA, Keller BB, and Clark EB (1992). Cardiac mechanics in the stage-16 chick embryo, *J. Biomech. Eng.* **114**, 427-434.

Taber LA, Sun H, Clark EB, and Keller BB (1994). Epicardial strains in embryonic chick ventricle at stages 16 through 24, *Circ. Res.* **75**, 896-903.

Taber LA, Yang M, and Podszus WW (1996). Mechanics of ventricular torsion, *J. Biomech.* **29**, 745-752.

Timoshenko S and Goodier JN (1969). *Theory of Elasticity,* 3rd Ed., McGraw-Hill, New York.

Tozeren A (1983). Static analysis of the left ventricle, *J. Biomech. Eng.* **105**, 39-46.

Truesdell C (1991). *A First Course in Rational Continuum Mechanics,* 2nd Ed., Academic Press, New York.

Truesdell C and Noll W (1965). The non-linear field theories of mechanics, in *Handbuch der Physik*, S Flugge, Ed, Springer-Verlag, Berlin, 1-602.

Truesdell C and Noll W (1992). *The Non-Linear Field Theories of Mechanics,* 2nd Ed., Springer-Verlag, Berlin.

Villarreal FJ, Lew WYW, Waldman LK, and Covell JW (1991). Transmural myocardial deformation in the ischemic canine left ventricle, *Circ. Res.* **68**, 368-381.

Vossoughi J, Hedjazi Z, and Borris FS (1993). Intimal residual stress and strain in large arteries, in *Proc. Summer Bioengineering Conference*, NA Langrana, MH Friedman, and ES Grood, Eds, New York, ASME, 434-437.

Waldman LK, Fung YC, and Covell JW (1985). Transmural myocardial deformation in the canine left ventricle. Normal in vivo three-dimensional finite strains, *Circ. Res.* **57**, 152-163.

Waldman LK, Nosan D, Villarreal F, and Covell JW (1988). Relation between transmural deformation and local myofiber direction in canine left ventricle, *Circ. Res.* **63**, 550-562.

Woods RH (1892). A few applications of a physical theorem to membranes in the human body in a state of tension, *J. Anat. Physiol.* **26**, 362-370.

Yin FCP (1981). Ventricular wall stress, *Circ. Res.* **49**, 829-842.

Yin FCP, Strumpf RK, Chew PH, and Zeger SL (1987). Quantification of the mechanical properties of nonconducting canine myocardium under simultaneous biaxial loading, *J. Biomech.* **20**, 577-589.

Young AA and Axel L (1992). Three-dimensional motion and deformation of the heart wall: estimation with spatial modulation of magnetization–a model-based approach, *Radiology* **185**, 241-247.

Zheng QS (1994). Theory of representations for tensor functions — a unified invariant approach to constitutive equations, *Appl. Mech. Rev.* **47**, 545-587.

Index